4-99
CW01655913

DISTRICT HEATING AND COOLING FOR ENERGY CONSERVATION

District Heating and Cooling for Energy Conservation

R. M. E. Diamant and David Kut

THE ARCHITECTURAL PRESS : LONDON
HALSTED PRESS
JOHN WILEY & SONS : NEW YORK

First published in 1981 by The Architectural Press:
London
Published in the U.S.A. by Halsted Press,
a Division of John Wiley & Sons, Inc.,
New York

© R. M. E. Diamant and David Kut 1981

British Library Cataloguing in Publication Data
Diamant, R. M. E.
District heating and cooling for energy conservation
1. Heating from central stations
2. Energy conservation
I. Title II. Kut, David
697′.57 TH7641
ISBN 0–85139–154–0 (UK)

Library of Congress Cataloging in Publication Data
Diamant, Rudy, 1925
District heating and cooling for energy conservation.
Bibliography: p.
Includes index.
1. Heating from central stations. 2. Air conditioning.
3. Buildings—Energy conservation.
I. Kut, David. II. Title.
TH7641.D44 1981 697′.03 81–4110
ISBN 0–470–27182–5 (USA) AACR2

All rights reserved. No part of this publication may be reproduced, stored in a retrieval system, or transmitted, in any form or by any means, electronic, mechanical, photocopying, recording or otherwise, without the prior permission of the publishers. Such permission, if granted, is subject to a fee depending on the nature of the use.

Set in Monotype Imprint
by Ronset Limited, Darwen, Lancashire
Printed in Great Britain by Mackays of Chatham Ltd

Contents

Preface

The growing scarcity and rising costs of the convenience fuels – gas and oil – is turning attention both towards the achievement of optimum efficiencies in the use of such fuels and towards the exploitation of less convenient but more abundant fuels and energy resources, such as coal, refuse and nuclear energy.

Environmental effects generally rule out the location of coal fired boiler plants, refuse incinerators and nuclear energy complexes in the heart of city centres or in the immediate vicinity of residential areas.

This book shows, at both theoretical and practical level, how district heating can play a major part in optimising the use of oil and gas and exploiting coal, refuse and nuclear energy by locating the heat stations at a distance from the consumers. Indeed district heating applications can be a major new source of growth in the use of coal for the provision of heat and hot water.

The authors similarly indicate the manner in which district cooling systems can provide greater efficiency and lower capital costs of individual cooling and air conditioning.

The book is generously illustrated, but the use of complicated mathematics is avoided, giving the reader a clear understanding of the broad aspects and details of district heating principles, methods and procedures. Copious references are provided to the literature on the subject to facilitate further research.

The authors are hopeful that this book will make a major contribution towards the understanding and expansion of district heating and district cooling. It should be useful also at the practical level, where so many more theoretical treatises have ill-served the plant owner, operator and design engineer by paying insufficient attention to critical matters, such as the correct insulation and protection of underground mains, water treatment etc.

Rudy Diamant, MSc, DipChemE, FInstE
David Kut, BScEng, CEng, FIMechE, FCIBS, FInstE, MConsE, FCIArb

Acknowledgements

The authors acknowledge with thanks the valuable assistance they obtained from the following firms in the matter of photographs and technical information:

Adsco Corporation, Buffalo, New York, USA
Advanced Thermal Systems Inc, Lancaster, New York, USA
AGA Heating Division, Stockholm, Sweden
Amalgamated Power Engineering Ltd, Bedford, England
Applied Air Conditioning Equipment Pty Ltd, Melbourne, Australia
Allen Ygnis Boilers Ltd, Tipton, England
Aquametro AG, Basle, Switzerland
Arkla Ltd, London, England
Associated Heat Services, London, England
Atomenergi AB, Studsvik, Sweden

Barco-Yates Ltd, Marple, Cheshire, England
Borsari & Co Ltd, Zurich, Switzerland
BP Oil Ltd, London, England
Bruun & Sorensen A/S, Aarhus, Denmark

Cadillac Meters Corporation, Detroit, Michigan, USA
Cepi Ltd, Purley, Surrey, England
Combustion Engineering Corporation, New York, USA
Corrall-Montenay Ltd, London, England
Clorius Ltd, Keynsham, Bristol, England
College of Fuel Technology, London, England
CTC AB, Gothenburg, Sweden

Danfoss Ltd, Middlesex, England
District Heating Association, Guildford, Surrey, England

Ebco Piping Systems Inc, Glenview, Illinois, USA
Engineering Appliances Ltd, London, England
Essoheat Ltd, London, England
Eternit AG, Berlin, West Germany

Ferranti Ltd, Manchester, England

Garrett Air Research Inc, San Diego, California, USA
General Electric Corporation, New York, USA
General Engineering Ltd, Radcliffe, Lancashire, England
Grunzweig & Hartmann AG, Ludwigshafen, West Germany

Haden-Carrier Ltd, London, England
Hitaveita Reykjavikur, Reykjavik, Iceland
Holmgrens Armaturfabrik AB, Gnosjo, Sweden

ICCO Boulogne-Billancourt, France
Iceland Tourist Bureau, London, England
International District Heating Association, Pittsburgh, Pennsylvania, USA
Ishikawajima-Harima Ltd, Tokyo, Japan
IWKA AG, Augsburg, West Germany

Kabel und Metallwerke AG, Hanover, West Germany
Kamstrup-Metro A/S, Copenhagen, Denmark
Kelit GmbH, Linz, Austria
Hans Knudsen GmbH, Flensburg, West Germany
KWF GmbH, Neuss, West Germany
Kymmene AB, Helsinki, Finland

Logstor Rorindustri A/S, Logstor, Denmark
Lubo AB, Stockholm, Sweden
Lundsberg Fabrik AB, Stockholm, Sweden

Macawber Engineering Ltd, Doncaster, England
Malmo Kraftvarmeverk AB, Malmo, Sweden
Maskinaffaren Generator AB, Partille, Sweden
Menzenbach KG, Duisburg, West Germany
Meier Schenk AG, Regensdorf-Zurich, Switzerland
Mollerpipe Ltd, London, England
Montfors & Reiners GmbH, Munchengladbach, West Germany

National Coal Board, London, England
D. J. Neil Ltd, Macclesfield, Cheshire, England

Permapipe Ltd, Watford, England
Permutit-Boby Ltd, London, England
Pittsburgh Corning Ltd, London, England
Pollux GmbH, Ludswigshafen, West Germany
Protexulate Ltd, Esher, Surrey, England

Ranger Ltd, London, England
Ric-Wil Ltd, London, England
Rorkonsult for Fjarrvarme AB, Orebro, Sweden

Samsons Control Ltd, London, England
Satchwell Controls Ltd, England
C. F. Scheer & Cie, Stuttgart, West Germany
Stal-Laval Ltd, London, England
Stuart-Forbes Insulations Ltd, London, England
Sulzer Bros AG, Winterthur, Switzerland
Sur Engineering Ltd, Mitcham, Surrey, England

Teknoterm Systems AB, Linkoping, Sweden
Teplo International AB, Upplands-Vasby, Sweden
Thermo-crete Isolation GmbH, Ureberach-Hessen, West Germany
Tubuflex KG, Hamburg, West Germany

United Flexible Metallic Tubing Ltd, London, England
United Kingdom Atomic Energy Authority, Risley, Lancashire, England

Vasteras Stads Kraftvarme AB, Vasteras, Sweden
Verein Deutscher Elektrizitatswerke, Frankfurt, West Germany
Vereinigte Kesselwerke GmbH, Dusseldorf, West Germany
Volund A/S, Copenhagen, Denmark

Wenit GmbH, Wanne-Eikel, West Germany

CHAPTER 1

Introduction

The progress of group and district heating schemes represents a revolution in traditional methods of providing heat and hot water to individual consumers. The basic case for this revolutionary change may be simply stated: energy conservation and the clamour to reduce atmospheric pollution. The former is achievable by economies in scale and by the optimum use of automatic control; the latter can be attained by reducing the number of chimneys and by the ability to obtain maximum combustion efficiency with strict control over chimney smoke and smut emission. The technology of group and district heating is complex because it must be fully related to the requirements of the individual consumer of heat.

CLASSIFICATION

It is essential to have a clear understanding of the terms employed to describe systems of heat supply and distribution which cater for multiple users. The three main classifications are central heating, group heating and district heating.

Central heating

This is a mode of heating a block of dwellings or similar buildings which uses a single heating appliance in a central position and employs either hot water, steam or air as a medium to distribute the heat to the various parts of the building.

Group heating

This description applies when the heating systems of several houses and/or other buildings are collected together and serviced from a central position. It does not really matter how many dwellings are connected in this way as long as there are two or more. In group heating practice there is usually only one single heating centrale which serves each group, with the possible addition of a standby station. It is usual for all the dwellings within a given group heating complex to be administered by the same authority. **Plate 2** shows the area of the Rowlett's Hill group heating scheme.

Group heating is quite widely used in the United Kingdom: by the middle of 1975 there were over 150 group heating schemes fired by coal and several hundred more in which oil and gas were used as the prime fuels. The schemes ranged from small ones, with perhaps two dozen council houses, to quite massive ones such as the Nottingham scheme through which 7000 homes are served by a combined coal and refuse incineration heating centrale. All schemes currently operational in the United Kingdom can only be properly referred to as group heating: district heating still lies in the future.

District heating

District heating systems must have the following basic characteristics:
(a) they should be large networks with an absolute minimum of about 5000 consumers, but usually of the order of 50000 or more
(b) the consumers are normally independent customers of the central district heating authority, purchasing their heat from this authority in metered quantities. In 'free economies' the consumers must be free to be connected or not to be connected to the district heating network, which should compete directly with other forms of heat supply ie individual oil, gas or electric heating. Only normal market forces must apply and there should be no compulsion. These criteria do not all or generally apply to district heating systems in 'controlled economies'
(c) the most important feature of all true district heating systems is that the district heating authority collects together all possible supplies of waste heat within the community, such as industrial waste heat, heat from refuse incinerators or sewage works, geothermal heat etc, and adjusts the heat supplied to the heat required by means of one or more total energy (combined heat and power) plants. These are either steam/electric power stations or gas turbine stations in which the ratio power/heat can be adjusted instantaneously in such a way that the balance of 'heat supplied to heat needed' is maintained.

ADVANTAGES OF DISTRICT HEATING

In so far as group heating can be considered a halfway house between the individual heating of dwellings and full district heating, it exhibits some of the advantages of the latter. The full advantages of fuel economy are not however realised until a system has been fully developed.

Avoidance of air pollution

There is evidence to show that there is a close relationship between air pollution and bronchitis death rates. District heating eliminates air pollution because the combustion appliances are re-located from centres of population to industrial areas where it becomes possible to site chimneys in such a way that flue gases blow away from the urban areas. Furthermore, when large scale combustion appliances of any kind are used, be they industrial boilers, electric power stations or even refuse incinerators, it becomes technically feasible to operate them at the very highest degree of combustion efficiency ie to obtain virtually complete conversion to carbon dioxide and water vapour. At the same time it becomes an economically viable proposition to remove dust and grit from the flue gases by the use of cyclone and electrostatic precipitators capable of removing well over 99.9 per cent of solid impurities from flue gases. Although the removal of sulphur dioxide from flue gases has not as yet emerged from the pilot plant stage, it could become feasible in the future once economic conditions warrant it. With small scale combustion appliances it is difficult to control the relatively copious emission of harmful flue gas components. It can be calculated that, for every thirty homes which are connected to a district heating system instead of being individually fired, atmospheric pollution is reduced by one tonne per annum.

The enormous expense of having inefficient combustion appliances, manifested in loss of human lives as well as in direct financial losses borne by the state and the community at large, should be taken into consideration when evaluating the overall cost of a district heating network. This is acknowledged in some countries by the provision of low interest rates for state loans to enable district heating to be established.

Avoidance of fire risks

When district heating is employed, heat is supplied to the consumer only in the form of steam or hot water. Fire risks are thus very much reduced in comparison with buildings where combustion appliances are under the control of the consumer and may be accessible to children. It is therefore no longer necessary to insist upon such strict fireproof construction conditions as those required for houses where normal combustion appliances are used. For example the use of open fires in timber houses is very dangerous and, quite rightly, most local authorities tend to discourage the building of such houses by insisting upon very strict spacing rules. If houses of this type could be constructed without any direct fuel combustion appliance it would probably be possible to rely for suitable fire protection on the impregnation of timber with fire resisting materials. The fire risk is accentuated in multi-occupation dwellings where the unwise action of a single person can cause the deaths of many and the destruction of much property.

Fire risks cannot be totally eliminated. People will still smoke or use kitchen appliances carelessly and there is the ever present danger of children playing with matches. But if it is possible to get rid of heating appliances on domestic premises there should be a reduction of fires by 50 per cent or even more.

UNITS OF SIZE EMPLOYED IN DISTRICT HEATING

Under the SI system (see Appendix A) one should refer to the size of a district heating station in terms of megawatts (MW) of heat produced.

However there is little uniformity about this. Continental practice refers to heat production in terms of G calories/hour (1 gigacalorie or G calorie = 10^9 calories). Elsewhere the Btu or therm still reigns supreme (particularly in the USA). The values are as follows:

1 Gcal/hour = 1.162 78 MW
1 million Btu/hr = 10 therms/hr = 0.293 072 MW

Quantities of heat provided are given under SI rules in either gigajoules (10^9 J) or terajoules (10^{12} J). Conversions from other units are as follows:

1 Tcal = 4.186 8 TJ
1 Gcal = 4.186 8 GJ
1 million Btu = 1.055 06 GJ
1 million therms = 105.506 TJ

Some complication is caused when heat is expressed in quantities of steam. It is then necessary to know the exact temperature and pressure of the steam and look up in steam tables how many joules of energy are obtained when a given quantity of steam at the temperature and pressure in question is turned into water.

Example: Determine the amount of heat which is given off when 10 000 kg of steam at a temperature of 450°C (842°F) and a pressure of 12 bars are condensed to water at 70°C (158°F). Consulting the steam tables (Appendix C below), the enthalpy of the superheated steam is given as 3.368 MJ/kg, while the enthalpy of the condensed water equals 0.293 MJ/kg. The difference in enthalpy is therefore 3.368 − 0.293 = 3.075 MJ/kg or 30.75 gigajoules for the 10 000 kg of steam involved.

OPTIMUM CONDITIONS FOR DISTRICT HEATING

District heating shows the greatest economic advantages when
(a) prime fuel is expensive
(b) the thermal density (the district heat requirement) of the area to be supplied is high

The following table illustrates German experience of thermal density in urban areas:

TABLE 1.1 THERMAL DENSITY IN GERMAN URBAN AREAS

Type of development	Thermal density (MW/km^2)
City centre development mainly 8 storeys and above	100 to 170 or more
Areas with mainly 6 to 8 storey buildings	100 to 110
Old parts with narrow roads and 3 to 6 storey buildings	55 to 100
Areas with 4 to 6 storey buildings	45 to 55
Areas with mainly 3 to 4 storey blocks	20 to 35

It used to be assumed that the economic lower limit for the installation of district heating in Germany was about 45 MW/km^2 for existing urban areas and 28 MW/km^2 for new areas. With increases in fuel costs the economic limit is coming down steadily so that today even areas with less than 14 MW/km^2 can be provided economically with district heat.

Further conditions under which district heating shows the greatest economic advantages are:

(c) when it becomes possible to borrow money at fairly low interest rates. This was the case during the periods when most of the continental district heating systems were installed. Inflation also favours capital intensive projects because money, once spent, no longer depreciates in value

(d) when the installation of district heating networks can be combined with city replanning or the construction of new towns: installation in existing towns is always very much more costly

(e) when a large section of the consumer network is under the jurisdiction of the operating authority, as one then has a built-in consumer circle. This is why district heating is such a success in east European countries. In western countries one always tries to find areas where many dwellings are municipally owned ie where a high percentage are so-called 'Local Authority dwellings'

(f) under climatic conditions where short periods of extremely low temperatures do not occur. The equable climate found in the United Kingdom is far more favourable for the economic operation of district heating systems than the more extreme conditions found in many parts of Europe and the United States

(g) in areas where the large district heating mains can be routed with relative ease.

DISTRIBUTION OF HEAT

The advantages of using hot water rather than steam as a distribution medium for heat are that hot water is obtained much more cheaply; there is no wastage of water (as there is with steam, where waste water has to be expensively purified and the condensate often has to be run into the sewers); hot water is also safer and simpler to handle, and it is easy to vary the capacity of a given network merely by increasing or decreasing the temperature difference between the flow and the return mains.

The advantages of using steam as a distribution medium instead of hot water are found when it becomes necessary to supply high-lying districts, operate refrigeration plant and supply industrial consumers.

In European systems where the bulk of the heat comes from heat or thermal (power) stations which can produce steam only under unfavourable conditions, hot water is by far the most popular heat exchange medium. In the USA on the other hand steam predominates. Because fuel prices were relatively low there in the past, less emphasis was then put on the need to obtain heat at optimum efficiency from total energy sources. Steam was found to be of value in driving centrifugal chillers for the provision of air conditioning and most space heating systems in the USA also use steam. Condensate returns are usually not fitted because it is necessary to use expensive stainless steel in their construction as it cannot be guaranteed that the condensate is free from dissolved oxygen and carbon dioxide.

With the current increasing fuel costs in the USA there are indications that even there steam lines may be on the way out, to be replaced by high temperature water lines in which flashed off steam can be utilised for driving the chillers.

CHECK LIST

1 Basic characteristics which favour district heating
2 Advantages arising out of reduced pollution and reduced fire risk
3 Optimum conditions for district heating
4 Relative advantages of hot water and steam networks distribution
5 Progress made to-date.

REFERENCES

Fernwärme International (articles are in German and are listed chronologically)
K. H. Orth, 'How a CHP station protects the environment', 2 (1973) 6, pp 153–162
'The development of district heating in the Unichal member countries', 3 (1974) 1, pp 9–10
R. Schaumann, 'The Hamburg district heating network', 3 (1974) 2, pp 36–38
H. Stumpf and E. Orth, 'District heating in West Germany', 3 (1974) 3, pp 64–70
V. A. Peer, 'District heating in the USSR', 4 (1975) 5, pp 36–38
F. J. Molter, 'District heating in Germany', 4 (1975) 5, pp 165–172
J. Wojcicki, 'District heating in Poland', 5 (1976) 2, pp 28–31
J. Halzl and C. Ignacz, 'District heating in Hungary', 5 (1976) 2, pp 42–47
H. P. Winkens and others, 'Economic prospects of German district heating systems', 5 (1976), 5, pp 134–152
K. Beck, 'An international comparison of district heating', 7 (1978) 5, pp 168–173

Other sources

E. Avers, 'District heating progress', Proceedings of the International District Heating Association New York, 65, 1974, pp 49–52
'Canada edges towards district heating', *Energy International*, 12 (6) 44, June 1975

'Denmark leads in district heating', *District Heating*, 60 (1) 1974, pp 9–10
R. M. E. Diamant, 'District heating and district cooling', Symposium on Fuels and the growing community, Canberra, 1st–3rd November 1972
R. M. E. Diamant and J. McGarry, *Space and district heating*, Butterworths, London, 1976
R. Ecabert, 'An energy concept for Switzerland', *H & V News*, February 1976, pp 42–49
R. Ecabert and others, 'Technical and economic possibilities of district heating in Switzerland', *District Heating*, 61 (1), 1976, pp 24–31
J. Olsen, 'Rational employment of energy in Denmark', Heizung, Lüftung, Haustechnik, 5th May 1976, pp 168–170 (in German)
A. E. Postlethwaite, 'National plan for heat', *Energy Digest* 8, June 1979, pp 26–31
B. V. Sazanov, 'Efficiency of district heating installations', *Teploenergetika*, 22 (2), 1975, pp 49–54 (in Russian)
E. Y. Sokolov and S. Y. Belinskii, '50 years of Soviet district heating', *Teploenergetika*, 21 (11), 1974, pp 2–7 (in Russian)
W. Wasilewski, 'In the Polish manner', *District Heating*, 60 (4) 1975, pp 14–19

CHAPTER 2

Assessment of heat requirements

ENERGY BALANCE

When planning the connection of premises to a district or group heating network, the heat requirements of the premises must be accurately established. It is necessary to establish the maximum heat demand, the minimum heat demand and the overall annual heat demand.

Human activities vary widely, even in a uniform housing estate, and the actual heat requirements are governed largely by individual living styles. Such calculations can therefore only be basic guidelines which must be modified by local circumstances.

It is necessary to draw up an energy balance between the building heat loss and the heat gain. In accordance with the first law of thermodynamics (energy can neither be created nor destroyed) the losses and the gains must balance absolutely and should indicate the actual heat requirement.

SOURCES OF HEAT GAINS

Heat supplied by the district or group heating system

This is the unknown quantity which has to be determined.

Heat supplied by electrical machinery or lighting

This item is assessed by site inspection. (1 kWh = 3.6 megawatts).

Heat emitted by occupants

Each adult contributes about 100 to 150 watts of thermal energy to the area occupied.

Radiant heat through walls

During the day heat is absorbed by radiation at the outer wall surface and transmitted by conduction and convection to the inside. At night the reverse occurs. On clear nights radiant heat exchange occurs with the upper atmosphere, which may be some 50°C (90°F) or more below the temperature of the ambient external atmosphere. The net effect is that the night heat losses through walls are greater than those during the day, even at the same external ambient temperatures.

Radiant heat through window panes

This can be a major factor, even during the winter months. Table 2.1

indicates the solar radiation which falls on vertical windows, table 2.2 provides correction factors for other latitudes and table 2.3 indicates the average and maximum solar radiation through horizontal roof lights. The tabulated data are only marginally affected by the use of multiple glazing but are lowered drastically when the windows are coated with dirt or are heavily curtained.

TABLE 2.1 SOLAR RADIATION THROUGH WINDOWS IN WATT/METRE^2, LATITUDE 56°N

Compass direction of window	North	North East	East	South East	South	South West	West	North West
Month								
January	5	5	30	75	75	30	5	5
February	10	10	60	120	105	35	10	10
March	15	35	105	155	110	35	15	15
April	20	65	135	160	95	35	20	20
May	35	100	165	160	80	35	20	20
June	40	110	170	145	75	35	25	25
July	30	100	160	155	80	35	20	20
August	20	60	130	155	95	35	20	20
September	15	30	95	145	105	35	15	15
October	10	10	55	110	100	35	10	10
November	5	5	30	70	70	35	5	5
December	3	3	15	50	55	25	3	3

TABLE 2.2 CORRECTION FACTORS AS A PERCENTAGE

Month	Jan. Feb.	Mar. April	May June	July Aug.	Sept. Oct.	Nov. Dec.
Latitude						
52°N (Southern England)	+36	+16	−5	−7	+8	+37
53°N (Midlands)	+27	+12	−4	−5	+6	+28
54°N (Lancashire and Yorkshire)	+18	+8	−3	−4	+4	+18
55°N (Newcastle–Carlisle belt)	+9	+4	−1	−2	+2	+9
56°N (Edinburgh–Glasgow belt)	0	0	0	0	0	0
57°N (Aberdeen)	−9	−4	+1	+1	−2	−9

TABLE 2.3 RADIATION THROUGH HORIZONTAL ROOF LIGHTS

	Average solar radiation (watt/metre²)	**Maximum solar radiation (watt/metre²)**
January	15	155
February	35	310
March	70	480
April	105	635
May	135	720
June	150	750
July	135	720
August	105	625
September	65	465
October	35	300
November	15	150
December	10	100

SOURCES OF HEAT LOSSES

Ventilation

It is difficult to forecast closely the heat loss by ventilation as individual habits vary so much in this respect. Table 2.4 below indicates likely rates of air changes for different buildings.

Ventilation heat loss is calculated from:
$0.34\ AC \times V\ (t_i - t_o)$ watts, where
AC is the air change rate (h^{-1})
V is the volume of the premises (m^3)
t_i is the internal air temperature (°C)
t_o is the external air temperature (°C)
Example: A building has a volume of 320 m³, an air change rate of 0.4 per hour, the internal temperature is 20°C (68°F) and the external temperature is 5°C (41°F).
Heat loss $= 0.34 \times 0.4 \times 320 \times (20 - 5)$ watts $= 653$ watts

TABLE 2.4 RECOMMENDED RATES OF AIR CHANGE

Type of building	**Recommended air changes per hour**
Canteens	2
Museums	1
Public garage	Normally 3 to 6 depending on local authority and garage layout
Ablution blocks	2
Small banking hall (height over 4.3 m)	2

Type of building		Recommended air changes per hour
Large banking hall (height over 4.3 m)		$1\frac{1}{2}$
Residence	living room	1 to $1\frac{1}{2}$
	bedroom	1
	bed/sitting room	$1\frac{1}{2}$
School	classroom	2
	assembly hall	$1\frac{1}{2}$
	gymnasium	$1\frac{1}{2}$
	common rooms	2
	dormitories	2
Warehouses	storage space	1
Offices	general	2
	private	$1\frac{1}{2}$
Factories – unpartitioned		
	up to 280 m^3	$1\frac{1}{2}$ to $2\frac{3}{4}$
	280 to 2800 m^3	$1\frac{1}{4}$ to $2\frac{1}{2}$
	2800 to 8400 m^3	1 to $2\frac{1}{4}$
	8400 to 17 000 m^3	$\frac{3}{4}$ to 2
	Over 17 000 m^3	$\frac{1}{2}$ to $1\frac{3}{4}$

The degree of exposure is included in the value of the surface resistance and in the design rate of air change. When heating tall spaces some height allowance must be made to compensate for the temperature gradient, ie build-up of temperature at the upper levels resulting from the buoyancy of warm air.

Interface systems generally

The components to heat losses through an interface system are:
Internal boundary layer heat conductance (hi) expressed as 11.5 W/m^2K for horizontal surfaces, where heat travels upwards, 7.5 W/m^2K for horizontal surfaces, where heat travels downwards and 9.5 W/m^2K for vertical surfaces.
External boundary layer heat conductance (h_o) which varies with the prevailing wind conditions from values similar to the internal ones in still wind conditions to very much greater values in high wind conditions. The through-wall conductance (hw) (or floor or ceiling) unit depends on the construction detail.
Overall conductance (U-value) is calculated from the relationship:
$1/U = 1/h_i + 1/h_w + 1/h_o$

The heat loss is calculated from the equation:
$HL = U \times A \times (t_i - t_o)$ watts, where
U is the U-value of the appropriate building section (W/m^2K).
A is the area of the particular wall floor or ceiling (m^2)
t_i is the internal ambient air temperature (°C)

t_o is the effective external ambient air temperature after taking account of radiation heating and cooling (°C).

Walls and roof

Tables 2.5 and 2.6 are published below by courtesy of the Building Research Establishment. Crown copyright is reserved. Normal wind and weather conditions are assumed with h_o as 18.2 W/m²K for walls and 22.2 W/m²K for roofs.

TABLE 2.5 STANDARD U-VALUES FOR WALLS AS QUOTED BY THE BUILDING RESEARCH ESTABLISHMENT UNDER NORMAL WIND AND WEATHER CONDITIONS

Construction		U-value W/m²K	Material Bulk dry density kg/m³	Conductivity (k) W/m K
Masonry construction				
Brickwork				
1 Solid wall, unplastered	105 mm	3.3	Brick 1700	0.84
	220 mm	2.3		
	335 mm	1.7		
2 Solid wall, with 16 mm plaster on inside face				
(a) with dense plaster	105 mm	3.0	Plaster	0.50
	220 mm	2.1		
	335 mm	1.7		
(b) with lightweight plaster	105 mm	2.5	Plaster	0.16
	220 mm	1.9		
	335 mm	1.5		
3 Solid wall, with 10 mm plaster-	105 mm	2.8	Plasterboard	0.16
board lining fixed to brickwork	220 mm	2.0		
with plaster dabs	335 mm	1.6		
4 Cavity wall with 105 mm outer and inner leaves with 16 mm plaster on inside face				
(a) with dense plaster		1.5	Brick 1700 (outer leaf)	0.84
(b) with lightweight plaster		1.3	Brick 1700 (inner leaf)	0.62
5 As 4 but with 230 mm outer leaf and 105 mm inner leaf				
(a) with dense plaster		1.2	Plaster 1300	0.50
(b) with lightweight plaster		1.1	Plaster 600	0.16

Construction		**U value W/m²K**	**Material Bulk dry density kg/m³**	**Conductivity (k) W/m K**
Brickwork/lightweight concrete block				
6 Cavity wall (unventilated) 260 mm with 105 mm brick outer leaf, 100 mm lightweight concrete block inner leaf and with 16 mm dense plaster on inside face		0.96	Concrete block 600	0.19
7 As 6, but with 13 mm expanded polystyrene board in cavity		0.70	Expanded polystyrene	0.003
Lightweight concrete block				
8 Solid wall, 150 mm aerated concrete block, with tile hanging externally and with 16 mm plaster on inside face		0.97	Aerated concrete block 750	0.22
9 Cavity wall with 75 mm aerated concrete block outer leaf, rendered externally, 100 mm aerated concrete block inner leaf and with 16 mm plaster on inside face, 50 mm cavity		0.84	Aerated concrete block (outer leaf) 750	0.24
			(inner leaf) 750	0.22
Concrete				
10 Cast	150 mm	3.5	Concrete	
	200 mm	3.1	2100	1.40
11 Cast, 150 mm thick, with 50 mm woodwool slab permanent shuttering on inside face finished with 16 mm dense plaster		1.1	Woodwool slab 450	0.09
12 As 11, but 200 mm thick		1.1		
13 Pre-cast panels, 75 mm thick		4.3		
14 As 13, but with 50 mm and sandwich lining panels composed of 5 mm asbestos-cement sheet, 25 mm expanded polystyrene and 10 mm plasterboard		0.80	Asbestos-cement sheet 1500	0.36
15 Pre-cast sandwich panels comprising 75 mm dense concrete, 25 mm expanded polystyrene and 150 mm lightweight concrete		0.72	Concrete 2100	1.4
			Lightweight concrete 1200	0.38
16 Pre-cast panels 38 mm on timber battens and framing with 10 mm plasterboard lining and 50 mm		0.62	Glass fibre	0.035
			Timber	0.14

Construction	**U value W/m²K**	**Material**	**Bulk dry density kg/m³**	**Conductivity (k) W/m K**
glass fibre insulation in cavity. 10% of area of glass fibre bridged by timber				
Curtain Wall construction				
1 Composite cladding panels, comprising 25 mm expanded polystyrene between 5 mm asbestos-cement sheets set in metal framing, 50 mm cavity, 100 mm lightweight concrete block inner wall, finished with 16 mm plaster rendering on inside face 5% of area bridged by metal framing	0.81			
2 Obscured glass, 38 mm polystyrene, cavity, 100 mm lightweight concrete back-up wall, dense plaster	0.51	Lightweight concrete	800	0.23
3 Stove-enamelled steel sheet, 10 mm asbestos board, cavity, 100 mm lightweight concrete back-up wall, dense plaster	1.1	Asbestos board	700	0.11
Framed construction				
1 Tile hanging, on timber battens and framing with 10 mm plasterboard lining, 50 mm glass fibre insulation in the cavity and building paper behind the battens. 10% of glass fibre area bridged by timber	0.65	Clay tiles		0.84
2 Weatherboard on timber framing with 10 mm plasterboard lining, 50 mm glass fibre insulation in the cavity and building paper behind the boarding. 10% of glass fibre area bridged by timber	0.62	Weather-boarding		0.14
Corrugated sheeting				
3 asbestos-cement 5 mm thick	5.3	Asbestos cement sheeting		0.36
4 As 3, but with cavity and aluminium foil-backed plasterboard lining	1.8			

Construction	U-value W/m²K	Material Bulk dry density kg/m³	Conductivity (k) W/m K
5 Double skin asbestos-cement with 25 mm glass fibre insulation in between	1.1		
6 As 5, but with cavity and aluminium foil-backed plasterboard lining	0.78		
7 Aluminium			
(a) Bright surface outside and inside	2.6		
(b) Dull surface outside, bright surface inside	2.8		
8 As 7, but with cavity and aluminium foil-backed plasterboard lining			
(a) Bright surface outside	1.8		
(b) Dull surface outside	1.9		
9 Plastic-coated steel	5.7		
10 As 9, but with cavity and aluminium foil-backed plasterboard lining	1.9		

TABLE 2.6 STANDARD U-VALUES FOR ROOFS UNDER NORMAL CONDITIONS

Construction	U value W/m²K	Material Bulk dry density kg/m³	Conductivity (k) W/m K
Flat or pitched roofs			
1 Asphalt 19 mm thick or felt/bitumen layers on solid concrete 150 mm thick	3.4	Asphalt 1700 Concrete 2100	0.50 1.4
2 As 1, but with 50 mm lightweight concrete screed and 16 mm plaster ceiling	2.2	Lightweight concrete 1200 Dense plaster	0.42 0.50
3 As 2, but with screed laid to falls, average 100 mm thick	1.8		
4 Asphalt 19 mm thick or felt/bitumen layers on 150 mm thick autoclaved aerated concrete roof-slabs	0.88	Aerated concrete 500	0.16

Construction	U-value W/m²K	Material Bulk dry density kg/m³	Conductivity (k) W/m K
5 Asphalt 19 mm thick or felt/bitumen layers on hollow tiles 150 mm thick	2.2	Hollow tile R = 0.27	
6 As 5, but with 50 mm lightweight concrete screed and 16 mm plaster ceiling	1.6		
7 As 6, but with screed laid to falls, average 100 mm thick	1.4		
8 Asphalt 19 mm thick or felt/bitumen layers on 13 mm cement and sand screed, 50 mm woodwool slabs on timber joists and aluminium foil-backed 10 mm plasterboard ceiling, sealed to prevent moisture penetration	0.90	Cement/sand 2100 Woodwool slab 560 Plasterboard	1.28 0.10 0.16
9 As 8, but with 25 mm glass fibre insulation laid between joists	0.60	Glass fibre	0.035
10 Asphalt 19 mm thick or felt/bitumen layers on 13 mm cement and sand screed on 50 mm metal edge reinforced woodwool slabs on steel framing, with vapour barrier at inside	1.4		
11 As 10, but with cavity and aluminium foil-backed 10 mm plasterboard ceiling below steel framing	0.90		
12 Asphalt 19 mm thick or felt/bitumen layers on 13 mm fibre insulation board on hollow or cavity asbestos-cement decking, decking, with vapour barrier at inside	1.5	Fibreboard	0.050
13 As 12, but with 25 mm glass fibre insulation in cavity, with vapour barrier	0.73		
14 Felt/bitumen layers on 25 mm expanded polystyrene on hollow or cavity asbestos decking with vapour barrier	0.87	Expanded polystyrene	0.033
15 Asphalt 19 mm thick or felt/bitumen layers on 13 mm fibre insulation board on metal decking, with vapour barrier			

Construction	U-value W/m²K	Material	Bulk dry density kg/m³	Conductivity (k) W/mK
(a) Cavity within decking	1.6			
(b) No cavity within decking	2.2			
16 Felt/bitumen layers on 25 mm expanded polystyrene on metal decking, with vapour barrier				
(a) Cavity within decking	0.9			
(b) No cavity within decking	1.1			
Pitched roofs 35° slope				
1 Tiles on battens, roofing felt and rafters, with roof space and 10 mm plasterboard on joists	1.9	Tiles Roofing felt		0.84 0.19
2 As 1, but with glass fibre insulation between joists				
25 mm glass fibre	0.71			
50 mm glass fibre	0.51			
75 mm glass fibre	0.38			
3 As 1, but with aluminium foil-backed plasterboard	1.5			
4 As 1, but with boarding on rafters	1.7			
5 As 1, but with aluminium foil-backed plasterboard and boarding on rafters	1.3			
6 Corrugated asbestos-cement sheeting	6.1	Asbestos cement sheeting		0.36
7 As 6, but with cavity and aluminium foil-backed 10 mm plasterboard lining	1.9			
8 Corrugated double-skin asbestos-cement sheeting with 25 mm glass fibre insulation between	1.1			
9 As 8, but with cavity and aluminium foil-backed 10 mm plasterboard lining; ventilated air space	0.80			
10 Corrugated aluminium sheeting	3.8			
11 As 10, but with cavity and aluminium foil-backed 10 mm plasterboard lining	1.9			
12 Corrugated plastic-coated steel sheeting	6.7			
13 As 12, but with cavity and aluminium foil-backed 10 mm plasterboard lining; ventilated air space	2.0			

Floors

In normal circumstances it can be assumed that the ambient soil temperature at about 1 metre below ground is about 5°C (41°F). The heat loss from the floor depends on the nature of the floor construction. Damp building materials have a k-value of about 1.8 W/m K; dry building materials of 0.1; expanded polystyrene and similar insulating materials with large interstellar voids of 0.04:

$1/h_W = d_1/k_1 + d_2/k_2 + d_3/k_3$ etc

where d_1, d_2, d_3 etc are the thicknesses (m) and k_1, k_2, k_3 etc are the k-values of the different layers.

Example: Evaluate the h_W value of a floor consisting of 800 mm wet building materials, 50 mm of expanded polystyrene and 150 mm of dry building materials:

Answer: $$\frac{1}{h_W} = \frac{0.8}{1.8} + \frac{0.05}{0.04} + \frac{0.15}{0.1}$$

or $h_W = 0.313$

Assume the internal h-value (h_i) at 7.5 W/m²K when heat is travelling downwards:

$$\frac{1}{U_{floor}} = \frac{1}{7.5} + \frac{1}{0.313}\text{, or}$$

$U_{floor} = 0.300$ W/m²K

The heat loss, assuming temperatures of 20°C (68°F) and 5°C (41°F) respectively is:

$0.300 \times (20 - 5) = 4.5$ watts per m² of floor.

Calculating total heat loss

For each of these heat loss constituents the related U-value is established, the internal temperature is specified and the areas are calculated. The heat losses are based on the lowest temperature which can be reasonably expected in the particular area. (Information from the local meteorological office or weather station.) These heat losses indicate the maximum heat load on the system.

For annual heat loss (or fuel consumption assessment) use is made of the degree-day figures published in the United Kingdom by the Department of the Environment for each specific area of the country.

The instantaneous heat loss on a specific day is therefore:

$U \times A\,(t_i - t_o)$ watts, where

U = U-value (W/m²K) of the specific building component

A = area of specific building part (m²)

t_i = internal temperature (°C)

t_o = external temperature (°C)

For the annual heat requirements:

$0.864\ U \times A\ (DD + 250(t_a - t_d))$ megajoules, where
DD is the published annual degree-day figure for the area concerned
t_a is the actual average internal temperature (°C)
t_d is the internal design temperature as given in the degree-day figures (base temperature).
Example: The annual number of degree days is 2557, the base temperature is 18°C (64.4°F) and the average internal temperature is 20°C (68°F). Then the heat loss during a year per m² of wall unit with a U-value 0.48 W/m²K is:
$0.864 \times 0.48\ (2557 + (250 \times 2))$ megajoules $= 1268$ megajoules.

ASSESSMENT OF GAINS AND LOSSES

It is necessary to consider the various types of premises which are to be heated and to evaluate the total heat losses likely to be incurred under conditions of maximum heat requirements and over the entire year.

The constituent heat losses (ventilation and fabric losses) are calculated and the sum of these is expressed in terms of watts for instantaneous loading and in terms of megajoules for annual loading.

The heat gains are similarly evaluated.

To arrive at the maximum winter heat demand from the district or group system heat gains are deducted from the heat losses.

The minimum summer heat must also be calculated to arrive at the assessment of the required base-load capacity.

CONDENSATION AT ROOFS AND GABLES

Condensation can only occur within an insulated or uninsulated roof or gable structure when the rate of ingress of water vapour from the warm internal area exceeds that to the outside to such an extent that the dewpoint is reached and water is shed within the insulation or at a surface. Condensation cannot occur in a structure with air space which incorporates an effective vapour barrier on the inside and an adequate degree of ventilation towards the outside.

Often inadequate attention is paid to vapour barriers; sometimes these are deficient or are altogether omitted. For example, if foil-backed gypsum panels are butt-jointed and the joints are not taped with a water vapour proof sealing tape, water vapour will penetrate through the joints and may nullify the effect of the vapour barrier. In many existing dwellings, no vapour barrier has been provided: it is then necessary to calculate the required ventilation rate to balance the ingress of water vapour.

It can be demonstrated that the effectiveness of even very efficient vapour barriers is at risk unless all joints are carefully prepared and sealed.

The rate at which water vapour enters the roof or gable from the room is given by the product of the diffusance value for the material and the difference in the water vapour pressure between the air close to the ceiling and that close to the vapour barrier. The very considerable molecular speed of water molecules ensures that extensive air spaces soon achieve a uniform water vapour concentration or vapour pressure.

The coldest temperature is always that against the inside surface of the outside leaf of the structure, but under certain circumstances this temperature may lie below the external ambient temperature: condensation will then occur however good the internal vapour barrier and the external ventilation. For example on cold and clear nights the external surface temperature of a structure can be some 25 to 30°C (45 to 54°F) lower than the ambient external air, due to radiation cooling.

In general the quantity of moisture which is condensed due to radiation cooling is small but it can nevertheless cause some condensation trouble. In the UK completely cloudless night skies occur only infrequently and are of short duration, particularly in winter. It is advisable however to avoid uninsulated external surfaces for roofs and gables because these tend towards low internal surface temperatures, even when radiation cooling does not apply.

Example: Consider a roof construction in which gypsum plaster boarding is covered with 100 mm of mineral wool of k-value 0.04 W/m K and diffusivity of 1.5×10^{-10} s. The external leaf consists of 20 mm of chipboard of k-value 0.12 W/m K and 10 mm of roofing felt with k-value 0.10 W/m K. The internal boundary layer of air is specified at 8.5 W/m²K and the air gap has a value of 11.5 W/m²K. The specified internal temperature is 20°C (68°F).

To evaluate the surface temperature on the internal surface of the external gable leaf, the values of U_{inw} (the U-value of the entire structure towards the inside of the room) and U_{outw} (the U-value towards the outside surface) must be established.

$1/U_{inw} = 1/8.5 + 0.012/0.20 + 0.1/0.04 + 1/11.5 = 2.765$
therefore $U = 0.361$ W/m²K
$1/U_{outw} = 0.020/0.12 + 0.010/0.10 = 0.2\,667$
therefore $U = 3.75$ W/M²K
The external boundary layer is ignored.

$$\frac{t_s - t_{es}}{t_i - t_s} = \frac{U_{inw}}{U_{outw}} = \frac{0.3617}{3.75} = 0.0965, \text{ where}$$

t_i is the internal temperature (°C)
t_s is the surface temperature of the inside of the external leaf (°C)
t_{es} is the external surface temperature (°C)
thus,

$$t_s = \frac{0.0965\, t_i + t_{es}}{1.0965}$$

If $t_i = 20°C\ (68°F)$

$$t_s = 1.760 + \frac{t_{es}}{1.0965}\,.$$

The rate of water ingress from the room to the roof or gable under the worst possible conditions (ie of 100 per cent relative humidity) is:
$(2330 - 695)\triangle_i \times 24$ kg/s, in which
$1/\triangle_i = 1/1.067 \times 10^{-10} + 0.1/1.5 \times 10^{-10}$, or
$\triangle_i = 9.9614 \times 10^{-11}$
Substituting for $\triangle_i$, the ingress of water vapour into the roof space is established at 3.909×10^{-6} kg/s.

To avoid condensation this quantity of water vapour generation must be dissipated to the outside and must take place at a rate equal to the above.

CONDENSATION AND VENTILATION

If the roof is impervious to moisture and the water vapour cannot transfer through it, adequate ventilation must be provided. During periods of radiation cooling, the rate of moisture ingress rises to:
$2330 \times 9.9614 \times 10^{-11} \times 24 = 5.5704 \times 10^{-6}$ kg/s

In such circumstances the equilibrium water vapour pressure at the inside of the external roof surface must be assumed as close to nil pressure; condensation is then bound to arise which is re-evaporated when the radiation cooling has ceased. In daytime the external surface temperature is far higher than the ambient air temperature (due to radiation heating) even on the coldest days in winter. The small amounts of moisture which are condensed during periods of radiation cooling do not usually penetrate deeply into the insulation materials.

Water vapour which has penetrated into closed gables and roofs can only be effectively dissipated if the building construction allows for adequate ventilation. This is best achieved by the provision of a ventilation gap in the construction. It can be shown that the area of the required gap (A) at the perimeter of the roof can be obtained from:
$A = 0.0143\ m^2$ or $1.43\ dm^2$
In practice, a factor of safety of about 3 to 10 is applied to the theoretical requirement (a theoretical gap of 1 mm becomes an actual provision of between 3 mm and 10 mm). Most building specifications specify air gaps of 1000 mm per m of roof perimeter. For highly insulated roofs figures of 10 000 mm^2 per metre run are often specified. However such figures which are not based upon proven scientific theory could be unsound, as they do not allow specifically for the size and shape of the roof. These factors are of great importance because increased ventilation must be provided for structures which incorporate a considerable distance between the unventilated roof centre and the ventilated perimeter. Roofs without an

efficient water vapour barrier at the ceiling should be more thoroughly ventilated than roofs with a good vapour barrier.

Condensation of water vapour within the insulation material is theoretically possible if very porous building materials are used on the inside of the insulation layer, combined with high internal relative humidity and poor ventilation outside the insulation materials. In practice, condensation within the insulation material is unlikely. Most condensation troubles with roofs occur where water vapour condenses against the uninsulated outer weathering leaf and then drips on to the insulation layer; in due course, the moisture then penetrates into the insulation layer and traverses the ceiling board by capillary movement. The adoption of a reasonably well insulated weathering layer for the roof structure, with adequate ventilation gaps designed to cope with the worst possible conditions, should prevent the occurrence of condensation.

CHECK LIST

1 Extent of heat gains
2 Extent of heat losses
3 Establish energy balance
4 Assess heat requirements for district heating demands and for summer base load
5 Precautions required to avoid condensation
6 Ventilation rate required to avoid condensation.

CHAPTER 3

District heating as a means to combat condensation

AVOIDING CONDENSATION

Many housing schemes in the UK suffer to some extent from condensation. Where condensation is not related to high external humidity conditions it can be prevented by the application of correct thermal designs, basically by adopting these three design criteria:

Adequate ventilation

Adequately high standard of thermal insulation

Maintenance of adequate internal ambient temperatures.

These three factors are interrelated. District heating (especially combined heat and power) provides heat more cheaply than other systems, making it economically feasible to maintain the higher internal ambient temperatures required to prevent condensation.

Air currents carry a certain amount of water vapour, the quantity depending upon temperature. Table 3.1 below lists the water vapour pressure and the moisture content related to the temperature of the air.

For example, at 20°C (68°F) air which is fully saturated with water vapour has a water vapour pressure of 2330 pascals. As this is a partial pressure one can evaluate the partial pressure of the air by subtracting the water vapour pressure from the total pressure. At a total pressure of 101 325 pascals (one atmosphere) this is 101 325 − 2,330 = 98 995 pascals.

The molecular weight of water is 18 and at NTP the molecular weight of any gas occupies 22.414 litres. Therefore the weight of water carried by one cubic metre at 20°C (68°F) is

$$\frac{18 \times 1000 \times 23330 \times 273.15}{22.414 \times 101\,325 \times 293.15} = 17.20 \text{ g/m}^3$$

If, for example, 1 cubic metre of air is cooled from 20°C (68°F) to 10°C (50°F) the maximum carrying capacity of the air for water vapour is reduced from 17.20 g/m³ to 9.07 g/m³. If the air at 20°C (68°F) was fully saturated, the difference, ie 8.13 g of water vapour, would be condensed as water of condensation.

The relative humidity of a particular sample of air is the fraction:

$$\frac{\text{Actual water vapour pressure at a given temperature}}{\text{Maximum theoretical water vapour pressure at that temperature}} \times 100\%$$

Example: The water vapour pressure at 20°C (68°F) is 1900 pascals, the maximum theoretical water vapour pressure at 20°C (68°F) is 2330 pascals, the relative humidity is then:

$$\frac{1900}{2330} = 0.8155 \text{ or } 81.55\,\%.$$

The temperature at which this condensation begins is that at which the saturation water vapour pressure is 1900 pascals, ie at 16.8°C (62°F). This temperature is termed the dewpoint for the given condition. To avoid condensation, the surfaces must be maintained at a temperature above this dewpoint.

Figure 3.1 diagrammatically illustrates the sources of condensation inside a dwelling and **3.2** shows a case study of condensation and indicates successful remedial action.

EVAPORATION

Standing surfaces

The rate of evaporation from standing areas of water is given by the following equation:

Weight of water evaporated per hour =

$$\frac{0.318 \times A \times K_t \times (P_w - P_a)}{P_a} \text{ grammes, where}$$

A is the area of the water surface (m^2)

K_t is the temperature constant (Pa/K)

(K_t (by empirical equation) $= 92.397 - 7.452\,t + 0.7873\,t^2 - 0.017587\,t^3 + 0.0001729\,t^4$)

P_w is the water vapour pressure (pascals)

P_a is the water vapour pressure in air in pascals

This is a temperature dependent value and is equal to the values given in table 3.1 for temperatures up to 25°C (77°F). Table 3.2 extends these values for temperatures up to 100°C (212°F).

Example: Calculate the hourly rate of evaporation of water from a standing area of 1.5 m^2 of water at 40°C (104°F) if the air temperature is 20°C (68°F) and the relative humidity of the air amounts to 60 %.

$K_t = 371$ Pa/K (by calculation)

$P_w = 7400$ Pa while $P_a = 0.6 \times 2330 = 1398$ Pa

Substituting in the master equation, the rate of evaporation from this surface of water is

$$\frac{0.318 \times 1.5 \times 371 \times (7400 - 1398)}{1398} \text{ g/hour}$$

= 760 grammes of water vapour per hour.

Combustion appliances

Hydrocarbon fuels release water during combustion. In correctly installed

TABLE 3.1 WATER VAPOUR PRESSURE OF SATURATED AIR AND THE WEIGHT OF WATER VAPOUR WHICH CAN BE CARRIED BY 1 m³ OF AIR AT VARYING TEMPERATURES

Temperature in °C	(°F)	Water vapour pressure (pascals)	Weight water per m³ (g)
−10	(14)	286	2.35
−9	(16)	309	2.53
−8	(17.5)	336	2.74
−7	(19)	361	2.91
−6	(21)	390	3.14
−5	(23)	421	3.38
−4	(25)	454	3.62
−3	(26.5)	488	3.89
−2	(28)	526	4.19
−1	(30)	567	4.50
0	(32)	610	4.83
1	(34)	655	5.17
2	(36)	703	5.55
3	(37.5)	755	5.94
4	(39)	812	6.40
5	(41)	870	6.83
6	(43)	932	7.23
7	(45)	1000	7.73
8	(46.5)	1070	7.90
9	(48)	1132	8.36
10	(50)	1228	9.07
11	(52)	1310	9.67
12	(54)	1400	10.33
13	(55.5)	1496	11.00
14	(57)	1590	11.74
15	(59)	1695	12.52
16	(61)	1810	13.37
17	(63)	1930	14.25
18	(64.5)	2055	15.17
19	(66)	2185	16.13
20	(68)	2330	17.20
21	(71)	2470	18.24
22	(71.5)	2620	19.35
23	(73.5)	2790	20.60
24	(75)	2980	22.01
25	(77)	3160	23.34
100	(212)	101 325	587.86

appliances, the flue gases are vented directly to the outside and do not contribute to the pressure of water vapour inside a building. The usual domestic gas appliances however may not have flue vents. Paraffin stoves are commonly used without any flue venting to outside and thus discharge water vapour into the dwelling. It is fairly simple to calculate the amount of water produced from the combustion of different fuels.

Town gas is methane which has a calorific value of 37.2 MJ/m³ at 20°C (68°F) and 1 bar pressure.

A therm corresponds to 100 MJ and liberates 4023 grammes (say 4 kg) of water.

Assuming that domestic kerosene (paraffin) has an average formula of C_9H_{20}, then each kg liberates on combustion 1.41 kg of water.

Kitchen and water boilers

The latent heat of water at atmospheric pressure is 2259 kJ/kg, so that each MJ of heat supplied to boiler water liberates 443 grammes of water vapour. 1 kWh equals 3.6 MJ; hence, each boiling unit of one kWh capacity can produce 1595 grammes of water vapour/hour.

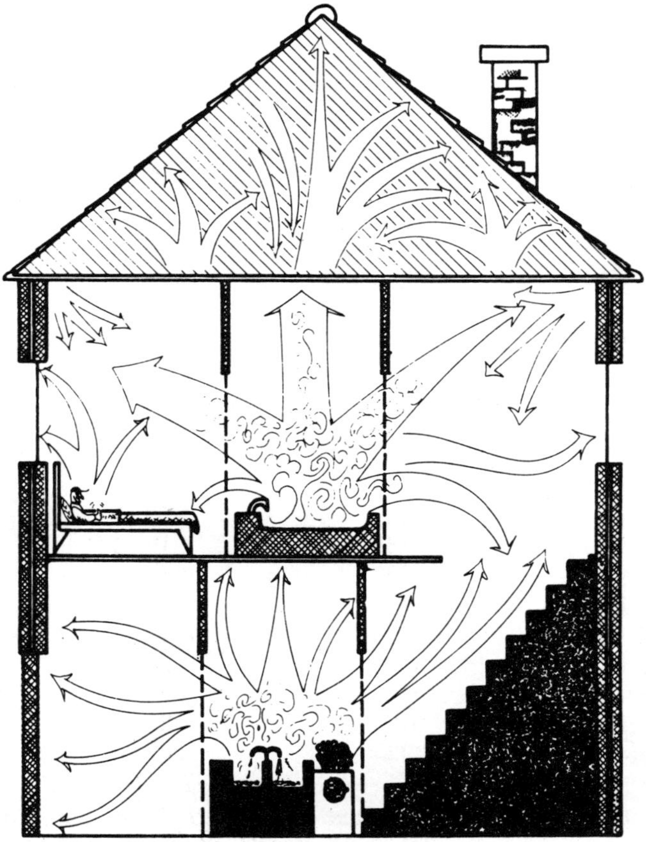

3.1 Source of condensation inside a dwelling

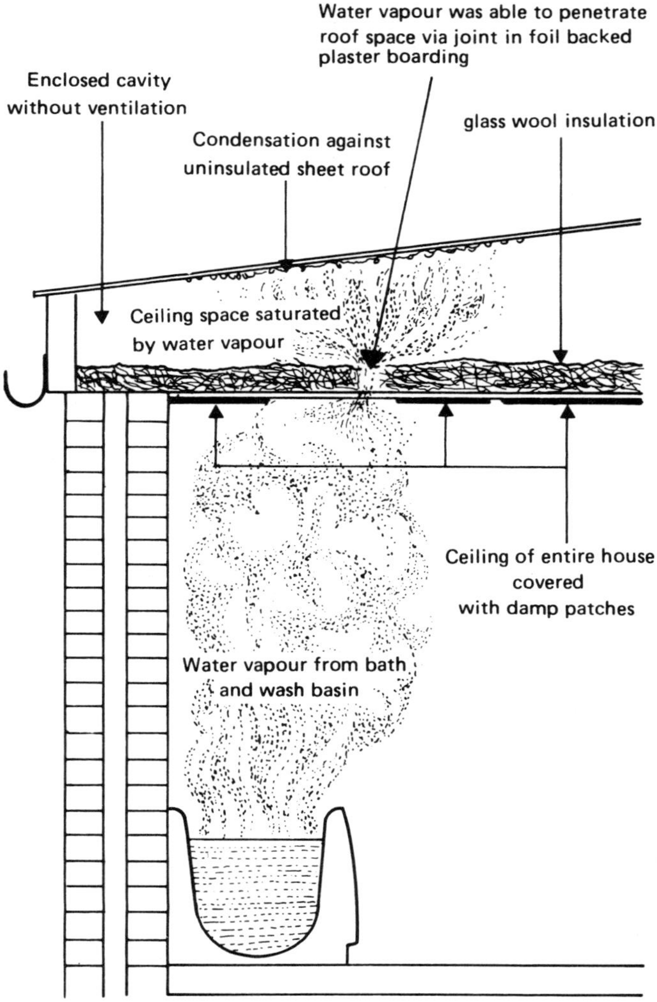

3.2 Faulty design of houses at Harlow, Essex caused condensation on the ceiling. The fault was cured by ventilating the roof space to the outside

Human and animal sources

All living creatures take in carbohydrates and other food and employ atmospheric oxygen to burn this into carbon dioxide and water vapour, the difference in enthalpy between the food and the waste products sustaining the living processes. On average, an adult person at rest releases 700 grammes of water vapour per 24 hours, more when taking strenuous exercise; when asleep he or she releases about 80 grammes/hour. Most domestic animals have a much higher water vapour output/body weight than humans. The

TABLE 3.2 EQUILIBRIUM WATER VAPOUR PRESSURES AT DIFFERENT TEMPERATURES (PASCALS)

Temperature in °C	(°F)	Water vapour pressure (pascals)
25	(77)	3316
30	(86)	4245
35	(95)	5600
40	(104)	7400
45	(113)	9580
50	(122)	12 350
55	(131)	15 710
60	(140)	19 900
65	(149)	25 000
70	(158)	31 160
75	(167)	38 540
80	(176)	47 340
85	(185)	57 800
90	(194)	70 100
95	(203)	84 510
100	(212)	101 325

steady output of water vapour from sleeping persons greatly affects condensation in bedrooms, particularly when the ambient air temperature in the bedroom is low and ventilation is inadequate.

Clothes drying

One of the major sources of water vapour (and condensation) in dwellings arises from clothes washing. Table 3.3 lists experimental and appropriate guidance data relating to the residual moisture content of clothes after washing and spin-drying.

TABLE 3.3 WATER CONTENT OF SPIN-DRIED CLOTHING

Type of clothing	Weight of residual water
Pair of stocking tights	15 g
Woman's blouse	60 g
Baby's nappy	100 g
Child's dress	120 g
Man's shirt	160 g
Towel	220 g
Man's heavy sweater	300 g
Dressing gown	480 g

These data permit an assessment of the amount of water vapour likely to be liberated in a given dwelling, school, hospital or public building. To obviate condensation problems, such calculations must be carried out during the design stage. It is stressed that the condensation troubles which tend to beset buildings in the UK are not a consequence of the damp nature of the British climate: the offending water vapour is generated and retained through local and internal activities.

VENTILATION CALCULATIONS

General considerations

The diffusion of moisture through the fabric of a dwelling is usually totally inadequate to dissipate the water vapour, particularly from kitchens and bathrooms. The bedroom condensation problem is not quite so serious. Each person gives out some 80 g of water vapour when asleep. If the bedroom is kept reasonably warm, natural diffusion through walls and ceilings should be able to dissipate nearly all this water vapour, with the balance drifting through the house and disappearing through gaps at windows and doors.

In an average bathroom up to 1000 g of water vapour per hour can be released at certain times of the day; in kitchens 3500–4500 g/hour. Natural diffusion cannot cope with such quantities. Ventilation is therefore necessary. One solution is to leave windows open to ventilate the water vapour. However, this loses expensive heat. The amount of ventilation necessary to avoid condensation against walls, windows, floors and ceilings depends very largely on the standard of thermal insulation of the dwelling. The better the insulation, the lower the required ventilation rate, so that well insulated dwellings with double glazing, effective wall and ceiling insulation etc require the lowest rates of ventilation.

In practice, inadequately insulated dwellings are often badly ventilated as well (to conserve heat), with consequent heavy condensation. It is no mere accident that those dwellings most badly affected by condensation at the present time are the various low-cost public housing units which were subsidised out of public funds as inexpensively as possible. Condensation in public dwellings has often been so serious that very expensive remedies and repairs have become necessary. In some cases fairly newly constructed blocks of flats have been so badly affected by condensation that they have had to be evacuated.

Another cause of condensation in low-cost dwellings has been the cost of heating directly borne by tenants. To reduce this, many tenants use paraffin heaters which cause much condensation.

Calculation of dew point on wall interfaces

The first step is to calculate the U-value of the wall, window or ceiling in the

conventional way, using the relationship:
$1/U = 1/h_o + d_1/k_1 + d_2/k_2 + \ldots$, where
U is the U-value of the entire structure ($W/m^2/K$)
h_o is the external boundary layer heat transfer coefficient (W/m^2K), which depends almost entirely upon the prevailing wind conditions and the nature of the surface. (Table 3.4 gives some actual values of h_o.)
d_1, k_1 etc give the thicknesses (m) and the thermal conductivities (W/m K) of the various layers of which the wall is constructed. Where the wall cavity exceeds 10 mm in width one assumes a d/k value of 0.18 for horizontal structures where the heat flow is upwards (roofs or walls) and a value of 0.21 for floors. If either one, or both, sides of such cavities are coated with a reflective layer (aluminium foil or metallised plastic foil), then the d/k values are increased to 0.34 and 1.05, respectively.

TABLE 3.4 h_0 VALUES WITH DIFFERENT WIND SPEEDS AND EXTERNAL SURFACES

Wind speed (m/s) parallel to surface	**Nature of surface and heat transfer coefficient (W/m^2K)**				
	Glass or metal	*Plaster or timber*	*Concrete*	*Brick*	*Stucco*
Wind still	8.5	8.5	8.5	8.5	8.5
2	13	14	15	15	17
5	27	33	42	48	61
10	47	56	75	84	107
15	64	77	102	115	147
20	81	98	131	147	188

The internal boundary layer heat transfer coefficient in every case equals 8.5 W/m^2K, as the assumption is made that still wind conditions apply indoors.

To establish the internal surface temperature, the following equation is used:
$8.5\,(t_i - t_{is}) = U_{wio}\,(t_{is} - t_o)$, where
t_i is the internal room temperature (°C)
t_{is} is the internal surface temperature of the wall or other interface (°C)
t_o is the external temperature (°C)
$U_{w,o}$ is the U-value of the wall section and the external boundary layer (W/m^2K).
Rearrangement of equation gives us:

$$t_{is} = \frac{8.5\,t_i + U_{wio}\,t_o}{(8.5 + U_{wio})}$$

The water vapour pressure in the air, at which condensation begins, is obtained from table 3.1.

Example: A wall consists of 100 mm of concrete of k-value 0.9 W/m²K, an air gap, 50 mm of expanded polystyrene of k-value 0.035 W/m K, 80 mm of timber with k-value 0.185 W/m K and finally 15 mm of plaster with k-value 0.65 W/m K. The external temperature is −5°C (23°F) and the wind speed is 10 m/s. The internal temperature is 24°C (76°F). At what vapour pressure does condensation commence?

Answer: The value for the wall (1_{wio}) is calculated from:

$$1/U_{wio} = 1/84 + (0.100/0.9) + 0.18 + (0.050 \times 0.035) + (0.080 \times 0.185) + (0.015 \times 0.65)$$

which gives $1/U_{wio} = 2.1871$ or $U_{wio} = 0.457$ W/m²K

Substituting:

$$t_{is} = \frac{(8.5 \times 24) + (0.457 \times (-5))}{(8.5 + 0.457)} = 22.52°C\ (73°F)$$

Interpolating from table 3.1, it is established that condensation against such a wall starts when the water vapour pressure exceeds 2708 pascals.

Now establish the temperature and vapour pressure at which water vapour pressure condensation would commence if, instead of the well insulated wall a mere 100 mm thickness of concrete of K-value 0.9 W/m k is provided; $1/U_{wio}$ now equals:

$1/84 + (0.100/0.9) = 0.12302$

or $U_{w,o} = 8.129$

Substituting:

$$t_{is} = \frac{(8.5 \times 24) + (8.129 \times (-5))}{(8.5 + 8.129)}$$

which gives us $t_{is} = 9.82°C$ (49°F)

Interpolating from table 3.1, it is established that in these different circumstances condensation commences whenever the water vapour pressure inside the room exceeds 1211 pascals, ie condensation occurs on the poorly insulated walls when the air carries only 44.7 per cent (1211/2708) of the water vapour which would cause condensation against the better insulated wall.

Internal surface temperatures

To reduce condensation it is necessary to ensure that the temperatures of the inside surfaces of walls, windows, ceiling and floors approximate closely to that of the internal ambient air.

Assuming an external temperature of 0°C (32°F) and an internal temperature of 20°C (68°F), then the internal surface temperatures required to avoid condensation, for a wide variety of U-values, are given in table 3.5

TABLE 3.5 INTERNAL SURFACE TEMPERATURES FOR WALL AND CEILING SYSTEMS, ASSUMING AN EXTERNAL TEMPERATURE OF 0°C (32°F) AND AN INTERNAL TEMPERATURE OF 20°C (68°F).

U-Value (W/m^2K)	Temperature ratio	Internal surface temperature °C	°F
0	Infinity	20.0	68
0.1	83.998	19.8	67
0.3	27.3330	19.3	67
0.4	20.2500	19.06	67
0.5	15.9998	18.82	65
0.6	13.1666	18.59	66
0.7	11.1428	18.35	66
0.8	9.6250	18.12	65
0.9	8.4444	17.88	64
1.0	7.4999	17.65	64
1.2	6.0833	17.18	64
1.4	5.0714	16.71	62
1.6	4.3125	16.24	61
1.8	3.7222	15.76	60
2.0	3.2500	15.29	59
2.2	2.8636	14.82	59
2.4	2.5417	14.35	57
2.6	2.2692	13.88	57
2.8	2.0357	13.41	55
3.0	1.8333	12.94	55
3.2	1.6562	12.47	54
3.4	1.5000	12.00	54
3.6	1.3611	11.53	53
3.8	1.2368	11.06	53
4.0	1.1250	10.59	52
4.2	1.0238	10.12	52
4.4	0.9318	9.65	49
4.6	0.8478	9.18	50
4.8	0.7708	8.71	48
5.0	0.7000	8.24	48
5.5	0.5454	7.06	46
6.0	0.4167	5.88	42
6.5	0.3077	4.71	41
7.0	0.2143	3.53	39
Infinity	Zero	0.00	

Table 3.5 shows noticeably the inter-relationship between surface temperatures and U-values. The table can be adapted for other internal and external

temperatures by applying the temperature ratio value (TR):

$$TR = \frac{t_{is} - t_o}{t_i - t_{is}}$$

which can be transformed in favour of t_{is} (the inside temperature):

$$t_{is} = \frac{t_i\,TR + t_o}{1 + TR}$$

Example: Given a wall system of U-value 0.8, so that TR is 9.6250, internal temperature of 25°C (77°F), external temperature of −5°C (23°F),

$$t_{is} = \frac{(9.6250 \times 25) - 5}{10.6250}$$

$= 22.18°C$ (74°F)
equivalent to a limiting water vapour pressure value of 2650 pascals.

CONDENSATION AGAINST WINDOWS

The thermal resistance of the glass in single glazed windows is negligible and the only major resistance to heat transfer is offered by the twin boundary layers on the inside and on the outside of the window pane respectively. If for example at a wind speed of 10 m/s, h_o is 47 (table 3.4) giving a U-value of the pane of:
$1/47 + 1/8.5 = 7.198\ W/m^2K$
At an external temperature of 0°C (32°F) and an internal one of 20°C (68°F) the internal window surface temperature is only 3.06°C (37.5°F), corresponding to a water vapour pressure of 756 pascals. Even under still conditions the internal surface temperature of single glazing is only 10°C (50°F), corresponding to a water vapour pressure of 1228 pascals. Still wind conditions seldom apply; it is thus virtually impossible to avoid condensation against single glazed surfaces with the usual ventilation rates. Neither net curtains nor thick cloth curtains can materially influence condensation. Because of the rapid motion of the water vapour molecules, the vapour pressure against the surface of the glass is that which exists in the room itself, even though there are curtains between room and window. Vapour pressures always achieve equilibrium, unless some impervious layer intervenes between them. Thick curtains improve the thermal insulation of a window system but they have no effect on the degree of condensation which occurs at the window. Multiple glazing of the window will prevent condensation or reduce the rate of condensation. Table 3.6 lists the U-values for a number of multiple glazing systems (information from Pilkington Brothers Ltd of St Helens).

Figure 3.3 shows the effect of double glazing upon the temperature of the room and the associated liability to condensation. **Figure 3.4** illustrates the temperature distribution with single glazed and with double glazed windows.

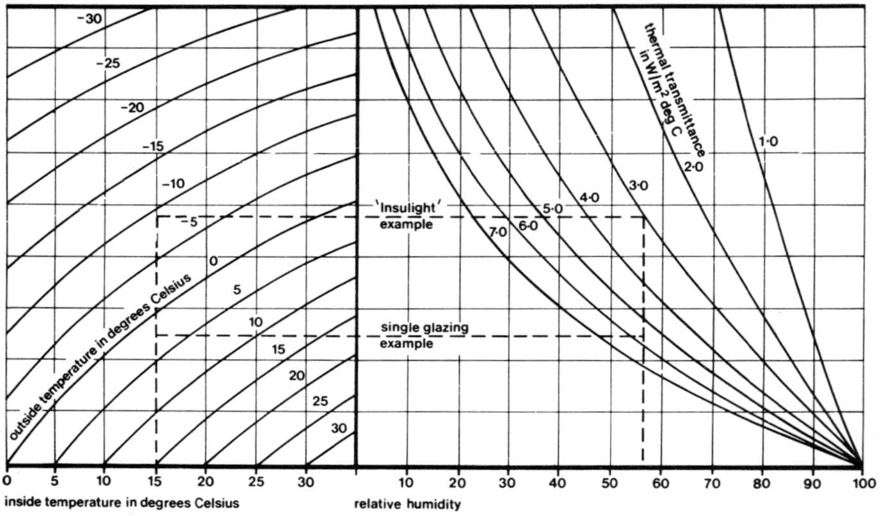

3.3 The effect of double glazing upon temperature inside room and the associated liability to condensation

TABLE 3.6 U VALUES OF VARIOUS MULTIGLASS SYSTEMS (W/m²K)

Air space mm	Type of window system	Sheltered location	Average location	Severe wind exposure
5	Double	3.3	3.5	4.0
	Triple	2.5	2.6	2.8
	Quad	2.0	2.1	2.1
	Quin	1.6	1.7	1.8
	Sext	1.4	1.4	1.5
6	Double	3.2	3.4	3.8
	Triple	2.3	2.5	2.6
	Quad	1.8	1.9	2.0
	Quin	1.5	1.6	1.6
	Sext	1.3	1.3	1.4
9	Double	3.0	3.2	3.5
	Triple	2.1	2.2	2.3
	Quad	1.6	1.7	1.8
	Quin	1.3	1.4	1.4
	Sext	1.1	1.2	1.2
12	Double	2.8	3.0	3.3
	Triple	2.0	2.1	2.2
	Quad	1.5	1.6	1.6
	Quin	1.2	1.3	1.3
	Sext	1.0	1.1	1.1
19	Double	2.8	3.0	3.2
	Triple	1.9	2.0	2.1

Air space mm	Type of window system	Sheltered location	Average location	Severe wind exposure
6	Insulight sun Cool double	2.7	2.8	3.1
12	Ditto	2.0	2.1	2.2
12	Insulight sun Cool triple	1.5	1.6	1.6

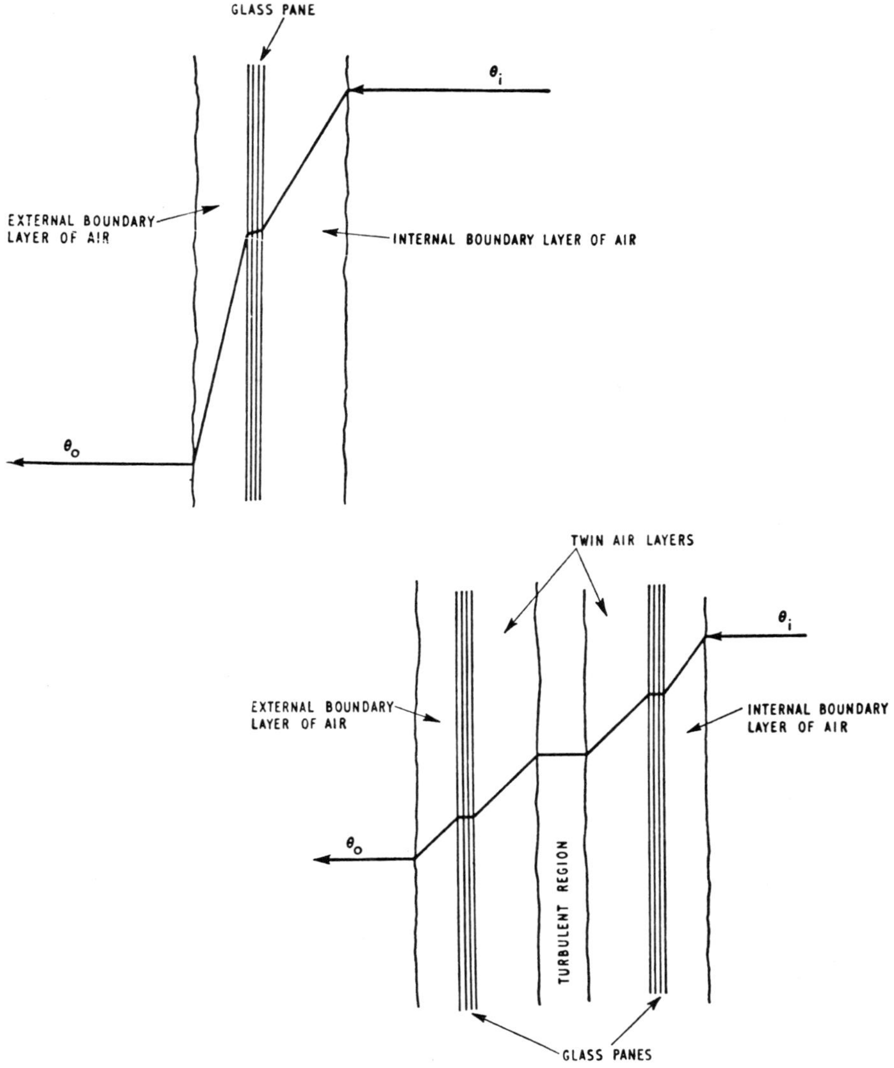

3.4 Temperature distribution with single and double glazed windows

The great improvement in the condensation situation due to multiple glazing can be shown by considering the following example:
Severe conditions; external temperature 0°C (32°F); internal temperature 20°C (68°F).
The internal surface temperature then is 3°C (36.8°F) on the inside window surface with single glazing. The water vapour pressure is about 756 pascals. Condensation against such a window surface begins whenever the vapour pressure inside the dwelling exceeds 756 pascals.
Double glazing, which incorporates an air gap of 6 mm, raises this internal surface temperature to 10.6°C (51.1°F), equivalent to 1277 pascals.
With triple glazing (9 mm air gaps) the internal surface temperature is 14.6°C (58.3°F), corresponding to an internal water vapour pressure of 1650 pascals.
Clearly, the risk of condensation and the quantity of water run-off are materially reduced when multiple glazed windows replace single glazed ones.

Three distinct types of multiple glazed windows are currently available: sealed units, coupled sashes sealed on the inside and having the space between the panes vented to the outside and various types of 'converted' double glazing units.
Sealed units have very dry air sealed between the panes which excludes the moisture-laden air inside the room from contact with the external cold pane of glass. However, if the sealing is imperfect (eg a pinhole is left in the junction) moisture-laden air can be drawn from inside the room. This cools inside the glazing unit and condensation occurs against the inside surface of the outer pane. As the water vapour pressure of the air within the sealed unit is lower than that of the warm room air, there will be a steady flow of moisture inwards, keeping the window permanently misted up. This used to be a frequent occurrence with sealed double glazed units of the early types. Modern improved adhesives and sealing techniques have virtually eliminated this difficulty which was caused mainly by the breakdown through ageing of the plastic material of the sealant. If a double glazed sealed unit mists up a major repair or even replacement is required.
Coupled sashes are particularly popular on the continent of Europe. They do not generally suffer from condensation between the panes. While the internal vapour barrier against the ingress of damp air into the cavity between the panes is inferior to that in sealed units, ventilation of the cavity to the outside is vigorous. Coupled units usually have a gap of about 1 mm wide around the perimeter; this allows the water vapour pressure inside the cavity to equalise with the external water vapour pressure. As the temperature inside the cavity is always higher than that of the outside air (in winter), the dewpoint is not reached and the internal surface of the outer pane remains clear.
Converted double glazing systems are sometimes designed to avoid condensation. The best of these systems have the extra pane mounted on the outside

of the original window, similar to the storm windows popular in North America. The frames of some of these makes are well sealed and have holes drilled in the surrounds of the original window frame to provide ventilation of the air space: such systems do avoid condensation. In other systems where the glazing units are sealed towards the inside of the dwelling (which has a higher temperature and, in consequence, a higher water vapour pressure than the colder outside even on the wettest of days) the protection is inadequate. This is because warm moist air drifts into the space between the panes and causes condensation against the inside surface of the outer pane of glass. Improvement can be effected by better sealing combined with a ventilation facility towards the outside.

Regulating ventilation rate to prevent interior condensation

Water is evaporated inside a building by a variety of agencies, the relative importance of which depends on local circumstances. For example, in a bedroom the major source of water vapour is from the occupants (asleep or active), in kitchens from the evaporation from boiling liquids, the use of washing machines, dish-washers, clothes dryers etc, and in bathrooms through evaporation from hot standing areas of water. Similar criteria arise with buildings other than dwellings, eg indoor swimming pools have suffered greatly from condensation. Surprisingly much more trouble has arisen with modern building structures than with numerous Victorian public baths and wash-houses where the engineers appear to have had a good understanding of the underlying causes of condensation.

When the total amount of water vapour is released in a given building structure is known (generally by calculation), the minimum number of air changes required to combat condensation can be calculated from the following equation:

$$n = \frac{(W - w)(P_i - P_s)}{V\,h\,(P_s - P_o)}, \text{ where}$$

n is the number of air changes (h^{-1})
W is the weight of water released per hour
w is the weight of water which is able to diffuse through walls, ceiling, cracks at windows etc (g/hr)
V is the volume of the room (m^3)
h is the carrying capacity of the air (m^3) for water vapour, obtained from table 3.1
P_i is the water vapour pressure (pascals) inside the room
P_s is the saturation water vapour pressure at the temperature of the critical wall, window, floor or ceiling surface (pascals)
P_o is the water vapour pressure of the external air (pascals)

W is fairly easy to work out from first principles, by summing up the likely sources of water vapour inside the room area

w is calculated by summing up $\triangle A (P_i - P_o)$ for the walls, windows, ceiling and floor concerned where
$\triangle$ is the diffusance constant of the interface sections (s/m)
A is the area of the interface concerned (m^2)
P_i is the internal water vapour pressure (pascals)
P_o is the external water vapour pressure (pascals).
The dimension of the results of these values is

$$\frac{s \times m^2 \times kg\,m}{m \times s^2 \times m^2} = kg/s$$

To obtain the value of w, which is quoted in g/hr, it is necessary to multiply $A (P_i - P_o)$ by a factor of 3.6×10^6.

Example: Determine the minimum number of air changes required to ventilate a room measuring 8 m × 4 m × 3 m if a quantity of water vapour of 500 g is released each hour. Assume that the loss of water vapour by diffusion processes through walls, ceiling, floor and imperfections at doors and windows is 900 g/hr. The external air temperature is 0°C (32°F). The room has single glazing with a U-value of 6.0. It is intended to change to double glazing with U-value 3.0. Determine the number of air changes necessary to avoid condensation. The internal air temperature is 20°C (68°F). The internal relative humidity should not exceed 65 per cent. The relative humidity of the external air is taken as 100 per cent.

The surface temperature of the single glazed window can be worked out at 5.88°C (42.6°F), corresponding to a water vapour pressure of 925 pascals, and an h-value of 7.2 g/m³. Condensation will take place against the external panes whenever the relative humidity inside the room exceeds $\frac{925}{2330} = 0.397$, where the water vapour pressure equivalent to a temperature of 20°C (68°F) = 2330 pascals.

To prevent such condensation the minimum required air change rate is:

$$n = \frac{(3500 - 900)(2330 - 925)}{96 \times 7.2 \times (925 - 610)}$$

$= 3.87$ air changes per hour

If double glazing with a U-value of 3.0 W/m² is used, then the internal surface temperature rises to 12.94°C (55.3°F) equivalent to a water vapour pressure of 1490 pascals and a carrying capacity for water vapour of 10.95 g/m³.

Condensation against the windows will then only occur when the relative humidity of the air inside the room rises to

$$\frac{1490}{2330} = 0.639 \text{ and}$$

$$\frac{(1500 - 900)(2330 - 1490)}{96 \times 10.95 (1490 - 610)} = 0.54 \text{ air changes per hour.}$$

In both cases it is assumed that ventilation proceeds by exhausting the water vapour as it is produced, at 100 per cent relative humidity.

It has been demonstrated that far lower ventilation rates are required to avoid condensation against cold surfaces if the thermal insulation standards are improved. The major methods available for minimising the air change rates are therefore:

(W − w) should be a minimum. Good natural ventilation reduces W (the evaporation of water in a given area) by increasing the value of w.

(P_i − P_x) is kept at a minimum by raising the surface temperature of the interface through improving the standard of thermal insulation.

V must be kept as large as possible: large rooms are less subject to condensation than small ones, all other factors being kept equal.

h should be as large as possible. Adequately heated rooms (where the carrying capacity of the air inside the room for water vapour is high) are less liable to suffer from condensation troubles than poorly heated rooms.

(P_s − P_o) should be as large as possible: there should be the largest possible difference in temperature between the cold external air and the warm internal air at the wall or window interface, implying efficient thermal insulation.

When the moisture cannot be extracted at source, a higher air change rate is required than derived theoretically. If the moist air is extracted at a higher temperature than that of the ambient air prevailing in the room (so that it carries a higher percentage moisture than calculated), a lower air change rate can be safely adopted.

Clearly excessive air change rate is undesirable as this involves considerable loss of sensible heat.

Figure 3.6 illustrates condensation-prone building construction and indicates remedial action and **figure 3.5** shows interstitial condensation in insulated walls.

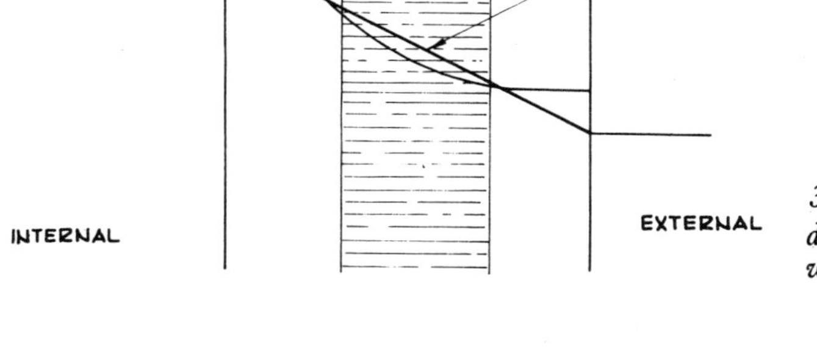

3.5 Interstitial condensation in insulated walls

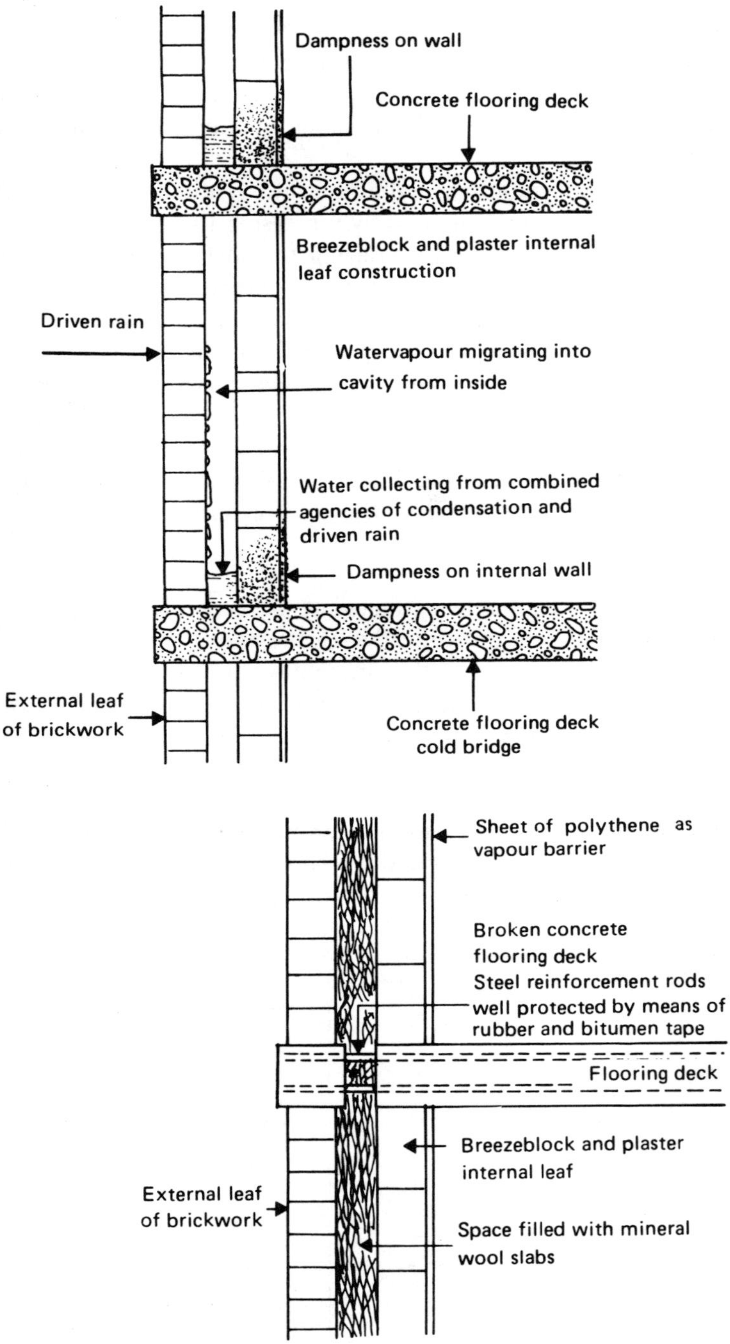

3.6 Faulty design of certain Scottish apartment blocks caused excessive condensation: methods which can be employed to remedy the situation

CHECK LIST

1 Basic requirements to avoid condensation
2 Total vapour release into dwellings from all sources
3 Ventilation requirements to avoid condensation
4 Dew point on surfaces
5 Conditions which lead to condensation against windows
6 Advantages of multiple glazed windows relating to condensation
7 Possible difficulties arising from the provision of multiple glazing
8 Required air change rate to prevent condensation.

CHAPTER 4

Group heating

ADVANTAGES AND DISADVANTAGES

Group heating schemes are always much smaller and more compact than comprehensive district heating schemes. The fuel costs in group heating schemes are only slightly lower than they would be if each dwelling were heated individually. It is therefore essential that pipeline costs are kept low. Even so, it is not uncommon for the actual cost of heating by a group centrale to be higher than the cheapest alternative individual heating. Gas fired central heating, whether by hot air or by hot water heaters, is often cheaper than group heating; coal and oil fired individual central heating compete very strongly with it.

Group heating can in fact only compete outright with electric heating, whether on or off peak. Financial considerations are not, however, the only factors which dominate decisions to adopt or reject group heating. The other factors are:

(a) Gas individual heating is banned altogether in tower blocks in the UK because of the potential explosion risk. (The Ronan Point disaster was caused by the fracture of a gas pipe high up in a tower block of flats.)

(b) The burning of oil and coal in individual dwellings in blocks of flats is generally unsuitable because of fuel storage difficulties and the inherent fire risks.

(c) Individual combustion appliances should not be specified for dwellings accommodating old persons or handicapped persons.

(d) Where individual combustion appliances are fitted it is necessary to expend capital on special areas for such appliances, on chimneys and on storage spaces for the fuel (and ash for solid fuel), standing areas for vehicles and access roads.

(e) Where appropriate it is desirable to design and build group heating schemes initially arranged to suit subsequent development into an urban or inter-urban district heating scheme. Nearly all large district heating schemes were developed by joining together a number of group heating schemes which had been established as a halfway stage. District heating is economically viable under nearly all circumstances.

(f) Even in a well-constructed district heating scheme it is necessary to have stand-by boilers to provide the heat demands during abnormally cold weather conditions or periods when the main thermo-electric plant is shut down because of strikes, repairs or servicing.

It is generally accepted that pure group heating schemes seldom survive unaided and commonly have to be subsidised. Comprehensive district heating schemes do not, however, require to be subsidised as most of them are very profitable.

GENERAL LAYOUT

Because of the need to restrain the network costs, the heat demand per unit length of pipeline mains must be as high as possible. With group heating stations, the target is a minimum connected capacity of around 1 to 1.5 megawatt or an average supply of 10 to 15 terajoules per kilometre of mains per annum. If the heat supply is much less than this, the system will have to be subsidised heavily by the operating authorities. To achieve such a large thermal density, it is usually necessary to have captive consumers, such as municipal or co-operative housing estates, public buildings, schools, hospitals, universities etc. District heating schemes which use waste heat from power stations, refuse incinerators etc can function profitably with much smaller thermal densities.

Plates 4 and **5** illustrate a coal fired boiler house serving a group heating scheme and a municipal swimming pool connected to such a scheme.

DIMENSIONING

A comprehensive study of group heating has been conducted by the South West and South Wales branch of the District Heating Association (UK) with reference to an assumed town called 'Thermville' (see **figure 4.1** for layout plan), using the following design parameters:

TABLE 4.1 CONNECTED CAPACITIES OF DWELLINGS (UK)

Type of dwelling	Connected heat to each dwelling
1 bedroom dwellings	5.85 kW
2 bedroom dwellings	7.30 kW
3 bedroom dwellings	8.80 kW
4 bedroom dwellings	10.25 kW
Primary school	586.00 kW
Secondary school	880.00 kW
Shops	404.00 kW
Estimated hot water requirements	1125 litres/dwelling/week at a temperature of 50°C (122°F)

The equivalent annual number of hours at full heat load in the UK are estimated at 3185, representing 35 weeks at 20 hours per day, using a weather

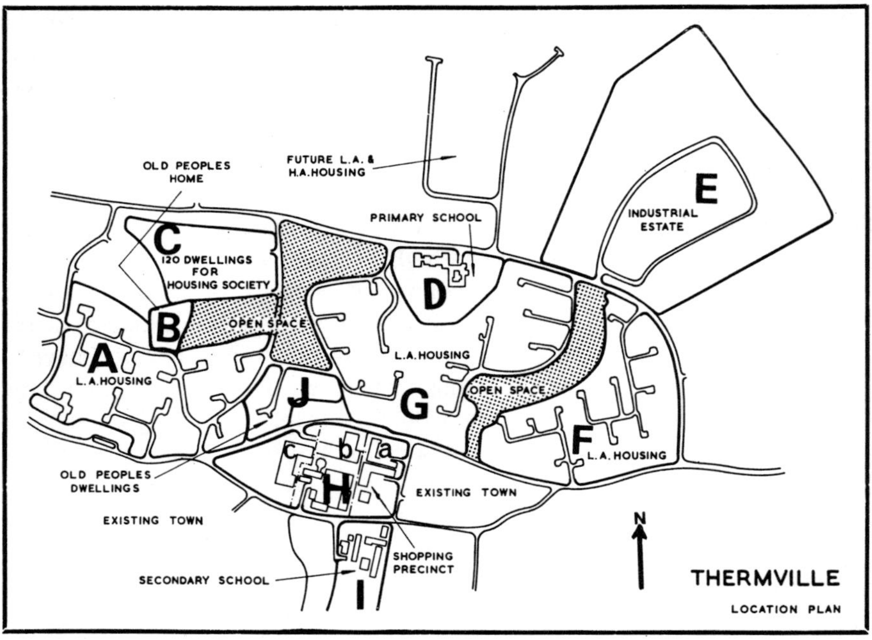

4.1 Layout of Thermville

factor of 0.65. The group heating scheme for the Thermville project caters for a total of 814 local authority dwellings and ancillary premises and requires 120 825 gigajoules of heat per annum, made up as follows:

TABLE 4.2 ANNUAL HEAT REQUIREMENTS OF THERMVILLE

Usage	Heat requirement
Hot water for district heating purposes	96 500 GJ
Domestic hot water	10 900 GJ
Mains losses in dwellings (10%)	10 740 GJ
External mains losses (2.5%)	2 685 GJ
Total:	120 825 GJ

EFFICIENCY

The study of Thermville compares the efficiencies of large group heating boiler plants with individually fired appliances with the following results:

TABLE 4.3 EFFICIENCY OF VARIOUS TYPES OF APPLIANCES

Fuel	Group heating centrales	Individual appliances
Electricity (off peak)	n/a	90%
Gas	80%	75%
Coal	70%	60%
Oil	78%	70%

The savings in efficiency with respect to gas, coal and oil in group heating centrales over efficient individual systems are not very great. However, group heating centrales can utilise much lower (and cheaper) grades of oil or coal than the small consumer. Savings are also made by buying fuel in bulk. For example, the cost of coal delivered in bulk to the district heating plant only costs 39 per cent of what the householder would have to pay buying on his account; that for oil about 70 per cent. Thus, there are some savings to be made, out of which group heating can be financed. However, given availability and suitability, it is seldom possible to make a viable case for the use of gas in group heating, as gas can be burned at an efficiency of over 75 per cent in domestic boilers.

Electricity, as a space heating fuel, is of necessity ruinously expensive. After all, when electricity is produced from prime fuel, only about a third of the energy in the fuel is converted to electric power and it requires very capital intensive plant to do this. Hence even if domestic electric heating appliances are taken as running at 100 per cent efficiency, the overall thermal efficiency is less than 33 per cent, much lower than the combustion efficiency of any other prime fuel.

The use of heat pumps is now gaining ground in applications where direct on-peak electric heating must be used. As such pumps work at a coefficient of performance (COP) of between 2.5 and 3.5, some of the thermodynamic energy loss inherent in electricity generation can be recovered. **Figure 4.2** illustrates heat pump operation.

The actual efficiency of heat production, using heat pumps is:

$$\frac{\text{Coefficient of performance of heat pump}}{\text{Efficiency of central power generation}}$$

Example: a heat pump operates at a COP of 2.13 to convert electricity into low grade heat. Determine the overall thermal efficiency, if the electricity used is generated at an efficiency of 34.5 %.

Overall efficiency of heat generation $= 2.13 \times 34.5 = 73.5\,\%$

Heat pumps do not produce something for nothing; they merely convert electric power more efficiently to low grade energy than nichrome wires. They are therefore only economic to use when it is necessary to dispose of surplus or waste electricity or when there is no alternative to electric space heating.

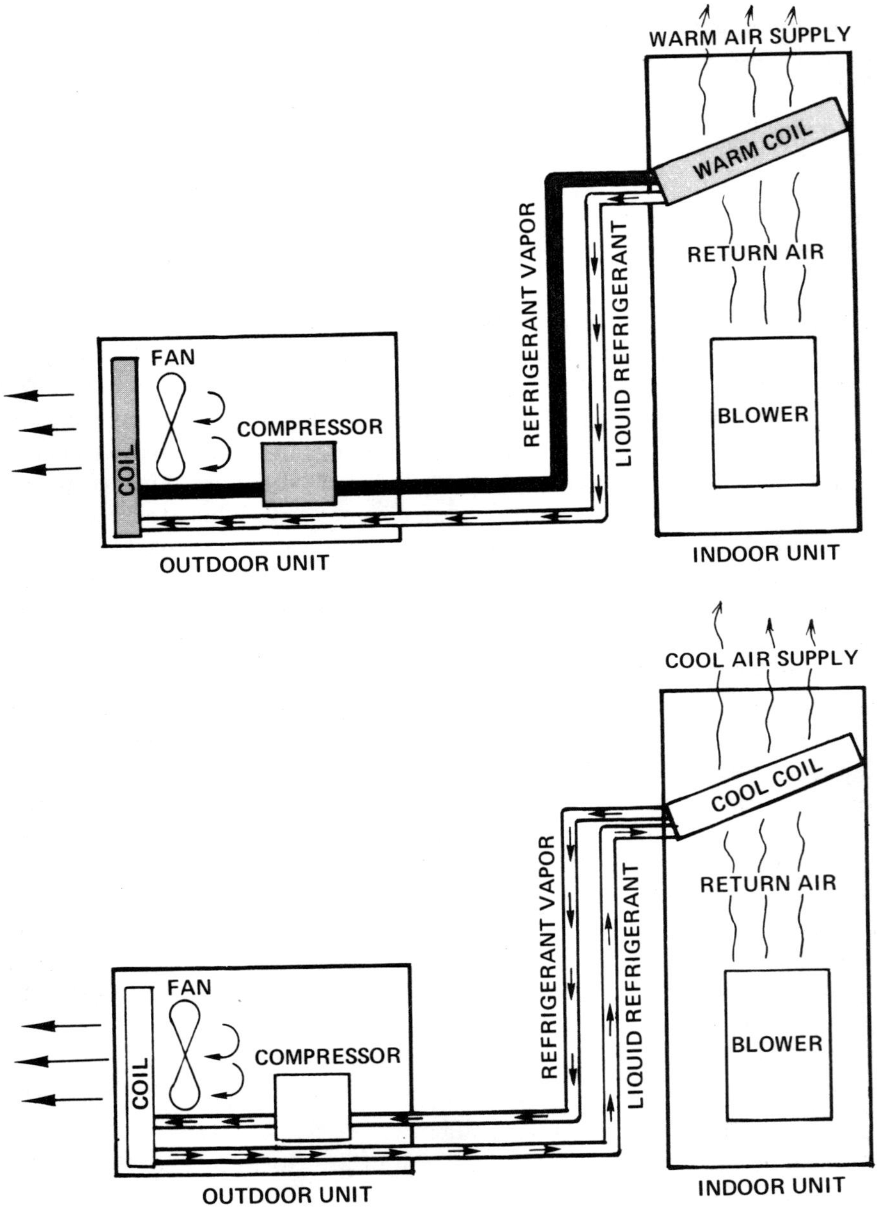

4.2 Illustration of heat pump operation

COMPARATIVE COSTS

It is of interest to compare the initial capital cost to the individual of domestic central heating systems with his share of boiler house capital costs and house connections in group heating schemes. 'House connections' includes everything from the entry of the pipes into the house or block of flats ie valve systems, meters, radiators, internal pipe runs, thermostats etc.

The following data were obtained by the Thermville study group and have been adjusted to make them directly comparable to the cost of a gas fired central heating system (index of 100). The index approach is useful because of the likely effect of inflation which would invalidate monetary data quite quickly.

TABLE 4.4 CAPITAL COST OF APPLIANCES

Cost to individual of domestic heating systems	
Gas	100
Coal	103.8
Oil	113.5
Electricity	93.8
Cost to individual of his share in group heating scheme*	
Coal	105.1
Oil	102.7

*Share of boiler house costs and house installation only – no mains costs.

It will be noted that plant costs for individual central heating systems and for group heating schemes are roughly of the same order of magnitude. Group heating schemes must however bear the following additional expenses:

capital costs of constructing mains and auxiliaries
capital costs of pumping and sub-stations
repair and maintenance costs for distribution system
pumping costs
heat losses in mains system
cost of metering the heat supplied and billing of customers including the cost of bad debts
loss of revenue during periods when the system is being repaired or extended
supervision
administration costs, insurances etc.

It has already been shown that there are no material savings in capital costs by using central boiler houses instead of individual appliances. The above expenses can therefore only be recouped by savings made in fuel costs. In most group heating schemes interest charges constitute the major expense. For example, assuming a life of 50 years for buried mains and an annual repair and maintenance charge of ½% (normal in European district heating networks), the annual cost of the mains is then 10 per cent of the initial outlay if the applicable interest rate is 7.5 per cent, and 12.5 per cent of such outlay if the interest rate is 10 per cent. Such a difference can be the decisive factor in proving the likely viability of a particular group heating scheme.

VARIATION OF HEAT LOADING

The heat requirement (loading) of a group heating system over the year varies with the relevant climatic conditions. For example, in central Europe there is a far wider spread of demand between summer and winter than in the UK, which enjoys a more equable climate. In the UK a given group heating system is likely to operate with the following loading figures for the following number of days in a year:

100% loading	100 days per annum
66% loading	83 days
33% loading	82 days
10% loading	100 days
	365 days

These loadings are only approximate and vary from year to year and also with different localities, but they serve for approximate costing purposes.

TYPES OF FUEL

Figure 4.3 indicates diagrammatically the relative complexity of coal, oil and gas fired schemes.

Gas

When gas is used in group heating schemes, it can be purchased in one of two ways:
(a) on a basic charge with guaranteed supply, or
(b) on an interruptible basis, the Supply Authority retaining the right and option to cut off the gas supply whenever it chooses (usually subject to some notice) during periods of maximum thermal demand. Such a supply arrangement is the cheaper, but the operating authority then must have a fall-back alternative, such as an oil burning plant.

Natural gas as a fuel for group heating stations is very favourable. Boiler plants can be straightforward and compact; the number of chimneys is reduced because of the Clean Air Acts and their height can be kept to a minimum, subject to standard and supply companies' codes. Gas firing at a group heating station is safer than individual gas fired central heating systems, and it is also more feasible to provide a full range of safeguards, such as gas leak detection.

Oil

In the recent past, oil fired group heating centrales were cheaper to run than

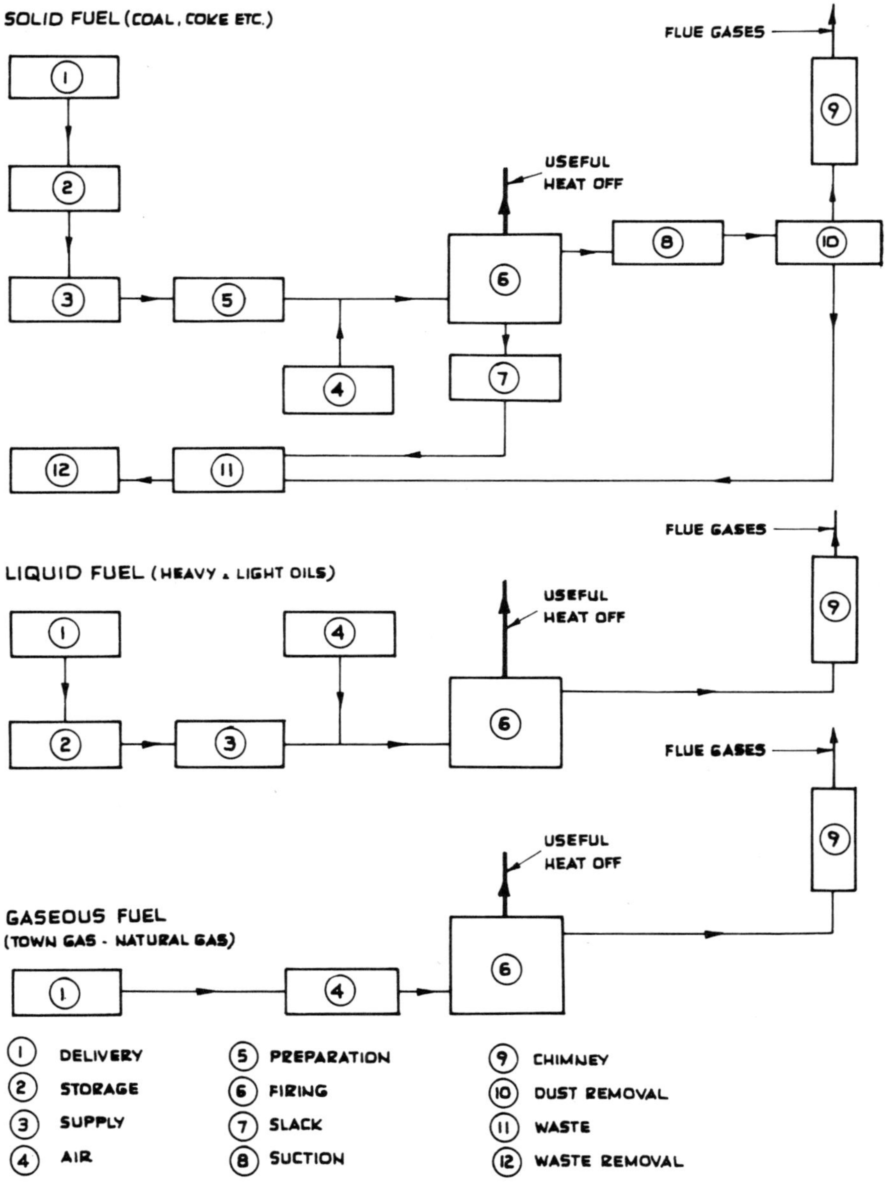

4.3 Diagram showing relative complexity of coal, oil and gas fired group heating schemes

coal fired systems. Oil has a calorific value of between 42.3 and 45.4 J/kg (with a density of approximately 930 kg/m³, a cubic metre of oil contains approximately 40 gigajoules of heat). The calorific value of coal is much lower, between 32 and 36 MW/kg, and it cannot be stored as neatly and cleanly as oil. The plant required for the storage, conveying and combustion of oil is simpler, requires less maintenance and can be operated more easily than an equivalent coal plant. On the other hand proven oil reserves are much smaller than those of coal, and there is an enormous competing demand for oil in respect of petrol, paraffin, plastics etc. In consequence the current price of oil per unit of heat content is far higher than that of coal, and this trend is likely to continue.

Oil is used to the greatest advantage in group heating schemes which are in the interim period before connection to a comprehensive district heating system. In large group heating plants the cheapest grade (3500 seconds viscosity) oil can be used, if there is a worthwhile cost benefit, allowing also for the expense in heating the oil.

Coal

Coal fired boilers have to be dimensioned about 15 per cent larger than the equivalent gas or oil fired units, and they operate in conjunction with costly delivery, storage and ash disposal plant. Modern coal fired group heating plants are highly automated and are no longer labour intensive. Three alternative coal supply arrangements are used with regard to large group heating plants (see also chapter 13):

(a) the coal is delivered to a storage depot at ground level and then conveyed by a bucket elevator and conveyor belt to overhead storage hoppers – one hopper per boiler;

(b) the coal is delivered to separate storage compounds at ground level, and from these it is supplied automatically to each boiler plant by an Archimedian type screw conveyor;

(c) the coal is pumped via a pneumatic pipeline system directly from the supply truck to the overhead bunker for each boiler plant; this arrangement is the most expensive, but it is cleaner and less labour intensive.

Coal ash is discharged automatically from the furnace using either a submerged belt system which removes water-quenched ash from the bottom of the furnace and conveys it to an overhead hopper from which it can be discharged easily into a collection truck, or a pneumatic conveyor system, which removes the dried ash from the bottom of the furnace and blows it into overhead hoppers. A furnace burning 30 tonnes of coal produces about $2\frac{1}{2}$ tonnes of ash. In general, the ash hoppers are designed for being emptied twice a week.

Most coal fired group heating plants burn washed smalls, a grade of coal which has a fairly low ash content and ash diameter of between 20 mm and 50 mm.

Coal fired group heating plants should be designed to store not less than one month's supply of the fuel close to the boilers so as to guard against fuel supply interruptions due to strikes and bad weather.

Currently, coal fired group heating schemes are the most viable because of their considerable saving in cost. Many of the old traditional markets for coal (eg railways or open domestic fires) have largely disappeared never to return, and it is therefore desirable to open up new markets such as in the field of group and district heating, where the economics scale favours coal. In West Germany, for example, coal fired group heating plants benefit specifically from various fiscal advantages and inducements.

CORNER TUBE BOILERS

Types available

These small boilers are widely used for group heating. They can be supplied in the form of packaged boiler houses, comprising hot water boiler with its burner and feed system, primary boiler circuit with shunt pump, distribution pump, pressurising system, expansion vessel, cold water controls and all associated electric circuits. Such packaged boilers tend to be between 5 and 10 MW in capacity; the maximum is around 160 MW. The typical weight for a 10 MW corner tube boiler unit is 11 tonnes when empty, 15 tonnes when filled with water. This is less than half the weight of an equivalent three-pass shell boiler system.

Corner tube boilers can be fired by gas, oil, coal, wood waste, peat or refuse, or they can function as waste heat boilers in conjunction with gas turbines.

Figure 4.4 shows the basic arrangement of a corner tube boiler, **4.5** shows sections through a large boiler of this type and **4.6** sections through a smaller one.

Basic design

The corner tube boiler was invented by Dr Vorkauf and consists basically of an all-welded cage system of corner tubes with interconnecting longitudinal and transverse headers. There are welded-in heating surfaces, such as furnace walls and convection heat banks. The corner tubes ensure that the cooling system is self-supporting.

For the smaller boiler, the furnace is made completely gas-tight by being enclosed in a sheet metal casing which expands and contracts together with the tubes. The larger models use integral fins on the furnace tubes.

A corner tube boiler can be designed to function both as a hot water boiler and as a steam boiler. One popular design is the combined steam/hot water boiler in which 30 per cent of the output is available as line steam. When using the standard hot water boiler the steam generated in the drum which is mounted directly above the boiler can be used for oil heating, soot blowing and oil jet atomizing.

Whilst it is practicable to use corner tube hot water boilers for direct connection into a group or district heating system, an alternative method

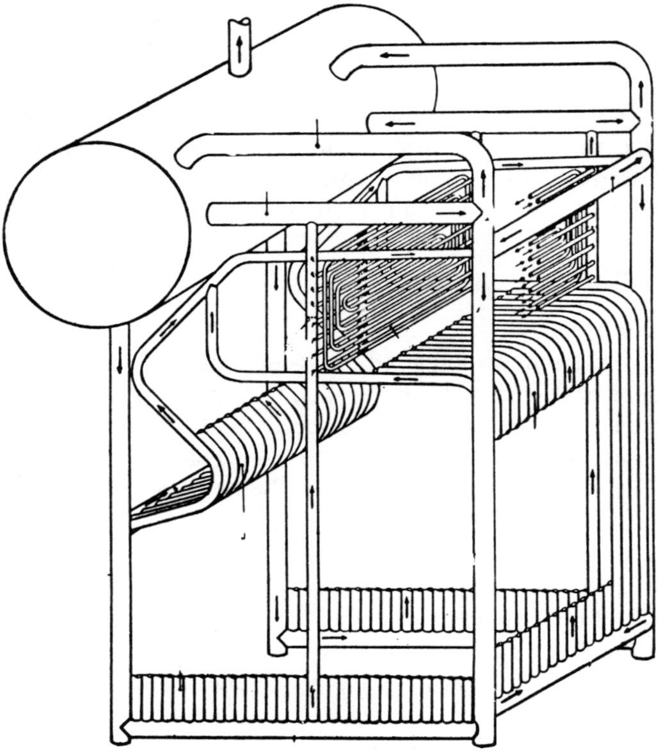

4.4 Schematic arrangement of a corner tube boiler

employs an arrangement in which the boiler water is circulated in a closed primary loop by a separate circulating pump and the heat then transferred to the network via a heat exchanger. The latter method gives better control of the pressure and temperatures in the primary circuit, avoids risks of low temperature corrosion on the boiler gas side and maintains the security and treated condition of the water in the boiler. A dual circuit (steam and hot water respectively) must of course, be maintained when a combined steam/hot water boiler is employed.

The corner tube boiler contains no refractories, except the burner quarls; this saves weight and maintenance. The gas-tight membrane wall avoids low temperature corrosion problems within the furnace.

Uses and efficiency

Currently, corner tube boilers have the lion's share in European group and district heating schemes. They are used for the following purposes:

(a) portable hot water boiler houses for temporary use in the initial stages of group heating schemes;

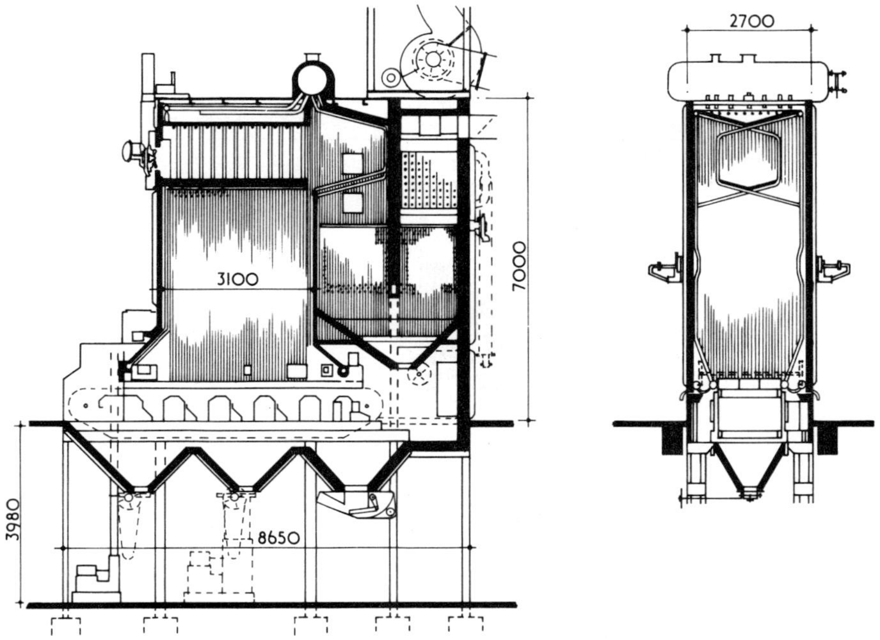

4.5 Sections through a large coal fired corner tube boiler as used in West Germany as a peak demand station in a district heating system: output 10 MW

(b) large permanent hot water boilers for established group heating schemes;
(c) hot water boilers for stand-by and peak duties in large district heating schemes, providing security of supply in the event of breakdown at the thermo-electric station;
(d) waste heat recovery boilers for use with municipal refuse incineration plants in conjunction with group and district heating schemes;
(e) extraction of waste heat from the flue gases of gas turbines;
(f) stand-by steam generators for the supply of medium pressure steam (specially designed corner tube boilers are capable of producing steam at a pressure of up to 105 bars and 520°C (968°F)).

The claimed combustion efficiency of a corner tube boiler is higher than that of an equivalent multiple-pass shell boiler. The largest dimension boilers, with a capacity of 160 MW, measure 6.5 m × 6.5 m × 11.5 m in height. Burner turn-down is to 3 : 1. Efficiencies are claimed of 90.5 per cent (nett calorific value) at 33 per cent load, 91 per cent at 50 per cent load and 89.5 per cent at full load.

A coal fired corner tube boiler is de-rated by about 60 per cent of its output when burning oil or gas, due to the very much lower calorific value of coal. One can employ either a chain grate stoker or adapt the system for pulverised fuel firing. It is necessary to provide appropriate fuel bunkers and

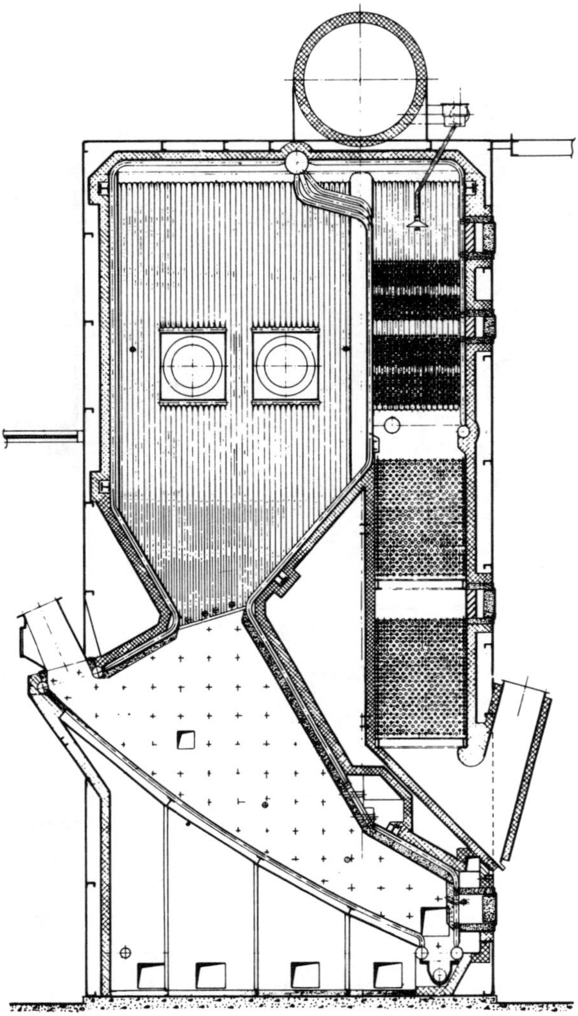

4.6 Section through a smaller corner tube boiler

ash handling plant. Corner tube boilers are well suited for use as waste heat recovery from large-scale refuse and wood waste incinerators and some 130 of them are currently known to operate in this manner. Some corner tube boilers are used for dual fuel firing, it being claimed that a change-over from oil to solid fuel firing (and vice-versa) can be effected in less than 24 hours.

SMALL MOBILE GROUP HEATING PLANTS

When a district heating system is being constructed, it is necessary to position temporary small boiler houses strategically so as to serve the various individual sections until the overall system has achieved sufficient capacity to be connected to the main combined heat and power stations. Typical of such packages are the small containerised heating plants manufactured by Witermo Oy of Finland of capacities between 160 kW and 5.9 MW, measuring between 5.3 m × 2.8 m × 3.8 m and 9.5 m × 3.7 m × 3.8 m. Weights are between 7.8 and 44 tonnes. **Figure 4.7** shows one such unit in diagram form. **Plate 8** is a view of a packaged boiler plant.

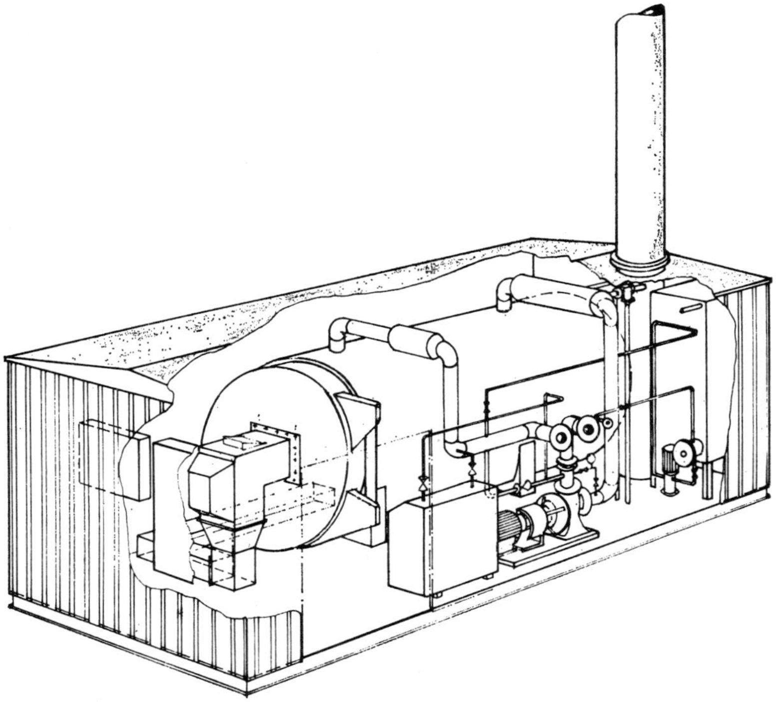

4.7 Diagram of a packaged heating plant

Each packaged boiler house consists of an all steel welded space frame system with insulated walls and roof. Floors consist of chequer steel plate. The fuel used is either heavy fuel oil or gas. The flue gases discharge into a twin-barrel guyed steel stack, in which the external barrel takes the stack weight and the wind forces; the maximum height of such a stack is about 30 metres. The packaged system includes the high pressure pumping system and all necessary valves and controls. Transportation to the site is on a low loader truck. Once arrived at site, it is only necessary to lift the assembly on to prepared foundations, erect the prefabricated chimney and connect the oil and gas supply. The unit is joined into the group or district heating supply mains and can be made operational in a very short time.

These boiler packages occupy much less ground space than traditional types of boiler houses. If so desired, they may be located on roofs of blocks of offices or flats. They can be easily moved from place to place as required.

Plates 6 and **7** show a packaged coal fired boiler plant and a view of a group heating centrale.

WASTE HEAT RECOVERY FROM SMALL SCALE INCINERATORS

It is practicable to operate a small group heating scheme utilising the heat produced by a simple continuously fuelled refuse incineration furnace. A typical plant of this type is the Consumat system (**4.8**) in which the waste is

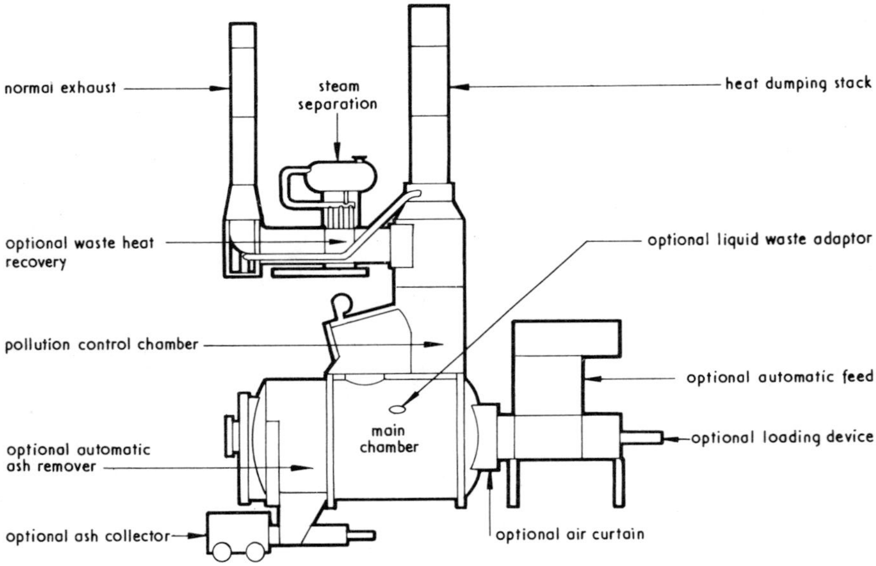

4.8 Consumat refuse incineration plant suitable for group heating

first of all heated in a refractory-lined combustion chamber with the aid of small auxiliary gas burners in an atmosphere of starved air incineration (or virtual pyrolisis). This process is autothermic and proceeds spontaneously as soon as an operating temperature of about 800°C (1472°F) has been achieved. The cracked combustion gases are then passed into the oxidation chamber, where they react with excess air until a temperature of 1000 to 1200°C (1832 to 2142°F) is achieved. The hot exhaust gases can be passed to a waste heat boiler to produce steam. The feeding of the waste into the incinerator is by automatic loader. Ash removal is made simple. All the Consumat units consist of standard elements which are each capable of handling up to 1 tonne of fuel per hour. Each module can provide about 3.6 tonnes of low pressure steam per hour. When no heat is recovered, the flue gases are diluted with cold air and discharged to atmosphere.

COMBINED HEAT AND POWER PLANT (CHP)

Suitability

In the USA and in Canada there exist numerous circumstances where comprehensive district heating systems are neither desirable nor practicable. Group heating with small total energy plants is used for such projects which include large privately owned apartment complexes, office and shop developments and industrial 'parks'. In many instances it is cheaper to supply heat to such complexes by small total energy generators than to purchase power from the electricity utility companies.

The advantages claimed for such schemes are:

(a) Fuel can be utilised at a far higher efficiency in central power stations which operate condensing turbines.

(b) The cost of long transmission lines and of auxiliary switchgear is very high: total energy operation avoids these costs.

(c) The disruption of power supply caused by adverse weather conditions (tornadoes, hurricanes, excessive snowfall, floods etc) can be avoided, and total energy installations are also immune to major breakdowns and to labour disputes at central power stations.

(d) Mass produced small scale gas turbines, gas engines etc are employed; these cost less per unit output than a massive central power station equipment which is generally a 'one off' installation; total energy operation is, however, not invariably cheaper than purchase of electricity from power utility companies combined with on-site heat generation in central boilers.

The prime essentials for successful total energy operation are:

(a) A constant supply of cheap fuel: in the USA this is normally provided by natural gas and/or waste gas from oil refineries.

(b) Minimal variation in electricity demand: some fluctuation is unavoidable over a 24 hour period, but taken over a year such variations should be as small as possible; this can be achieved by coupling industrial and domestic power loads and by using summer power loads for air conditioning or refrigeration purposes.

(c) Adequate utilisation of the heat produced in the total energy system: the dissipation of waste heat in cooling towers is very wasteful of resources and expensive. In the UK, and in other countries with similar climates, it is, in general, very difficult to achieve, unless all-year-round industrial processes are included as heat load. In North America the winter and summer heat demands are of similar magnitude due to the widespread provision of air conditioning.

(d) Efficient equipment: it is necessary to select only equipment which has proved itself to be reasonable, trouble-free and able to operate reliably with minimum supervision.

Total energy schemes tend to be most economic at times of low interest rates, high basic fuel costs and conditions of rapid price inflation. Privately owned total energy installations may be less cost-effective when cheap power can be supplied by nuclear energy than when power is provided by expensive fossil fuel.

Types of CHP plant

Small scale steam turbines are used mainly in factories where there is a steady demand for heat as low pressure steam for process and space heating. It may then be more economic to supply this steam from a back pressure turbine and obtain electricity as a by-product, rather than to employ a conventional boiler plant. Such a system tends to produce excessive heat and inadequate electricity and it is therefore usually operated with supplementary power supply from the grid system. It is common practice to use steam pressures of up to 45 bars only, as small boiler plants with higher operating pressures are disproportionately costly. The practical upper limit of temperature is about 450°C (842°F). Small scale steam turbines can also be used in conjunction with gas turbines and diesel or gas engines.

In open-cycle gas turbines air is drawn from atmosphere, compressed, heated at constant pressure by the fuel/air mixture and expanded in the turbine. The work produced by the turbine drives the compressor in the cycle, and the balance (usually about 33 per cent) provides the useful power output of the machine.

The major application of open-cycle gas turbines is in the aviation field. There are, however, circumstances where small open-cycle gas turbines can be employed in stationary positions, utilising the waste heat in the exhaust for the supply of heat. With conventional gas turbines, about 35 per cent of the net heat in the fuel is converted into useful work capable of generating electricity; a further 30 per cent is dissipated in engine coolers, oil coolers and radiation and convection heat losses and the remaining 35 per cent is available for producing heat.

Group heating schemes which are based on open-cycle gas turbines must have a steady heat demand which balances with the power demand. Such matching can be achieved when air conditioning is provided, preferably in conjunction with some other heat requirements, such as those of swimming pools, laundries, industrial process heating needs etc.

Closed-cycle gas turbines differ from open-cycle gas turbines in that the operating gas (nitrogen or helium) is entirely enclosed. Closed-cycle gas turbines are much more expensive than open-cycle gas turbines, but they are more efficient for power generation and can be fired by almost any fuel. CHP operation is on the same lines as with open-cycle sets.

Diesel engines have a very high thermal efficiency and can convert up to

38 to 39 per cent of the net energy in the fuel into useful mechanical energy. Gas engines are rather similar in design. Heat is abstracted in either case from the exhaust gases, the cooling jacket and the oil coolers. About 37 per cent of the heat produced in these types of engines is recoverable, 20 per cent of this from exhaust heat, 12 per cent from cooling jacket heat and 5 per cent from oil cooler heat. The exhaust heat is utilised by employing waste heat boilers; the heat in the water jacket is extracted by flashing off steam from the water jacket. In general slightly less heat than power is produced by such engines.

Operating group heating schemes on CHP

Before deciding on the adoption of a total energy plant for the operation of a group heating scheme it is necessary to ascertain that there is always sufficient demand of heat to remove the waste heat produced by the prime mover. For most types of such plant the amount of heat produced balances the amount of generated electricity. CHP group heating plants cannot be effectively operated unless there is an adequate summer take-off of heat energy, either for air conditioning and/or for industrial uses. In North America there is usually little difficulty in this respect as a heavy air conditioning demand exists in most areas during the summer; the waste steam from the prime movers is then utilised to drive centrifugal chillers.

Typical of the operation of a total energy scheme is a plant installed at Regency Square, Jacksonville, Florida (**4.9**). All the power, heating and cooling requirements of this complex are supplied by nine standard 450 kW Caterpillar natural gas engines plus an additional 570 kW system to drive a large air conditioning compressor. At peak demand periods eight of the nine units are in operation; normally only five sets are needed by day and three

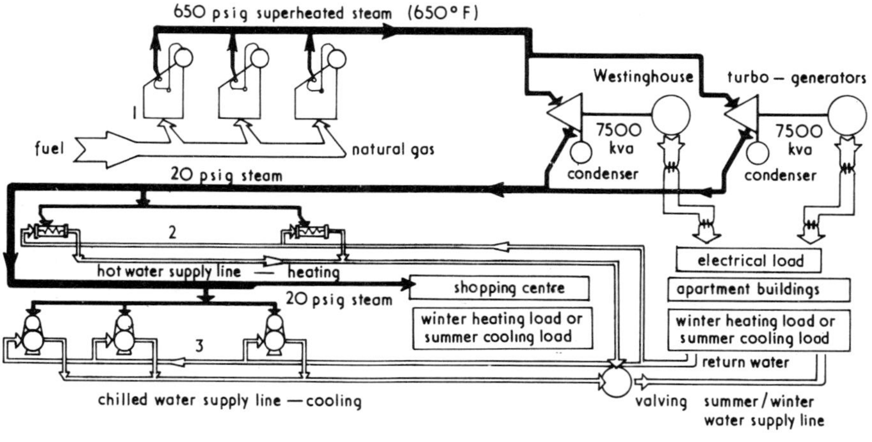

4.9 CHP scheme at Regency Square, Jacksonville, Florida, USA

at night. Additional sets are started and stopped automatically as power demand increases or decreases. The power is generated at 480 volts, but it can be distributed at varying voltages, depending upon specific requirements, using transformers. Heat is extracted from the exhaust and from the jackets of the gas engines. When an engine is operating at full load it is possible to extract 2540 MJ of heat per engine per hour in the form of steam at 1.8 bars (absolute). A four-pipe distribution system feeds the chilled water from the absorption chiller and the hot water produced in a steam/water heat exchanger to the various locations in the group scheme. The chilled water pipes have a diameter of 355 mm; the hot water pipes of 202 mm. The system is economically very successful because care has been taken to match power and heat.

Economic viability of CHP

The most important criterion in assessing viability is the pay-out (or pay-back) period of a system (in years). This is derived as follows:

$$\text{Pay-out} = \frac{(C_e - C_m)}{H\,(E - (M + F - H_r))} \text{ years, where}$$

C_e is the first cost of a total energy plant (£/kW)
C_m is the first cost of electric motors (£/kW)
H is the number of hours of operation (hours/year)
E is the cost of electricity (£/kWh)
M is the engine maintenance cost (£/kWh)
F is the fuel cost (£/kWh)
H_r is the price of the recoverable heat (£/kWh)

A more sophisticated approach to the costing of group heating plants which operate on total energy (instead of buying in electricity and generating steam or hot water locally) has been developed by the Energy Economics Section of the Northern Natural Gas Co, Omaha, Nebraska, USA.

In this, the Uniform Present Worth Factor (UPWF) is employed. It is expressed as follows:

$$\text{UPWF} = \frac{I}{GS + HRS - ME - FE - IT}\text{, where}$$

I is the investment ((a) below)
GS is the generator saving (b)
HRS is the heat recovery saving (c)
ME is the maintenance expenses (d)
FE is the fixed expense item (e)
IT is the income tax rate (f)

Paragraphs (a) to (f) below demonstrate how all these data can be evaluated separately before being introduced into the above master equation.

(*a*) *Investment factor* (**I**)

$I = (\text{max } D) \times IR \times (1 + SP) \times CI$, where

(max D) is the maximum power demand (kW)

IR is the instantaneous peak ratio which gives the ratio between peak and average demands
SP is the capacity of extra stand-by plant needed given as a fraction of the installed power generating capacity
CI is the installed cost (money value/kW)
In practice, in the USA, IR works out at 1.1 and SP at 0.33.
These figures would not however apply to countries (such as the UK) which possess no significant summer air conditioning load where the value IR would consequently tend to be higher. If power demand is steady over the years, however, figures close to those experienced in the USA may apply.
(*b*) *Generator saving* (**GS**)

$$GS = kWh\left(ER - \frac{3600 \times FR}{Eff \times NCV}\right), \text{ where}$$

kWh is the number of kilowatts of power used each year
ER is the cost of electricity as bought in from the public utilities (money value/kWh)
FR is the fuel rate or the cost of fuel in terms of money per unit weight or volume
Eff is the efficiency of the generator plant expressed as a fraction
NCV is the net calorific value of the fuel used (kJ per unit weight or volume) (The factor 3600 correlates this value with the kWh used, for the cost of power.)
(*c*) *Heat recovery saving* (**HRS**)

$$HRS = \frac{Kwh \times PHR \times PRU \times FR}{NCV \times Eff \times BEff}, \text{ where}$$

PHR is the fraction of heat which can be recovered
PRU is the fraction of recoverable heat used
BEff is the boiler efficiency expressed as a fraction.
(*d*) *Maintenance expenses* (**ME**)
ME = kWh × MR, where
MR is the maintenance rate in terms of money per kWh of power produced.
(*e*) *Fixed expenses* (**FE**)
FE = I × FER, where
I is the investment costs
FER is the fixed (overhead) expenses such as administration costs etc given as a fraction of the investment cost. In the USA this is often assumed as being 0.03 but it is always necessary to evaluate it for a particular case.
(*f*) *Income tax rate* (**IT**)

$$IT = ITR\left(GS + HRS - ME - FE - \frac{I \times DF \times INT}{2} - I/EL\right), \text{ where}$$

ITR is the income and other tax rate as a fraction
EL is the economic life of the plant
DF is the debt fraction
INT is the interest rate as a fraction.

Income tax rates are likely to be complex due to the various base rates of taxation, corporation tax, value added taxes, exemption limit grants etc; therefore this part of the equation is best evaluated by an accountant.

When the above six items have been computed, they are substituted into the master equation to provide the pay-out period in years. For example, if the UPWF is 8.33 years the anticipated return on capital is 12 per cent. Obviously if loans are available below 12 per cent the project can be deemed viable.

CHECK LIST

1. The various factors which decide on the adoption or rejection of group heating
2. Whether the minimum connected capacity is available
3. Relative efficiencies of group heating centrales and individual appliances
4. Scope for installing a heat pump and any limitations it may have
5. Comparative capital costs between creating group heating schemes and relying on individual appliances
6. Likely variations of heat loading of group heating scheme over a full year
7. Most suitable fuel
8. Most suitable boiler type (consider corner tube boiler and packaged boiler)
9. Scope for waste heat recovery using small scale incinerators
10. Application of combined heat and power plant for group heating
11. Suitable plant for CHP schemes
12. Economic viability of CHP schemes.

REFERENCES

H. Buck, 'Operation of a 25 MW pure heating station', *Fernwärme International*, 3 (1974) 3, pp 79–82 (in German)

'Coal fired district heating systems: the National Coal Board's contribution', *Building Services Engineer*, 46 (3), June 1978, pp A11–A13

R. M. E. Diamant and J. McGarry, *Space and District Heating*, Butterworths, London, 1968

S. Haal, 'Comparison of different methods of heating small houses', *VVS*, 6/7, 1974, pp 21–26 (in Swedish)

R. F. Hampton, 'Heating schemes for council house developments', *Heating and Ventilation Engineer*, June 1974, pp 533–538

E. Nickel, 'Hot water boilers with oil and/or gas firing', *Fernwärme International*, 2 (1973) 5, pp 115–121 (in German)

T. Ojima and K. Nezu, 'Waste heat by utilities and industries in urban areas of Japan', *District Heating*, 60 (2), 1974, pp 26–33

T. C. Petersen, 'Natural gas fired energy plant serves the entire city centre', *VVS*, 2, 1974, pp 41–43 (in Swedish)

'Supply of district heat from pure heating stations', VWEW Frankfurt, 70, 1973 (in German)

'The district heating of Thermville', *Heating and Air Conditioning Journal*, June–September 1974
'The Tredegar Road district heating scheme', *Consulting Engineer* 43, May 1979, pp 38–39
K. Waterking, 'The corner tube hot water boiler', Heizung, Lüftung, Haustechnik, 6 May 1955, pp 170–175 (in German)

CHAPTER 5

Combined heat and power generation

COST OF HEATING SCHEMES

A district heating scheme will prove economically viable when its overall costs compare favourably with the burning of fuel in individual appliances. The decision to embark on such schemes usually involves a great commitment of financial and engineering resources. It is essential that all the relevant factors are closely examined and analysed with a view to arriving at correct conclusions and plant specification.

Such consideration should take account of the fact that district heating offers advantages other than cost advantages, eg reduction of air pollution, centralised maintenance, space savings, improvement in safety of dwellings and an acceptable method of burning solid fuel on a large scale.

Examining, for example, the prices of fuels as quoted in the UK in October 1980, domestic electricity was the most expensive at £8.11 per GJ, followed by off-peak electricity at £3.78 per GJ, gas at between £2.60 and £2.85 per GJ but domestic heating oil at £4.01 per GJ. Commercial oil was somewhat cheaper at about £2.92 per GJ. As regards coal, while domestic coal varied in price between £2.00 and £2.50 per GJ and smokeless solid fuel cost up to £3.02 per GJ, commercial grades were much cheaper at about £1.58 on average. Coal fines and coals with high ash content, which can be burned efficiently in modern large scale pulverised fuel furnaces, cost still less: in the order of £1.00 per GJ.

All fuel prices fluctuate in relation to each other as technology changes or new political factors arise and particularly, in response to an imbalance in the supply and demand situation. Fuel prices tend to be more responsive to inflationary pressures than most other common commodities.

To justify group heating schemes on pure economic grounds it is necessary to produce heat in a central station cheaply so as to compensate for the annuity costs of the pipeline system. Examination of relative costs indicates that this condition is not achieved very easily.

A modern domestic gas fired central heating system operates at an average combustion efficiency of 75 per cent. Hence the net cost of heat to the consumer paying a retail price of, say £2.60 per GJ delivered, is only £3.47 per GJ for the actual heat obtained. Admittedly the individual domestic plant has to be purchased and installed, but the overall cost is far less than that of providing a group heating scheme.

If commercial oil is burnt in a central plant at an efficiency of 82 per cent (which is quite difficult to achieve) resulting in an actual heat cost per useful GJ of £3.68, then it will actually cost more to supply heat from a central oil fired boiler house than to produce the same heat output by small domestic gas fired boilers. All this has to be taken into account before any money can be spent on pipelines or wages.

Profitability is somewhat better with coal fired stations. If coal costing £2.20 per GJ at 75 per cent efficiency is burned in a modern boiler plant the net cost for heat obtained is £2.93 per GJ ie a saving (compared with gas) of a mere 54 pence per GJ to pay for everything.

These prospects are not very encouraging and, however one may try to juggle with the figures, it is obvious that direct fuel combustion cannot possibly justify the huge cost of installing district heating networks.

COST OF COMBINED SCHEMES

To consider the economic problems in a different way: let the heat be generated during power production. It has been established that during simultaneous heat and power production a sacrifice of only one GJ of electricity produces about five to six GJ of heat, so that the process of obtaining heat as a by-product from power generation can be considered to be between 500 and 600 per cent efficient. The fuel consumed in Utilities power stations is relatively cheap amounting to only £1.00 per gross GJ. Even at its most conservative, therefore, waste heat from power generation should not cost more than 25 pence per useful GJ.

Comparing the cost of this heat with that produced by efficient domestic gas fired central heating systems (stated above as about £3.47 per useful GJ) it can be seen that heat costs as a by-product from power generation are quite insignificant. The difference of £3.27 per GJ is adequate to pay for district heating networks, pumping costs, wages, administration etc and allows an operating profit. It is now possible to understand why the district heating systems in Germany, Sweden, USSR etc are so very successful. If heat is produced as a by-product of electricity generation using nuclear power the cost picture is still more favourable because nuclear fuel costs are quite negligible. **Figure 5.1** indicates the costs of operating various types of combined heat and power stations.

The above argument has deliberately taken into consideration only the actual fuel cost directly involved in the production of heat supplied to the district heating system and has not allowed for the capital cost of turbine plant which is one of the main items involved in determining the cost of electricity. The capital cost of constructing power stations is not governed by their average power requirement but by their maximum power need which may occur for only a few hours each year. For most of the remaining time the power plant runs well below its maximum capacity. It is at present

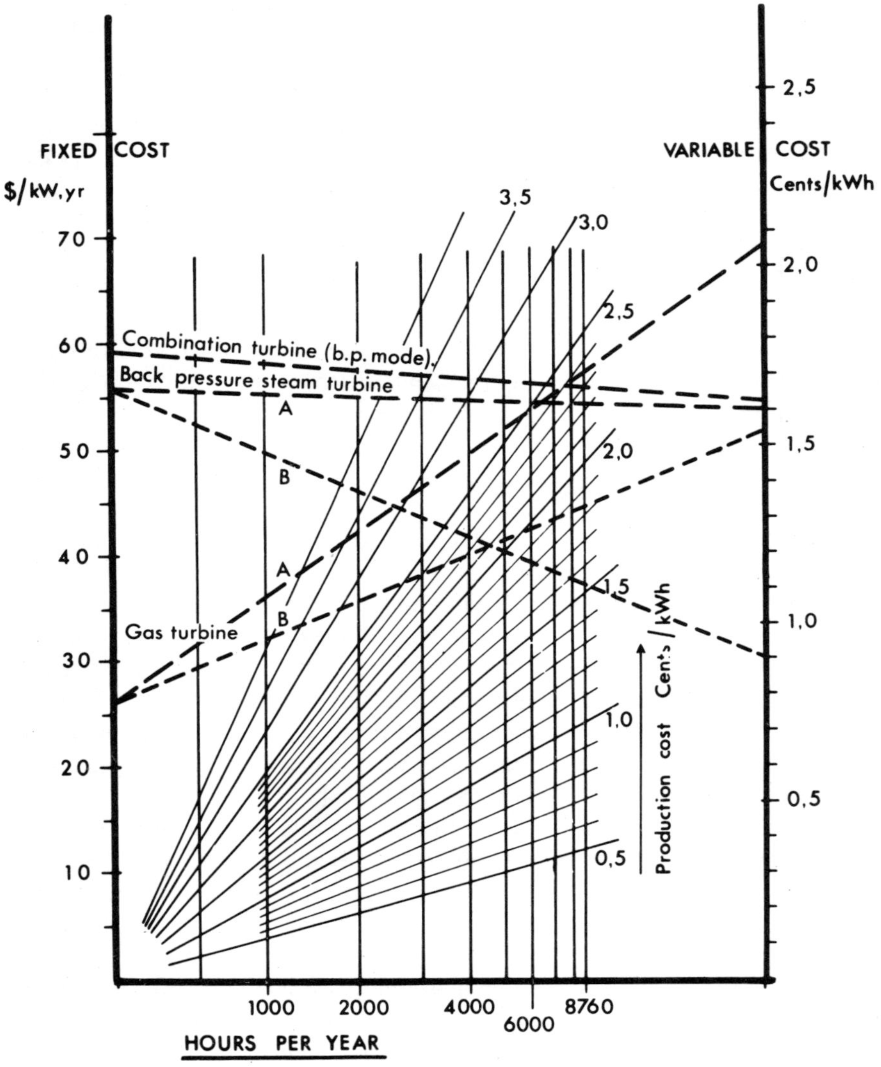

A Credited 0,355 c/kWh$_{th}$ for heat

B —"— 0,75 —"— —"—

5.1 Costs with the various types of combined heat and power stations

impossible to store electricity (except by means of ruinously expensive and very doubtful schemes such as the Dinorvic pumping basin) but there is no difficulty whatsoever in storing hot water either in well insulated pipeline systems or in efficiently insulated hot water accumulators. It is practicable in this way to operate turbines with maximum heat bleed at night when power requirements are low and run them normally during periods when the power demand is high. In consequence no additional turbine investment is necessary to run intermediate take-off condensing (ITOC) stations of this kind except for small initial modifications. Moreover district heating now takes care of some of the peak electric power load because it is possible to use waste heat generated at night instead of having to generate extra electricity for electric heaters. Consequently the power demand curve is flattened out to such an extent that the electricity supply industry can be run with less plant than previously. If the plant is already installed the older less efficient equipment can be taken out of service with considerable savings in labour, maintenance and administration and without risking inadequate power generation capacity during periods of maximum power demand.

Capital savings achieved in this way can pay for the turbine modifications required to enable plant to operate as combined heat and power stations.

ADVANTAGES OF ITOC STATIONS

ITOC stations produce heat in the form of hot water or low pressure steam at a very small cost; they also offer other advantages. By the operation of a single valve it is possible to adjust the output of a turbine instantaneously from producing electricity but no heat (at an efficiency which is almost identical to that obtained with normal condensing turbines) to producing both electricity and heat up to the maximum heat output possible with a given design of plant. The heat output of such a plant can be adjusted, second by second, to accord accurately with the heat demand of the system without varying the fuel throughput of the furnace because such adjustment merely alters the amount of electricity generation sacrificed to produce heat. With this arrangement between five and six units of heat are obtained on average for each unit of electricity output sacrificed.

No other method of equating the heat demand and heat supply in large district heating systems shows anywhere near the same efficiency. Although it is possible, in directly fired district heating systems to vary the fuel supply to the furnace, this is a wasteful process without immediate effect upon the circulating heating medium and it often becomes necessary to dissipate excess heat to the atmosphere in dump condensers. In ITOC turbine systems an automatically controlled valve adjusts the output instantaneously.

Figure 5.2 illustrates the effect of an ITOC turbine in flattening electricity demand. In practice the turbines are operated at night with maximum back pressure operation and the surplus heat is stored in well insulated accumulators and within the pipeline network in the form of sensible heat. During the morning peak demand period for electric power the bleed valves for steam are closed and the turbine then operates as a full condensing turbine

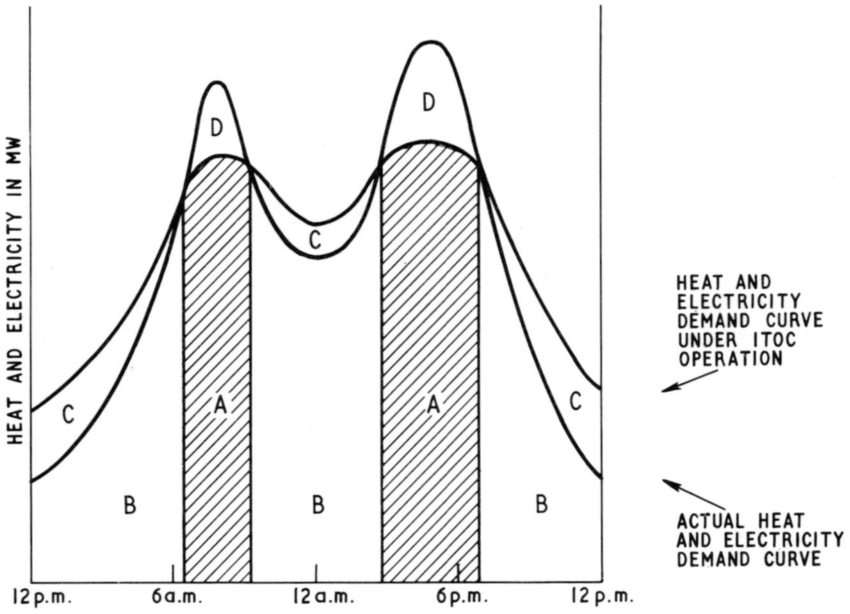

5.2 The operation of an ITOC turbine flattens down demand
A *Turbine under condensing operation only*, **B** *Turbine under partial back pressure operation*, **C** *Extra heat produced is stored against peak demand*, **D** *Heat demand supplied from stored hot water*

giving its maximum rated output of electricity but no heat. The hot water which has been heated up overnight and stored then provides space and water heating and relieves the power load by lessening the demand for peak current. The turbines are operated on maximum condensing cycle until the power demand lessens around lunchtime. The back pressure valves are then opened somewhat to boost the water temperature to meet the heating load during the afternoon and evening peak power demand periods when the turbines are again operated on full condensing cycle. This procedure offers the following very considerable advantages:

(a) loading of turbines is spread much more evenly over the year because the heat load, which normally causes the greatest fluctuation in power demand, is supplied by bleed steam and not by electric power. Thus a given power consumption network needs investment in power plant for the achievement of complete security of supply because the power plant is used more fully and for much longer periods. It is consequently now feasible to construct only high efficiency plants without any need for a reserve of lower efficiency plants which would have to be operated expensively only a few hours each year to supply peak needs.

(b) by operating the turbine system fully night and day, yielding mainly heat at night and the maximum amount of electricity during the day, a far more even plant operation is achieved which has considerable operational advantages: for a given expenditure on plant, labour and overhead charges such equipment produces a much higher output of usable energy with worthwhile financial benefits.

(c) when operating at night on back pressure very much better fuel utilisation is achieved than by running plant on condensing operation only, so that it is no longer necessary to sell electricity at an uneconomic 'night tariff' because in ITOC systems excess power is converted into heat at a very favourable rate and stored in the form of hot water. The heat losses from well insulated modern accumulators and pipelines are relatively small and do not greatly influence the overall cost analysis.

Advantage (c) is probably the most vital aspect of ITOC turbines.

Figures 5.3 and **5.4** show, respectively, the practical loading distribution with a typical town's district heating network and the total energy balance of the Linkoping district heating system.

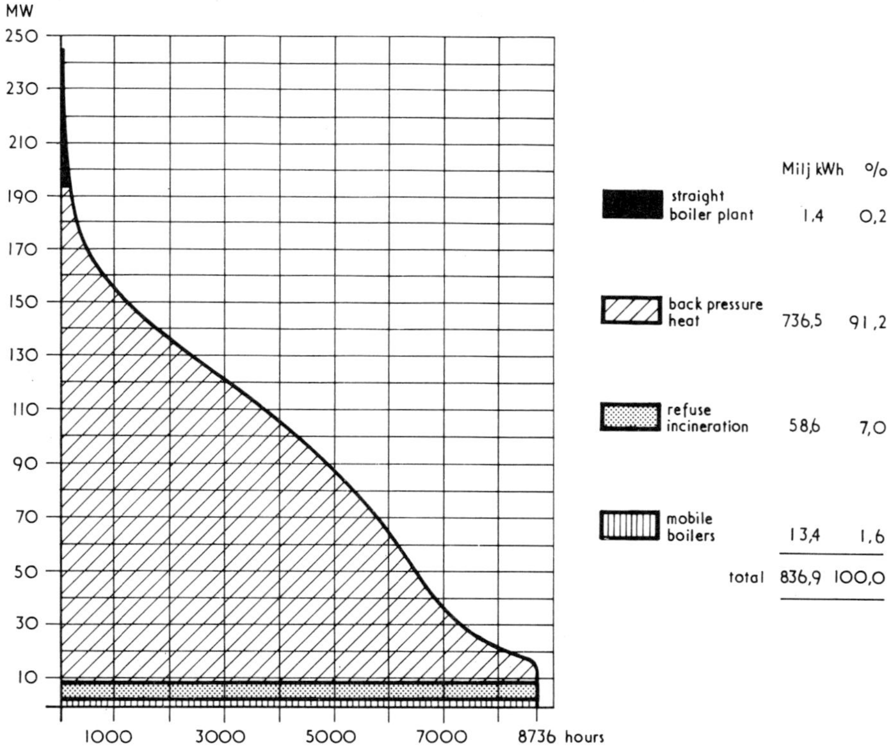

5.3 Practical loading distribution with a typical town's district heating network

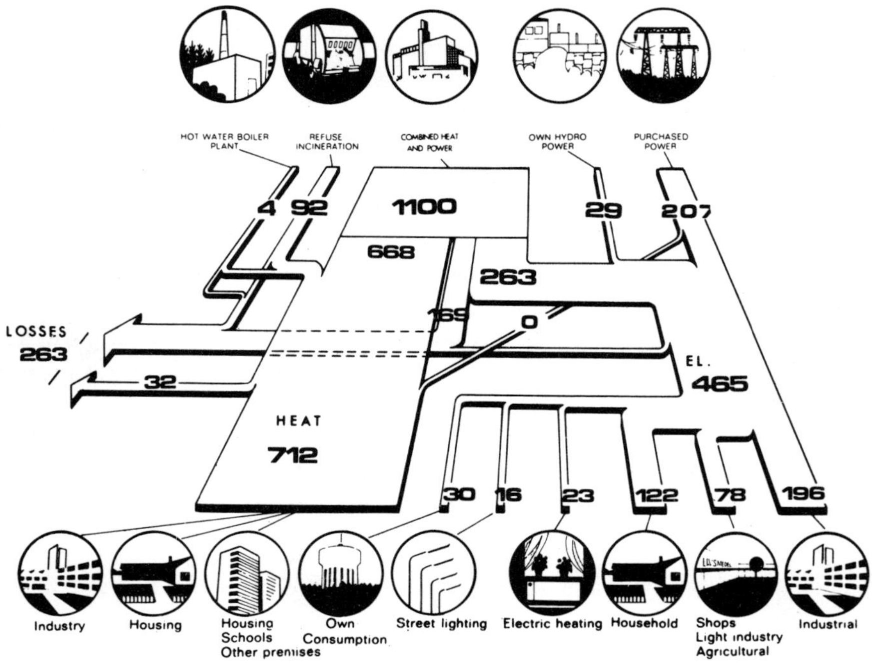

5.4 Total energy balance Linkoping district heating system

THEORY OF STEAM TURBINES

Turbines convert heat energy into mechanical energy and steam turbines operate at their greatest efficiency when designed for very large outputs. Sets of up to 1100 MW driven at 1800 rpm have been made but perhaps a more common size is around 650 MW. Steam turbines are not efficient unless operated at steam conditions well above the critical temperature for water which is 374.15°C (705°F).

According to Carnot's law the efficiency of a gas engine, in terms of the ratio of work done/heat input at the fluid feed temperature, cannot exceed the value of:

$$\frac{T_1 - T_2}{T_1} \text{ where}$$

T_1 is the top temperature and T_2 is the exhaust temperature, both measured in kelvins (Kelvin temperature = Centigrade temperature + 273.15).

In practice no engine, however perfect, fed with a medium at a temperature of T_1 and exhausted at a temperature of T_2 can yield more than the above fraction of the energy input as useful work. With steam turbines, this efficiency is much lower as steam is not a perfect gas. In practice, circulating water is heated up by steam bleeds to a temperature practically the same as that at which the water is likely to boil. This feed water is then fed into the boiler where latent heat is added to generate saturated steam. The application of additional heat in a superheater heats the saturated steam above the boiling point which relates to the steam pressure. When the superheated steam is fed into the turbine it performs work and in consequence its temperature drops. Partial condensation takes place. It is not permissible to have more than 12 per cent condensed water in the steam flow of the turbine because an excess causes cavitation of the turbine blades. For this reason the partially condensed steam is returned to the boiler for a so-called 'reheat' when more heat is given up to it to enable it to become superheated steam once more.

CONDENSING TURBINES AND BACK PRESSURE TURBINES

A condensing turbine is operated, with the object of achieving the highest possible electric current-producing efficiency, by employing a virtual vacuum on the outlet side. This makes it practicable to achieve a pressure of around 4 kPa on the condensing side of the turbine where the steam is extracted and condensed by means of circulating water. At a pressure of 4 kPa (absolute) the boiling point of water is only about 29.2°C (84.5°F). Enormous quantities of lukewarm water at around this temperature have to be cooled down to operate the condensers and unless the power station is situated near a river, lake or the sea it is necessary to provide the associated large cooling tower capacity to enable such quantities of low grade heat to be dissipated to the environment. There is no economically feasible method by which this can be utilised. Most district heating systems require hot water at minimum temperatures of between 90 to 100°C (194 to 212°F), though higher hot water temperatures are usually desirable to reduce the costs of heat transmission.

Assuming that, instead of a condensing turbine with outlet pressure of 4 kPa, a back pressure turbine with outlet pressure of 100 kPa is employed, and also assuming frictionless bearings, perfect electricity conversion and a number of other similar impossible criteria, then

$$\frac{\text{Electric power obtained}}{\text{Heat added to steam}} = \frac{h_1 - h_2}{h_1 - h_3}, \text{ where}$$

h_1 is the feed enthalpy of the steam
h_2 is the exhaust enthalpy of the steam
h_3 is the enthalpy of the condensed water.
(Items such as the work done by the feed pump, the effect of reheat cycles etc are ignored in the above.)
Assuming further that steam is used at 170 bars and 500°C (932°F), the steam tables (see Appendix C) show that the enthalpy of steam (h_1) at that temperature and pressure is 3.396 MJ/kg. At a back pressure of 4 kPa

the exhaust temperature is 29°C (84.2°F), and the enthalpies for steam and water at this temperature at 2.554 MJ/kg and 0.121 MJ/kg, respectively (h_2 and h_3). The ratio of electricity produced to heat supplied to generate steam is therefore:

$$\frac{3.396 - 2.554}{3.396 - 0.121} = 0.257 \text{ or } 25.7\%$$

At an exhaust pressure of 100 bars the enthalpy of steam at that pressure and at a temperature of approximately 100°C (212°F) is 2.676 MJ/kg (h_2) while that of water is 0.420 MJ/kg (h_3). The ratio of electricity supplied to heat used in the steam generator is then:

$$\frac{3.396 - 2.676}{3.396 - 0.420} = 0.241 \text{ or } 24.1\%$$

and the sacrifice of electricity is:

$$\frac{(25.7 - 24.1)}{25.7} \times 100\% = 6.22\%$$

representing an enthalpy in the incoming steam of 0.211 MJ/kg. However in exchange for this loss of heat 2.676 − 0.420 = 2.256 MJ of useful heat at 100°C (212°F) is now obtained which can be employed for district heating, yielding a theoretical efficiency of heat generation in this case of about 1000 per cent. In practice small combined heat/power generating plants usually yield six to eight units of heat for each unit of electricity sacrificed.

The above reasoning is a considerable simplification of what actually happens. Modern high pressure and temperature steam power stations operate with steam pressures in excess of 300 bars and temperatures of 650°C (1202°F) and more, and complex systems of feed water preheating, multiple reheat cycles etc enable overall efficiencies in excess of 35 per cent to be achieved in terms of the energy in the electricity produced relative to the energy input in the fuel. Steam feed enthalpies in excess of 4.00 MJ/kg steam are achieved and, in such power stations, only a fraction of the heat produced can be taken off for district heating purposes: the station has to be operated partly on a condensing cycle and partly on back pressure ie it becomes an ITOC station.

It can be shown that relationship of useful heat is improved by operating at higher feed pressure.

EFFICIENCY

In the foregoing calculations it has been assumed that the power generation plant operates at 100 per cent efficiency. In practice the case for district heating with waste heat from power stations is not quite as strong as stated because power generation proceeds at an average efficiency of only around

Our experience at your service.

STAL-LAVAL produces steam turbines with electrical outputs from 50-1000 MW for combined heat and power generation systems. Extensive modularization and a variety of proven designs simplifies selection to suit any application and power. STAL-LAVAL also produces gas turbines for district heating.

STAL-LAVAL has manufactured turbines for district heating schemes for over 50 years, and has been a world leader in turbinology for almost a century. Consulting us about your energy and environmental problems may be the decisive step for their solution.

STAL-LAVAL

STAL-LAVAL Turbin AB
S-61220 FINSPONG
Sweden

Tel: 0122/810 00
Telex: 64045 STAVAL S

Subsidiaries in: Benelux, France, Finland, Norway, UK, USA.

Still pooh-poohing solar energy?

This new book might just make you change your mind

Small Solar Buildings in Cool Northern Climates
David Oppenheim

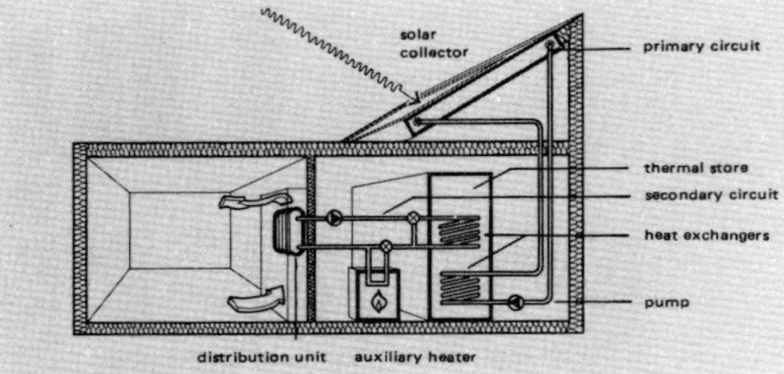

31 LIQUID FLAT PLATE COLLECTOR

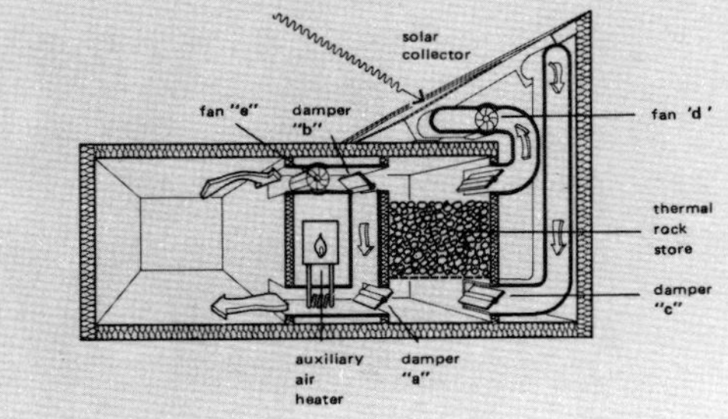

32 AIR FLAT PLATE COLLECTOR

Making efficient use of solar energy is now well-established in hot, sunny climates. But the techniques and experience gained in these areas is not directly applicable to cool temperate zones.

Here is a new book which concentrates solely on the problems and possibilities of solar buildings in cooler climates.

144 pages illus 210×297mm
£12·95 paper

- Practical design information for architects on solar buildings
- A unique fully illustrated catalogue of every known building in the UK exploiting solar energy for space heating
- Full glossary of technical terms
- Lists of solar designers to be approached for specialist advice
- A book full of practical ideas by a specialist in the field and invaluable to designers looking to the future

Order form

Please send me copy/copies of ***Small Solar Buildings in Cool Northern climates*** (ISBN: 0 85139 596 1). Please add 10% for postage and packing. Orders over £30 for UK delivery, post free.

The Architectural Press Ltd, 9 Queen Anne's Gate, London SW1H 9BY Registered in England: 1176599

Name

Address

0.82. In modern turbine plants the various efficiencies are of the following order:

Generator efficiency	0.98
Mechanical efficiency	0.98
Internal turbine efficiency	0.85

By multiplying these factors with each other an overall engine efficiency of 0.82 is obtained. Heat exchangers operate at far better efficiencies than this because they possess no moving parts.

The efficiency of heat generation as a by-product of electricity production is only expressed in terms of electricity lost. It should be mentioned that modern steam generators are able to utilise well over 90 per cent of the net calorific value of the fuel used for the production of steam. Small boilers are often not more than 75 per cent efficient.

It is often stated that electricity can be converted into heat at 100 per cent efficiency. This may be true for electricity at the point of use. It is not true for power which has to be supplied from the power station. Step-up and step-down transformers, switchgears and transmission lines all waste a considerable amount of expensive energy. There are also pipeline heat losses when hot water is pumped over long distances.

A typical insulated twin pipeline hot water system with a flow temperature of 100°C (212°F) and a return temperature of 60°C (140°F), employing a flow velocity of 3 m/s and an internal pipeline diameter of 600 mm (ie supplying something of the order of 141 MW of heat) has a heat loss of approximately 100 kW per kilometre run, representing about 0.71 per cent of its energy content per kilometre pipe line length. This is admittedly higher than equivalent electricity transmission losses, but hot water is far cheaper than electricity.

The relative inefficiency of a steam turbine process implies that the amount of waste heat produced is quite enormous. It has already been shown that out of 4.00 MJ/kg of steam produced a condensing turbine can only utilise about 1.48 MJ/kg for electric current generating purposes. Hence $4.00 - 1.48 = 2.52$ MJ/kg of low grade heat steam generated must be dissipated in the condensers of the power station.

The cooling capacity required in a typical condensing power station is roughly 1.7 times its electricity output. For example in a power station with power output of 2000 MW the cooling capacity must be of the order of 3400 MW; this is why cooling towers are such a prominent feature of all power stations. When operating back pressure power stations, the heat evolved is even greater. When steam is extracted at 110°C (230°F) roughly twice as much heat is produced as electricity, eg a 650 MW turbine set operated on the back pressure principle will produce 1300 MW of heat night and day, winter and summer.

It is possible to see how this relates to actual district heating practice by mentioning the district heating system in Hamburg (West Germany). This is the most advanced system of district heating in Western Europe and its entire connected load is only just over 3000 MW (although in summer the heat demand may be as low as about 300 MW).

A very high percentage of all domestic industrial and commercial consumers in Hamburg is connected to the municipal district heating network which has been expanding steadily for more than half a century. By comparison the largest British district heating network, the Nottingham scheme, has a present day capacity of about 132 MW. In summer, utilisation is probably about 20 to 30 MW.

There can only be a few cases where a straight forward back pressure turbine plant of appreciable size can be used for district heating purposes. When the West German district heating grid has been fully extended, its capacity will be approximately 25 000 MW; back pressure turbines of adequate size will then obviously be a matter of consideration for base load purposes, and this also applies in the Soviet Union where the Moscow system has a capacity of well over 10 000 MW of heat.

Is a solution available? At Pimlico (London) an attempt was made to use a diminutive back pressure turbine connected to the Battersea power station with a capacity of just $1\frac{1}{2}$ MW to supply the local district heating system. However with such relatively tiny installations the current-producing efficiency of the plant is ridiculously low: a mere $11\frac{1}{2}$ per cent. Capital costs and labour and repair costs then loom prohibitively high in comparison to the minute quantity of electricity available. Yet in the summer even a very small steam driven turbine like this produces far too much heat. The Battersea back pressure turbine produces roughly 6 MW of heat for each MW of electricity so that the plant delivers $1.5 \times 6 = 9$ MW: far too much for summer operation of the Pimlico district heating system. In consequence the turbine has to be shut down because it is uneconomic to use on back pressure. This adds to the overhead costs.

What is therefore the viable solution? It is the current view that new power producing turbines should have a minimum capacity of about 300 MW; preferably they should be larger. Units of this size are employed widely for district heating purposes employing ITOC operation ie used in the variable mode between conditions of full condensing operation and about 20 per cent back pressure. For example at its maximum back pressure operation a 300 MW turbine unit which is quite economic to build and maintain would yield about 120 MW of heat, roughly the requirement for the peak operation of a district heating system such as that at Nottingham. Used at maximum condensing operation the heat output of such a set would be zero. Assuming an overall operating efficiency of 37 per cent when on full condensing operation this would fall to around 34.5 per cent at maximum heat extraction. 120 MW of heat can therefore be obtained for a sacrifice of 15 MW of electricity.

Where there are existing turbine units in urban stations, requiring no capital investment requirement except for design modifications, sets as small as 60 MW can be operated economically on an ITOC cycle.

USES OF BACK PRESSURE TURBINES

In back pressure turbines steam is fed in at a high pressure, but instead of being exhausted at the normal low temperature of at about 40°C (104°F) as is the case with condensing turbines, the steam is taken off at a temperature which is still useful for heating purposes. The combined effective output of such turbines is therefore all the heat input in the fuel, less such unavoidable factors as fluegas losses, radiation losses, bearing losses etc. It is found that a back pressure turbine can be operated at a combined heat plus power efficiency of 80 per cent plus. **Figure 5.5** is a diagram showing the arrangements of a turbine for back pressure operation.

Back pressure turbines are, however, limited in their applicability. They require a constant or near constant very high heat demand, a condition which may be met in industrial plants but is seldom found with urban district heating systems. In practice a back pressure plant gives far too much heat for its output of electricity ie it is relatively inflexible.

It can be shown that in a back pressure plant (**5.6**) with a typical overall efficiency of 85 per cent, the difference between condensing operation and back pressure operation at an outlet steam pressure of 1 bar and a temperature of 100°C (212°F) which is normal for most district heating plants, is as follows:

Condensing plant: for an input of 100 kW of fuel 34.8 kW of electricity and no useful heat.

Back pressure plant: for an input of 100 kW of fuel 27.84 kW of electricity and $85 - 27.84 = 57.16$ kW of heat.

The loss in electricity suffered is thus $34.8 - 27.84 = 6.96$ kW.

If the electricity produced in normal condensing operation were to be converted into heat by using an electric fire or similar electrical appliance exactly 6.96 kW of heat would be obtained for 6.96 kW of electricity. Yet by sacrificing 6.96 units of electricity through operating a turbine on back pressure rather than on the condensing principle no less than 57.16 kW of heat can be obtained for a loss of 6.96 kW of electricity. The process is therefore $\dfrac{57.16 \times 100}{6.96} = 821.3\%$ efficient.

Since the heat output/power ratio for a given design of turbine cannot be varied, a back pressure turbine can only be employed for carrying the base load of a district heating system and this greatly restricts the size of turbine which can be used for such a purpose.

Back pressure turbines are however particularly suitable for use with refuse incinerators. The heat output of such an incinerator relates entirely to the amount of refuse fed into it at any given time, and this in its turn

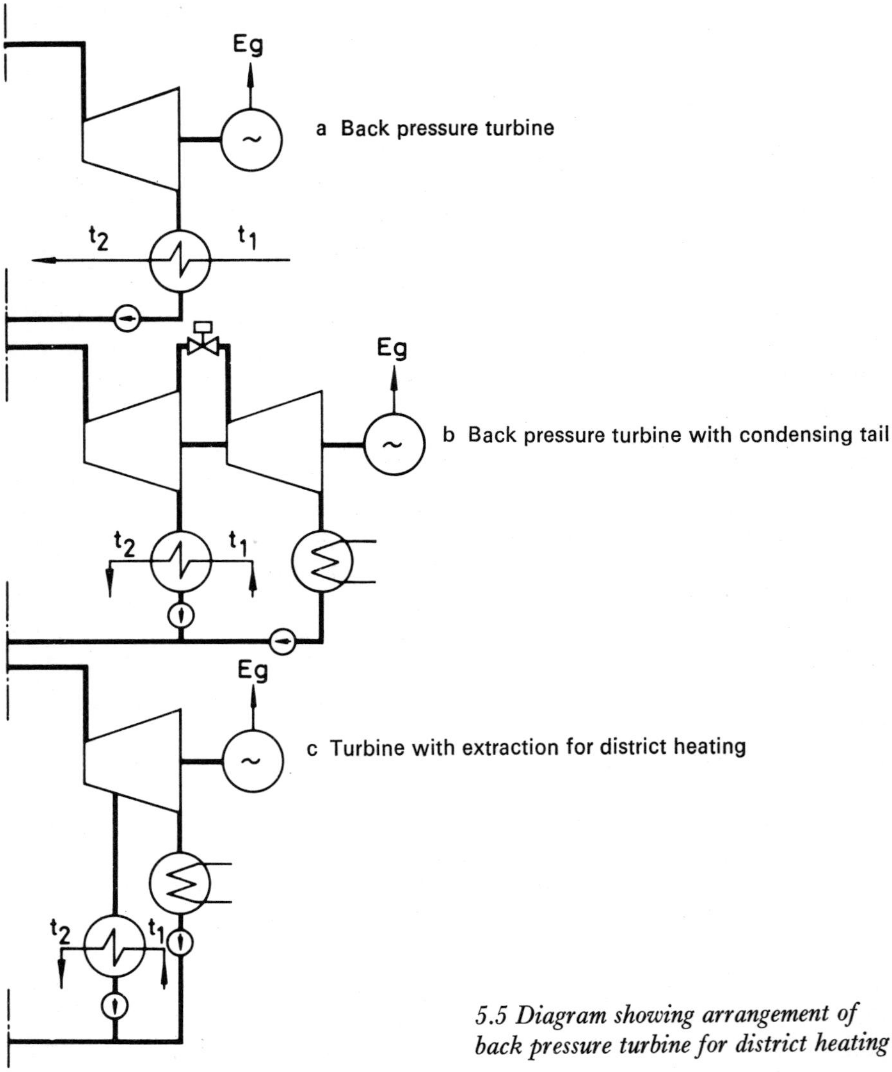

5.5 Diagram showing arrangement of back pressure turbine for district heating

depends upon the quantity of refuse which the community happens to be throwing away at that time. Refuse cannot be stored for more than a few days, making the operation of a refuse incinerator rather inflexible.

Back pressure turbines can be fed reasonably economically with low pressure steam generated in the waste heat boiler of an incinerator plant. It is necessary to incorporate an ITOC turbine somewhere in such a system to balance the supply of heat with the demand as the back pressure turbine cannot do this.

The common dimensioning temperatures for the inlet and outlet to the

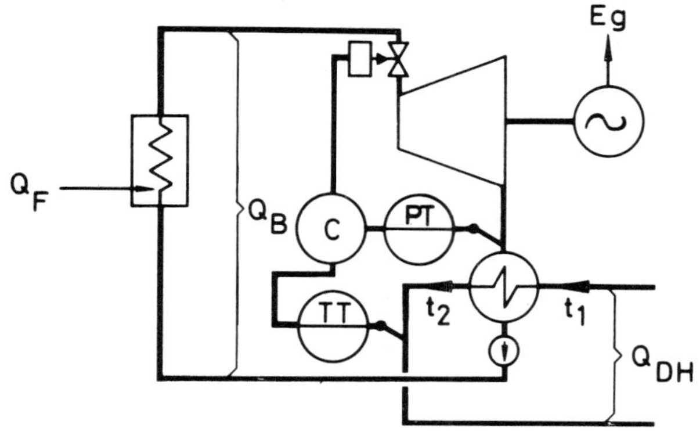

5.6 District heating back pressure turbine

district heating network are 50°C (122°F) and 110°C (230°F) respectively. With a temperature rise across the system of 40°C (104°F) it is desirable to split the heating into two stages with larger plants which then increases the power yield by about 5 per cent. The economic viability of such an arrangement depends upon the extra cost of the turbine and heating condensers and on the extent by which the improvement in power yield is available with varying loads. The simplest arrangement consists of an unregulated extraction for the second stage. This however only permits the division of flow to be at its best at one given design point. If the heat load varies the system tends to resemble single stage heating because at the high loads most of the heat is taken off at the higher temperature and at the low loads at the lower one. This can cause difficulties with a refuse incinerator back pressure turbine which is inherently subject to load fluctuations.

For cases involving a considerable load variation the assymetric double flow turbine can be used (**5.7**) In this the steam expands to two different pressures from the central feed point. The ratio of the high pressure and low pressure outflow rates is therefore kept constant whatever the loading. The disadvantage lies in increased cost and a slight reduction in turbine efficiency. Such turbines are usually only viable for plants with outputs in excess of 200 MW. Splitting the turbine up into two cylinders is carried out for the reheating of feed water or for other reasons.

The Stal-Laval company has designed a back pressure turbine for the 60 to 200 $MW_{(el)}$ range which is cheaper than a conventional assymetric double flow turbine. It has a concentrically divided exhaust with a different number of stages in each, giving two different exhaust pressures as required by the two condensers. The steam flow is split concentrically a few stages upstream of the first exhaust stage by dividers on the fixed and moving blades, dividing up the heat load effectively at varying turbine loadings as with assymetric flow turbines.

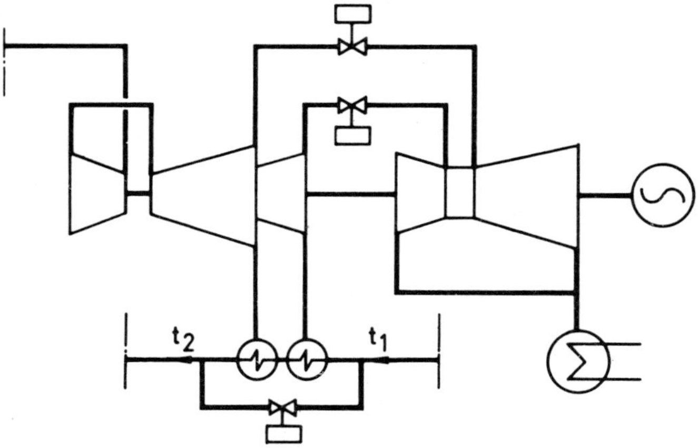

a Non-reheat DH-extraction condensing turbine

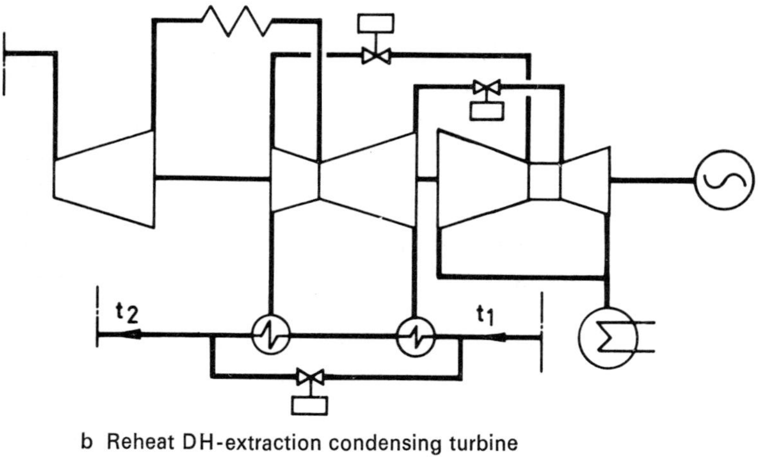

b Reheat DH-extraction condensing turbine

5.7 Non-reheat types of extraction: condensing turbines

It is also possible to obtain two-stage hot water heating by connecting two separate back pressure turbines in series.

The Stal-Laval company makes a number of standardised back pressure district heating turbines. Table 5.1 gives some of the applicable operating conditions.

The largest of these units is only suitable for use in very extensive district heating systems with a design load of at least 1000 MW; it must never be forgotten that back pressure turbines are unsuitable for variable load conditions.

TABLE 5.1 STAL LAVAL'S STANDARDISED BACK PRESSURE DISTRICT HEATING TURBINES

Type	Counter-rotation radial-flow			Axial flow
Steam pressure (bars)	80	105	110	140
Steam temperature (°C)	495	530	530	535
Feed water temperature (°C)	180	220	240	250
Generator power (MW_e)	22	48	70 to 80	100 to 120
Heat power (MW_{DH})	50 to 55	95 to 105	130 to 150	180 to 220
Number of heating stages	1	1	2	2
Outgoing hot water temperature (°C)	80 to 90	80 to 90	80 to 90	80 to 90

OPERATION OF ITOC TURBINES

Intermediate take-off condensing systems of combined heat and power generation are all characterised by being able to vary the ratio of electricity/heat extracted. One or more ITOC turbines are essential features of all major heating systems. They have been found to be the only acceptable and economic method of balancing the demand with heat supplied in a system. **Figures 5.8** and **5.9** show, respectively, an ITOC turbine plant working from a nuclear energy plant and the arrangement of a Sulzer ITOC plant.

Basically ITOC turbines can be operated within the following modes:

(a) When little heat is required by the district heating system, the ITOC

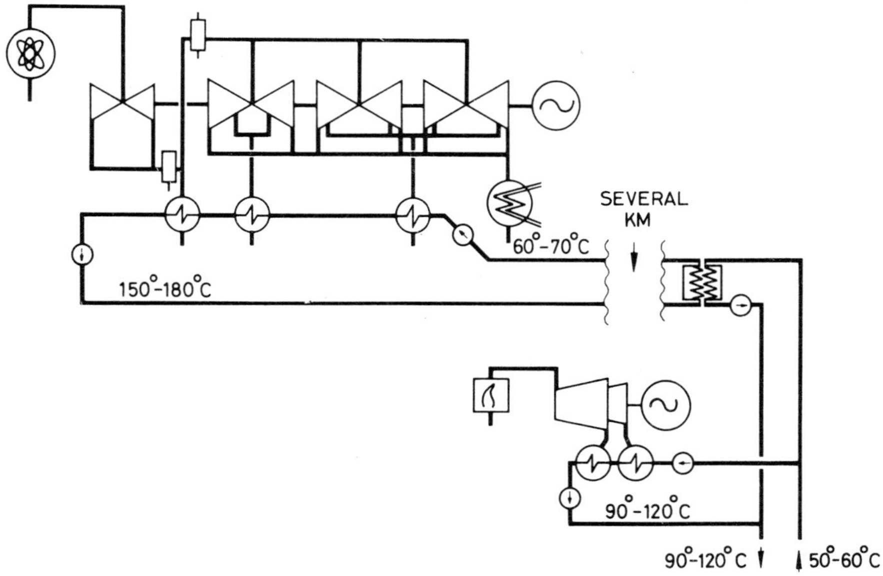

5.8 ITOC turbine plant working from a nuclear energy plant

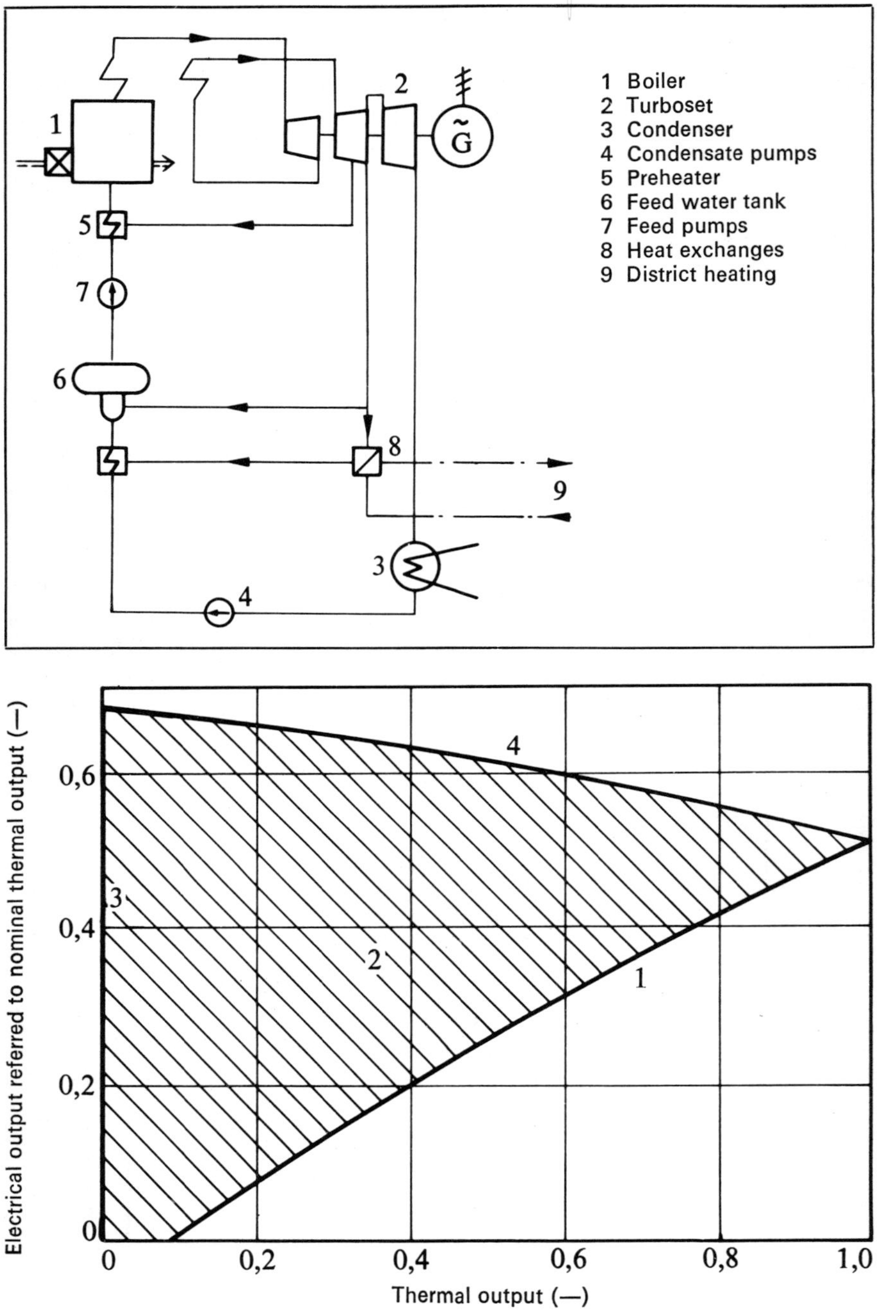

1 Pure back-pressure operation
2 Possible operating rang
3 100% condensation
4 Constant maximum live steam flow

5.9 Sulzer fossil fuelled ITOC plant

turbine is switched to full condensing operation and the heat load is carried by the base load heat generators (the back pressure turbines of the refuse incinerators, heat obtained from burning sewage methane, sewage sludges, geothermal heat, waste heat from industry etc).
(b) As the demand for heat rises the ITOC turbine is switched gradually to partial back pressure operation and, in some designs, even to full back pressure operation. Sophisticated electronic valve systems are incorporated in the system for this purpose, and it is essential that these are under the control of the district heating operating authority which has to decide, second by second, the manner of turbine operation and the balance of heat/electricity generation.
Current British legislation has rested the control of power production in a state monopoly, so that the necessary balancing control cannot be exercised. For this reason alone ITOC turbines and, in consequence, viable district heating schemes cannot at present be operated in the United Kingdom.

The back pressure ITOC turbine with condensing 'tail' is probably the type most commonly employed with small and intermediate size stations. Its advantages are that it is simple to instal and operate and that it can easily be produced by modifying existing back pressure turbines. On the other hand such turbines tend to be not quite as efficient in operation as the more complex condensing turbines for district heating extraction extensively used in some of the very big district schemes in West Germany, Sweden and Eastern Europe.

The simple back pressure turbine described previously can be provided with a non-symmetric low pressure cylinder (as shown in **figure 5.10**). For larger installations steam from the first back pressure section is reheated as shown in **figure 5.11.** Under such conditions the turbine is likely to be of the non-symmetric double-flow type for optimum two-stage water heating. The low pressure turbine is then also a non-symmetric double-flow plant. For larger plants a reheater can be used which improves the power yield by up to about 10 per cent but can also increase the plant capital costs quite considerably. The installation of reheat arrangements improves the power yield but not the total heat plus power yield of a given plant. Careful costing is there-

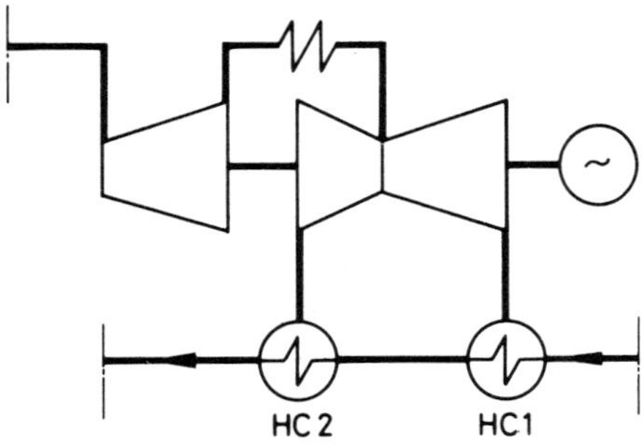

5.10 Non-symmetric low pressure cylinder for back pressure turbine

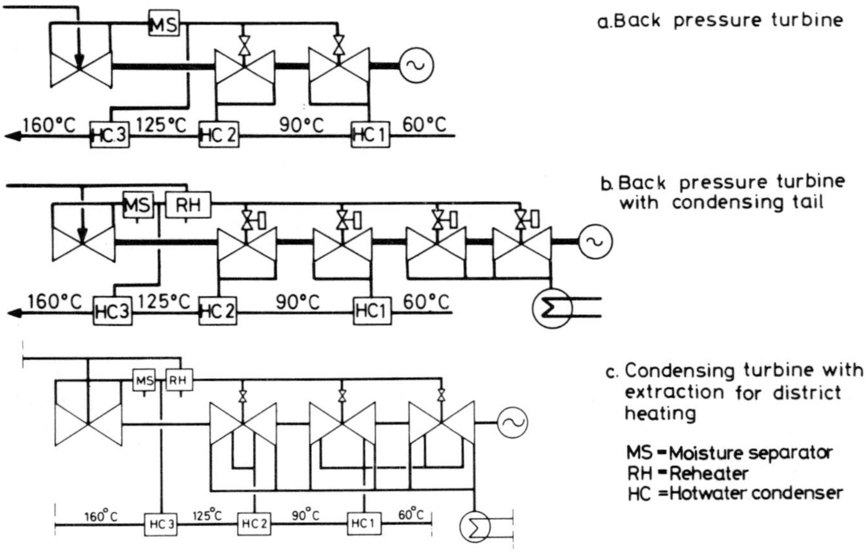

5.11 Various types of large steam turbines for combined production of heat and electricity

fore necessary with ITOC turbines to establish whether the extra plant cost for reheat equipment (somewhat in excess of 10 per cent of the total plant cost) can be justified.

At high heat load the outgoing hot water temperature is regulated by throttling the steam flow to the low pressure turbine. The valves in the transfer pipes between the intermediate pressure and low pressure turbines are fully closed for pure back pressure operation which is carried out during the night when the power demand is at its lowest, but heat can be extracted at minimum cost in the form of hot water and stored in the system for use during the period of maximum heat demand on the following day. A certain minimum flow of cooling steam is always needed for the low pressure turbine section and this is provided by arranging a fixed leakage through the valves even when these are completely closed. The ITOC turbine can therefore never have the same high power and heat efficiency as a back pressure turbine. The common limit of efficiency is about 70 per cent as against over 80 per cent for a back pressure turbine.

At low heat load and high electrical load the temperature of the outgoing hot water is higher than required and a proportion of the hot water is shunted around the hot water condenser to achieve the required temperature through mixing.

For a given heat load, there is only one operating condition when neither throttling in the transfer pipe nor shunting of the hot water is required. At all other loads, one or other procedure must be applied, leading to losses.

Very large plants

There is a tendency today for power stations to be situated at a distance from urban areas, close to coal fields or oil installations and where their presence does not unduly conflict with the environment or safety aspects (such as when electricity generation is by nuclear energy). In countries where population densities are low the distance between a power station and the urban areas it serves can be 200 km and more. In Great Britain a number of large power stations are well within 50 km of industrial areas. This distance is considered technologically feasible for associated district heat supply, so that construction of the long hot water mains from the station to the urban user is more easily arranged. **Figure 5.12** shows typical annual operating curves with combined equipment.

Power stations constructed in conjunction with nuclear energy installations are always very large, 2000 to 3000 MW being usually considered to be the minimum acceptable size. It is not feasible to operate stations of this type on back pressure operation because an impossibly large connected district heat load of 5000 to 7500 MW all the year round would be required to utilise even one such station.

The best technical and economic solution is the provision of a standard condensing turbine with district heat take-offs; these are generally unregulated and are extracted from the low pressure turbines and behind the high temperature turbines. The maximum heat flow which can be taken off in this way is about half that which would go to the condenser in normal condensing turbine operation.

For example, it would be possible to design a 3000 MW (thermal) coal or nuclear energy fuelled power station to deliver about 1000 MW of electricity

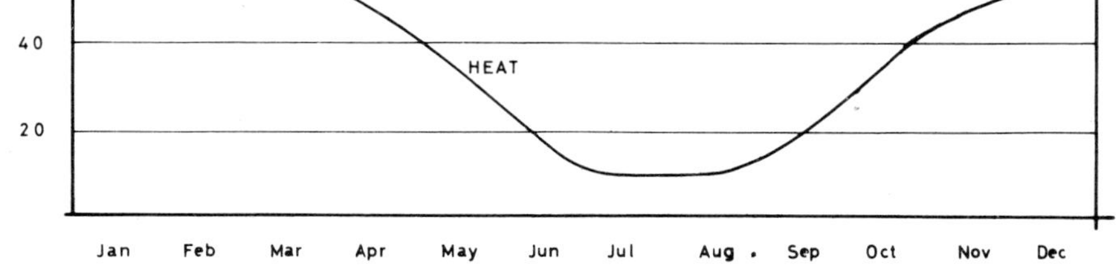

5.12 Annual operating curves with combined equipment

and 1000 MW of heat. It is not, in this case, quite so easy to vary the ratio of electricity/heat produced in such turbines as it is with the smaller back pressure turbines with condensing tail. But this can be done by throttling down one or other of the sidestream take-offs when heating demand is low, as for example during the summer months. A station of this sort can be operated on a complete condensing cycle if needed, although this is not advisable for technical and economic reasons.

A large turbine plant designed to yield an appreciable amount of hot water for district heating purposes requires a smaller exhaust area and less cooling capacity than a corresponding condensing turbine. The exhaust area and cooling capacity provided is optimised with reference to the variations in heat load and cooling water temperature during the year. Naturally, if the exhaust area is reduced considerably the electrical output of the plant is lower under conditions of full condensing operation. However even in summer there is always some need for hot water for domestic, industrial and commercial purposes. Therefore the system should be planned so that it should never be necessary to operate it on a complete condensing cycle. The savings in exhaust condenser and cooling tower capacity considerably reduce the enormous cost to the consumer authority of installing hot water mains from the station. If it were necessary to run the plant on a complete condensing cycle for even a few days in the year the complete complement of exhaust condensers and cooling towers would be required.

Operating the system with extraction for district heating slightly reduces the overall power production efficiency of the plant. The magnitude of such reduction depends on a number of factors such as quantity of heat extracted, temperature level of extracted heat, number of heating stages etc. The power loss lies within the range of 120 to 150 watts of electricity for each kilowatt of heat extracted eg the process of heat extraction from a very big ITOC turbine such as the one described ranges in efficiency between 667 and 833 per cent.

It is desirable for economic reasons for large turbines to operate at maximum load throughout the year and from the point of view of power production very large coal fired and nuclear energy fuelled plants have always provided the base load of the public electricity output. On the other hand district heat output requirement fluctuates considerably and can be varied either by adjusting the hot water flow at constant outgoing temperatures or by regulating the temperature at constant flow by by-passing the hot water sides of the heating condensers so that they are progressively short-circuited from the highest level to the lowest. This method results in the minimum loss of electrical energy from the turbine per unit of extracted heat over the whole year of operation but requires constant pumping power to convey the hot water. When the hot water flow is governed at constant outgoing temperature the pumping power requirement drops with heat load while the loss of electricity per unit of heat extracted remains constant. Which of these two methods of control is the more advantageous depends upon a number of factors such as the length of the heating mains network and the nature of other heat suppliers to the district heating system.

One of the major advantages of systems using non-regulated extraction for district heating is that this method involves only a minimum number of modifications to existing turbine systems. High pressure and medium pressure turbines need not be modified at all. Only low pressure turbines require some slight blading modifications to enable steam extraction to proceed (see **figures 5.13, 5.14** and **5.15**).

GAS TURBINES

Of the two basic types of gas turbine (open-cycle and closed-cycle) open-cycle are the more widely used. In these air is taken from the atmosphere, compressed and heated at constant pressure by the combustion of the fuel. The compressed gases are then conveyed through the turbine blades to produce work. The exhaust gases are discharged at a very high temperature and can be used to produce more mechanical energy by coupling with a small steam turbine or employed for the production of thermal energy (**plate 9**).

As regards waste heat abstraction gas turbines compete with ITOC tur-

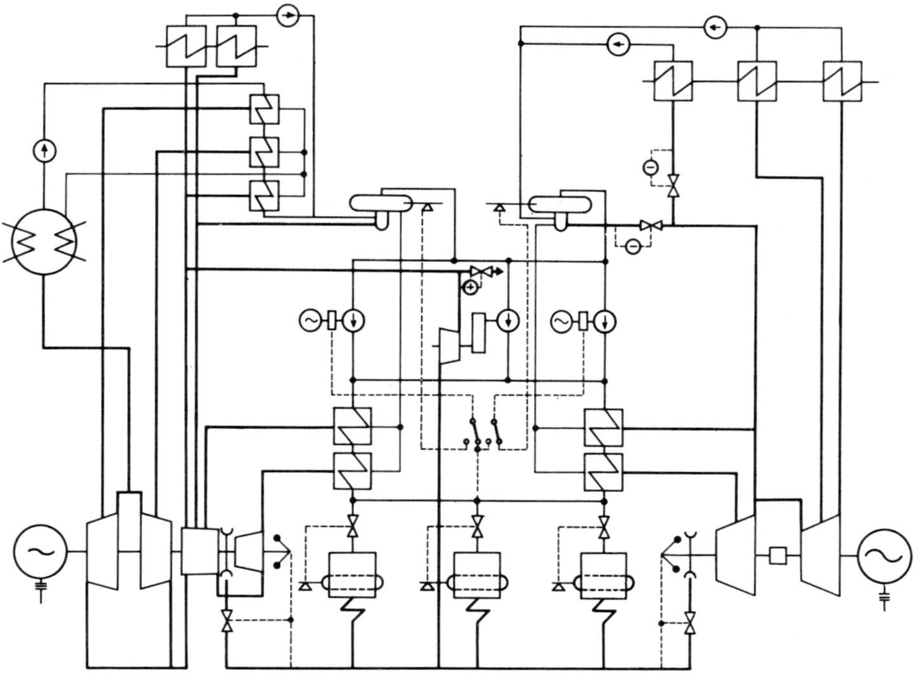

5.13 Layout of modern ITOC station

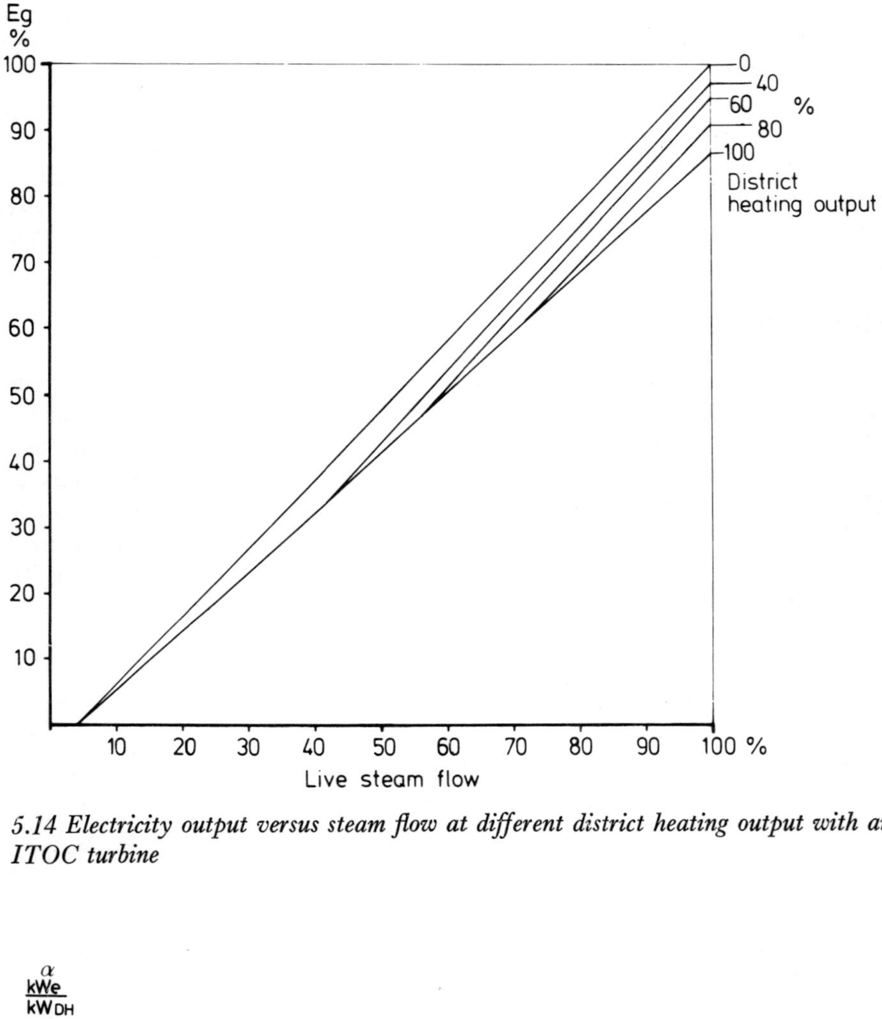

5.14 Electricity output versus steam flow at different district heating output with an ITOC turbine

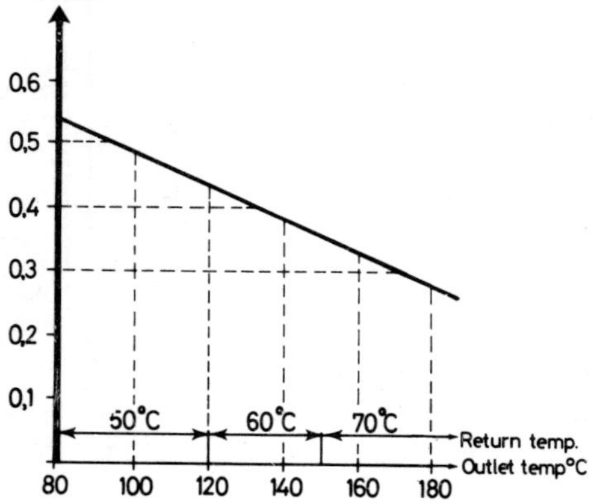

5.15 Power yield versus hot water outlet temperature with a 70 MW ITOC turbine using 2-stage hot water heating (530°C, 111 bars)

bines and offer some important advantages. The inlet and exhaust temperatures of gas turbines are for example very much higher than those associated with steam turbines. Operating temperatures in excess of 800°C (1472°F) are common with waste gas temperatures between 200°C (392°F) and 400°C (752°F). Combined gas turbine/steam turbine plants can be constructed which have overall efficiencies of power generation of up to 43 per cent: much higher than can be achieved with even the most modern steam turbines. Efficiencies of power generation well over 35 per cent and heat plus power efficiencies in excess of 80 per cent can be obtained by passing the exhaust through a steam turbine. It is practicable and economical to heat district heating flow water by the exhaust to a temperature of 150°C (302°F) and more without lowering the efficiency of power generation of the gas turbine plant appreciably. Indeed the cost of waste heat from a gas turbine plant, in terms of loss of electricity and cost of obtaining heat from the plant in question, is by far the lowest of any combined heat/power generation plant. Gas turbines can also be put on to load very rapidly so as to be fully operational from cold within ten minutes.

On the other hand gas turbines also have serious disadvantages compared with ITOC plants. Their capital cost per unit of output is far higher and they do not possess the very considerable flexibility of ITOC steam turbines. They are also not very adaptable in electricity generation as they have to be operated at a constant load and their efficiency falls off considerably when this load is reduced. Also, their fuel costs are high because they have to use prime fuels whereas steam power stations can use the lower grades of coal and oil. In consequence they are usually only employed for peak power generation when a sudden need for extra electricity arises. To operate them without attendant heat recovery is very wasteful as considerable quantities of high temperature flue gases are evolved and are exhausted to atmosphere.

It is necessary for gas turbines to be operated at maximum temperature, limited essentially by the types of alloys available for blading (modern alloys can operate at 800°C (1472°F) for long periods of time). They can only work efficiently at very high temperatures when using either distillate oil or natural gas as fuel. If heavy oil is used the content of vanadium or sodium salts limits the operating temperature to below 670°C (1238°F). It is possible to drive gas turbines with low calorific value gases (less than 19 MJ/m^3) but problems then arise with regard to the added volume of gas to the compressor. Certain grades of heavy oil are available without sodium and vanadium content: these are intermediate in price between distillate and crude oil and are useful for running some high temperature gas turbines. Other types of gas turbines are designed to operate on combined gas/heavy fuel oil.

Gas turbines, like all other heat engines, comply with Carnot's cycle described earlier in this chapter. The inlet temperature (T_1) is very high indeed, up to about 1100 K with some of the most modern units equipped with blade cooling etc. On the other hand the exhaust temperature (T_2) is very high also (300 to 400°C) (572 to 752°F) and only a small fraction of this heat can be utilised for such purposes as air feed preheating.

One system utilises the waste heat to produce steam for a boiler plant to be used in conjunction with the gas turbines. The steam produced under such circumstances is likely to be at a temperature of about 15 bars and 350°C (662°F) offering quite poor thermodynamic characteristics particularly when used on back pressure. It is, however, estimated that a steam turbine connected to an open-cycle gas turbine only raises the efficiency of the latter as a power producer by a mere 1.5 to 2 per cent and it is therefore questionable whether the expense of recovering such small quantities of electricity is worthwhile. But the hot water obtainable from the back pressure end can of course be employed for district heating purposes provided that use can be made of low grade heat. There have been instances where open-cycle gas turbines have been used in conjunction with conventional steam boilers to perform the basic function of extracting some energy from the burning fuel prior to its use for steam raising.

It has been found best to operate open-cycle gas turbines in conjunction with waste heat boilers to produce hot water for direct use in district heating, offering the following advantages:

(a) There is virtually no drop in the electricity production efficiency when heat is abstracted in this way because, unlike steam power stations, it is impossible to operate open-cycle gas turbines with a very low exhaust temperature.

(b) Waste heat boilers or similar heat exchange equipment are relatively cheap to install and simple to operate.

(c) Hot water can be supplied to the district heat network at a fairly high temperature, saving on pipeline costs, pumping charges etc. Using steam power stations for the supply of hot water, the sacrifice of electrical energy becomes appreciable when high water flow temperatures are required.

Altogether it is very economical to employ gas turbines for combined power and heat generation because virtually all the residual heat in the exhaust gases can be abstracted (whereas ITOC turbines always need appreciable cooling capacity as well).

With gas turbines use is made only of heat which would otherwise have been wasted by exhaust to the atmosphere and this accounts for the low unit cost of the heat produced in this way. However gas turbines are commonly only brought into use in public electricity generation when the power system demands electricity in excess of that produced by the base load stations to meet a peak demand. Waste heat is then produced as a by-product. However it will probably not be available at the time when it is required. It can therefore only be used as a small part of the total heat available to the system. It is in fact impossible to base the operation of a district heating system on waste heat produced by a gas turbine unless that system includes an effective heat control appliance ie an ITOC steam turbine to adjust the heat demand to the heat supply. Waste heat from a gas turbine is very useful and cheap, but it can only be used within the general framework of a well controlled and organised district heating network.

A typical example of the effective use of gas turbines for district heating exists in West Berlin where three gas turbine sets with a total electric power output of 210 MW were installed in 1975 in the Charlottenburg power station. They are able to contribute up to 160 MW of heat to the district heating which has a total connected load of 2500 MW. A further gas turbine system at the Berlin Heizkraftwerke Wilmersdorf will eventually yield a further 210 MW of heat.

The above observations have been made entirely in the context of the well proven open-cycle gas turbine (**5.16**). In a closed-cycle gas turbine (**5.17**) the circulating fluid is heated in a type of 'boiler', is compressed and then performs work in a turbine. After abstraction of the sensible heat for district heating the fluid is returned to the 'boiler'. The working fluid is helium and, although there are technical advantages, such systems are much more expensive to construct then the relatively simple open-cycle gas turbine plants.

For closed gas turbine systems plant costs per unit of power or of heat

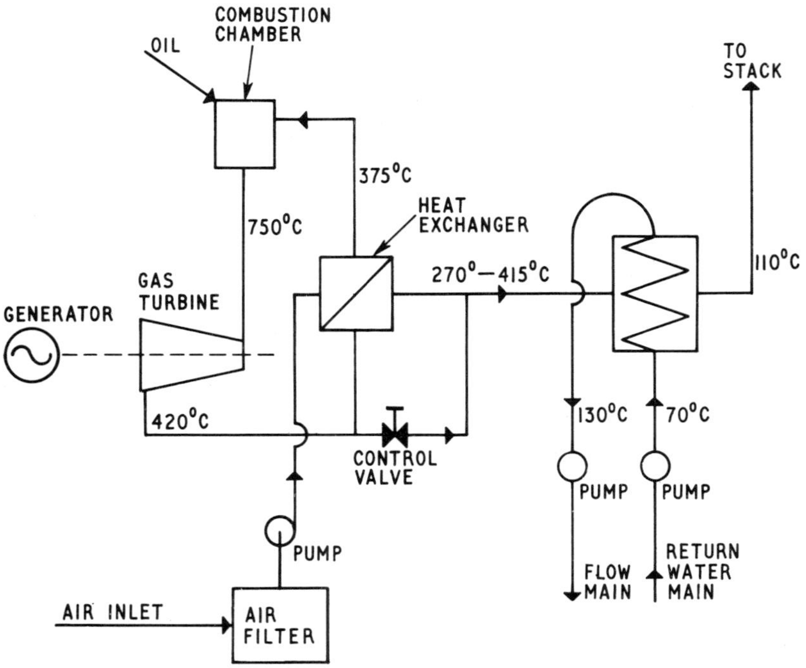

5.16 An open-cycle gas turbine circuit used for district heating supply

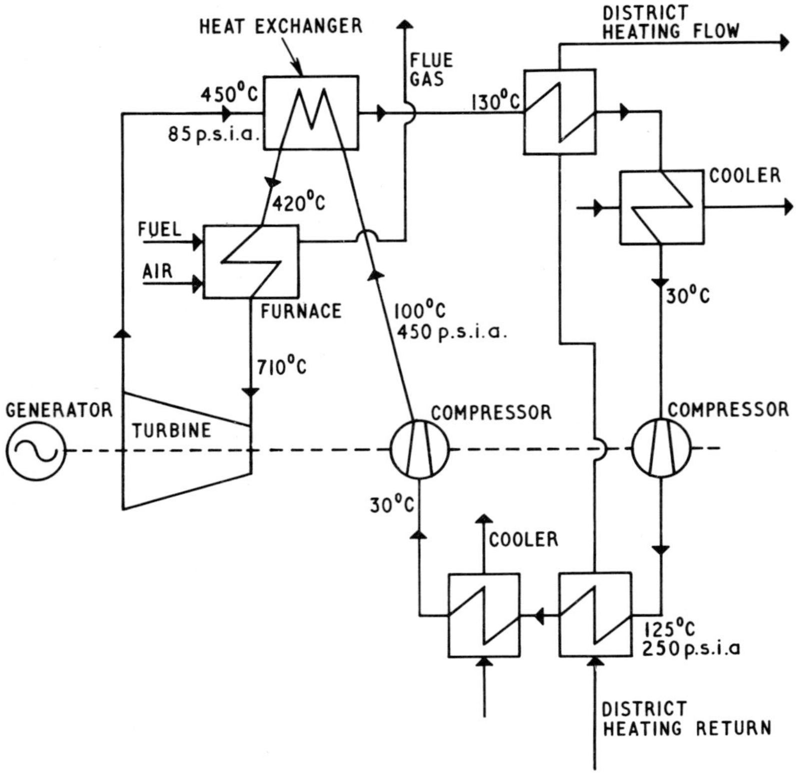

5.17 A closed-cycle gas turbine circuit used for combined heat and power generation

are currently many times higher than those of the highly successful large steam ITOC stations on which district heating is generally based.

ECONOMICS OF CHP GENERATION IN WEST GERMANY

Experience of different turbine systems

Climatic conditions in West Germany are roughly similar to those in Great Britain and there are also other similarities. The only pronounced differences are in interest rates charged for capital (lower) and in wage rates (higher).

Five main types of combined turbines are being used successfully in West Germany:

(a) Back pressure turbines with capacity of up to 100 MW
(b) ITOC steam turbines with capacities of up to 700 MW
(c) Open-cycle gas turbines with capacities of up to 100 MW
(d) Combined gas turbine steam plants with capacities of up to 130 MW
(e) Diesel and gas engine plants with capacities of up to 20 MW.

Of these, back pressure turbines have the highest overall efficiencies: 88 per cent for combined heat and power generation has been quoted. Open-cycle gas turbines combined with direct heat extraction or with steam plants have combined efficiencies of up to 83 per cent. Diesel and gas engine aggregates have the same maximum efficiencies as the ITOC turbines (around 80 per cent). Table 5.2 summarises the main features of turbine systems used in district heating production in West Germany.

TABLE 5.2 WEST GERMAN EXPERIENCE OF TURBINE SYSTEMS FOR DISTRICT HEATING PRODUCTION

Back pressure turbines	
Characteristics	Designed mainly for supply of heat with electricity as by-product
Advantages	Cheap and can use unconventional fuels such as refuse, excellent overall thermal efficiency
Disadvantages	Fixed heat/electricity output ratio. Poor electricity yield in operation. Expensive for small units
Main uses	To cover base load of heat for large systems; useful in connection with refuse incinerators
Back pressure turbines with condensing tail	
Characteristics	Designed mainly for supply of heat with electricity as by-product, but able to vary output ratio of heat/electricity
Advantages	Excellent thermal efficiency for obtaining heat, but able to obtain electricity at reasonable efficiency when heat load is low
Disadvantages	A very high basic heat demand is always needed for this turbine to run efficiently. At very low heat demand electricity production is uneconomic. More expensive than basic back pressure turbine
Main uses	To cover base load of heat for large systems where fluctuations in demand are expected. It is unwise to dimension such turbines to cover an excess of 50 per cent of the expected peak demand
Condensing turbines with district heating extraction (ITOC)	
Characteristics	Designed mainly for power production but permits take-off of steam at around 1 to 1.5 bars pressure for district heating purposes
Advantages	Steam take-off can be adjusted instantaneously between zero and the maximum of about 20 per cent of turbine output. Existing condensing turbines can be modified at moderate cost. Efficiency of electricity production very high
Disadvantages	Limitation of quantity of heat which can be abstracted without adversely affecting running of turbine
Main uses	Most common of all district heating turbines for large schemes. Able to act as a control valve to balance heat output and heat requirements. Used for nuclear district heating schemes and others where relatively long supply lines are used
Gas turbines	
Characteristics	Designed for putting into operation at very short notice to supply

	peak electricity. Combined with heat exchangers and small back pressure turbine to step up electricity supply and to obtain waste heat for district heating purposes
Advantages	Suitable for very small group heating undertakings and for small power take-off. High electric and overall thermal efficiencies
Disadvantages	Expensive in capital investment and maintenance charges. Open-cycle gas turbines require high grade fuel, while closed-cycle turbines are even more expensive than open-cycle ones. Difficult to vary heat/electricity ratio
Main uses	Supplying heat and electricity to very small schemes. Useful as a peak station for the supply of both heat and electricity
Diesel engines with waste heat extraction	
Characteristics	Diesel and gas engines extract heat from (a) exhaust, (b) cooling jacket, (c) oil coolers to supply either steam or hot water
Advantages	Making the best use of prime fuel when electricity is being generated in small quantities by diesel engines, by application of CHP principle
Disadvantages	Use very expensive fuel. Capital outlay quite prohibitive for larger installations. Poor overall thermal efficiency. Heat/electricity output ratio invariable
Main uses	Supply of heat and electricity to small group heating schemes. Often used in conjunction with supply of chilled water for refrigeration purposes

Power loss

For optimum efficiency West German experience shows that ITOC turbines should be operated with the lowest practicable district heating flow temperature, as the power loss per unit of heat supplied is appreciably higher when the flow temperature of the district heating system is elevated. The effect, in terms of power loss, for various combinations is as follows:
Gas turbine and diesel (gas engines): The power loss per GJ of heat supplied is between 1.2 and 2.4 kWh or 4.32 to 8.34 MJ per GJ of heat giving a heat generating efficiency of 11500 to 23000 per cent. This is hardly affected by changes in the flow temperature of the water to the district heating system as the operating conditions of such plant are virtually unaltered when heat is being extracted. Such plants can be started very rapidly from cold: diesel and gas engines require a mere two minutes to be put into operation from cold; gas turbines about seven to ten minutes. Open-cycle gas turbines are, however, not particularly effective for operation at variable heat output, as the efficiency reduces rapidly when the turbines are operated at partial load.
Gas turbines in combination with steam generation: Such plant offers thermal efficiency of extraction of between 10 and 25 kWh per GJ of heat supplied when the circulating water is at 80°C (176°F), corresponding to an extraction efficiency of between 1100 and 2800 per cent. There is a loss of 39 to 45 kWh of electricity per GJ of heat when the heat is extracted at the much

higher temperature of 160°C (320°F): equivalent to extraction efficiencies of 620 to 710 per cent. This type of plant can be commissioned from cold in between 30 minutes and one hour. The heat output can be varied to suit the heat requirements.

Back pressure turbines and various types of ITOC steam turbines: The efficiency of operation of these turbines relates to the temperature of the hot water being supplied. At a flow temperature of 80°C (176°F) the power loss per GJ of heat supplied varies between 16 and 30 kWh per GJ of heat: the heat generating efficiency lies between 925 and 1740 per cent. It falls off rapidly when the outflow temperature is raised. Thus, at a flow temperature of 160°C (320°F), the power loss per GJ of heat supplied is between 45 and 60 kWh: the heat generating efficiency lies between 460 and 620 per cent. Back pressure turbines are useful for carrying the base load of large district heating networks; the ITOC turbines are employed for the variable load. Both types require between one and four hours to move from cold to full-load operation.

Costs of production and distribution

The figures in the following section are given in West German currency (Deutschmarks), and applied in January 1975.

Fuel costs worked out as follows:

Coal and heavy fuel used in ITOC turbines	DM 5.20 per GJ
Fuel for peak boilers	DM 5.80 per GJ
Fuel for gas turbines	DM 6.20 per GJ
Nuclear fuel	DM 1.10 per GJ

It is assumed in these figures that the operating company funds 40 per cent of its own capital and borrows 60 per cent at 8 per cent. When taxes are included, the average interest payment is assessed at 12 per cent. The annual unit production costs for electricity at a unit production cost of DM 0.0562 per kWh equals DM 190, assuming 3380 hours of power output per year.

If combined heat and power stations are considered, which are required anyway to meet power requirements, then the costs of heat production are as follows:

Nuclear stations	DM 2.50 to 4.00 per GJ
ITOC stations fuelled by fossil fuels (150 to 600 MW)	DM 3.00 to 4.50 per GJ
Gas turbines	DM 5.50 to 7.00 per GJ

When special ITOC or other stations have to be provided in circumstances where the electric power could be supplied from existing sources, the cost situation is that large ITOC stations produce heat at DM 3.00 to 4.50 per GJ while the cost of heat produced by small back pressure plants is, as the following figures show, very high:

5 MW back pressure plant	DM 8.00 to 10.00 per GJ
10 MW back pressure plant	DM 6.00 to 7.00 per GJ

30 MW back pressure plant	DM 5.00 to 6.00 per GJ
20 to 100 MW gas turbine plant	DM 5.00 to 7.00 per GJ

Heat produced in a combined gas turbine/steam plant costs only DM 1 to 2 per GJ. That produced by employing diesel motors is much more expensive, amounting to DM 8 per GJ. With gas engines the cost can be as high as DM 13 per GJ. Heating by direct on-peak electricity in Germany costs between 10 and 12 pfennige per kWh, DM 28 to 33 per GJ: very much more expensive than any of the alternative methods of heat production.

It has therefore been concluded in West Germany that district heat supply from large ITOC stations, whether conventional or nuclear, is much cheaper than the supply from small stations, even though it may be necessary to convey the heat across long distances. Small combined heat/power stations or central boiler houses are in fact uneconomic under many circumstances and cannot easily compete with individual house heating using gas or oil. District heating from nuclear power stations costs about DM 0.5 to 1.0 per GJ less than from conventional stations. In current economic conditions, the additional heavy expenditure on the long supply lines associated with nuclear heating centrales cannot as yet be justified. This evaluation may change as fossil fuels become ever more expensive and nuclear power (hopefully) cheaper.

As regards distribution wide ranging studies have been conducted in West Germany into the costs of transporting hot water. Recent developments have greatly reduced these. Given modern types of pipelines, the transmission cost for heat, assuming an annual usage of 3000 hours and a temperature difference between supply and return lines of 80°C (176°F), is set out in table 5.3 below.

TABLE 5.3 COST OF TRANSPORTING HOT WATER

Quantity of water (m^3/h)	Cost (including pumps and pumping costs) (DM per GJ/km)
3000	0.11
5000	0.095
10 000	0.080
15 000	0.065
20 000	0.050

Heat losses average to about 0.5 per cent per km pipeline, depending upon the quantity of heat in transport.

Specific heat transport costs in Germany, when conveyed over distances of 50 km, are between DM 4 and DM 9 per GJ depending upon the quantity of heat transported. The higher figure applies to heat quantities below 240 MW, the lower to heat quantities exceeding about 1800 MW. These figures include the cost of heat losses.

The further the distance between the heating/power station and the supply water system is as low as possible. The optimum temperature difference between the flow and return mains.

ITOC turbines operate at their best when the supply temperature to the water system is as low as possible. The optimum temperature difference between flow and return mains is 60°C (say 120°C (248°F) flow and 60°C (140°F) return) at a supply distance below about 10 km rising to around 100°C with 50 km long supply lines.

When supply lines in excess of 60 km are required and circumstances are appropriate, it is best to bleed very hot water into the twin urban supply system and to use the excess lukewarm water for purposes such as process hot water in industry, laundries, swimming baths etc. A district return main is then not necessary.

In January 1975 the West German consumer paid DM 14.30 per GJ for district heating. At various types of stations the cost of heat production, and the margin between this cost and the price paid by the consumer, were (in DM per GJ):

	Cost of heat production	*Margin for costs of distribution administration and profit*
Heat-only station	10.70	3.60
Small urban combined heating/power station	8.30	6.00
Medium sized ITOC station at edge of city	4.10	10.20
Large nuclear or other station some distance away from city	3.10	11.20

At all such stations the cost of administration was DM 1.20 per GJ and the cost of house stations and pumping DM 0.50 per GJ.

Assuming the cost of heat transportation from small pure heating stations or small urban combined heating/power stations and the cost of heat lost to be DM 0.18 GJ per kilometre, profitability can be achieved when the supply lines are shorter than 20 km for the heat-only station and less than 33 km for the ITOC station.

Assuming, for medium sized ITOC stations, the costs of heat transportation and the heat losses to be DM 0.12 GJ/km, profitability can be achieved with supply lines less than 85 km long.

For the very large power stations, where transportation costs for heat approximate at DM 0.08 GJ/km, profitability is achieved if the supply lines are shorter than 140 km.

The cost to the consumer of DM 14.30 per GJ equates to about 5 Pfennige per kWh; if this heat were supplied instead by electricity, this would cost 10 to 12 Pfennige per unit. It is of interest to compare this figure with the price the domestic consumer pays for competing fuels. For example, the price of fuel used in domestic oil fired appliances at 27 Pfennige per litre corresponds to about DM 14.33 per GJ calorific value of the fuel. Fuel oil cannot be burned in a domestic installation at an efficiency much exceeding 75 per cent: therefore district heat supplied at DM 14.30 per actual GJ of useful heat represents an attractive proposition indeed. Although the householder has still to supply his own radiator system and auxiliaries, he need neither buy nor maintain a boiler plant. At the time of writing, domestic gas central heating in West Germany was costing more to operate than oil fired central heating, giving district heating an even greater advantage.

Advantages to the public

From the individual consumer's point of view district heating using ITOC turbines (as employed in West Germany and elsewhere) gives very much lower national heat prices than those prevailing in the UK.

From the national point of view it has been estimated that the total amount of space heat which can be covered by district heating throughout West Germany is 450 000 to 625 000 TJ each year. The present installed capacity of ITOC stations there is calculated to save around 9 to 13.4 million tonnes of coal equivalent each year. This is due to rise to about 16.6 million tonnes when the present extension programme has been completed. Nuclear stations planned to carry some of the heating load will increase this saving by a further 2.1 to 2.8 million tonnes of coal equivalent each year.

It has also been estimated that it is necessary to spend between DM 55 and DM 65 in the form of ITOC stations, pipelines etc for a saving of 1 GJ of heat valued at about DM 15, therefore giving a return of around 23 to 27 per cent based on present day fuel prices. By comparison the return on improved thermal insulation (over the very demanding German standards) would only amount to about 6.25 to 9.0 per cent, so that increasing the already well expanded West German district heating network offers far better fuel savings than could be achieved by better thermal insulation. (In West Germany, 100 mm thick void insulation in walls, ceilings and underneath the floor is already mandatory in most areas and double glazing is a standard facility.)

COSTING CALCULATIONS FOR COMBINED SCHEMES

Numerous technical papers have been written about apportionment of savings engendered by combined, as opposed to separate, generation of heat and power. This is less important when the same authority supplies heat and power but much more so when different administrations are involved.

Rotislav Hromek of the Czechoslovakian Ministry of Fuel and Energy has correlated the different techniques of apportionment and the following section is based upon his work.

Costs are always a sum of the fixed costs (C_f) and the variable costs (C_v).

The total costs (C) therefore equal the sum of these ie:

$C = C_f + C_v$

Fixed costs in combined heat and power generation schemes are usually higher than in schemes for the separate production of heat and power. Designating the costs of separate production of heat and power by the symbol C(S), combined production by C and the suffixes E and H for electricity and heat respectively the following relationships are obtained:

For separate heat and power generation:

$C(S) = C(S)_E + C(S)_H$

For combined heat and power generation:

$C = C_E + C_H$

The savings in cost ($\triangle C$) are thus:

$C(S) - C = \triangle C$

Limiting values: It is possible to credit all the savings to power generation. Under such circumstances,

$C(S)_H = C_H$, but

$C_E = C(S)_E - \triangle C$

It is also possible to credit all the savings to heat generation:

when $C(S)_E = C_E$

then $C_H = C(S)_H - \triangle C$

All other systems of apportioning heat and power costs fall within these limits.

Calorific technique

The total production costs are subdivided into those applicable to power production and those applicable to heat production by evaluating the ratio of heat energy in the fuel needed to produce steam for the two final products.

If the hot water represents Q_H joules of heat and the electricity Q_E joules, while the original fuel contains Q joules, then the cost of heat is assessed as:

$$C_H = \frac{Q_H}{Q} \times C$$

and the cost of electricity as:

$$C_E = \frac{Q_E}{Q} \times C$$

In this method the variable and the fixed costs are kept down for the power production side of the business. It can be argued that this way of calculating does not reflect the higher grade of energy represented by electricity.

Thermodynamic methods

Enthalpy method

H_{st} is the enthalpy of the steam produced in the boiler

H_b is the enthalpy of the steam at the bleed take-off,

H_c is the enthalpy of the steam at the condenser of the turbine.

Equating the relative costs in a thermodynamically correct manner produces the following formulae:

$$C_H = \frac{H_b - H_c}{H_{st} - H_c} \times C, \text{ and}$$

$$C_E = \frac{H_{st} - H_b}{H_{st} - H_c} \times C$$

This method assesses the value of the steam in terms of its thermodynamic capability of doing work.

Entropy method

The difference in capability of doing work can also be expressed in terms of entropies. If

(S_{st}) is the entropy of the boiler-house steam, and

(S_b) is entropy at the bleed take-off, then

$$C_H = \frac{S_{st} - S_b}{S_{st}} \times C, \text{ and}$$

$$C_E = \frac{S_b}{S_{st}}$$

Commercial method

Separate ratios are calculated for the cost of producing heat and for that of producing electricity (ie heat generated in a directly fired boiler house and electricity generated in a condensing turbine):

$$C_H = \frac{C\,(S)_H}{C\,(S)_H + C\,(S)_E} \times C, \text{ and}$$

$$C_E = \frac{C\,(S)_E}{C\,(S)_H + C\,(S)_E} \times C$$

This method gives the total production costs of heat and electricity, rather than, as is the case with the other systems, only the costs of fuel.

Omega system

All the above methods have some inbuilt inaccuracies, Hromek therefore suggests the 'Omega' method, a rather complex one which combines the main criteria of the other methods:

(a) the savings in the cost of heat due to ITOC operation are first of all credited to power generation

(b) the increase in investment costs due to combining heat and power generation, in contrast to supplying heat and power separately, is then debited to power generation

(c) the cost of heat generation is assessed by subtracting the cost of power generation from the total costs.

It can be shown mathematically that, if this calculation is carried out in

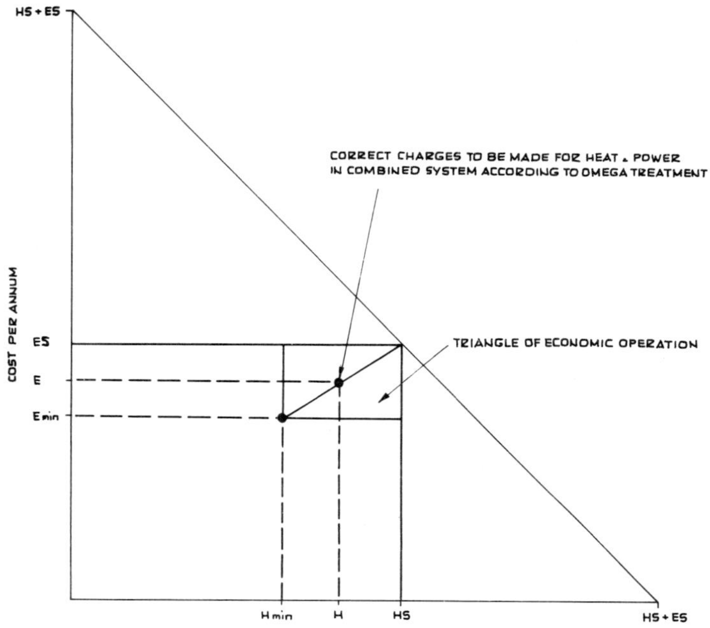

5.18 Graphical representation of correct price for heat and electricity in combined plant

reverse ie if one credits the savings and debits the additional investment costs to heat generation and then subtracts the cost of heat generation from the total costs to obtain the cost of power, exactly the same results are obtained.

The costs of heat and power are expressed by the Omega method as follows:

$$C_H = \frac{Q_H C_Q}{Q} + (I - (I (S)_H + I (S)_E))$$

$$C_E = \frac{Q_E C_Q}{Q} + (I - (I (S)_H + I (S)_E)), \text{ where}$$

Q is the heat contained in the fuel
Q_H and Q_E are the heat contents of hot water and power, respectively
C_Q is the fuel cost per unit heat content
(These values, like the following ones, are quoted on an annual basis)
I is the annual capital cost of the ITOC plant,
$I (S)_H$ and $I (S)_E$ are the annual capital costs of plants producing only heat or only electricity, respectively, divided by the annual output of heat and electricity.

The advantage of the Omega treatment for apportioning heat and power costs is that it represents accurately both the fixed and variable parts of the relative costs of heat and power generation.

CHECK LIST

1 Basic conditions for economic viability
2 Function of intermediate take-off condensing (ITOC) stations
3 Thermodynamics of steam turbine systems
4 Turbine types suitable for use with district heating
5 Machine efficiencies
6 Energy transmission losses
7 Quantities of heat evolved
8 Major relative advantages/disadvantages of condensing and back pressure turbines for use with district heating
9 Uses of very large ITOC turbines
10 Nuclear energy generation with turbine operation for district heating
11 Advantages/disadvantages and uses of gas turbines for district heating applications
12 Economic assessments relating to combined heat and power generation
13 Operating information relating to the various types of turbines used for district heating
14 Operating experience regarding the cost of transporting district heating hot water
15 The national or public advantages of using district heating.

REFERENCES

Fernwärme International (*in German*)

E. Dern, 'Improving CHP stations by the use of twin extraction points', 2 (1973) 3, pp 41–54

'Thermal power stations in Sweden', 2 (1973) 3, pp 57–67

J. Korek, 'The heat pump and district heating', 3 (1974) 2, pp 53–55

H. Muhlhausen and others, 'Various systems of operating CHP turbines', 4 (1975) 1, pp 15–19

K. D. Schellenberg, 'Gas turbine waste heat for district heating', 4 (1975) 1, pp 20–23

R. Schach and F. Linder, 'CHP operation in built-up areas', 4 (1975) 4, pp 139–148

H. P. Winkens, 'Economic aspects of CHP stations', 5 (1976) 4, pp 48–56

S. Andrzejewski, 'Coupling heat and electrical energy', 5 (1976) 4, pp 111–114

'Study of district heating in West Germany', 6 (1977) 2, pp 24–35

Other sources

E. A. Bruges, *Available energy and the second law of thermodynamics*, Butterworths, London, 1959

A. Chuter, 'Case for combined heat and power generation', *Engineer*, 249, 2nd–9th August 1979

H. Cohen and G. F. C. Rogets, *Gas turbine theory*, Longmans, London, 1958
E. F. Crawley, 'The application of medium speed diesel engines', Diesel Engineers and Users Association No 373, 1976
'Computer energy control system can optimize CHP output', *Electrical Review*, 204 (8), 7th February 1979
R. M. E. Diamant, *Total energy*, Pergamon, Oxford, 1970
R. M. E. Diamant, 'District heating with combined heat and power generation', *Heating and Air Conditioning Journal*, January 1977 et seq
A. A. Field, 'District heating and total energy', *Energy World*, February 1975, pp 11–15
H. A. Freeman and A. G. Horsley, Total energy evaluation and feasibility studies Paper 184, 8th December 1975, p 660
H. Harboe, 'District heating and power generation', Nottingham Symposium, 14th November, 1975
H. Harboe, 'Challenge of electricity', *Consulting Engineer*, 40, January 1976, pp 22–23
L. Jarfas and M. Torma, 'Limits of heat/power coupling with district heating in Hungary', Congress on district heating, Paris 1975, pp 9–11
J. F. Lee, *Theory and design of steam and gas turbines*, McGraw-Hill, New York, 1954
J. Maccadam, 'How power stations can heat Great Britain', *Electrical Review*, 200, 21st January 1977, pp 16–17
T. N. Melentev, 'District heating in the Soviet power industry', *Teploenergetika*, 21 (11), 1974, p 10 (in Russian)
I. Oliker, 'Cogeneration plants serve district heating systems', *Mechanical Engineering*, 100, July 1978, pp 24–29
G. S. West, 'The electricity supply industry's role in district heating', *Electric Power Engineering*, 57, March 1975, pp 46–47
S. V. Wildman and T. J. Goldsmith, 'Application of steam turbines to energy recovery', Energy Digest 4, October 1975, pp 20–30
A. C. Valentine, *Total energy*, Amalgamated Power Ltd, Bedford, 1977
H. Wilson, 'District heating from old power stations', *DHA Journal*, 21, March 1978, pp 19–21

CHAPTER 6

District heating using refuse or geothermal energy

REFUSE INCINERATION

In the developed countries refuse of all types discarded per head of population amounts to about 300 kg per annum. Assuming that the average calorific value is 6.2 MJ/kg, Great Britain (population of 55 million) disposes of about 16.5 million tonnes of refuse each year, representing about 100 million gigajoules of heat (2.9 million tonnes of coal equivalent).

Serious efforts have been made in many countries to utilise the calorific content of refuse by incinerating it for district heating purposes. Cost effectiveness is not always the deciding factor in the adoption of incineration for this purpose eg there may be a shortage of landfill sites for tipping rubbish. Moreover burned refuse converts into a much smaller volume of slag which may be of some value to the construction industry.

Small municipal incinerators (those handling less than 9000 tonnes of refuse per annum) are not particularly economical to operate as it is impossible to recoup a major part of the cost of disposal by efficient heat recovery. The operation of very large refuse incinerator units in which most of the heat can be recycled for district heating purposes is however a viable and attractive means of disposal and recovery.

COMPOSITION OF REFUSE

The content and calorific value of refuse differ according to the country and even the town from which it comes. The composition of municipal refuse does, in fact, reflect the life style of the population and changes from time to time: as standards of living improve so do the quantity and quality of the discards.

The following figures are a typical example of the composition of domestic refuse:

TABLE 6.1 COMPOSITION OF DOMESTIC REFUSE

Paper, cardboard	39.8%
Coal, coke, etc.	1.1%
Cloth and fabric	1.7%
Wood, fibre, straw	3.4%
Rubber, leather, plastic	0.4%

Bones and other organic materials	7.6%
Incombustible materials	12.3%
Slack and ashes:	
up to 9.5 mm diameter	15.8%
between 9.5 mm and 16 mm diameter	9.3%
between 16 mm and 25 mm diameter	8.6%
Total	100.0%

Such refuse contains: 37% combustible matter
25% incombustible matter
38% water

The average calorific value of typical municipal refuse is about 6.2 MJ/kg.
It contains 0.2 to 0.7% of sulphur
0.2 to 0.4% of chlorine
0.02 to 0.06% of fluorine

The presence of noxious materials (such as sulphur and chlorine) very materially limits the permissible temperature of flue gases.

Urban communities also produce large quantities of sewage sludge, filtered off or sedimented from sewage plants. In the past this was used in agriculture as a fertiliser. Nowadays, however, such use is discouraged because of the high lead content of the sludge and incineration is considered a safer method of disposal. Sewage sludge has a very high moisture content which can amount to 60 to 70 per cent of its weight; specific expertise and technique are therefore required if it is to be disposed of by incineration.

TYPES OF FURNACE FOR REFUSE INCINERATION

In general it is difficult to operate a furnace in conjunction with heat recovery unless it has a capacity exceeding 3 tonnes of refuse per hour. Attempts have been made to use conventional oil or coal fired furnaces for refuse incineration by mixing a quantity of refuse with conventional fuel. It has been found from these that the presence of refuse in the furnace lowers the overall combustion efficiency to such an extent that very little benefit is obtained over firing by conventional fuel alone. Much extra expense is involved in handling the refuse and the ash produced and maintenance and plant costs are higher. For these reasons it is better for such furnaces to be purpose-designed for refuse incineration alone. **Figure 6.1** shows a typical fixed grate refuse incineration system.

The main difficulties encountered in burning refuse are:

(a) it is generally very damp, containing up to 40 per cent water
(b) it is very variable in size, moisture content and calorific value
(c) it is bulky with an average density of only 300 kg/m^3 and its density varies according to its composition

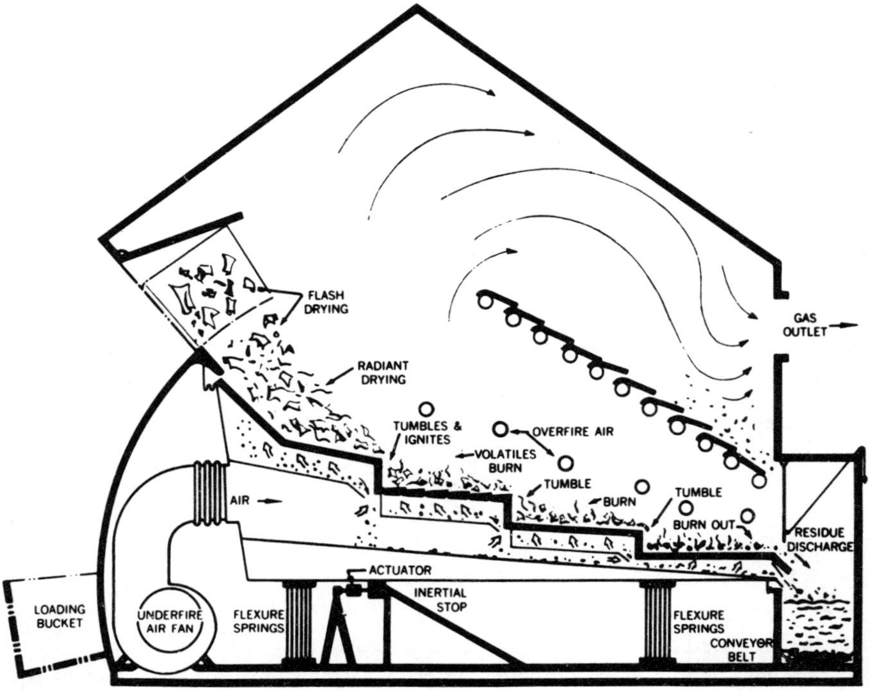

6.1 A typical fixed grate refuse incinerator

(d) it is a mixture of very soft and very hard components and contains substances varying from dust to large furniture sections, steel springs, wires etc which may be difficult to separate magnetically because they stick to other components
(e) it may contain noxious components such as plastics, rubber etc
(f) it may include constituents (particularly glass) which possess a low ash fusion (or melting) point, leading to rapid clogging and failure of the fire bars and of other furnace parts.

Furnaces for refuse incineration are designed so that the refuse is well dried before combustion and turned and agitated in such a way that every part of it can be completely incinerated. The following types of furnace are available:

Fixed grates: These operate on the step-grate principle. A number of separate grates are placed one above the other, and the refuse is loaded on to the top grate where it is dried. After drying, it is pushed on to the next grate where it commences combustion, eventually reaching the bottom grate as ash. A typical plant of this type is shown in **figure 6.2.**

Chaingrate stokers without mixing action: Some refuse incineration plants use conventional chaingrate stokers for burning either coal or refuse. In general

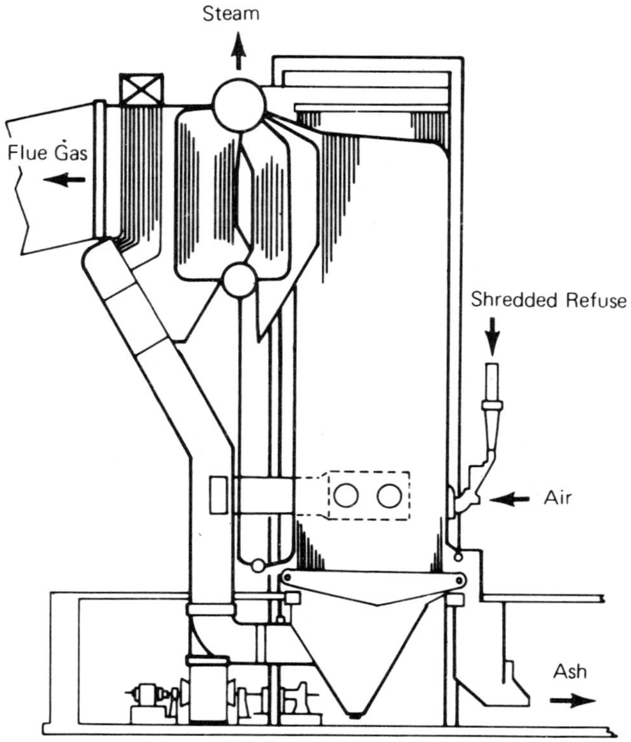

6.2 Steam generator for processed municipal waste at Hamilton, Ontario, Canada

such furnaces are not particularly suitable for refuse incineration owing to the heterogenous nature of the material.

Chaingrate stokers with mixing action: Some systems of this type have been purpose-designed for burning refuse (**6.3**). For example, the Karnap system comprises a number of individual chaingrate units which are positioned one behind the other so that the partially burned refuse is turned over and allowed to continue burning. The speed of each of the individual chain grates can be varied to maintain the height of the refuse and of the slack bed constant. In the Babcock system, on the other hand, the chaingrate units themselves move in relation to each other so as to agitate the refuse, while in the Düsseldorf system the chaingrate moves up and down as it passes along a number of large rollers and the refuse is agitated sufficiently to prevent part of it from being only partly burned.

Agitating grates: With these the refuse is moved around and reduced in size without allowing any holes to form in the fuel bed (**6.4**). The best-known system of this type is the Von-Roll system, in which the material is agitated constantly by hydraulic means in preference to mechanical operation so as

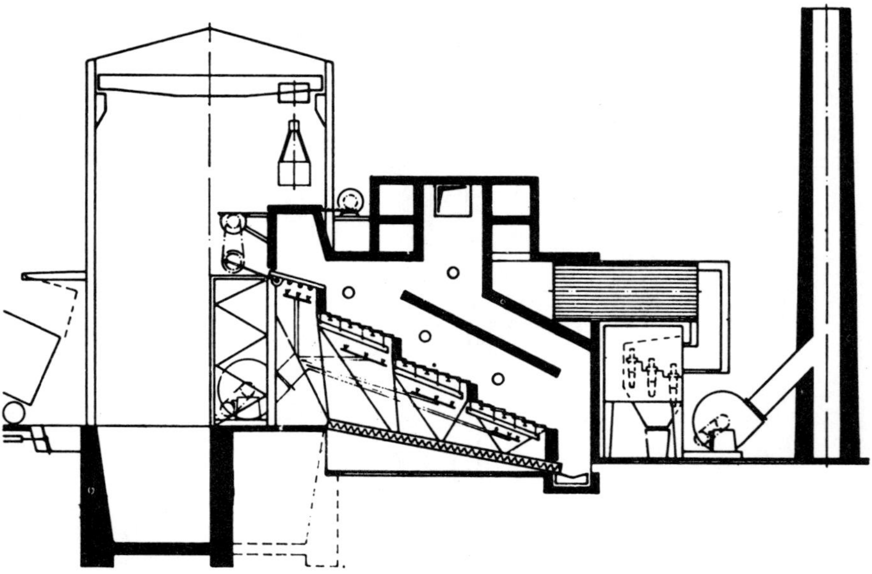

6.3 Forward-feed grate incinerator

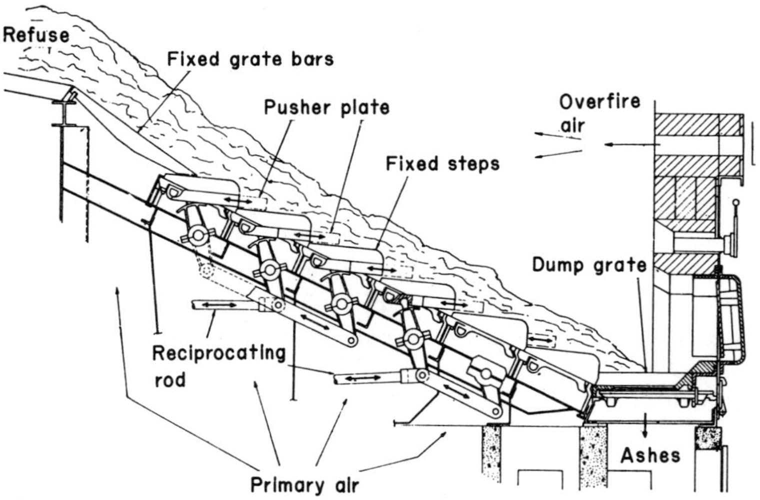

6.4 Agitating grate system

to minimise the risk of components breaking when solid materials become wedged.

Grateless systems: In these the refuse is incinerated in rotating ovens. The best known systems of this type are the Vølund system (see **plate 10**) (in

which an inclined rotating drum is used) and the Fellner and Ziegler system (which employs two drums, one above the other).
Flame chamber and fluidised combustion systems: These burn the refuse suspended in air either by itself or mixed with ground coal so as to permit complete combustion to take place.
Cyclone furnaces: The refuse enters the circular (or cylindrical) furnace tangentially and is spun around the furnace. These units give a much longer dwell time to the material being incinerated, thus promoting efficient pre-drying and complete combustion.

It is a common essential feature of all the various refuse incineration plants that the combustion space provides for a maximum furnace heat release of only 1.5 MJ/m^3 h, far lower than in furnaces which burn conventional fuels.

The refractory materials used in the construction of refuse incineration plant must be of very high quality because they have to resist considerable attack by the slag. They require to be carefully selected to suit the varied operating conditions inside the furnace and the characteristics of the materials being incinerated. The incinerator must be thermally insulated to retain its heat.

Wet refuse must be pre-dried. Such drying is carried out in large undertakings by the use of special drying grates in which the hot combustion gases pass through the damp refuse. This process also serves to reduce the temperature of the flue gases to about 350°C (662°F) which is necessary to obtain adequate operation of the dust collector used prior to discharge into the chimney. Much of the ash residue is in the form of flyash and techniques have been developed to sinter this afterwards to produce a material which can be used in the building industry.

As the calorific value of refuse is likely to vary a great deal it is common practice to provide auxiliary oil or gas burners for use when a batch of lower than average calorific value is fed into the furnace.

ECONOMICS OF REFUSE INCINERATION

Since the grate temperature in refuse incinerators is much lower than in either coal or oil fired installations, the combustion efficiency of a refuse incinerator is also much lower. Therefore its furnace plant must be far larger per unit thermal output than plant designed for efficient burning of fossil fuels (see **figures 6.5, 6.6** and **plates 11, 12**). To achieve reasonably complete combustion the quantities and proportions of secondary air supply must be strictly controlled and directed, and in order to obviate the emission of black smoke and achieve thorough mixing of combustion air with the refuse fuel relatively large amounts of secondary air must be fed in to the incinerator. The proportion of CO_2 in the flue gas of an efficient incinerator is of the order of only 8 to 12 per cent. The large amount of slag which is

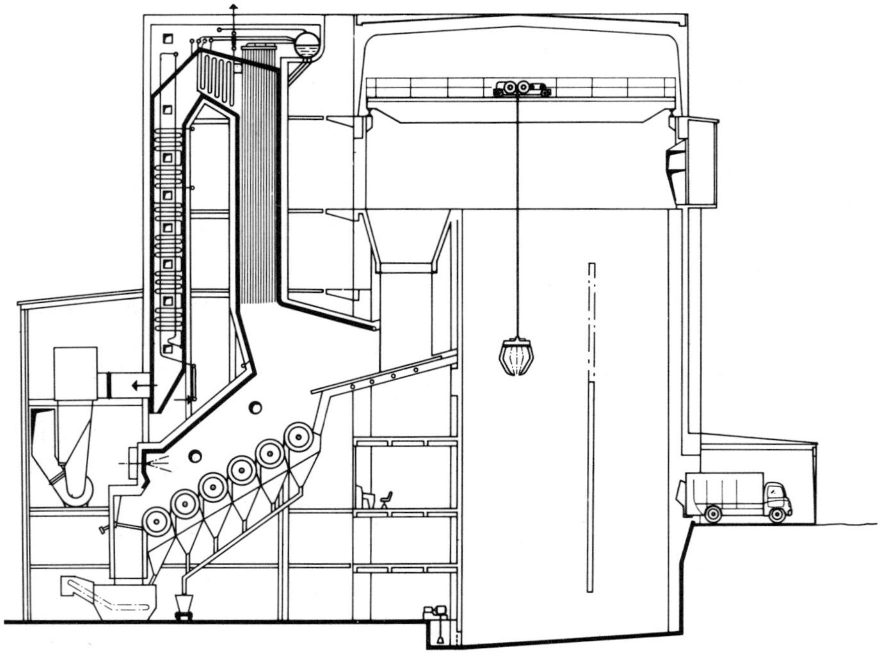

6.5 Layout of large urban incinerator which operates on the Düsseldorf system

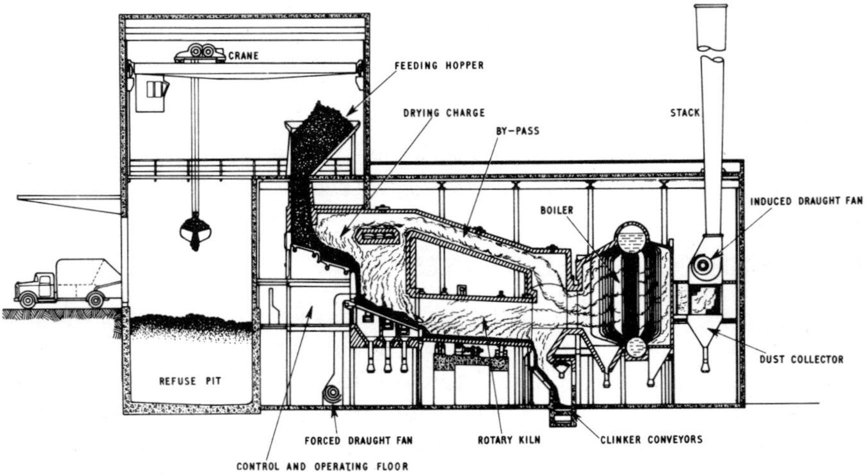

6.6 Layout of large urban incinerator plant which operates on the Volund principle

produced (between 25 and 30 per cent by weight and about 15 per cent by volume of the refuse fed into the furnace) carries away much sensible heat.

In consequence the combustion efficiencies of refuse incinerators are low, as is their heat recovery per unit size.

To summarise the losses of heat in operation, the following figures offer a general guide:

Loss of heat in flue gases	12 to 15%
Loss of heat in slag	10 to 18%
Loss of heat by radiation	5%
Other heat losses	5%

The overall average thermal efficiency is thus in the order of 60 to 70 per cent.

When refuse incinerators are linked to district heating they should carry the base load, the peak load being supplied by oil or coal fired furnaces. For example in Paris the heat recovered by refuse incinerators is the only heat supply to the municipal heating system from about March to October. During the winter months the thermo-electric plants are also on-line, the conventional heating plants being used only in severely cold weather.

Due to the low efficiency of refuse incineration plants, and because they are expensive to build, maintain and operate, it is seldom possible to make the sale of heat pay for all the costs which are involved. In almost every incinerator plant so far built the running costs have to be subsidised by the authority wishing to dispose of its refuse.

It is generally agreed that the combustion of refuse in large district heating installations is the cheapest method of disposing of it. There are hopes that in future it may be possible to achieve a 'break-even' operating cost through the profits made from the sale of steam (or hot water) and clinker.

USE OF HEAT FROM REFUSE INCINERATORS FOR DISTRICT HEATING

The incineration of refuse reduces its original volume to about 10 per cent and produces a non-putrescible ash with some valuable properties for use within the building and other industries. The extraction and recycling of metals is possible and is achieved in some existing large plants. For district heating purposes it is necessary to use incinerators of very high output (a minimum capacity of some 5 tonnes per hour and preferably larger). The limit on the size of a municipal refuse incinerator plant complex is governed by the cost of transporting the refuse from points of collection (eg residences) to the plant complex.

Assuming that the calorific value of the refuse discarded by one average family of four members is about 10 GJ per annum, while the heating demands for such a family are between 40 and 60 GJ per annum, it can be seen that the refuse from between five and eight families can produce enough useful heat (after taking into consideration heat losses during combustion etc) to provide space heating and domestic hot water for one family.

Because of the difficulties often faced by local authorities in the disposal of their town refuse, established district heating incinerator plants tend to collect it from a ring of neighbouring communities. As the authority which operates the incinerator is paid for disposal of the refuse, this then constitutes a fuel with a negative cost.

The pattern of operation of a refuse incinerator is mainly governed by the presence of large quantities of chlorine, hydrogen chloride, sulphur dioxide and sulphur trioxide in the flue gases. The chlorine compounds originate from the combustion of PVC plastic components in the refuse.

Special attention must be directed to the high flyash content in flue gases which can lead to heavy deposits, the clogging up of heat exchanger surfaces and the erosion of boiler tubes. It is necessary to limit the gas velocities to below 5 to 8 m/s and thereby permit the flyash to settle in the dust collectors associated with the incinerator.

Boiler design should be generous to permit cleaning. The spaces between tubes should be at least 120 to 150 mm and the height of each tube bundle should not exceed 1500 mm. Flue gas temperatures must be kept below 700°C (1292°F) to prevent depositing of slag. In order to achieve sufficiently long operating cycles it is necessary to use soot blowing devices to clean the fireside surfaces. These are powered by air and/or steam and permit operating periods of between 4000 and 6000 hours between plant shut-downs.

Hydrogen chloride and oxides of sulphur attack steel and many of its alloys very rapidly at low temperatures. Exhaust gases have therefore to be maintained at above 220°C (428°F), appreciably higher than for other types of furnaces. The rate of corrosion increases dramatically at temperatures exceeding about 480°C (896°F) so that the practical limits of steam generation are about 43 bars and 450°C (842°F). Most plants operate at about 30 bars and 350°C (662°F). Steam at such relatively low temperature and pressure cannot generate much electricity even at very low exhaust temperatures. Furthermore the amount of steam produced per given quantity of refuse varies appreciably, as does the amount of refuse produced by each inhabitant. During the summer when demand for heat is at a minimum, more refuse is produced than in winter (the extra summer load consists of drink containers, ice-cream cartons, garden waste etc). The calorific value of refuse also tends to be higher in summer than in winter. It has been found that the average amount of steam obtained from each tonne of refuse is about 1.4 tonnes in winter and that this figure can rise to almost two tonnes in summer.

It has been explained that refuse incinerators have poor characteristics as the sole suppliers of waste heat. This has been proved by the fact that early projections and forecasts for the operation of such plant have been found impossible to achieve in practice. The most suitable arrangement is to employ refuse incinerators to supply electricity by a back pressure turbine and to use the waste heat from this turbine to provide only the summer load of the district heating system. The variable and peak loads for the system are then supplied by normal ITOC turbines fired by oil or coal.

Such systems have been very successfully applied. By adopting combined heat and power generation the maximum utilisation of the very expensive incinerator for the production of saleable energy is achieved. It then no longer matters how much or how little refuse passes through the plant or to what extent its calorific value varies. A large ITOC station run in parallel can adjust the total heat supply relative to the total heat demand instantaneously within very narrow limits by the operation of automatic valves.

In summer when refuse incineration plant operates at its peak capacity the ITOC station runs almost entirely in a condensing mode. This may therefore be the appropriate time for shutting it down for annual maintenance ond overhaul. The back pressure refuse incinerator turbine combined with one or two small gas turbines can very likely then meet the power requirements during July and August. In winter the bulk of the heat load can be carried by the ITOC station, with the refuse incinerator plant still contributing an appreciable quantity of electricity and heat from a fuel (the refuse) which is cheap (because the only cost involved is transport and because the operator is paid for burning it).

USE OF INDUSTRIAL WASTE HEAT

In any community the generation of electricity produces a large amount of utilisable waste heat. This is why ITOC turbines are so very popular. It is estimated that to supply electricity for a community of 10 million persons power stations consume 470 000 GJ of high grade energy per day of which about half is discharged in the form of waste heat. A high percentage of this waste heat can be, and is, utilised to provide the low grade heat requirements of the population. The major industries listed below discharge the following percentages of their total energy intake as waste heat:

Chemical works	92%
Oil refineries	50%
Food factories	25%
Non-ferrous metal plants	20%
Potteries	16%
Iron and steel works	10%
Gas works	8%

SOURCES OF GEOTHERMAL ENERGY

The idea and prospect of harnessing the almost limitless resources of energy beneath the earth's crust is very attractive. Appreciable difficulties are however involved in designing viable applications and schemes. Considerable work has been done in this field in Iceland, New Zealand, France, Hungary, Italy, Japan and elsewhere.

The earth's surface consists of a large variety of igneous and sedimentary rocks. Below a depth of about 35 km beneath the surface of the continents there is the Mohorovicic discontinuity, usually understood to be the lower limit of solid rock formations. In the case of the oceans the Mohorovicic discontinuity lies only about 5 km beneath the surface of their floor or, on average, 10 km beneath the surface of the waves. It can be seen therefore that the earth's crust is very thin indeed, varying between 5 km and 35 km. Beneath this crust, there is molten rock or lava in equilibrium with solid material. Because the crust is far thinner below sea areas than under continental land areas there is a greater tendency for earthquakes and volcanoes to arise close to the sea. At the Mohorovicic limit the basic composition of the semi-solid mantle is now assumed to be iron magnesium orthosilicate ($(Mg,Fe)_2SiO_4$).

It has been long known that, as a drill penetrates downwards, there is a growing temperature gradient the nature of which varies according to the thermal conductivity of the underlying rocks. On average the temperature rises by one degree Centigrade for every 100 metres bored beneath the surface. The average flow of heat upwards is approximately 50 mJ/m^2s or roughly 6.5 terawatts per annum over the whole surface of the globe. Some of this heat may be the original heat with which the earth was endowed during its formation, but scientists now speculate that heat is being produced continuously inside the earth by radioactive disintegration, ie that the interior of the earth, like that of the sun, is in fact a continuous atomic furnace. The temperature at the Mohorovicic discontinuity is generally taken to average 1400°C (2552°F).

The most promising conditions for extraction of heat from beneath the earth's surface exist where such heat is not only being conducted but also convected so that water flows upwards from the depths and carries heat with it. In volcanic regions, such as New Zealand and particularly Iceland, the water is heated to boiling point and emerges in the form of geysers often with a spectacular display. In the United States the geysers in Yellowstone National Park are well known.

UTILISATION OF GEOTHERMAL ENERGY IN DIFFERENT COUNTRIES

Iceland

Iceland has no deposits of fossil fuel and but few trees. It is however richly endowed with geothermal energy and much use has been made of this from the beginning of this century: see **6.7** and **plate 13**). The first attempt at town heating from geothermal sources took place in 1928 when a 3 km long pipeline was constructed to carry 14 litres of water per second at 87°C (189°F) to heat 70 houses, some swimming pools and other premises in the Reykjavik area. From these beginnings the exploitation of geothermal energy has proceeded very rapidly.

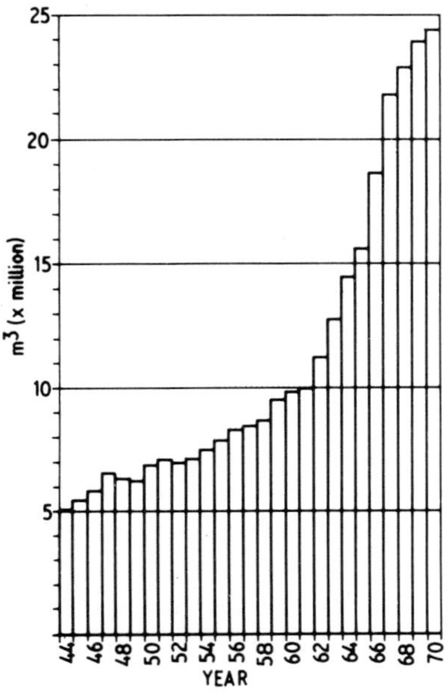

6.7 Pattern of development of hot water utilisation in Iceland, 1944–70

TABLE 6.2 USE OF GEOTHERMAL ENERGY IN ICELAND

Year	Geothermal energy used (terajoules per annum)
1960	2500
1962	3010
1964	3470
1966	4350
1968	5100
1970	9130
1972	9400
1974	10 800
1975	11 800

The total availability of geothermal energy in Iceland was estimated in 1968 at 293 000 terajoules per annum. By 1975 only about 4 per cent of the easily accessible resources had been utilised. The 11 800 terajoules of energy used in that year was employed as follows:

Reykjavik district heating system	6700
Other district heating systems	400
Power production	1600
Industrial users	2000
Heating of plant nurseries etc	1100
Total	11800

Reykjavik, a city of 85 000 inhabitants, had a 98 per cent connection to the geothermic district heating network in 1975. It is a city without chimneys. The hot water which it uses is obtained partly within the city area itself and partly from the Reykir field about 20 km away. There are ten large hot water accumulators on the Oskjuhlid hill close to the town and the distribution temperature of the hot water which these supply remains constant at 82°C (180°F). It is so plentiful that the system does not usually include return mains: the water is cooled down to about 40°C (104°F) and finally discharged into the sewers. Return mains are used, however, where the hot water is needed for such purposes as filling swimming pools etc.

The connected capacity of the Reykjavik district heating system was 261.2 MW in 1974 of which only 35 MW was provided by (peak use) oil fired plant. A particular advantage of this system is that there is a virtually even heat demand all the year round. The single pipelines are constructed from welded mild steel and are laid into concrete ducts, insulated thermally and with mineral wool. In August 1974 the consumer paid 79 pence per GJ of heat. In the context of the Icelandic cost situation this was a small charge. The following data indicate the size of the Reykjavik scheme:

Distribution area	11.3 km^2
Supply mains	29.1 km
Collecting mains	14.2 km
Street mains	125.2 km
House connections	120.2 km

The steam which emerges from the varying bore holes in Iceland can be at temperatures of up to 180°C (356°F). It is mainly used in a back pressure turbine with small power output (about 3 MW) utilising the waste heat produced for other purposes. In addition bore holes yield steam and water at varying temperatures. Table 6.3 summarises the main features of a comprehensive scheme of heat utilisation developed for industrial purposes in Iceland.

Currently a very large expansion of the geothermal energy facilities is taking place in Iceland mainly for provision of district heating to smaller communities.

TABLE 6.3 INDUSTRIAL USES OF GEOTHERMAL FLUIDS IN ICELAND

Temperature of fluid (steam or hot water)	Use
160–180°C (320–356°F)	Heavy water production using H_2S, ammonia absorption, refrigeration, digestion of paper pulp, Kraft process
150 to 160°C (302 to 320°F)	Fish meal production, drying of timber
140 to 150°C (284 to 302°F)	Alumina production using the Bayer process

Temperature of fluid (steam or hot water)	Use
120 to 140°C (248 to 284°F)	Evaporation of sugar solutions, extraction of salts, high rate evaporation for food factories
100 to 120°C (212 to 248°F)	Concentration of seawater for salt extraction, desalination processes
80 to 100°C (176 to 212°F)	Drying of stockfish, seaweeds, grass, vegetables, wool and other materials
70 to 80°C (158 to 176°F)	Operation of refrigeration plants
50 to 70°C (122 to 158°F)	Greenhouse heating
30 to 50°C (86 to 122°F)	Mushroom growing, soil warming to increase yield
20 to 30°C (68 to 86°F)	Fish farming, fermentation processes, de-icing

Hungary

The Hungarian scientist Boldissar estimated that the hot water at temperatures varying between 60 and 200°C (140 to 392°F) beneath the Hungarian plains has a recoverable heat content of no less than 10^{20} joules, equal to 50 per cent of the heat value of all the known petroleum reserves in the world. Since 1962 when the bulk of the fields were discovered considerable advances have been made in the use of this hot water for farm and domestic heating. By the end of 1970 for example 800 000 m² of hothouses were heated by geothermal hot water and considerable use was being made of the geothermal fluids for supplying district heating to the towns of Szeged, Hodmezovasarhely and Mako. The thermal waters of Hungary, produced by drilling to a depth of about 2000 metres, are at a temperature of about 90°C (194°F). The average yield per well is some 54 to 108 m³ per hour.

Italy

The main Italian geothermal field is in Tuscany some 50 km south of Pisa. This is the Larderello field and it yields superheated steam at a temperature of up to 260°C (500°F). Typical conditions are 5 bars and 225°C (437°F) at the wellhead with a maximum enthalpy of 2950 kJ/kg. The resources of the field are almost exclusively exploited for electric power production. Steam plant with a total capacity of 390 MW is available and in 1974 the total annual yield of electricity was more than 3 million kWh. It has been established that the well yield declines with time. New wells have therefore to be drilled on a continuing basis.

Japan

The potential output of the Japanese geothermal fields is estimated at several thousand megawatts. The main field is at Otake on Kyushu Island. A 13 MW plant has been operating there since 1967 and a new 180 MW plant is almost finished. Another geothermal power station of 20 MW capacity is situated at Matsukawa.

New Zealand

Although New Zealand is very plentifully endowed with underground hot springs (geysers), geothermal district heating is limited to the town of Rotorua where about 1000 hot water boreholes are connected to the heating systems of public buildings, hospitals, schools and some private houses. At Wairakei use is made of steam from underground boreholes to drive a 192 MW power station. This steam emanates from the volcanic region at a temperature of about 260°C (500°F).

France

A number of working schemes exist in France for using subterranean hot water for district heating purposes. One such scheme which is an undoubted success is at Carrières sur Seine near Paris, hardly a volcanic region (**6.8, 6.9**). Yet the heat contained in the hot water which is pumped up from a depth of 1580 metres has been supplied for over ten years to carry part of the heating load of a group heating scheme of 800 dwellings. The temperature of the water at the base of the well is 60°C (140°F). A flow rate of about 90 m³ per hour was obtained from this single borehole. By the time the water reaches the surface it has cooled down to about 57°C (135°F). This

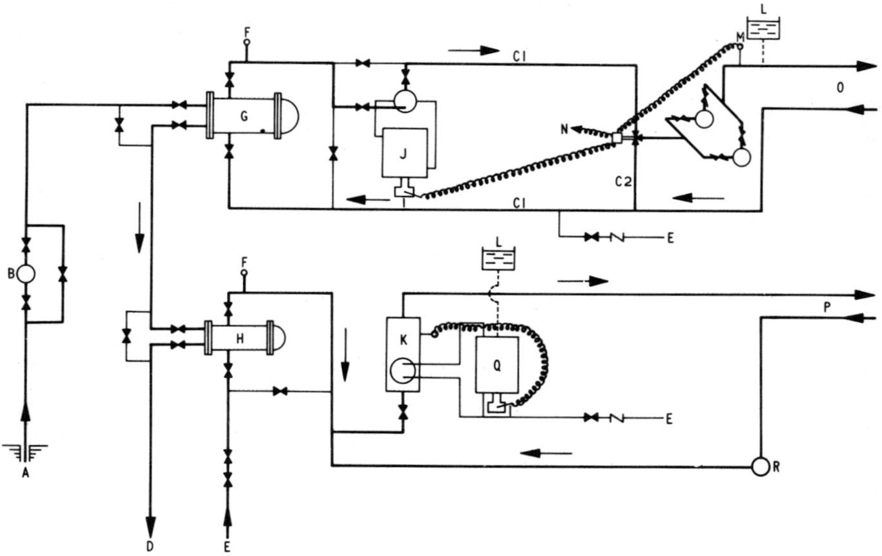

6.8 Layout of network at Carrières sur Seine

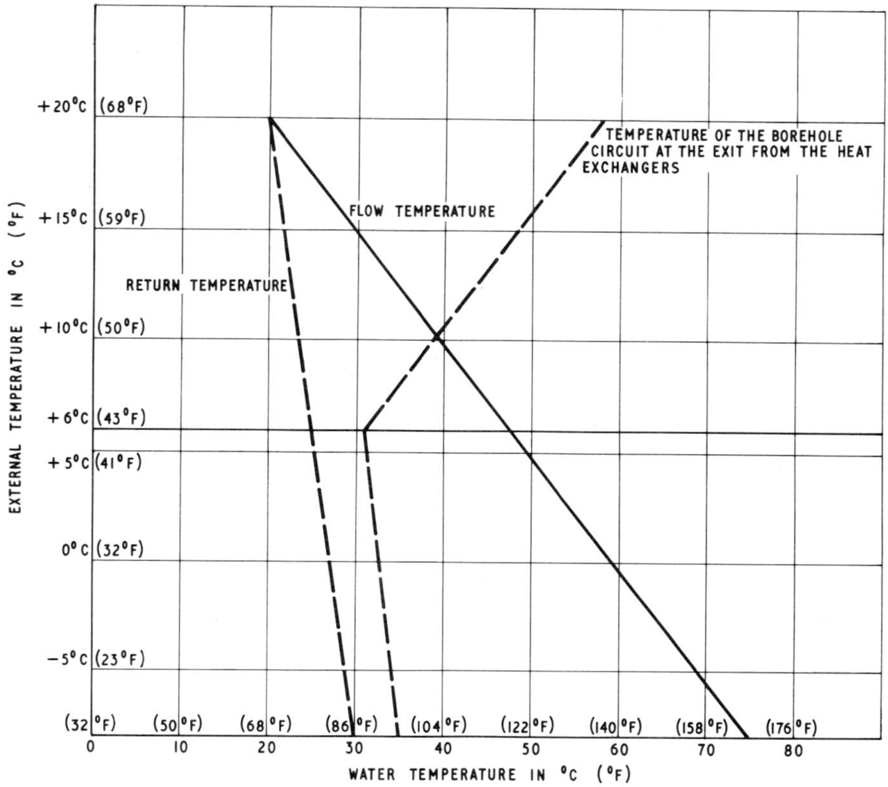

6.9 Typical conditions of the system at Carrières sur Seine

temperature has been found to be adequate for circulation purposes at external temperatures below 6°C (43°F). When the ambient air temperature is lower the geothermal heat has to be augmented by the output from an oil fired boiler. The geothermal water contains up to 15 000 parts per million of dissolved salts mainly sodium, potassium and magnesium chlorides. It is therefore necessary for the heat exchange to be effected by means of special heat exchangers equipped with high density polyethylene separators. The bore hole has had to be lined with stainless steel in order to make it withstand the corrosive effect of the hot saline water. When the geothermal water has exchanged its heat to the circulating water its temperature falls to 35°C (95°F). Some of the cooled water is discharged to the public sewer but a proportion of it is used as saline water for public open air swimming pools.

Considerable deposits of underground hot water have been found in France. In the Paris area it is found in great quantities at 75°C (167°F) at a depth of 2000 metres. In Aquitaine it occurs at 70°C (158°F) at 1800 metres, and in Alsace at 100°C at only 400 m. On the whole studies made in

France indicate that a specific geothermal heating system is only economic there provided that geothermal energy can supply at least 65 per cent of the total energy requirements of such a system. In 1976 four new projects were started there, utilising geothermal heat for district heating. **Figure 6.10** shows the locations of the resources of subterranean hot water so far located in France. Northern France has a similar geological structure to that in parts of Great Britain, so that French studies are of relevance to Great Britain (where heating by geothermal energy has not alas been practised since the Romans utilised it in the City of Bath almost 2000 years ago).

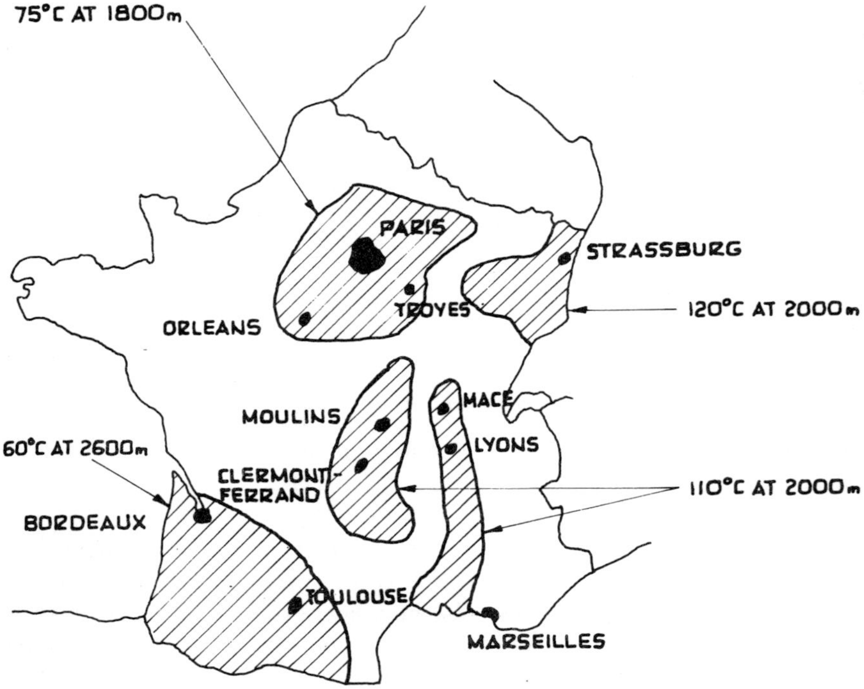

6.10 Map showing areas of subterranean hot water in France

Other countries

In the United States currently installed geothermal power plant amounts to a total of 195 MW with a further 220 MW planned. Most of these are at The Geysers in California.

Mexico utilises geothermal energy at Pathe and Mexicali with a total of 78.5 MW of installed plant.

In the Soviet Union much geothermal energy is used for agricultural purposes. In 1970 the total geothermal capacity of the Soviet Union was 970 MW, roughly four times the capacity of that at the Icelandic town of Reykjavik. Most of this capacity is used in Russia for district heating schemes.

Several other countries are actively taking part in geothermal exploration projects including Turkey, El Salvador, Chile, Ethiopia and the Philippines.

EQUIPMENT FOR EXPLOITING GEOTHERMAL ENERGY

Geothermal fluids are basically of two types: superheated steam (or hot water) caused by the flow of underground streams through hot subterranean regions, and plain hot water. Steam is generally only found in volcanic areas where the earth's crust is thin, particularly at intersections between the continental plates. Typical examples are the Larderello field in Italy, the Otake field in Japan and largest of them all, the Geysers field in California with a potential of over 4000 MW of exploitable steam resources. Specially designed turbines are used to convert such steam into electricity. Where there is no steam and only hot water is obtainable the best use for it is in the provision of district heat to cities and hot water supply to agriculture and industry.

The steam which emerges from a borehole may be dry, as in Larderello and The Geysers, or wet as at Wairakei (New Zealand), Mexicali and in Iceland. The equipment used for dry steam, which contains only gaseous contaminants, is similar to that used in conventional steam plants. Wet steam on the other hand contains large quantities of sodium chloride, silica and many scale forming substances and a steam/water separator is generally necessary whenever the amount of wetness exceeds 10 per cent. Condensing turbines are employed and where the initial temperature is below 100°C (212°F) flash-off steam has to be used for the feed. The most favourable feed pressure found so far is at Wairakei where the inlet is at 13 bars. Usually the maximum available pressure is about 7 bars. The largest units in operation are 30 MW (at Wairakei): these work with a feed pressure of 4 bars and a back pressure of 5 kPa. Modern materials have largely eliminated the corrosion trouble originally found. Considerable amounts of hydrogen sulphide are present in the steam and in consequence the tip speed of the turbine blades has to be limited to below about 300 m/s to prevent over-stressing which can lead to stress corrosion. Blades are inevitably of non-hardened 12 per cent Cr steel.

The condensers installed with geothermal plants are of the jet or spray type in which there is direct contact between the steam and the condensing water. There are three good reasons for this:

(a) condenser costs are far higher with geothermal plant than with conventional steam plant per unit output, owing to the very low inlet pressure

(b) it is unnecessary to keep the condensed steam clean for recycle to the boiler as with conventional plant
(c) copper alloys, which are used in the construction of conventional condensers, are particularly badly attacked by hydrogen sulphide in the condensate.
At The Geysers use was made of 316 clad steel with stainless steel piping, but it has since been found that coated mild steel is quite adequate to prevent corrosion by H_2S.

Geothermal plants require a copious supply of cooling water, about 300 kg of water being needed for each kWh of electrical output. They are unique in that they provide their own make-up water for the circulating cooling systems. In locations which have no readily available cold circulating water from a river or lake cooling towers can be used to cool the hot or warm water. Gaseous impurities in the condensing water such as H_2S, CO_2 etc are scrubbed off. Geothermal plant is best applied for base load duty because it has a high investment cost and a low operating charge in that fuel costs are zero.

Hot water, when it is separated from steam with wet steam feeds, can be used in the following ways:
(a) for refrigeration purposes (in absorption refrigerators)
(b) to drive a boiling water turbine, a system which has so far only been studied theoretically
(c) either for district heating schemes or for industrial purposes.

Attempts have been made to use fluids other than steam for geothermal power plants. At Ischia in Italy ethyl chloride has been used and at Kamichatka in Russia Freon 12 is employed, but the capacities of these plants are only small, 300 and 500 MW, respectively. Advantages claimed for this method are that it is possible to utilise hot water below 100°C (212°F) without having to go sub-atmospheric in pressure and that there is less risk of corrosion within the turbine.

USE OF GEOTHERMAL FLUIDS FOR DISTRICT HEATING

Most of the experience in the field of district heating derives from schemes in Iceland but some has also been gathered in France and also from the Soviet Union, where the largest scheme is one in Zgoudidi (Georgia) with a capacity of 58 MW.

The inherent characteristics of geothermal energy which must be considered when using this heat source for district heating purposes are:
(a) the thermal fluids in each bore hole usually have a constant temperature and thermal output
(b) the unit cost of energy produced is virtually entirely capital cost and not running costs, as is the case in the provision of hydro-electric power, wind power, wave power or solar energy

(c) the thermal fluids are highly corrosive and the distance through which they can be transported economically is limited
(d) the unit cost of the energy depends upon the capacity of the system.

The annual load factor of a system is given by the value LF and is commonly quoted as:

$$LF = \frac{G_h}{365\,(t_d - t_{min})}, \text{ where}$$

LF is the load factor
G_h is the number of degree days per annum on which heating above the temperature t_{min} °C is required
t_d is the design temperature inside the dwelling (degrees C).

If the geothermal system were designed to maintain the desired temperature to a minimum outside temperature of t_o only, where t_o is greater than t_{min}, the annual load factor would be:

$$LF_o = \frac{G_{ho}}{365\,(t_d - t_o)}, \text{ where}$$

G_{ho} is the annual number of degree days to a temperature of t_o °C.
The annual load factor is thus greatly increased, so that a plant with a lower capacity can supply the system and supplement the heat needed, when the temperature outside is between t_{min} and t_o, by using oil fired, coal fired or refuse incinerator plant. The selected temperature (t_o) depends on the relative capital costs of geothermal and conventional boiler plant and on the level of fuel costs. The equation used to determine this is:
$C_o/C_{min} = LF/LF_o \times F(H_o/H)$, where
C_o and C_{min} are the unit costs of heat when base temperatures of t_o and t_{min}, respectively, are considered
$F\dfrac{H_o}{H}$ is a function of the ratio of capital investment costs required to cater for a minimum temperature t_o and t_{min} respectively
To this must be added the cost of supplying supplementary heat on days when the external temperature lies between t_{min} and t_o

The following are approximate figures for exploration (in 1970 US dollar values):

The cost of drilling to a depth of about 1000 metres is \$60 000, or per successful bore about \$90 000.
Transmission costs average out at \$264 000 per borehole assuming an average distance between borehole and consumer not exceeding 1.5 km.
The average yield of heat from such a borehole is 190 GH/h or 53 MW.
Boreholes have an average effective life of about 10 years, pipelines etc of 25 years.
Add 2% for operation, repair and maintenance.

Assuming therefore a capital charge of 15 %, the cost of heat as supplied is:

$\frac{\$264\,000 \times 15}{100 \times 190\,h} = \frac{\$208}{h}$ per gigajoule supplied, where h is the number of hours of full operation per annum.

The actual costs of heat (per gigajoule supplied) at different fields were reported in 1970 to have been as follows:

Larderello	\$0.032
Rotorua	\$0.0336
Iceland	\$0.075 to \$0.121
Wairakei	\$0.032
The Geysers:	\$0.125

The cost of fossil fuels supplied during the same period averaged out at between \$0.25 and \$0.65.

The above geothermal fields all pay their way. It may be more difficult to argue the case for geothermal energy use in areas where conditions are not quite as favourable. On the other hand, with escalating costs of fossil fuels, a good case exists for thoroughly researching the feasibility of utilising the limitless resources of the geothermal energy beneath our feet.

CHECK LIST

1 Suitability of refuse incineration for the particular area concerned
2 Composition of local domestic refuse
3 Types of furnaces available for refuse incineration and their limitations
4 Economic aspects of operating refuse incinerators
5 Application of refuse incinerators to district heating
6 Application of ITOC turbines in conjunction with refuse incineration
7 Whether geothermal energy is available and the principles governing its use
8 Uses made of geothermal energy throughout the world
9 Application of geothermal energy for industrial purposes
10 Types of steam turbines suitable for power generation using geothermal energy
11 Conclusions in this chapter concerning district heating using geothermal fluids
12 Actual costs of using geothermal heat in various existing fields.

REFERENCES

K. Beck, 'Geothermal utilisation in Iceland', *Fernwärme International*, 4 (1975) 4, pp 116–119 (in German)

T. Brown, 'Thermal recovery from incineration processes', *Plant Engineer*, 19th May 1975, pp 11–13

R. M. E. Diamant, 'A case for geothermal energy', *Heating and Ventilation Engineer*, December 1976, pp 10–13

'Geothermal district heating in France', *Building Services Engineer*, 42, January 1975, pp 10–11

'Geothermal energy and its uses', UN Energy Department, 3.1–3.17 of Conference on world energy, Detroit, 1974

Geothermal energy, Review of research and development, Unesco Press, Paris, 1974

Geothermics: Proceedings of United Nations symposium on the development and utilisation of geothermal resources, Pisa, 1970

J. R. Holmes, 'Refuse into energy in the city of Sheffield refuse incinerator department', *Solid Wastes*, 65, December 1975, pp 576–581

A. Jover and others, 'Underground heating', *Révue Genèrale de Thermonique*, December 1975, pp 456–458 (in French)

'Refuse incinerator with practicable heat recovery', *Heating and Air Conditioning Journal*, November 1975, pp 10–14

R. S. Rochford, 'New and existing boilers for burning municipal refuse', *District Heating*, 61(2), 1975, pp 16–19

R. Singh, 'District heating from Sheffield's refuse', *Municipal Engineering*, 152, 21st February 1975, p 358

R. E. Sommerlad and R. H. Pai, 'Refuse; the world's wasted energy source', *District Heating*, 61 (3), 1976, pp 16–23

K. H. Thomen, 'Energy recovery by the incineration of solid waste', *Heating and Ventilation Engineer*, January 1977, pp 18–21

J. Zoega and G. Kristinsson, *The Reykjavik district heating system*, Reykjavik, 1970

CHAPTER 7

Nuclear power and district heating

ALTERNATIVES TO FOSSIL FUELS

Sooner or later (possibly sooner) our stores of the fossil fuels (oil, coal and natural gas) will come to an end. Mankind must prepare for the escalating costs of such fuel resources which will eventually become intolerably expensive. District heating, which can utilise fossil fuels so much more economically, can assist in delaying this day of reckoning. However district heating methods should also adapt to the usage of alternative energy sources.

Only three potential sources offer a promise of economic viability: geothermal energy, solar energy and nuclear energy.

Geothermal energy

This has been discussed in chapter 6. Geothermal fluids offer a very low energy grading so that although they can be utilised for space heating purposes they are not very suitable for purposes of power generation. Geothermal energy sources are usually confined to a particular locality and require much investment for their efficient exploitation.

Solar energy

The amount of heat which finally reaches the earth after solar radiation has passed through the atmosphere varies from about 200 W/m^2 to a maximum of about 1 kW/m^2. If it were practicable, say, to collect the solar energy from an area measuring 250 km $\times$ 250 km in central Australia at a collecting efficiency of 10 per cent this would theoretically meet all the current energy requirements on earth. No presently known methods can however collect solar energy at an efficiency approaching 10 per cent of incident solar radiation; 5 per cent is more likely, using the most costly and complex energy conversion systems as yet constructed. Biological systems can recover a mere 0.5 to 1.5 per cent of the incident solar radiation. It is also estimated that the investment necessary for the collection of solar energy would cost between 20 and 30 times more per unit of energy than that required for fossil fuel fired plant. It therefore seems unlikely that solar energy will ever contribute more than a small percentage of the total energy required by the world's growing population – and at a very high cost.

Nuclear power

Nuclear power appears to be the only realistic alternative to fossil fuels.

Unlike major schemes of harnessing solar energy (still very much at the science fiction stage) nuclear power has already proved itself to be a viable method of providing energy, with production costs lower than those which apply to fossil fuel. Its past record also shows that it is safer, cleaner and more desirable from the environmental point of view than power and heat production by fossil fuels.

THE PRINCIPLES OF NUCLEAR POWER

Nuclear power is based on the Einstein equation, which gives the relationship between mass and energy. Einstein modified the previously established dictum that matter can neither be created nor destroyed, by showing that matter can indeed be transformed into energy according to:

$E = mc^2$, where

E equals energy (joules)

m equals mass destroyed (kg)

c equals velocity of light (m/s)

To calculate the quantity of energy which is released by the destruction of one kilogram of matter, we introduce into the equation the speed of light (2.997925×10^8 m/s in vacuo) as c and calculate E at 8.987404×10^{16} joules, equivalent to about 90 000 terajoules of heat. The magnitude of this amount of energy can be placed into perspective by comparing it with the 114 000 terajoules of heat supplied annually by West Germany's 93 district heating companies, comprising a major portion of West German space heating demands. It can be calculated that the entire annual electricity output of the British Central Electricity Generating Board equates to a mass loss of about 7 kg.

There are two theoretical ways in which one can effect this mass loss and convert it into energy:

(*a*) *Nuclear fission:* start off with a fissile heavy chemical element, such as the 235 isotope of uranium (U^{235}), and convert it into a number of lighter elements such as lead, cadmium, lithium, etc. This is the basis of every current practical nuclear reactor.

(*b*) *Nuclear fusion:* start off with hydrogen, the lightest element of all, in the form of its isotope deuterium, and fuse several together to produce helium. The atomic weight of helium is 4.00260. Assuming that helium has been made from four atoms of hydrogen with atomic weight 1.0080, it can be seen that there is a weight loss of $4 \times 1.0080 - 4.00260$ ie 0.0294, or approximately 0.74 per cent – very many times more than the weight loss which arises during nuclear fission processes. Although nuclear fusion processes can be (and have been) induced in an uncontrolled fashion with so-called 'hydrogen bombs' the production of controlled quantities of energy by this method has not yet been achieved and presents major difficulties. Intense research is in progress.

THE THERMAL NUCLEAR REACTOR

Operation

The manner in which the different types of nuclear reactor function is remarkably similar **(7.1)**. The fuel rods, which consist of either normal uranium or of uranium in which the percentage of fissile U^{234} and U^{235} has been increased by various methods from the normal 0.72 per cent, are stacked in the reactor core. They are brought together into the moderator/reflector mass to an extent that criticality is reached and slightly exceeded. Special neutron absorber rods, able to slow down the nuclear reaction by sweeping up excess neutrons, are employed to ensure that the process does not get out of hand. These are made from lithium or boron alloys.

Currently the fuel rods are usually made from zirconium alloy in the form of cans which contain the uranium as dioxide UO_2. When criticality has been achieved heat is produced within the reactor and has to be abstracted in some suitable manner by either a gaseous or a liquid heat exchange medium.

Nuclear reactions produce extremely harmful radiation. It is therefore necessary to surround the core of the reactor with a biological shield constructed in heavy steel-lined concrete. For reasons of safety the operations within the nuclear core must be carried out by remote control.

The heat abstracted from the reactor is eventually converted into steam and is used to drive steam turbines for electricity generation; this part of the operation is essentially conventional.

Gas cooled reactors (GCR)

These were the first reactors used for commercial purposes **(7.2)**. The British Calder Hall reactor of this type went critical in 1956 and became the prototype of the whole series of nuclear power stations which now carry an appreciable portion of the base load of electricity produced in the UK.

The fuel consists of metallic uranium held in Magnox cans. The moderator/reflector is of graphite in the form of carefully engineered blocks. The heat exchange medium is virtually unpressurised carbon dioxide which passes the generated heat to the steam boilers. Such reactors operate with either totally unenriched or only very partially enriched uranium. There is now much accumulated operational experience.

While these reactors give little trouble, they suffer from three severe disadvantages:

(a) Everything about the stations is exceedingly massive. The reactor core of the 500 MW Transfynydd reactor, which is typical, weighs 1900 tonnes, the boilers are 35 metres high, and the 18½ m diameter pressure vessel is made of 90 mm solid steel. In consequence the costs of materials, civil engineering works etc are enormous

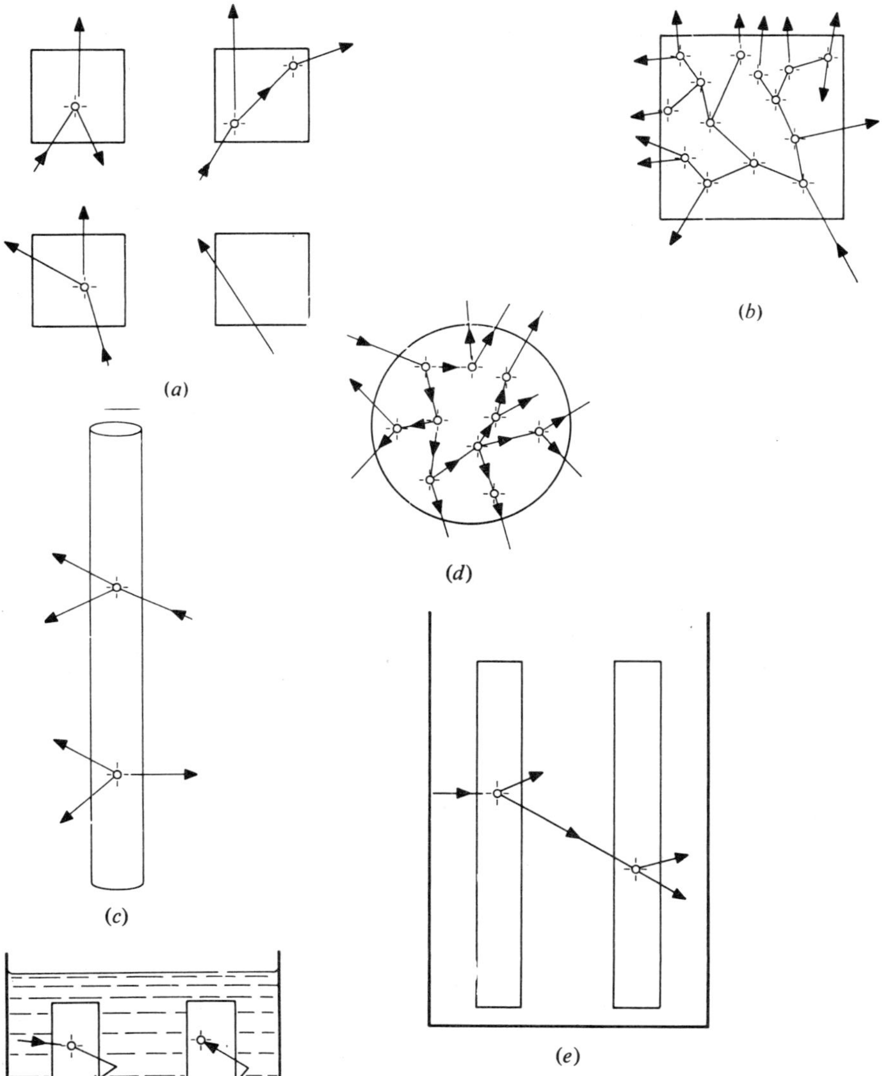

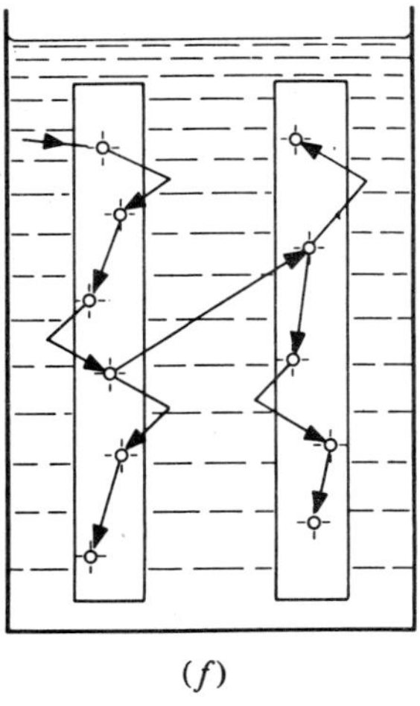

7.1 The mechanism of nuclear power: (a) and (b) illustrate the effect of critical mass upon the chances of neutrons being able to cause fission in nuclei; (c) and (d) show that nuclear fission is least likely to take place in a rod and most likely to take place in a sphere. This is why fuel elements are always in the form of thin cylinders. (e) and (f) show the effect of a moderator and reflector in inducing a higher intensity of nuclear fission

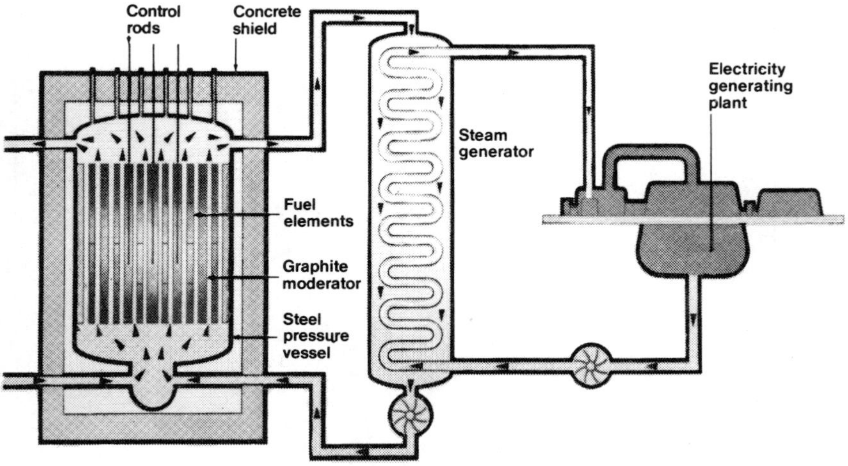

7.2 The gas cooled or Magnox reactor (GCR)

(b) Carbon dioxide is a very poor heat exchange medium so that feeder pipes of very large diameter must be used
(c) Even the most modern GCR plants cannot achieve steam temperatures in excess of 360°C (680°F); hence the thermal efficiencies of such plants are low.
The massive size of nuclear plants of this kind also tends to constitute a visual eyesore in the rural surroundings in which they have to be sited.

Advanced gas cooled reactors (AGR)
These have been developed from the well proved GCR. They use partially enriched (1.46 to 1.76 % U^{235}) fuel in the form of uranium dioxide capped in Zircalloy inside a concrete pressure vessel, clad on the inside with stainless steel. The core is pressurised to about 31 bars. The moderator is graphite and the coolant is highly pressurised carbon dioxide. The peak can-temperature which can be achieved is 800°C (1472°F) so that the associated boiler can generate steam at 565°C (1049°F) and 160 bars, which is close to the optimum steam conditions of any industrial boiler plant. AGR plants are moderately sized with small annular steam boilers within the biological shield. (**Figure 7.3**).

High temperature gas cooled reactors (HTGR)
Unlike the AGR, which can use only very slightly enriched fuel, the HTGR requires highly enriched fuel in which the percentage of U^{235} exceeds 90 per cent (**7.4**). The production of such fuel is exceedingly difficult and expensive. To obtain virtually complete separation requires an enormous and very costly plant and such large capacity installations are currently possessed only by the United States and the Soviet Union. In other countries there are

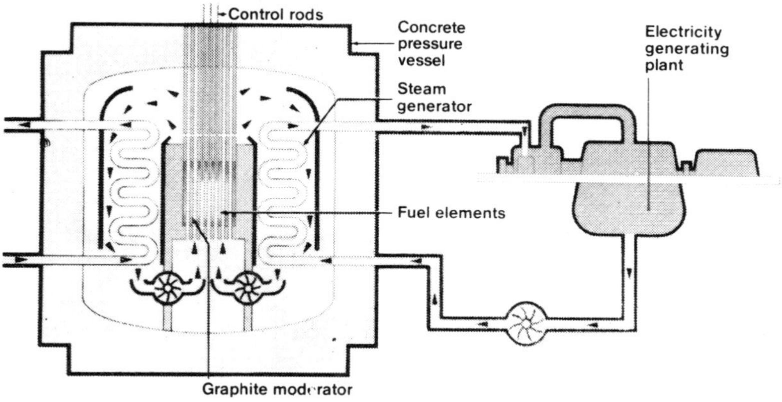

7.3 The advanced gas cooled reactor (AGR)

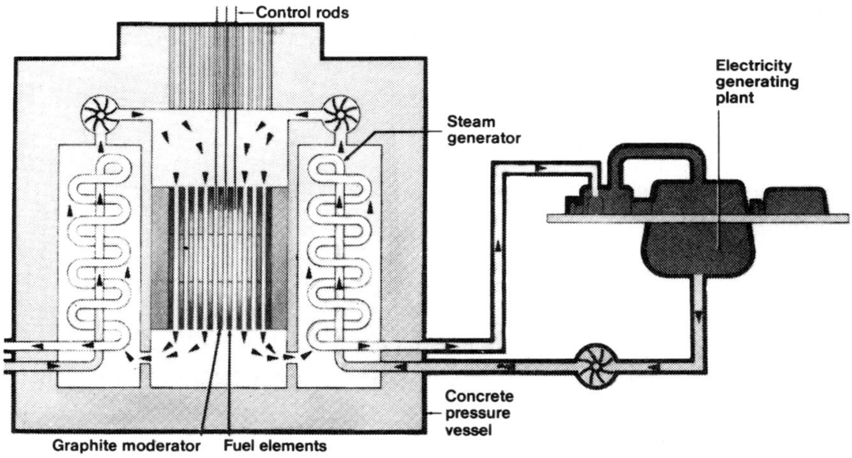

7.4 The high temperature gas cooled reactor (HTGR)

however a number of smaller plants able to process uranium for military purposes.

HTGR units use helium as a coolant and can heat this to 850°C (1562°F) or more.

It is most convenient to use such plant in conjunction with closed-cycle gas turbines using the waste heat generated on the exhaust side of the turbine for space heating purposes. The main disadvantages are difficulty in getting adequate quantities of highly enriched fuel and the enormous cost of the enrichment operation.

Pressurised water reactors (PWR)

The nuclear fuel used with these types of reactor is slightly enriched UO_2 in Zircalloy cans. Light (normal) water is used both as a moderator/reflector and as the heat exchange medium. The reactor vessels are made from high carbon steel clad internally with stainless steel and designed to withstand operating pressures of up to 175 bars. The water inside the pressure vessel is heated up by the nuclear reactions and transfers this heat via a heat exchanger to a secondary water circuit to generate steam at a pressure of 48 bars, which then drives turbines in the normal way.

This reactor system is popular in the United States, being comparatively cheap to build and operate. A typical reactor of this type, the Connecticut Yankee, with a thermal output of 1473 MW and an electrical output of 490 MW, uses a reactor vessel which is only 4 m in diameter and 10.5 m high with a 75.5 tonnes charge of fuel. The PWR is controlled by means of 20 silver/indium/cadmium rods, cased in stainless steel, to absorb the unwanted neutrons. PWRs have been operating for many years without much trouble. They have become probably the most popular of all reactor types and have the best cost economy figures. They are also safe. Even the Three Mile Island mishap, which involved a PWR, was far less serious than was thought at the time because even though many things went wrong, no harmful radio-active emissions were released outside the biological shield. PWR designs are the ones used for all marine applications such as nuclear submarines and merchant vessels such as the SAVANNAH, the LENIN and the OTTO HAHN.

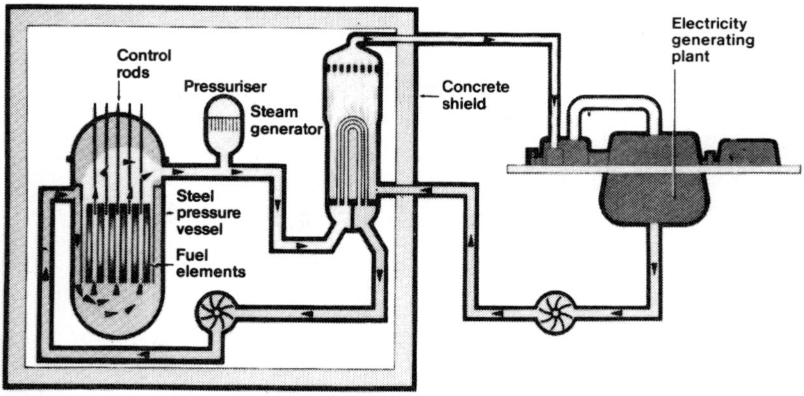

7.5 The pressurized water reactor (PWR)

Pressurised heavy water reactors (PHWR)

Normal light water is quite a good moderator and reflector but it suffers from the disadvantage that hydrogen can act as a neutron captor: H + n →

D (ie some of the neutrons are removed from the sphere of action and build up the hydrogen atom into a deuterium atom). Using heavy water D_2O instead of normal water the neutron capture rate would be very much reduced because the moderating and reflecting effect would be improved. The plant can be made very compact and be made suitable for positioning underground. The Swedish Agesta reactor is of this type. It was shut down in 1974 (as too small for economical operation). It had however been in continuous operation since 1964 and is discussed in some detail later in this chapter.

The Canadian CANDU reactor (**7.6**) operates on somewhat similar principles and plans are now afoot to use such a reactor for the supply of district heat in the Pickering area of Canada.

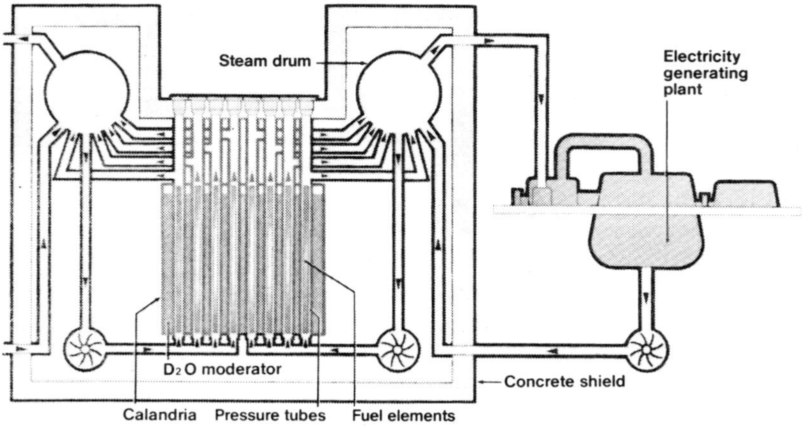

7.6 The CANDU reactor (*PHWR or SGHWR*)

Boiling water reactors (BWR)

These differ from the other reactor designs in that there is no integral heat exchanger (**7.7**). Ordinary water is pumped directly through the assembly of canned fuel rods where it acts as a moderator/oblique reflector and also removes the heat generated by the nuclear reactions. Part of the water is generated into steam. This is separated carefully, superheated to ensure that it contains no entrained water and then conveyed to the turbines. Although absolutely pure water does not become radioactive by being subjected to radiation it does become highly so in the presence of the slightest traces of impurities, particularly metallic ones, careful purification of the water used employing highly sophisticated ion exchange processes is thus an essential feature of the process. The fuel is in the form of slightly enriched (2.5 %) pellets of UO_2 capped in Zircalloy. Typical reactors of this type are the Oyster Creek reactor of the Jersey Power and Light Co. and the Dresden No 2 reactor (both in the USA).

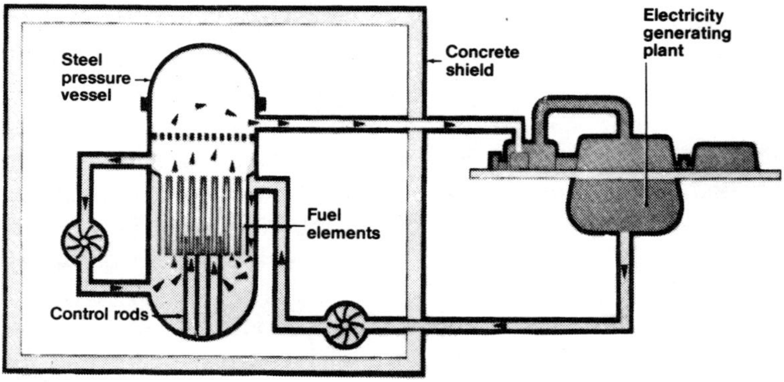

7.7 The boiling water reactor (BWR)

The 515 MW (e) Oyster Creek reactor uses a reactor vessel 5.5 m in diameter and 27 m high with 200 mm thick steel walls. Feed water passes through the fuel rods which are fed in from underneath and not from the top, as is usual with other kinds of plants. The reactor is controlled by neutron absorbing control rods. It is possible to inject sodium pentaborate into the circulating water which is able to absorb neutrons quickly and effectively, thereby bringing the reactor to a stop.

An advantage of the BWR over the PWR is the omission of the heat exchanger between the reactor core and the turbine which saves much capital expenditure and also reduces heat losses associated with the operation of a heat exchanger. Disadvantages are that pressures inside boiling water reactors are as high as in pressurised water reactors and that the same water circulates inside the reactor and the turbine making very extensive safety devices and monitoring techniques necessary to ensure that radioactivity is never released to the outside.

There are two projected heat only systems which are being studied currently from the point of view of district heating. Both are basically light water reactors, situated underground with automatic safety devices to shut down the reactor under any conditions of irregularity. They are the Swedish/Finnish Secure and the French Thermos projects.

NUCLEAR FUEL

Content of uranium ore

Uranium is one of the rarer metals of the earth's surface; ores seldom contain more than one per cent by weight of the metal (**7.8**). The main compounds of uranium in the natural state are:

Uranite	$(UO_2)_n$	$(UO_3)_m$
Carnotite	$K_2O . V_2O_5$	$(UO_3)_2 . 3 H_2O$
Tobernite	$CuO . P_2O_5$	$(UO_3)_2 . 8 H_2O$
Autunite	$CaO . P_2O_5$	$(UO_3)_2 . H_2O$

TABLE 7.1 COMPARISON BETWEEN THE PROPERTIES OF HEAVY WATER AND LIGHT WATER

	Heavy water (D_2O)	Light water (H_2O)
Density at 25°C (77°F)	1.108 kg/dm^3	1.000 kg/dm^3
Freezing point	3.81°C (39.9°F)	0.00°C (32°F)
Boiling point	101.41°C (214°F)	100.00°C (212°F)
Critical temperature	371.5°C (700.7°F)	374.2°C (705.6°F)
Critical pressure	221.44 bars	221.35 bars
Neutron absorption cross section	9.2×10^{-32}m^2	6×10^{-29}m^2

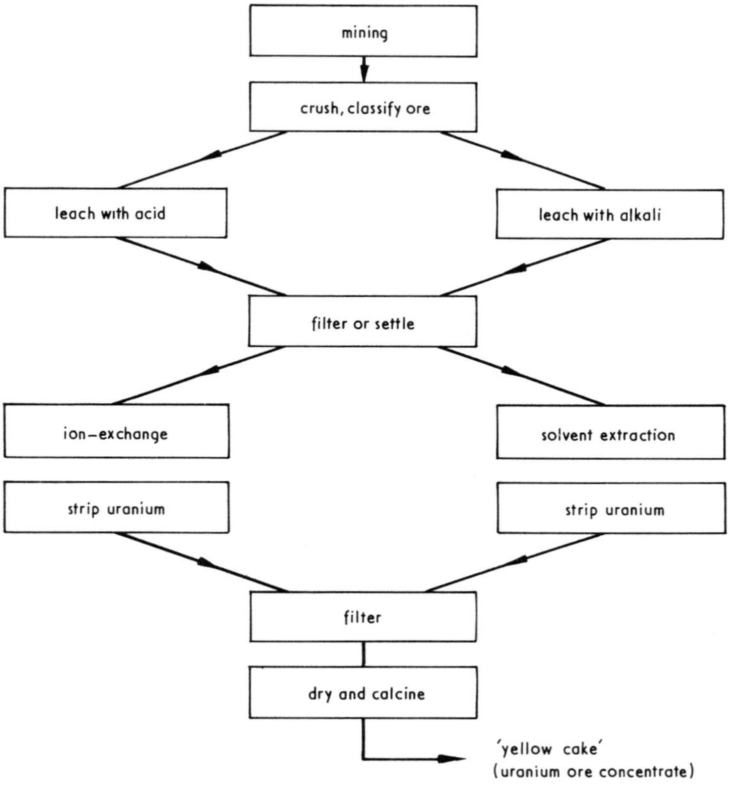

7.8 Flow sheet of production of uranium oxide from crude uranium ore

There are also deposits of oxides, hydroxides and titanates where uranium is associated with such metals as iron, cerium, vanadium, thorium etc. It is possible to work uranium ores containing as little as 0.1 per cent of the metal at reasonable costs. Obviously, as the richer uranium ore deposits are depleted, the less rich deposits will have to be utilised.

The ore is crushed and ground and some attempt is made to enrich the mud by classification and floatation. It is possible to use either acid or alkali leaching for extracting the uranium from the ground ore. If sulphuric acid is used, the uranium comes out as soluble $H_2UO_2(SO_4)_3$ or $H_2UO_2(SO_4)_2$. If the leaching agent is sodium carbonate the uranium is removed as the soluble $Na_4UO_2(CO_3)_3$. Acid leaching is more complete but requires more expensive equipment than alkali leaching. It is the more popular method. The uranium compounds are removed from the clear leaching liquors either by solvents or (increasingly) by ion exchange materials. The latest ion exchange methods are able to extract uranium from solutions which have not been filtered. When the uranium has been stripped out, it is roasted to form about 80 per cent pure UO_3. To obtain metallic uranium the crude UO_3 concentrate is reacted with hydrogen to form uranium dioxide:

$$UO_3 + H_2 \rightarrow UO_2 + H_2O$$

This then forms the following with hydrogen fluoride:

$$UO_2 + 4\,HF \rightarrow UF_4 + 2\,H_2O$$

UF_4 then reacts with magnesium metal to give uranium and magnesium fluoride:

$$UF_4 + 2\,Mg \rightarrow U + 2\,MgF_2$$

Metallic uranium has poor corrosion resistance coupled with a tendency to warp and twist under irradiation conditions. It must always be protected by canning in metal shells.

Enrichment

Most modern nuclear reactors require either slightly or highly enriched uranium ie uranium in which a higher percentage than normal consists of the fissile isotope U^{235}. Separation of U^{235} from U^{238} is very difficult because the chemical reactions of the two isotopes of uranium are naturally identical. Unlike the separation of hydrogen from deuterium, where there is a 100 per cent weight difference, the difference between the weight of U^{235} and U^{238} atoms is only 1.28 per cent.

Uranium tetrafluoride is converted into uranium hexafluoride, and a cascade diffusion process is used to enrich some of this in U^{235} and to impoverish the rest. After enrichment the uranium hexafluoride is reacted with steam and then with hydrogen to form ultimately uranium dioxide, UO_2. To produce this compound in a form that can be used effectively in nuclear reactors the UO_2 is made up into a slurry, spray-dried and pelleted. The pellets are heated up to 800°C (1472°F), sintered in hydrogen at about 1650°C (3000°F), washed and dried. For the newer types of water cooled reactors zirconium alloy tubes are used; these are loaded with the pellets, filled with helium and sealed by welding end caps into place.

Processing irradiated fuel elements

When the fuel elements are removed from a nuclear reactor they are very hot and extremely radioactive. It is necessary to let the rods cool down and then keep them under water for a few weeks or months to dissipate the short lived radiation.

Irradiated uranium always contains appreciable quantities of plutonium 239. This is formed as follows:

U^{238} + neutron $\rightarrow U^{239}$

$U^{239} \rightarrow Pu^{239}$ + beta radiation (electrons)

This reaction only takes place very partially in a conventional thermal reactor because the moderator slows down the neutrons too much to permit the process to proceed very fully. The change of uranium 238 to plutonium 239 is of very much greater importance in fast or breeder reactors which operate without moderators. The plutonium content of spent fuel in enriched thermal reactor systems is about 0.5 to 0.9 per cent by weight. When using enriched fuel there is also quite enough U^{235} present to make recycling worthwhile.

Processing of irradiated fuel is complicated by its intense radioactivity; in consequence all operations must be by remote control. After opening up the cans their contents are dissolved in nitric acid, oxygen being bubbled through and sodium nitrate being added to ensure that the uranium is present in the U^{6+} state and plutonium in the Pu^{4+} state. The solution is filtered off and brought into intimate contact with a suitable organic solvent, which dissolves the uranium and the plutonium. The operation is most usefully conducted by employing a packed tower, down which both the aqueous solution and the organic liquid are allowed to flow. The radioactive fission products remain in the aqueous solution. Plutonium and uranium can be separated from each other by partial reduction which enables the Pu^{3+} ions to be removed by water, washing off the solvent. By repeating all these processes several times 99.9 per cent separation of plutonium and uranium is achieved. The uranium has then to be passed through an enrichment process for re-use as a nuclear fuel. The plutonium is precipitated from the aqueous solution as the oxalate, which produces plutonium dioxide on roasting. If it is necessary to provide plutonium metal this dioxide is converted into a fluoride using hydrofluoric acid and the fluoride changed into the metal by reacting with magnesium. The plutonium produced by the above process has little immediate applicability. It can be used in modified light water reactors but its real value is as the primary fuel in fast or breeder nuclear reactors.

In contrast to much popular belief nuclear fuels are totally innocuous. They do not produce organic compounds during processes of energy delivery and all waste products can easily be contained. Radioactive materials are of course handled with the greatest care, employing all the most modern facilities technology can provide and under the strictest scientific and technical supervision. Radioactive substances disintegrate and the radiation danger therefore diminishes with the passage of time.

Radiation intensity and half life

Intensity of radiation is measured in curies, where 1 curie (Ci) $= 3.7 \times 10^{10}$ disintegrations per second. Radioactive materials decay according to the first order reaction:

$dN/dt = - k N$, where

N is the number of fissile atoms available at time t

t is the time (seconds)

k is the intensity of radiation

dN/dt is the rate of disintegration taking place.

On integration this becomes:

$N = N_0 e^{-kt}$, where

N_0 is the original quantity of available fissile atoms.

Radiation is thus an exponential process, starting fast and gradually diminishing, reaching zero at infinity time. The most useful method of expressing radioactive decay is by the half-life technique.

Reserves of fissile fuel

While nuclear fusion has unlimited quantities of fuel reserves available in the form of deuterium present in sea water, the future life and use of traditional nuclear reactors is limited by the accessible uranium resources. Each kilogram of uranium 235 produces the heat equivalent of 2800 tonnes of coal at a heat content of 35 GJ/tonne ie one kg of U^{235} is equivalent to about 100 TJ. Only about 0.72 per cent of natural uranium consists of U^{235}; the remainder consists of non-fissile U^{238}. It is well known that all the U^{235} cannot be extracted, nor can all the resultant heat energy be converted into electrical energy. In practice one tonne of natural uranium equates with only 250 terajoules of usable energy. A world population of about 4000 million has energy needs of about 220 million terajoules per annum, equivalent to 880 000 tonnes of uranium. Assuming that energy were supplied solely by standard thermal nuclear reactors, how long would the uranium last? The latest available information on uranium reserves at various extraction cost levels is given in Table 7.2:

TABLE 7.2 URANIUM RESERVES

Extraction costs (1976)	Global reserves (million tonnes)	Years' fuel supply (100% nuclear energy supply using thermal reactors)
Less than $25/kg	1.45	1.65 years
$26 to $40/kg	1.10	1.25 years
$40 to $100/kg	2.95	3.35 years
$100 to $200/kg	15.00	17.05 years
More than $200/kg (from sea water)	53.00	60.22 years

Ignoring uranium extraction from sea water (as no viable techniques for extracting heavy metals from sea water have as yet been developed) the total global uranium reserves stand at 23.3 years of total energy supply.

The thermal nuclear reactor is therefore certainly no long term solution to the depletion of the world's energy resources. At the very low rate of uranium utilisation of U^{235} the reserves would run out before oil or coal. Hence the vital importance of the fast or breeder reactor which can utilise all the uranium, both U^{235} and U^{238}, for power and heat production; if this were to be generally adopted the energy supply from accessible uranium reserves would be extended to hundreds or even thousands of years.

HOW NUCLEAR REACTIONS OCCUR

Natural uranium is composed of 99.283 per cent U^{238}, 0.711 per cent U^{235} and 0.006 per cent U^{234}. Only the last two isotopes are radioactive ie break up with the release of alpha rays, beta rays, gamma rays, heat energy and neutrons. Uranium 238 is virtually non-fissile, as it has a half-life of 4.67×10^9 years, uranium 234 has a much shorter half-life of about two million years while uranium 235 has a half-life of 7.13×10^8 years. All these elements are alpha ray emittors only (ie the products of decomposition are in all cases isotopes of thorium, which then decompose further into other elements). All isotopes of uranium are totally stable; whether uranium is in the form of its element or of a compound its degree of radioactivity is quite negligible by any standards.

The source of nuclear power lies within the neutrons released from secondary fission reactions after the U^{234} and U^{235} have been converted into thorium isotopes. These neutrons are capable of inducing a very much more rapid break-up of the U^{235} than occur in nature. If they hit the nucleus of an atom this can be smashed up completely, giving off heat, radioactivity and more neutrons. However not every neutron carries out such a nuclear reaction; the vast majority are either 'captured' (simply absorbed into the structure of the atom) or lost by escape through the geometric boundaries of the system. If a neutron hits an atom of uranium 238 it is not split up but is converted into plutonium 239. It is also a fact that not all atoms of uranium 235 and plutonium 239 are split up when hit by neutrons. This demonstrates that the nuclear power process is not at all easily achieved; nuclear reactions cannot take place by accident, so they have to be engineered. To promote an appreciable rate of nuclear reactions it is necessary to:

(a) have a big enough quantity of nuclear fuel together, so that the neutron losses along the geometric boundaries of the fuel are reduced relative to the number of neutrons produced inside the system, (the 'active mass'). To obtain 'criticality' (the stage in **figure 7.1** at which neutron generation balances neutron losses) the active mass must always constitute a given value;

(b) endeavour to mitigate neutron losses by slowing down the neutrons inside the reactor and reflecting them back against the fuel; this is the function of the moderator/reflector used in nuclear plants (which may consist either of solid materials such as graphite, or of liquids such as water, particularly heavy water (D_2O)).

The critical mass required to promote a nuclear reaction varies enormously according to such factors as the nature of the fissile material and its abundance in the nuclear fuel, the type, nature and quantity of moderator/reflector used and the shape of nucleus mass. The lowest ratio of surface volume for such a mass is given by a sphere, while the highest is provided by long thin rods. For example if a pure sample of uranium 235 is converted into a water soluble salt the critical mass will be as little as one kilogram. On the other hand the critical mass of natural uranium in the form of fuel rods within graphite moderator/reflector blocks is 30 tonnes. Natural uranium without a moderator/reflector can never go critical, however large the quantity stored, because the neutron losses will always far exceed the neutrons' production. If

$e^{-k\,t_{\frac{1}{2}}} = \frac{1}{2}$, then

$t_{\frac{1}{2}} = 1/k \log_e 2 = 0.6931/k$

The half-life is therefore inversely proportional to the intensity of decay. Materials with very long half-lives have a low radiation intensity and vice versa.

DISPOSAL OF RADIOACTIVE MATERIALS

Materials with very high radiation intensity

These substances emit alpha, beta and gamma rays at a very high rate. If human beings were exposed to them intense radiation burns would result. Such materials are handled very carefully by remote control behind strong shielding devices, and are kept either under water or in underground mortuary chambers for up to two years. At the end of that time, virtually all the short-lived radiation has disappeared and with it all handling dangers. For example, in the case of thorium 233, which has a half-life of 27.4 days, the original intensity of radiation is reduced to $\frac{1}{2^{26\cdot64}}$ or to just over 10^{-6} per cent after two years (730 days).

Materials with very low radiation intensity

The dangers arising from materials which have a very long half-life are frequently mentioned, particularly with reference to plutonium 239 (half-life of 24 400 years). However such materials also have an extremely low radiation intensity which is unlikely to be harmful. Natural background radiation in Great Britain is about 0.1 to 0.15 rads per year (0.03 from cosmic radiation, 0.025 from naturally occurring radioactive isotopes such as potassium 40, and the rest from gamma radiation of the earth). Relative to this, radioactive waste disposal increases the radiation exposure of an individual by only about 0.000 01 rads each year. Thus even if the British nuclear industry should increase its output a hundred times from its present 4500 MW to 450 000 MW (most unlikely) the increase of radiation caused by the disposal of long-lived nuclear isotopes would not exceed 0.001 rads, almost insignificant in relation to the natural background radiation.

Materials with intermediate radiation intensity

Materials with very short half-lives are easily made harmless by storage under water or in rock chambers until the harmful radiation has disappeared. Materials with very long half-lives can be diluted adequately so that the effect upon the environment is virtually nullified. Some materials, however, have intermediate half-lives making it difficult to store them long enough to make an appreciable impact on their radiation intensity; yet the intensity is too high for satisfactory disposal by dilution. For example such materials as strontium 90 (half-life 26 years) and caesium 137 (half-life 30 years) are best separated out by precipitation techniques and sealed in the form of a plastic/cement mass into stout steel drums which can then be shipped to regions of ocean trenches where the depth is 10 km or more. The drums take about one hour to sink slowly to the bottom of the ocean, and they are made and tested to be highly impact resistant so as to obviate a burst on impact with the sea floor to which they drop very gently through the water. Their contents cannot be leached out by the water and gas valves are fitted to permit the escape of gases (which are always non-radioactive) thereby preventing build up of excessive internal pressure. Steel is incapable of rusting in an oxygen-free sea water environment. No marine life exists at the great depth to which the drums sink and the technique is therefore also ideal for the disposal of radioactive constituents which can be concentrated by certain organisms.

At the present moment the main plan for long-term nuclear waste disposal is to fix radioactive constituents to form glass blocks and to store these underground in dense rock formations. This system, called the 'harvest' process, is not as yet practised as problems are not acute today.

The only radioactive waste material which still presents difficulty is radioactive krypton. However this only emits weak beta radiation and is an inert gas which cannot form a chemical compound with anything, it is therefore fairly harmless and can be disposed of by dilution in the atmosphere. Research is now under way into the encapsulation of radioactive Krypton in metal blocks.

Altogether the disposal of radioactive substances does not constitute a major problem. Harmful pollution is infinitely worse when carbonaceous substances are burned as organic carcinogens are then formed. Statistics in the United Kingdom have compared the number of deaths by cancer of persons working in or close to nuclear energy establishments with those of persons living in industrial areas: the former were shown to suffer lower cancer rates, presumably because they were working in an environment free from organic carcinogens caused by the combustion of carbonaceous fuels.

FUTURE DEVELOPMENT OF NUCLEAR REACTORS

The possibilities

The uranium deposits so far known to exist on earth are as previously stated very limited in quantity (although more may of course be discovered). The very low utilisation factor of the precious uranium mined therefore causes concern. U^{235} and U^{234}, the two fissile isotopes found in natural uranium are present to the extent of 0.72 per cent. The rest is inactive uranium 238. But during actual reactor operation not more than perhaps 0.2 to 0.3 per cent of the uranium is utilised; the rest is, in effect, a waste product. On the other hand if the entire metal content of crude uranium can be effectively used, utilisation will be bettered by at least two hundredfold and the basic reserves of uranium will last for many centuries, even after expansion of nuclear capacity has proceeded to the stage of carrying out all electricity generation. Beyond this lies the possibility of extracting uranium from sea water which, although ruinously expensive by present standards, becomes economical when obtaining the sort of utilisation possible with a fast breeder reactor.

It can be demonstrated that fast or breeder reactors alone, once they have been developed into efficient units, can provide all the electric power humanity will need for many hundreds of years. Looking beyond that period, unlimited electric power will undoubtedly be generated by means of fusion reactors.

Fast or breeder reactors are at present still in the pilot plant stage ie at a point of development when quite small quantities of electricity are being produced in enormously costly plant at great expense. Before fast reactors can become a commercial proposition much data has to be gathered regarding their operating characteristics and the necessary design changes will have to be made.

As far as fusion reactors are concerned experiments are in hand throughout the world but have not as yet been successful. Even the design of a working pilot plant is many years or even decades off. Fusion plants will eventually be able to produce enormous quantities of energy from minute quantities of heavy hydrogen or deuterium: as deuterium is present to the extent of 0.0017 per cent in sea water, and as it can be calculated that there are no less than 1396×10^6 km^3 of water in the oceans of the earth, the available energy potential is virtually unlimited.

Fast or breeder reactors

Fast neutrons colliding with the nuclei of normal uranium 238 can convert this quite inactive metal into plutonium 239 which is a fissile element,

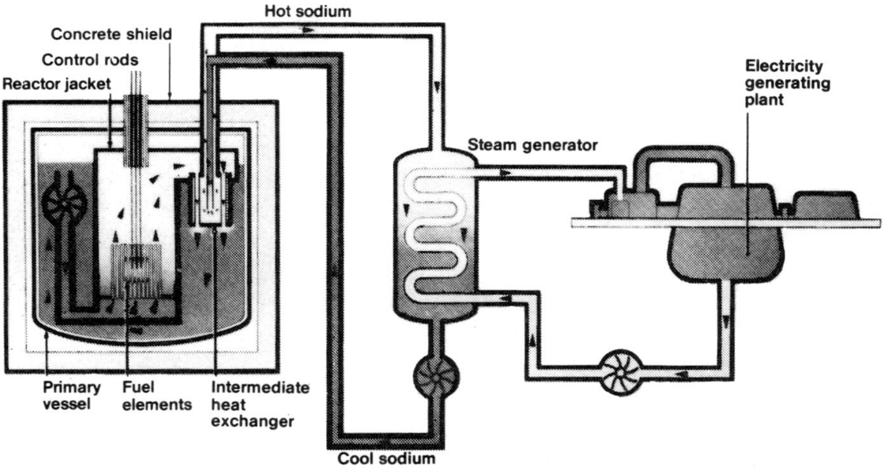

7.9 Dounreay prototype fast breeder reactor

though not quite as potent as uranium 235. In the conventional thermal reactor this reaction only takes place to a very slight extent because most of the neutrons are being slowed down by the moderators.

When an uranium atom is hit by a slow neutron the kind of reaction which takes place depends upon whether the atom is U^{238} or U^{235}. Slow neutrons have no effect whatsoever upon uranium 238; they just bounce off. When a slow neutron impinges upon uranium 235, the resultant reaction is:

$${}_{92}U^{235} + \text{neutron}_{(\text{slow})} \rightarrow {}_{56}Ba^{137} + {}_{36}Kr^{97} + 2 \text{ neutrons} + \text{energy (heat, radiation etc)}$$

The barium and krypton are themselves somewhat radioactive and disintegrate further. It should be noted that the above reaction is only one amongst many taking place and that a very wide spectrum of products is obtained during a nuclear fission reaction.

When the neutrons are fast ones uranium 235 reacts in more or less the same way although the reaction is not quite as efficient. For this reason the reaction in a conventional thermal reactor stops when the moderator is withdrawn. The major difference is the manner in which fast neutrons act upon the normally inactive uranium 238. The following reactions then take place:

$${}_{92}U^{238} + \text{neutron (fast)} \rightarrow {}_{92}U^{239} + \text{gamma rays}$$

The uranium 239 is very unstable and has a half-life of only 24 minutes. The reaction which takes place is then:

$${}_{92}U^{238} \rightarrow {}_{93}Np^{239} + \text{beta radiation (electrons)}$$

The neptunium is also rather unstable and has a half-life of 2.3 days. It disintegrates as follows:

$${}_{93}Np^{239} \rightarrow {}_{94}Pu^{239} + \text{beta radiation (electrons)}$$

The plutonium produced is stable and valuable as a nuclear fuel. Provided conditions are suitable it splits up giving off neutrons and energy. One reaction which takes place is:

$$ {}_{94}Pu^{239} + \text{neutron} \rightarrow {}_{56}Ba^{137} + {}_{38}Sr^{100} + 3 \text{ neutrons} + \text{energy (heat, gamma radiation etc)} $$

In conventional thermal reactors the percentage of uranium 238 which is converted into plutonium is very small, of the order of 0.5 to 0.9 per cent. However it is very much simpler to separate plutonium 239 from uranium 238 than it is to separate uranium 235 from 238 because the plutonium is a different element and is easily separated by purely chemical means. Uranium 235 can only be enriched by the use of highly complex gas diffusion and gas centrifuge processes operating on a cascade principle.

In fast reactors the main aim is to obtain as completely as possible a conversion of uranium 238 to plutonium 239 employing a highly compact reactor core (able to produce as many fast neutrons as possible) (**plates 15** and **16**). It is essential that moderators (ie materials which slow down the neutrons produced) are absent. The reactor core is surrounded by rods of uranium which, in the process, are converted by the fast neutrons into plutonium. Enrichment operations which are currently arranged to produce nuclear fuel with a percentage of U^{235} in excess of the natural 0.7 per cent, inevitably produce considerable quantities of uranium in which the amount of fissile uranium 235 is far less than 0.7 per cent. At present such uranium is a waste product, but when fast reactors have become a commercial proposition it will be possible to make the breeder rods entirely from uranium 238, the present waste material.

Coolants for fast or breeder reactors

The requirement of a very compact core with as little neutron retardation (moderation) as possible makes it impossible to use the conventional heat exchange media usually employed with thermal reactors (eg carbon dioxide or helium gas, water or heavy water). Virtually the only suitable coolants for fast reactors are molten sodium and molten potassium. Most of the fast reactors which are currently in use or projected use pure sodium as a heat exchange medium. This has a melting point of 98°C (208°F) and a boiling point of 883°C (1621°F), excellent heat transfer characteristics and a very low neutron retarding influence. Although a very reactive element it is compatible with the metals used in the construction of fast breeder reactors. Because of the high melting point it is necessary to use external heating to maintain the sodium fluid when the reactor is shut down for any length of time. Sodium/potassium mixtures are also used as heat exchange media in experimental reactors. One of the disadvantages of liquid metals relative to water, heavy water and gases, such as carbon dioxide and helium, is their property of becoming intensely radioactive when used as coolants inside nuclear reactors.

Sodium absorbs neutrons to produce the radioactive isotope sodium 24. It is a fissile element which gives off gamma and beta radiation. It has a half-life of 15 hours. Its final decomposition product is stable magnesium 24.

Potassium is converted by neutron bombardment into radioactive potassium 42, which has a half-life of 12½ hours and gives off alpha, beta and gamma rays to end up as stable calcium 42.

Nickel, Iconel, nichrome and various other nickel and chromium containing stainless steels are the most suitable metals for use in conjunction with molten sodium circuits. The presence of even the most minute quantities of oxygen must be avoided as otherwise the very highly corrosive Na_2O may be formed. Apart from causing corrosion troubles, sodium oxide may also encourage blockages in tubes.

Because the coolant used with fast reactors becomes intensely radioactive it must be kept entirely within the biological shield of the reactor. A second circuit of molten sodium is therefore used to abstract the heat from the primary circuit which carries the heat to the outside of the biological shield. A second system of heat exchangers then transfers the energy created within the reactor to the water circuit which produces steam at the desired temperature and pressure to drive the turbines. Provided that there is no fault in the primary circuit/secondary circuit heat exchanger, the sodium contained in the secondary circuit should be free from radioactivity.

Nuclear fusion

Nuclear fusion endeavours to reproduce energy in the way that it is being produced by the sun and other fixed stars **(7.10).** In this process hydrogen is converted into helium with the emission of positrons and energy as under:

$4\ H^1 \rightarrow He^4 + 2$ positrons $+ 4.28$ pJ

(pJ equals pico joule or 10^{-12} joules.)

Considering that there are 6.023×10^{23} atoms in the gramme atom the

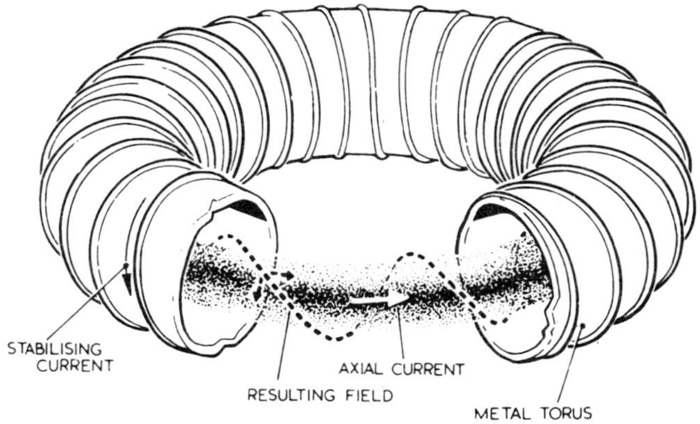

7.10 The Tokomak principle of plasmas containment to obtain nuclear fusion

quantity of energy thus obtainable is immense. 4 grammes of hydrogen yield $4.28 \times 10^{-12} \times 6.023 \times 10^{23}$ joules of energy ie 2.58 terajoules, which is equivalent to the energy obtainable from the combustion of about 60 tonnes of oil or 100 tonnes of coal.

It is extremely difficult to reproduce this fusion reaction exactly on earth. Therefore one generally has to rely on reactions not quite as efficient at producing energy as those which occur within the sun. It can be calculated that the reserves of energy in the form of deuterium contained in the oceans are 1700 million times greater than oil and gas resources, 300 million times greater than coal resources and 17 million times greater than uranium deposits (used only in fast breeder reactors).

The basic difficulty which has to be resolved before fusion power can become a practical reality is that of containing a gas of charged particles (a plasma) to about 100 000 000 kelvins sufficiently long to enable the various necessary reactions to take place. There are also a number of subsidiary problems:

(a) the tritium has to be produced by the reaction of lithium with neutrons because tritium is only very short-lived and does not occur in nature

(b) about 80 per cent of the energy produced in fusion reactors is carried away as kinetic energy in the production of high speed neutrons; this has to be converted into heat energy, and that energy must generate electricity

(c) on average the energy of the fusion neutrons is about seven times greater than that of the neutrons produced in a fission reaction: the design of suitable structures able to slow down these neutrons and to convert the associated energy is very difficult.

The helium-4 produced in a fusion reaction is not radioactive and can be dissipated quite harmlessly into the atmosphere.

Apart from the difficulties involved in starting a fusion reaction which generates more energy than that required to initiate the reaction, there are enormous ancillary problems involved in the shielding and containment of the radioactive particles produced during nuclear fusion. Cooling by means of liquid metals and helium would be required and a wall of lithium, beryllium and perhaps graphite (about one metre thick) would be needed to absorb about 97 per cent of the heat produced from the plasma. Even so there would still be some escape of neutrons and gamma rays beyond this shield requiring further really massive shields of lead and steel. The sheer complexity and size of a fusion plant relative to that of a fission reactor is so staggering because the energy density in a fusion reactor is so much greater.

Development of fusion reactors

Up to 1974 the US Atomic Energy Commission had spent about $500 million on fusion reactor research and another $1350 million was spent between 1975 and 1979. It is estimated that the cost of research needed before a viable power reactor using the fusion principle can be built is of the order of $6000 million. It is projected that the first commercial size fusion reactor (of about 1400 MW capacity) will be 'on-line' by the year 2005.

The amount of available deuterium is virtually unlimited. However fusion reactors require considerable quantities of tritium, which can currently only be made by the reaction of lithium with neutrons. Assuming that by the year 2020 there is an installed capacity of one million megawatt of fusion powered generating stations, 1100 tonnes of lithium will be required to provide the tritium fuel. In the United States alone there are at present known lithium deposits of over 6 million tonnes, sufficient for about 6000 years. The sea water content of lithium would be adequate to supply the total 1980 world electricity generating capacity for almost 300 000 years.

When fusion reactors do enter upon the energy scene they will gradually replace power stations supplied by fossil fuel although fission reactors will still continue to show a considerable expansion. By about the year 2025 fossil fuel fired power stations should have become extinct.

DISTRICT HEATING FROM NUCLEAR REACTORS

General considerations

Nuclear power is ideally suited for the supply of heat and power by means of ITOC turbines but the following properties of atomic power stations require special consideration:

(a) they are usually very large, delivering an enormous and constant supply of steam

(b) for reasons of safety (real and imagined) nuclear power stations are located at maximum distance from urban centres

(c) the fuel costs (particularly for the breeder reactor type) are quite negligible but the capital costs of building stations are enormous

(d) it is very costly to shut down a nuclear reactor after it has been made operational and it is difficult to vary its output.

All this must be considered in the overall design of nuclear power based district heating systems (**7.11** and **7.12**). Viewing the spiralling cost of the fossil fuels, it is almost inevitable that in the long run all district heating systems will draw on the waste heat discharged by nuclear stations.

Basically a district heating system fed from a nuclear energy system has to be very large and must cover a considerable number of consumers, (eg in West Germany, a district heating grid network which covers the entire country is being constructed). In the operation of such a district heating system it is necessary to vary the heat output drastically, yet keep the plant running at a constant capacity. This entails the selection of very sophisticated and very large ITOC turbines. In some cases the district heating hot water can be transported economically at very high temperatures in single pipeline networks.

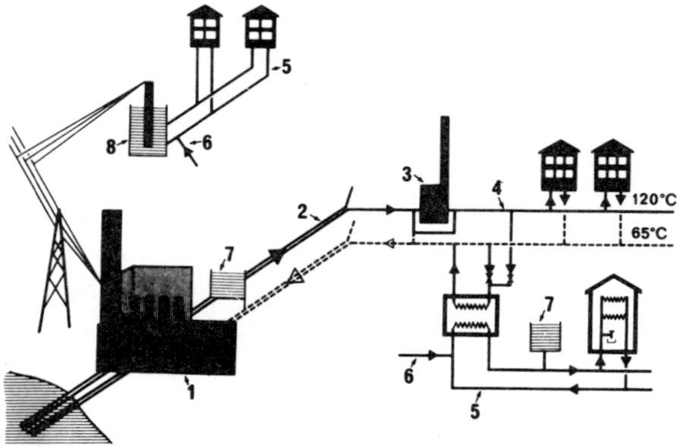

1. Nuclear Heat/Electric Station
2. Regional network of large pipes
3. Local fossil fuel plants for peak load and reserve
4. Local distribution system
5. Secondary low pressure distribution system of plastic pipes, common for space heating and hot tap water
6. Cold water make-up
7. Optional magazine
8. Hot water magazine for night electricity storage and continuous heat distribution to isolated housing estate

7.11 Simplified schematic diagram of a regional district heating scheme connected to nuclear power station

Current experience: the Agesta reactor

Up to the time of going to press, only two nuclear combined heating and power station systems have been operated. One of these is the Soviet Bilibino plant which went operational in 1973 and consists of four ITOC turbines, each producing about 12 MW of electricity and a maximum of 65 MW of heat, fed from a pressurised light water reactor. The other, the Swedish Agesta plant, operated highly successfully from 1964 to 1974 supplying district heating to the town of Farsta (population 40 000). It was closed down in 1974 as being too small for economic operation (all new reactors planned are far larger and will serve much bigger communities). It is nevertheless to be regarded as the prototype of all nuclear schemes in the district heating field and is therefore worth considering in some detail (see **7.13–7.17**).

Farsta is a 'new town' situated between two lakes, Magelungen and Drevviken, about 10½ km south of Stockholm. In 1955, when the streets were laid out, excavation of a rock tunnel under Lake Magelungen was started to serve for laying the insulated flow and return mains of the first nuclear energy heated district heating system in the world.

Work at the Agesta reactor, situated about 4 km from the centre of the new town and about 2 km from the nearest permanently occupied dwelling, was begun in 1957. A rock chamber, about 60 000 m³ in volume, was

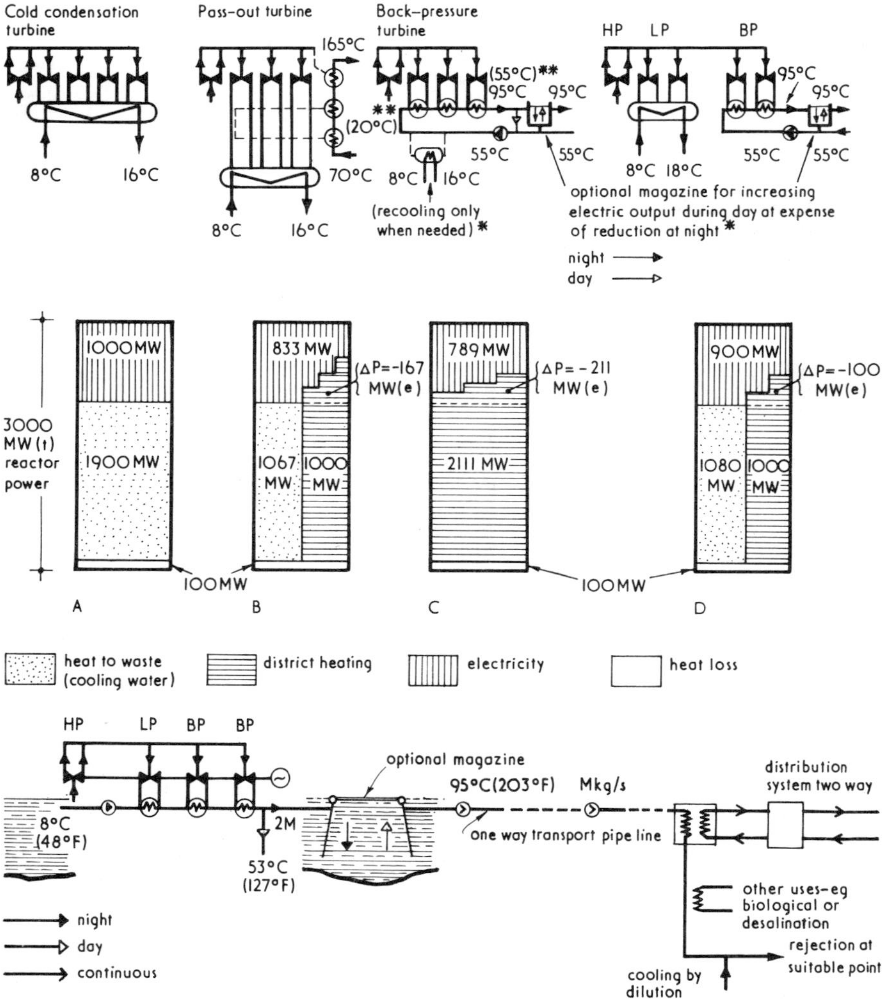

* *This figure applies when average Q_u during the day is less than total station heat rejection.*

** *Temperature in brackets applies when turbine is operated with recooler only during the day, while charging the heat store (magazine) during the night.*

7.12 Turbine schemes to be used in conjunction with a nuclear power station serving for heat and power production: capacity 3000 MW (thermal). **A** *Electricity production only: 1000 MW electricity, 1900 MW heat to waste (cooling water),* **B** *$Q_u/\triangle P = 6$, where $\triangle P$ is change in utilized power output relative to* **A.** *1000 MW electricity pass-out heat at high temperature,* **C** *$Q_u/\triangle P = 10$, where $\triangle P$ is change in utilized power output relative to* **A.** *2110 MW electricity back pressure heat at low temperature,* **D** *Ratio of district heating heat, and heat to waste varies between zero and maximum. Parallel low pressure/back pressure turbine—low temperature.*

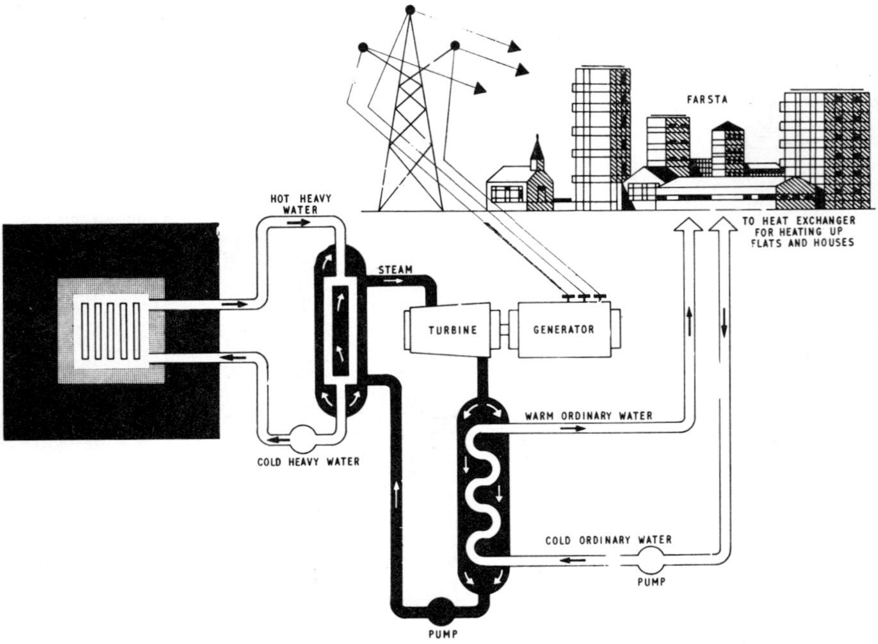

7.13 Simplified layout of the Agesta reactor at Farsta district heating network

blasted out of the granite giving the reactor chamber a covering of at least 60 m of solid rock. It was considered that such a reactor would be safe even under the most unlikely and extreme mishap. This arrangement also had the considerable advantage that, when the plant became obsolete, it would only be necessary to seal off the supply tunnels to the reactor leaving the whole site in place without its becoming a dangerous and unsightly eyesore.

The reactor which, in its first stage of development, yielded 55 MW of heat and 10 MW of electricity, was designed to provide only 60 per cent of the peak heat load required by the town of Farsta, the balance being carried by an oil fired boiler installed during the building of the town to supply the heat demand before the reactor achieved criticality. A large portion of the cost of the undertaking was undoubtedly due to the cost of blasting out the rock chamber. Savings were made as a result of the components being much smaller than in gas cooled reactors.

The reactor core consists of a steel vessel with internal diameter of 4550 mm and a total height of 5880 mm. The vessel is constructed of mild steel employing a wall thickness of 70 mm and clad internally with stainless steel sheeting. The top of the reactor tank consists of two plain steel plates, separated by a stiffening framework. The space between the plates is 1200 mm and is filled with light water which gives shielding and permits temperature control of the large metal masses. The top is fixed to the vessel by

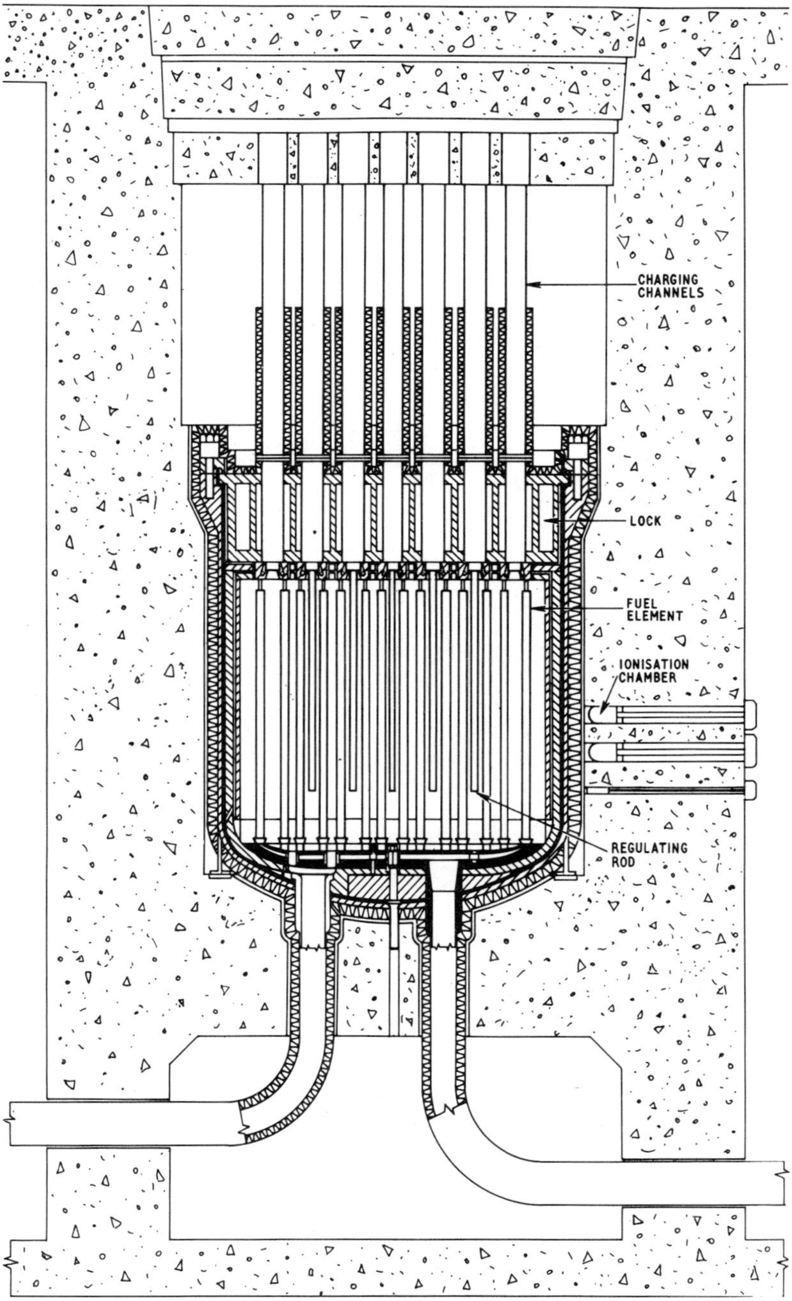

7.14 Section through the reaction vessel of the Agesta reactor

means of bolts, the joint being lined with a gasket. A weld lies outside this to provide an absolutely tight join.

Even after the reactor had been in operation for some time the gamma ray concentration on the outside was quite low. There is a heavy water reflector and a thick steel thermal shield of about 150 mm around the reactor core. The bottom of the tank is protected with a number of plates which also serve for the passage of the coolant. The top plate has 140 short tubes for spacing the 140 fuel elements. At the bottom of the tank there are four inlet

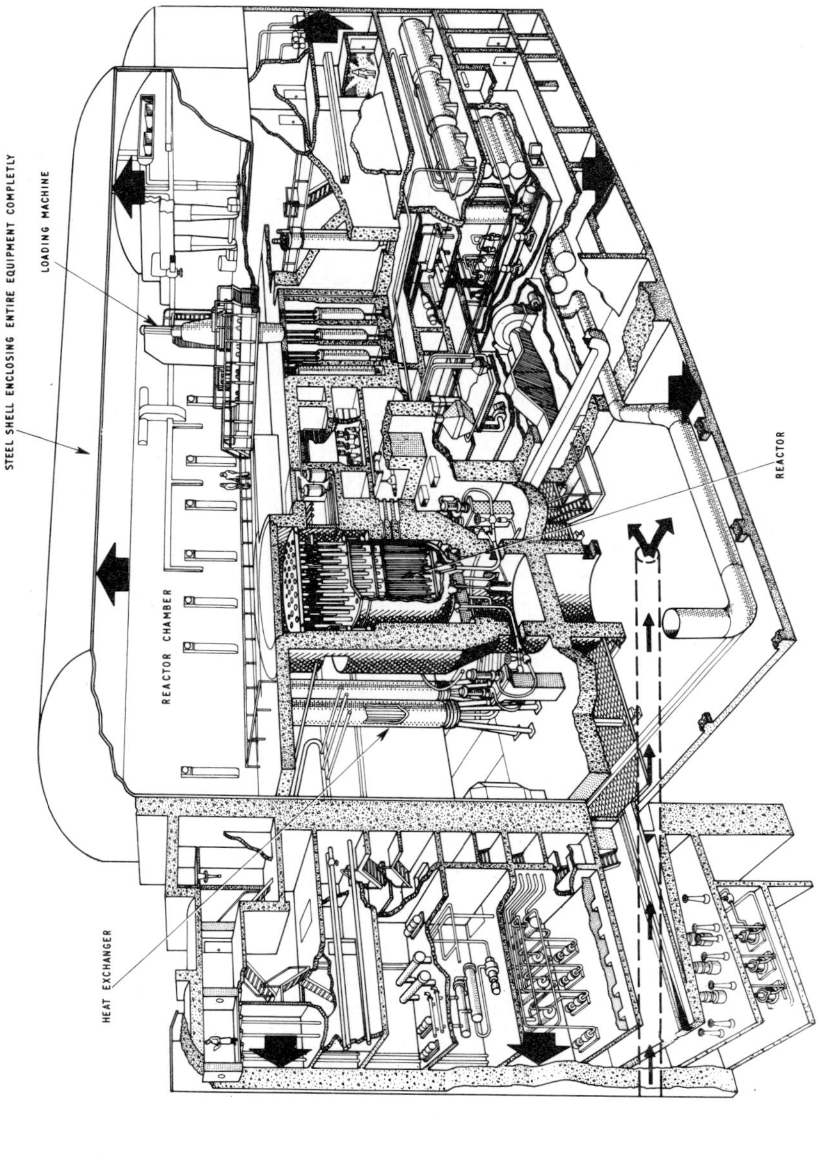

7.15 Layout of Agesta reactor

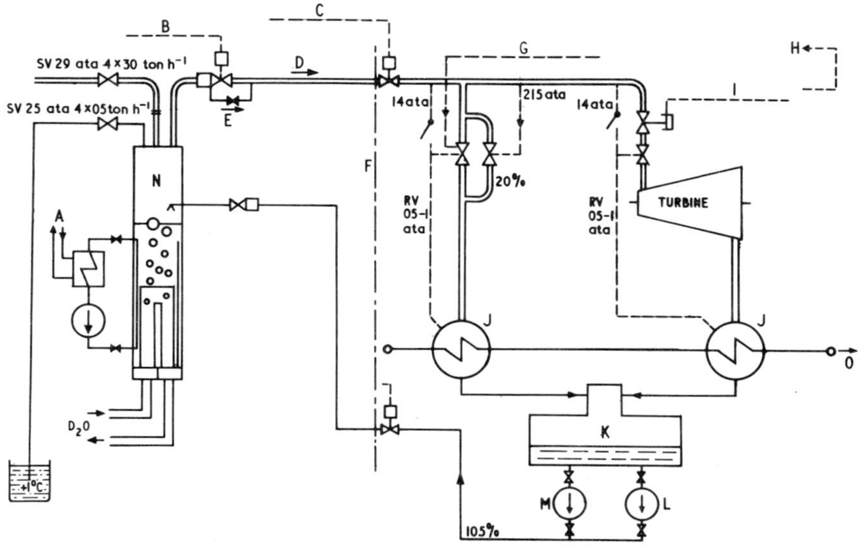

7.16 *Secondary side of Agesta reactor*

and four outlet pipes, each 350 mm in diameter, to permit the passage of the coolant to and from the four heat exchangers. The lid of the reactor has holes of 220 mm diameter for loading the fuel elements. There is also a pipe through the lid to the pressure control vessel.

The fuel elements used in the reactor are in the form of five bundles each consisting of 19 fuel rods which are arranged end-to-end inside a Zircalloy tube. The fuel consists of pellets of uranium dioxide canned in Zircalloy. The diameter of each rod is 17 mm and its length is 3600 mm with a distance between the centres of the rods of 22 mm. The 140 fuel elements in the reactor contain a total of 18 tonnes of uranium dioxide.

The pressure vessel and the internal stainless steel lining weigh 176 tonnes and 105 tonnes, respectively. The bottom and sides of the reactor core are covered in by a biological shield of heavy concrete which is cooled by means of air flowing between the insulation of the reactor tank and the inner concrete surface.

There is no solid moderator and reflector mass inside the core to slow down the fast neutrons and reflect them back on to the uranium rods. This function is achieved entirely by the charge of 70 tonnes of heavy water which is also used as the coolant to transfer the heat developed inside the reactor to the light water which drives the turbines. The complete reactor core is filled with heavy water under a pressure of 34 bars and thus remains liquid even under conditions of severe overload.

The heavy water passes from a duct at the bottom of the reactor vessel through the noses of the Zircalloy fuel elements and abstracts the heat

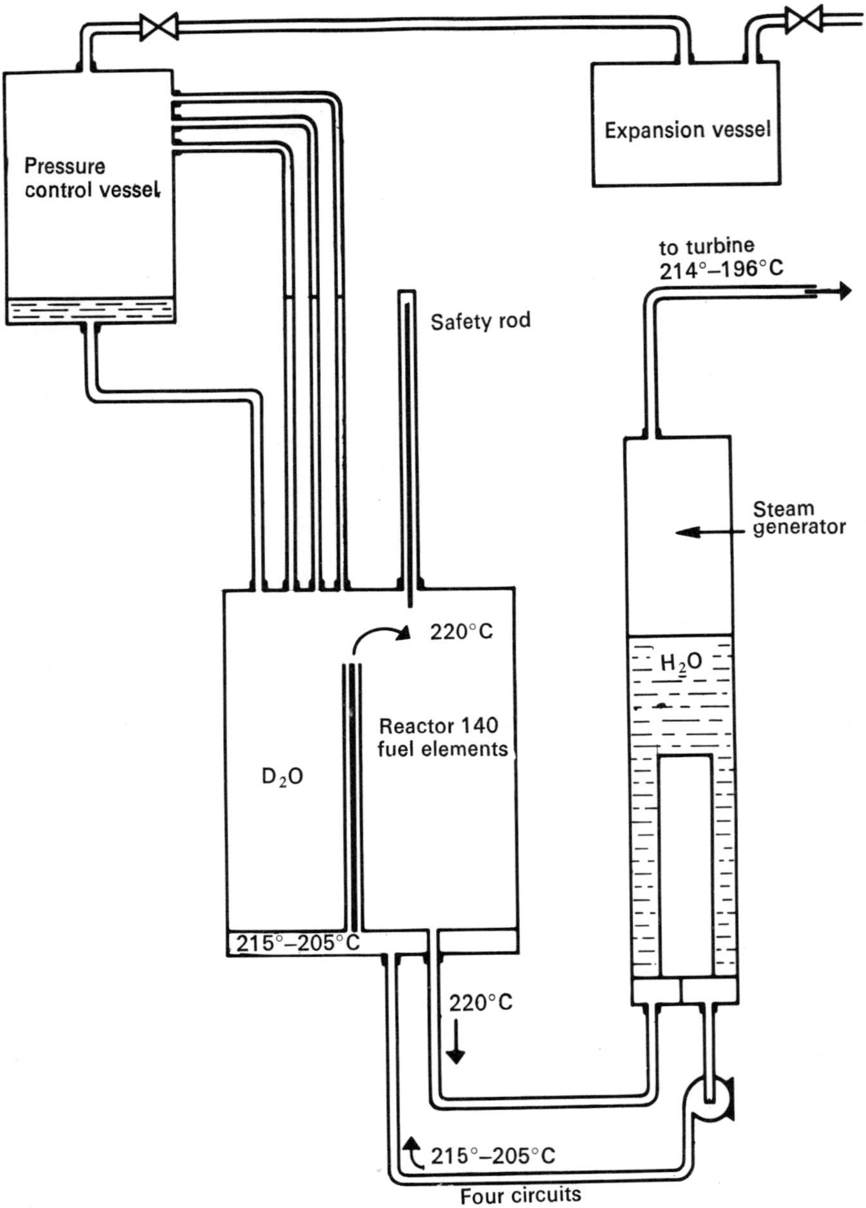

7.17 Primary heat exchanger circuit of Agesta reactor

developed in the uranium dioxide rods. The heavy water, heated from 205 to 220°C (401 to 428°F), then passes back into the bottom shell of the reactor vessel and thence to the four heat exchangers, each of which is 1.7 m in diameter and 10 m in height.

The heat is abstracted by feedwater which is converted into steam and drives the turbines. The steam on the back pressure side is then condensed and the condenser water, at a temperature between 70 to 100°C (158 to 212°F) depending upon the season of the year, is passed through the rock culvert to the town of Farsta, its flow controlled at the sub-station. During severe winter weather the water is further heated at this stage by conventional oil fired boilers. At the flats themselves or at other conveniently situated stations there is yet another heat exchanger system in which the water to be used for feeding into the central heating systems and into the taps for consumer water is heated by water/water heat exchangers. There are thus four completely independent flow systems:

(a) the heavy water circuit between the reactor core and the primary heat exchangers

(b) the turbine circuit

(c) the condenser water circuit which transfers the heat from the station to the sub-stations in the town of Farsta

(d) The final consumer circuit which employs water from the municipal waterworks.

The reactor and the heat exchangers together with other auxiliary equipment are situated in the main underground reactor chamber which has a volume of 22 500 m³ and is lined with steel plates capable of withstanding a positive pressure inside the chamber of two bars. On top of the reactor vessel and the biological shield is the charging machine for recharging and removal of the fuel elements from the reactor core. Also contained inside the main chamber are the pressure control vessel, electric boiler for starting off the reactor, as well as chambers for fuel elements, space for radioactive spent elements etc.

An adjoining rock chamber contains the expansion vessels for any D_2O (heavy water) which might be driven off from the pressure control vessel as a result of leaking safety valves, mishaps owing to overheating of the reactor core etc. There are also auxiliary chambers to deal with ventilation of the main reaction chamber.

A tunnel connects the reactor chamber to the turbine house above ground and contains the general switchboards, electric auxiliaries, steam pipes and a corridor for the personnel working in the plant. There is a second tunnel from the reactor chamber which is used for the movement of heavy equipment and fuel elements.

The fuel elements are inserted from the top by means of the loading machine. The loading channels, each of which serves four fuel elements, reach up through the top guard to the operating floor above.

The fuel elements are taken out in the following manner:

(a) the lid of the top of the loading channel is removed by hand and the loading machine is connected to it

(b) the shut off plug is unscrewed automatically and taken up into one of the chambers of the loading machine
(c) the magazine is turned until an empty channel is level with the loading channel
(d) a manipulator is lowered down from the top of the machine into the loading channel until it rests against a stop within the lid of the reactor tank
(e) the fuel element has a ring at the top by which it is supported on a hook on the underside of the tank lid. There is a movable hook on the manipulator which can engage in the ring of the fuel element. Once so engaged the fuel element can be lifted up into the chamber of the loading machine. A new fuel element can be inserted into the machine in a reverse way. Its bottom end is lowered against the tube of the cooling liquid compartment in the reactor vessel
(f) the plug is screwed back and the loading machine disconnected
(g) the used-up fuel elements are then dumped into the 'mortuary' section of the reactor chamber.

When the reactor is in operation the average temperature of the fuel elements is 600°C (1120°F) but the temperature in the centre of the uranium oxide rods in much higher, about 1400°C (2552°F). This is still however a comparatively low temperature for uranium oxide which has a melting point of about 2800°C (5072°F).

Probably the greatest advantage of this type of reactor, relative to reactors with solid moderators, is the self-adjusting effect of the reactor operation without the need to use either control rods or neutron absorbers. (These are also incorporated as an auxiliary system of reactor control.) In short, if the temperature inside the reactor should rise owing to inadequate removal of heat in the primary heat exchanger or for some other reason, the reactor slows down because of the negative reaction coefficient of the moderator. In the last resort, if the pressure inside the reactor should rise excessively, the reactor shuts down altogether, because heavy water vapour blows off into the expansion vessel and with the disappearance of the moderator and reflector from the reaction chamber the reactor is below criticality. Such activity does not take place during the ordinary running of the plant, only during an extreme emergency. Ordinarily the reactor is controlled by the operation of the negative reaction coefficients of the moderator.

The reactivity coefficients have been calculated at -1.5×10^{-5} deg C^{-1} for the fuel and at -32×10^{-5} deg C^{-1} and -15×10^{-5} deg C^{-1} for the moderator, under conditions of freshness and burn-out of the fuel elements, respectively. The reactivity of the fuel in the centre of the reactor has been calculated at 0.85×10^{-2}. Thus when the fuel is fresh, an increase of 27°C (80.6°F) in reactor temperature ($27 \times 32 \times 10^{-5} = 0.86 \times 10^{-2}$) means that the reactor is practically shut down. When the fuel is almost burned out it requires an increase of 56°C (100.8°F) above the working temperature (220°C or 428°F) of the reactor to bring the reactor to a stop. Adjustments are made with the control rods to allow for an increasing degree of burn-out of the fuel elements.

The working pressure inside the reactor is 34 bars and the standard working temperature is 220°C (428°F). As the boiling point of heavy water at this pressure is at 238°C (460.9°F) there is no tendency for any heavy water to turn into steam during the normal operation of the plant. There are however three safety valves capable of permitting the passage of 77 tonnes h^{-1} of steam at a pressure of 40 bars which open at the following respective pressures: 36 bars, 36.5 bars and 37 bars. These safety valves are placed between the pressure control vessel and the heavy water expansion chambers.

The heavy water from the moderator chamber passes along stainless steel tubes at a velocity of 7 m/s being pumped along by totally immersed electric motors with water lubricated bearings. Each of these pumps has an effect of about 180 kW at full power and 30 kW at half power. There are also special electrically operated shut-off valves on each pipe and special valves between the two halves of the single-pass heat exchanger; these come into operation if the pressure difference at the two ends of the U-tubes becomes excessive.

On the secondary side of the main boilers, the heat-exchange medium is very pure water which has been dosed with hydrazine to prevent differential aeration corrosion. Each of the four boilers contains 11 tonnes of water during operational conditions and is able to produce 27 tonnes h^{-1} of steam at 14 bars pressure. The boilers are however constructed for a working pressure of up to 32 bars. Owing to the very considerable volume of water in the boilers (about 45 tonnes) the system has considerable flexibility. Even if all the steam were shut off much time would elapse before the pressure inside the boilers would rise to 30 bars when the safety valves normally lift. The flow of water through the secondary system is monitored constantly to detect any leak of heavy water into the secondary system. In such event it would be a matter of seconds only before the valves of the primary and the secondary system were shut off automatically. As there are four independent boiler systems the remaining three could then be adjusted easily to take up the extra load.

Heat transfer on the secondary side is by natural convection only. The dome of each heat exchanger has a 200 mm steam outlet pipe. The four steam pipes lead into a main pipe 400 mm in diameter. The feed water is pumped in from a feed water tank at the turbine station, which has a capacity of 140 tonnes h^{-1}.

The steam from the secondary side next passes above ground via the pipes in the rock tunnel and enters the turbine which has a power output of 10 MW. In case of turbine failure it is possible to divert the steam directly into a special heat exchanger to abstract its total heat. The back pressure steam from the turbine is passed into the condenser where it is condensed at 105°C (221°F). This water is then returned to the boilers. The water emerging on the secondary side of the condenser next passes into the concrete underground culvert through one of the twin steel pipes at a temperature of about 100°C (212°F) (during winter). This water is then pumped underneath Lake Magelungen to the supply station at Farsta and the return water re-enters the station at about 60°C (140°F). The culvert is thermally insulated by encasement with aerated concrete of very low density.

After the Agesta plant had been in operation for a few years its thermal output was stepped up to 125 MW and its electric output to 18 MW when working on back pressure and to 31 MW during the summer when operating on the condensing cycle.

Projected schemes

Based on the experience gathered at Agesta several large schemes have been thoroughly costed and pre-planned, including Swedish schemes, one Finnish, one West German and one French.

There are proposed Swedish schemes at Stockholm, where it is proposed to construct two boiling water reactors, at Forsmark some 130 km to the north of Stockholm, and at Haninge about 40 km south. The Forsmark station will be the first to be completed and will have a power output of 1000 MW_e when operated on full condensing cycle, but capable of being operated on ITOC to provide the heat requirements of the conurbations of Uppsala and Stockholm. The main feeder main between Forsmark and Stockholm will be 130 km in length. In addition, the new Barseback nuclear reactor is under construction in the south of Sweden. When completed it will be capable of supplying 1000 MW of electric power and 960 MW of heat by means of ITOC turbines. It will serve the district heating networks of the towns of Halsingborg, Lund and Malmo. At Ringhals a nuclear reactor station is projected for the supply of heat and electricity to nearby Gothenberg. In Finland a nuclear district heating plant using ITOC turbines with a power output of 1000 MW_e and 1170 MW thermal load is planned for Helsinki. The designers for the Swedish and Finnish schemes are AB Atomenergi of Studsvik, Sweden (**7.18**).

In France it is planned to construct a nuclear reactor for the district heating network of Paris (CPCU). The reactor is to be located at Saclay about 20 km from the centre of Paris and is projected to yield 1100 MW of heat at a pressure of 31 bars and at a temperature of 270°C (518°F). This scheme will be most viable if 88 per cent of the district heating load of Paris can be supplied by nuclear power from the Saclay reactor. Pipeline diameters between the reactor and the consumer network would be of the order of 1000 to 1300 mm.

Another projected scheme is the West German Kirchgartshausen project which will have ITOC turbines with an electrical output of 1100 MW and a thermal output of 58 MW.

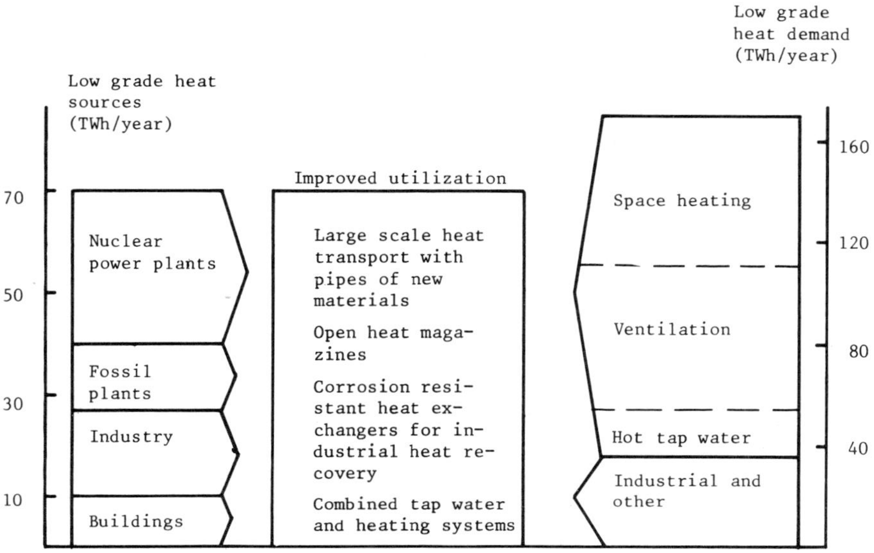

7.18 Projected supply and demand of waste heat in Sweden at the end of the 1980s

. Several studies have also been reported from various East European countries into the feasibility of nuclear energy utilisation for the supply of district heat.

Heat pumps

Heat pumps are basically refrigeration circuits able to convert high grade energy, such as electricity, into low grade heat at an efficiency between 200 and 300 per cent (ie at coefficient of performance of 2 to 3). There is no commercial logic in converting the heat potential of fossil fuel into electricity at an efficiency of say 30 per cent and then reclaiming this heat by heat pump, since the overall efficiency of the heat pump process will then only be between 60 and 75 per cent: a higher fuel/heat conversion efficiency could be achieved by the use of a conventional boiler plant.

On the other hand nuclear electricity costs are very low (ignoring the capital cost element). During periods of low power demand and high heat demand it would be more worthwhile to convert the waste or surplus electricity into heat via a heat pump at an efficiency of 200 to 300 per cent than to convert it into heat via an electric fire at 100 per cent efficiency.

Conclusions with respect to district heating schemes

It is generally accepted that nuclear district heating systems are only viable at a heat supply in excess of 1000 MW. To achieve the necessary flexibility of heat supply across the four seasons it is essential that heat and power generation should be combined using ITOC turbines. Nuclear stations cannot be used for heat supply only.

In order to make use of the large amounts of heat involved the consumer network must be very extensive before the decision can be reasonably taken to supply district heating from a nuclear reactor plant (**7.19**). In many potential applications it is only possible to assemble an adequate consumer network by the construction of very long pipelines. It is considered that pipeline lengths between 50 and 100 km are the limit of viability in current circumstances. As fossil fuel prices rise, while nuclear power generation and pipeline laying costs reduce relatively and pipeline techniques improve, the break-even pipeline lengths are bound to increase. Longer pipeline lengths can also be envisaged if the thermal density per unit length pipeline can be increased. The heat can be conveyed more cheaply by raising the supply temperature but then the cost of the heat is greater in terms of lost electricity.

The ultimate future role of nuclear power on the district heating scene is not as yet clear. The employment of thermal reactors which are only able to utilise either directly or indirectly a mere 0.72 per cent of the uranium fuel is obviously not a viable proposition because world resources of uranium will come to an end more quickly than oil and coal. Fast or breeder reactors (like Britain's Dounreay or France's Phenix) in size ranges of about 1000 MW_e appear to be the only types which will have a long term future, but fast reactor technology has not yet developed to the stage when such reactors can be safely located close to urban centres. Fast or breeder reactors are ruinously expensive. Yet nuclear heat supply by fast reactors with large ITOC turbines to balance heat/power requirements remains the only long-term prospect.

Fossil fuels are wasting assets and are bound to become progressively more scarce and more expensive. Once a given district heating network has been established it should however be feasible to convert it to nuclear heat supply when this becomes available.

CHECK LIST

1 Whether nuclear power is realistic alternative to fossil fuels
2 Principles on which nuclear power operates
3 Basic descriptions given concerning major nuclear reactors currently in use
4 Properties of nuclear fuel

7.19 Various systems of utilising reactor power for district heating. **A** *Nuclear power station with steam turbine,* **B** *Nuclear power station with gas turbine,* **C** *Nuclear heating and power station with steam turbine,* **D** *Nuclear heating and power station with gas turbine,* **E** *Pure heating station*

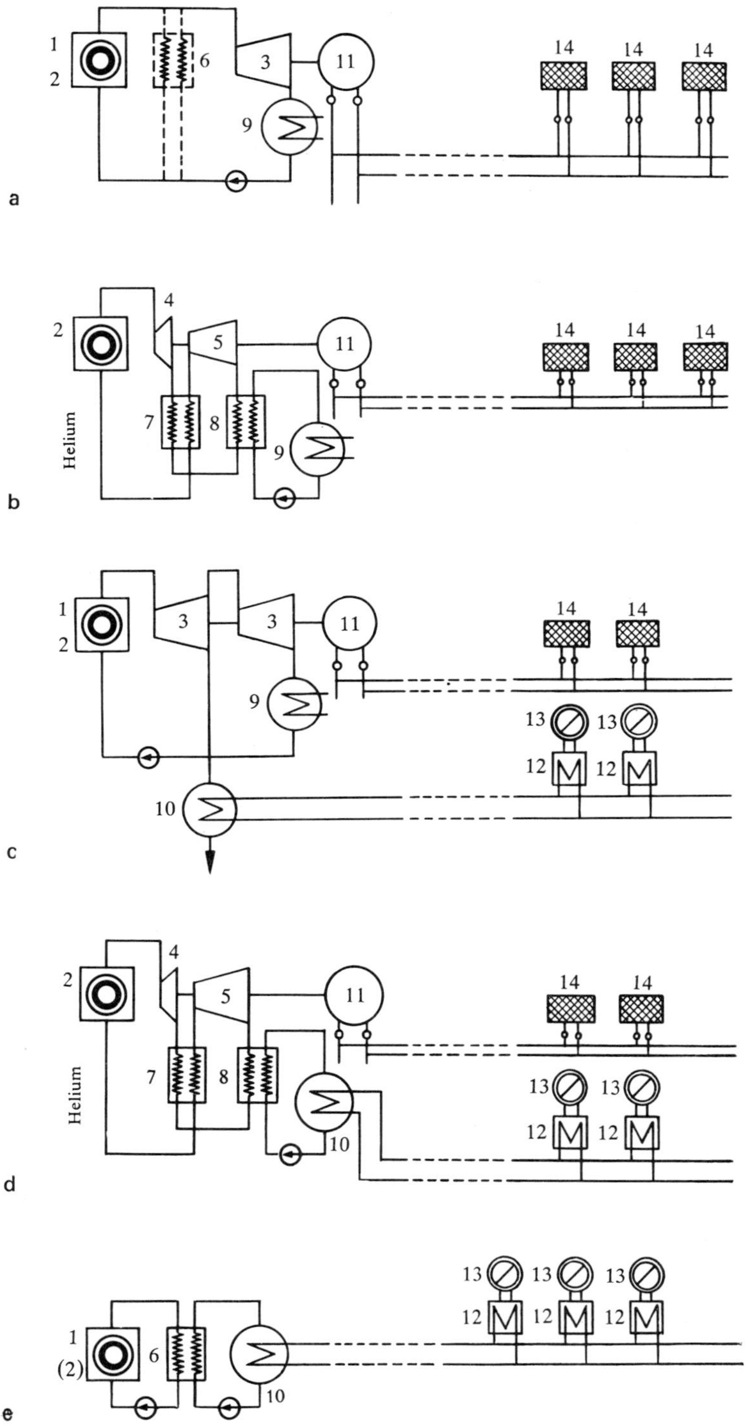
a
1
2
6
3
11
9
14 14 14
b
2
4
5
11
Helium
7
8
9
14 14 14
c
1
2
3
3
11
9
10
14 14
13 13
12 12
d
2
4
5
11
Helium
7
8
10
14 14
13 13
12 12
e
1
(2)
6
10
13 13 13
12 12 12

5 Methods of disposing of nuclear wastes
6 Principles of radiation intensity and half-life
7 Erosion of energy reserves by nuclear fission reactors
8 Working principles of nuclear reactors
9 Radioactive materials with different grades of radiation intensity
10 Prospects for thermal reactors
11 Prospects for fast or breeder reactors
12 Coolant for fast breeder reactors
13 Principles of nuclear fusion process
14 Impact on the economy of fusion reactors
15 Prospects for district heating using nuclear reactors
16 Current experience in the use of nuclear power for district heating
17 Projected schemes for district heating in conjunction with nuclear power generation
18 Heat pumps and their relevance to district heating by nuclear energy
19 Basic conclusions on the suitability of installing nuclear district heating systems

REFERENCES

Fernwärme International (*articles in German*)
G. Deuster, 'District heating from nuclear stations', 3 (1974) 4, pp. 110–114
G. Dietrich and others, 'District heating from nuclear stations', 4 (1975) 5, pp 155–162
G. Deuster, 'Nuclear energy for district heating', 4 (1975) 6, pp 200–203

Other sources
A. A. Field, 'Nuclear and geothermal energy for district heat supply', *Energy World*, December 1975, pp 5–8
A. A. Field, 'Heat from nuclear energy', *Heating and Ventilation Engineer*, December 1978/January 1979, pp 5–6
'Germany looks towards nuclear district heating', *District Heating*, 61 (2), 1975, pp 22–30
S. Glasstone and A. Sesonke, *Nuclear reactor engineering*, van Nostrand, New York, 1967
J. Hill, 'The nuclear fuel industry', *Atom* 207, January 1974, pp 2–35
G. Kulcinski, 'Fusion power: an assessment of its potential in the USA', *Energy Policy*, June 1974
P. Margen, 'The use of heat from nuclear stations for residential heating', ASME/ANS International Conference, Pittsburg, 13th–16th March 1976
T. N. Marsham, 'Nuclear power – the future', *Atom* 196, February 1973
T. N. Marsham, 'The UK view on fast reactors', *Atom* 201, July 1973
F. W. Mayes, 'Energy cascading in large district heating systems', *Nuclear Technology*, 40, October 1978, pp 23–29

H. Muhlhausen and others, 'Heating by means of nuclear energy', Technische Rundschau, *Sulzer News* 3, 1975, pp 1–8 (in German)
'Nuclear heating', *Heating and Air Conditioning Journal*, February 1976, pp 23–26
F. J. Pearson, *Nuclear power technology*, Oxford University Press, London, 1963
T. Schwarz, 'Four steps to nuclear district heating', *Nuclear Engineering International* 22, August 1977, pp 45–49
'The parameters of nuclear district heating schemes', *Teploenergetika*, 25 (12), 1978 (in Russian)
'Thermal power plants and nuclear engineering', Technische Rundschau, *Sulzer News*, 1, 1976

CHAPTER 8

Design of district heating mains systems

PIPELINE ROUTES

Choice of route and design

In a district mains system the financing charges of the pipe mains network are generally the major element in the final cost of heat to the consumer. It is therefore essential for the mains network to remain in good order for many years without requiring major or costly repairs. In virtually all the successful European district heating systems repair and maintenance charges are restricted to less than 0.5 per cent of the capital value of the mains. Careful and detailed design thus ensures the viability of the system.

District heating mains pipelines should if possible be run through the areas with the highest heat demand. Major savings in construction costs can also be made by routing the pipes via existing dry cellars, railway tunnels, other underground borings etc to simplify and reduce the cost of insulation and external protection sheathing.

It is often possible to route pipelines beneath bridges or along railway tracks and canals. Such routes are inferior to weathertight routes because they make it necessary to provide good surface protection: the pipelines should never carry cold water during periods of frost because heat losses may then be so great that they cause frost damage even though the insulation is good (trace heating may however offer adequate protection). If suitable enclosed spaces or good right-of-way runs locations are not available for pipelines then they are best placed inside trenches in parks or gardens where pressures on the pipes from traffic are only slight.

The most expensive form of pipe laying is beneath busy city streets. It is then necessary to use very expensive traffic pressure-proof pipelines to safeguard the mains against damage from heavy vehicles. Great difficulties may also have to be overcome relative to existing services in the ground such as sewers, water pipes, gas mains, power and telephone cables. Moreover maps showing the positions of the existing mains services may not be available in the centres of old cities. Designers of district heating networks must therefore try to avoid having to lay pipelines in such difficult areas. In new towns it is usually far easier to lay district heating networks.

One of the handicaps in the United Kingdom is that control over water, gas, electricity and telephone services is in the hands of separate nationalised industries, each with its own bureaucracy. This makes it difficult (or impossible) to adopt the techniques of pipeline laying used in many other countries (eg where the so called *Stadtwerke* (City works departments) control all the utility services and the sanitary departments and traffic organisation, etc). Where there is such unified control district heating pipelines can often be laid as part of a composite system together with water pipes, gas pipes and other services. This method is less costly and it entails less disturbance: two advantages which are essential if district heating is to be made a viable proposition (**8.1**).

Ground conditions

The cost of a district heating network is very dependent upon the nature of

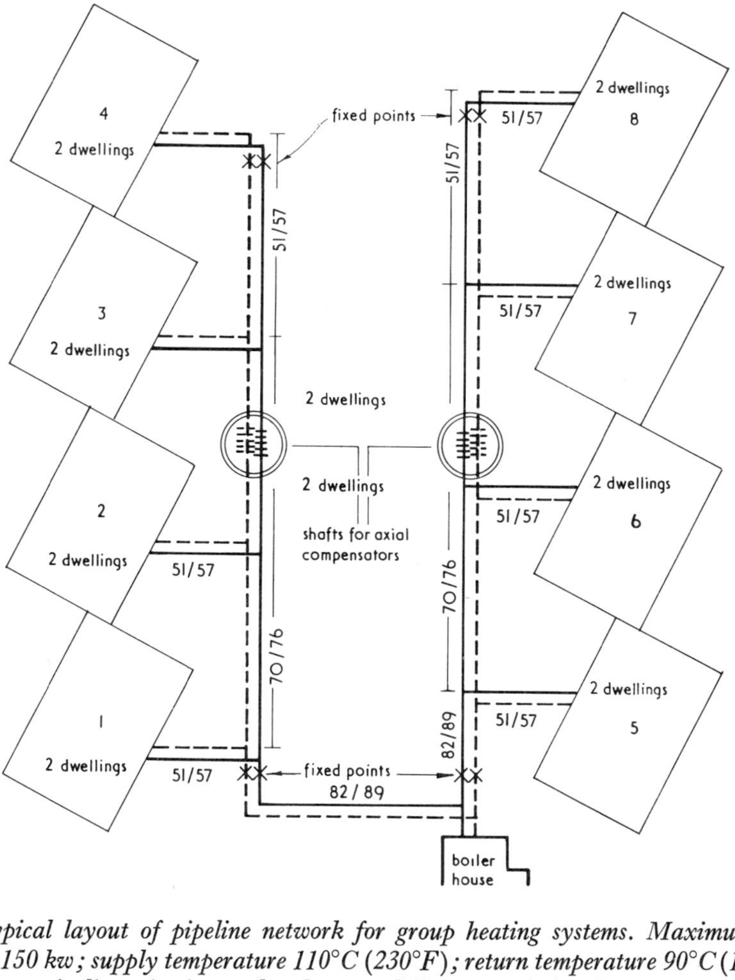

8.1 Typical layout of pipeline network for group heating systems. Maximum heat supply 150 kw; supply temperature 110°C (230°F); return temperature 90°C (194°F). Figures on pipelines give internal and external diameters of pipes in mm

the ground into which the pipes have to be laid. The best conditions are found in well drained areas of sandy soil where ground water never rises to engulf the pipes. If such areas are also free from traffic they are better still and it is then possible to adopt the cheapest forms of pipeline insulation. Other ground conditions may not be so favourable. The main hazards are:

Underground streams which may flow intermittently usually after periods of heavy rainfall: if there is the slightest risk of this happening loose fill insulation systems must not be used.

Variable ground water levels or inadequately drained ground which make it likely that the pipeline and insulation will be submerged at least part of the time: in such circumstances pipeline systems must have a waterproof mantle; gas concrete systems are inadmissible.

High traffic pressures which make it necessary to construct pipeline systems with a rigid mantle pipe: abestos cement systems and those with a tough mild steel mantle pipe are more suitable here than systems with a polyolefin or PVC mantle tube.

Liability to land slips which necessitates a certain degree of elasticity in the pipeline system: plastic piping is less likely to fail in such circumstances than more rigid systems.

Rocky areas which, although theoretically offering a good environment for district heating pipelines, usually make the costs of excavation prohibitive.

Proximity to other services

In city areas it is sometimes necessary to build district heating pipelines close to other services. The German VDE (*Verein Deutscher Elektrizitats-werke*) recommends the following minimum distances between these pipes and other services:

Services crossing district heating mains	
1 kV cables, telephone cables, test cables etc	300 mm
10 kV to 30kV cables	600 mm
Several 30 kV cables or single cables over 60 kV	1000 mm
Gas mains	300 mm
Water mains	300 mm
Services running parallel to district heating mains	
1 kV cables, telephone cables, test cables, etc	300 mm
10 kV to 30 kV cables	700 mm
Several 30 kV cables or single cables over 60 kV for distances shorter than 5 m	1000 mm
Several cables of 30 kV or single cables over 60 kV for distances greater than 5 m	1500 mm
Gas pipes	500 mm
Water pipes	500 mm

Direct-current railway lines or other earthed direct-current systems constitute a considerable corrosion danger to district heating pipelines because of the possible incidence of stray currents.

Traffic pressures

The determination of the depth at which district heating pipelines are best installed depends on the anticipated traffic conditions. The greater the depth, the lower the traffic pressure to which a given pipe is subjected but the static pressure from the earth cover is increased at a rate proportional to the depth. Care must therefore be exercised in evaluating each set of circumstances; the manufacturers standard recommendations should not be blindly followed (**8.2**).

Manholes

Manholes are positioned in the pipe network to house, and provide access to,

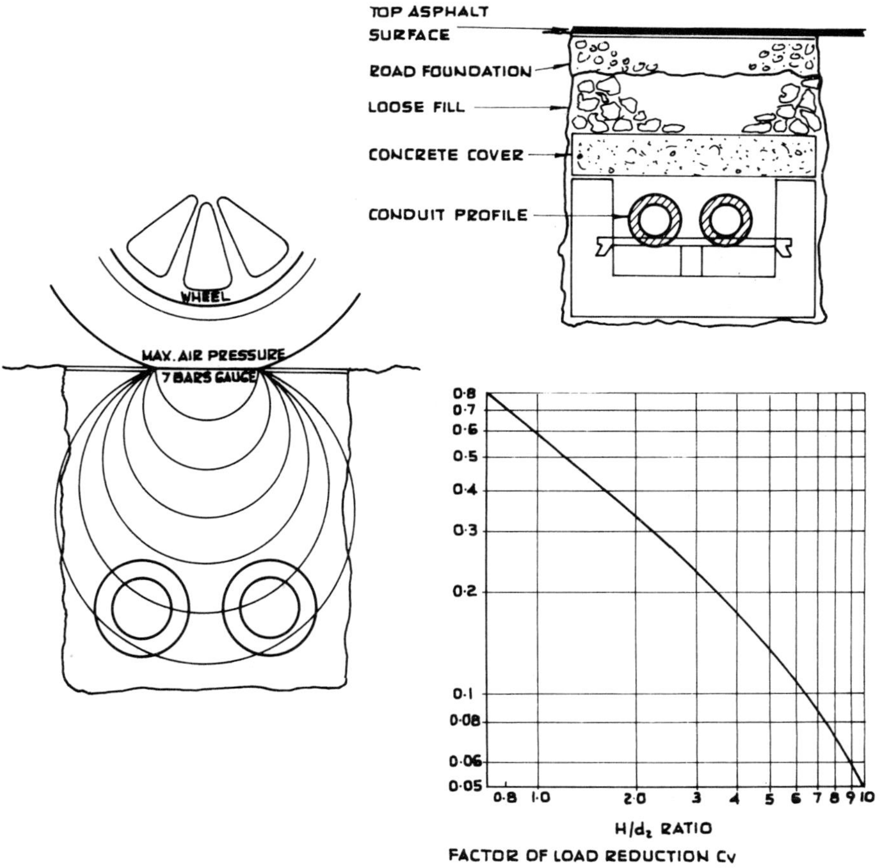

8.2 The effect of traffic loading upon buried pipelines

pipeline components which require routine inspection and/or occasional replacement, such as branch connections, temperature compensators, isolating/regulating valves, drain cocks, air vents, sump pumps, steam trap sets and electric isolators.

As the name implies it is necessary for men to work inside manholes: hence these must be dimensioned to permit safe access and adequate working space. **Figure 8.3** indicates the basic parameters of height and width. Stout access ladders should be securely fitted. The cover of the manhole must relate to the likely traffic conditions and loads and it must have a water-tight seal. Any inflow of water is most undesirable and a sump pump facility must therefore be installed. Various types of prefrabricated manholes are commercially available (**8.4** and **plate 14**).

Pipes and components must be kept well above any wetness inside manholes. They must also be protected with suitable moisture excluders and the connections into the manhole must be via puddle flanges.

To assist routine valve operation valves with extended spindles are often used inside deep manholes.

EXPANSION AND CONTRACTION

Anchorage

Pipelines are mostly anchored firmly to the ground at specific positions so as

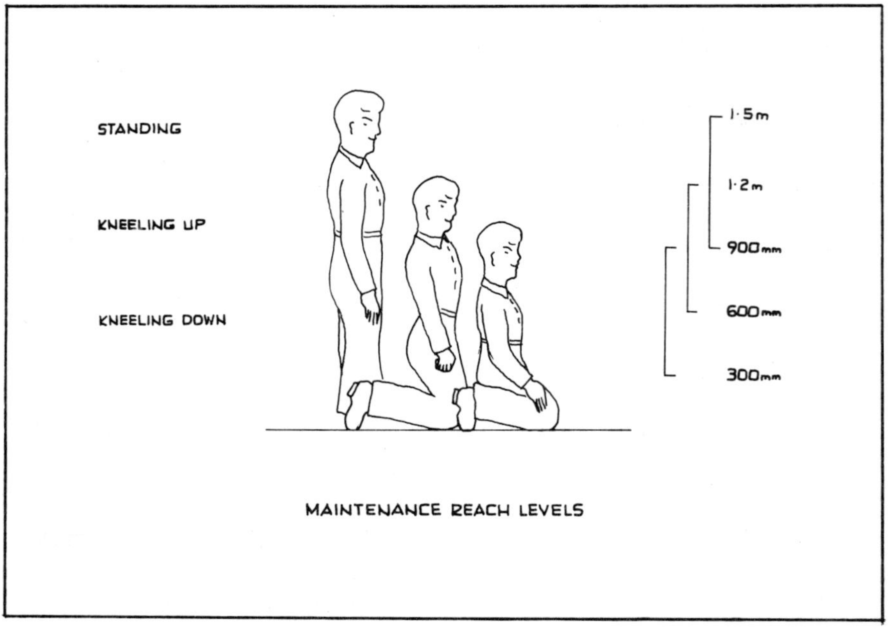

8.3 Dimensions for accessible manholes and inspection pits

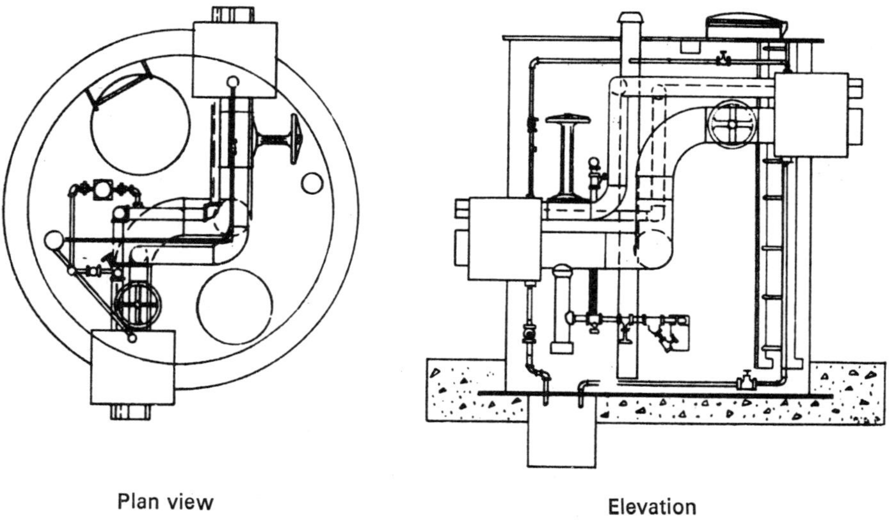

8.4 A prefabricated manhole assembly

to define clearly the direction and manner of their expansion and contraction under specific temperature conditions (**8.5**). In addition to these fixed points of anchorage they are also supported by bearings which permit linear movement and avoid movement stresses in the mains system. Expansion and contraction of the pipeline system is compensated by a variety of devices.

The forces acting upon the fixed point anchors are very considerable. They are caused by the following factors:

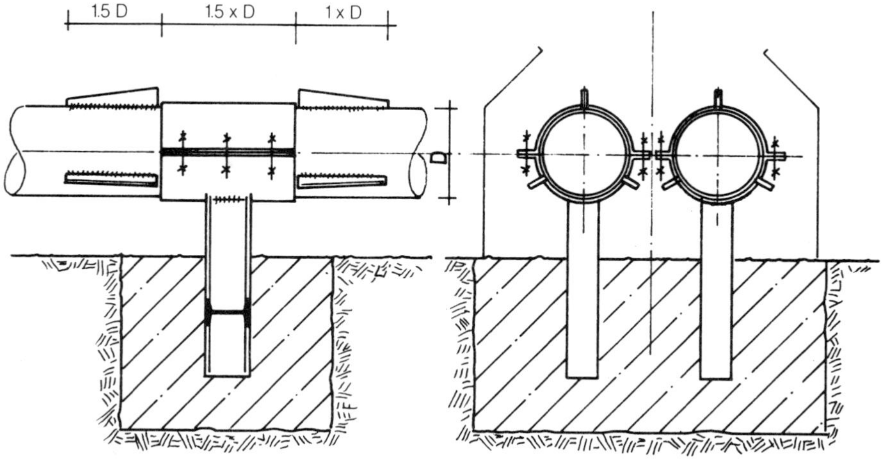

8.5 Fixed points along a straight pipeline

(a) reaction due to friction of a pipeline against its insulation and support bearings
(b) reaction of compensators which are often elastic in nature
(c) reaction of pipeline bends and U-compensators
(d) weight of pipeline, insulation and heating medium.

Various designs of fixed points are in common use: most of these depend on very solid reinforced concrete and rigid steel construction above the sole of the trench. Fixed points should be positioned as far as possible close to branch line connections, main valves or changes of pipeline direction.

Glide supports are required to permit smooth movement of pipelines (**8.6**). At locations where access to the pipelines may be difficult it is best to install sliding bearings only. (Practical experience with roller bearings, ball bearings etc has not been entirely successful.) The spacing between sliding bearings must be calculated in each specific case. It must never be so wide that the pipeline tends to sag between supports, but the more rigid the pipeline, the greater the permissible spacing. Where supports are placed close to pipeline bends it is also necessary to compensate for sideways movement. In vertical sections support should be given by sprung bearings or by a gliding support on an inclined plane (**8.7** and **8.8**).

If sideways movement is permitted it is likely to cause severe damage to compensators. Pipeline guides are therefore necessary to limit such movement. These should be placed at both ends of U-compensators, positioned at about three times the compensator movement. Unless other techniques have been employed to prevent sideways movement of axial compensators, guides should also be installed at both ends of these.

Compensation

It is necessary to incorporate suitable compensators to permit thermal movement of pipes in a longitudinal direction: changes in length mainly occur because of alterations in the temperature of the heating or cooling fluid which the pipes convey. When a pipeline is commissioned its linear expansion is of the order of 100 to 150 mm over 100 m; the exact extent of thermal movement depends on the expansion coefficient of the particular metal and on the temperature change. For example the expansion coefficient of steel is $12 \times 10^{-6}{}^{\circ}C^{-1}$ heated (or cooled) through a range of 100° C (180°F). The thermal movement is $12 \times 10^{-6} \times 100 = 1.20$ mm per metre length of pipe per °C rise in temperature.

Pipe movement is facilitated by the provision of certain fixed points along the pipe line at which the pipe is anchored securely to the base slab or to a firm post and by the use of rolling or sliding shoes and of compensators elsewhere. Thermal movement can most easily be taken up where pipelines change direction: bends can often be used as the simplest and cheapest form of compensator. The most common and least costly form of compensators used for long straight lines of pipes are all-welded U-shaped pipe

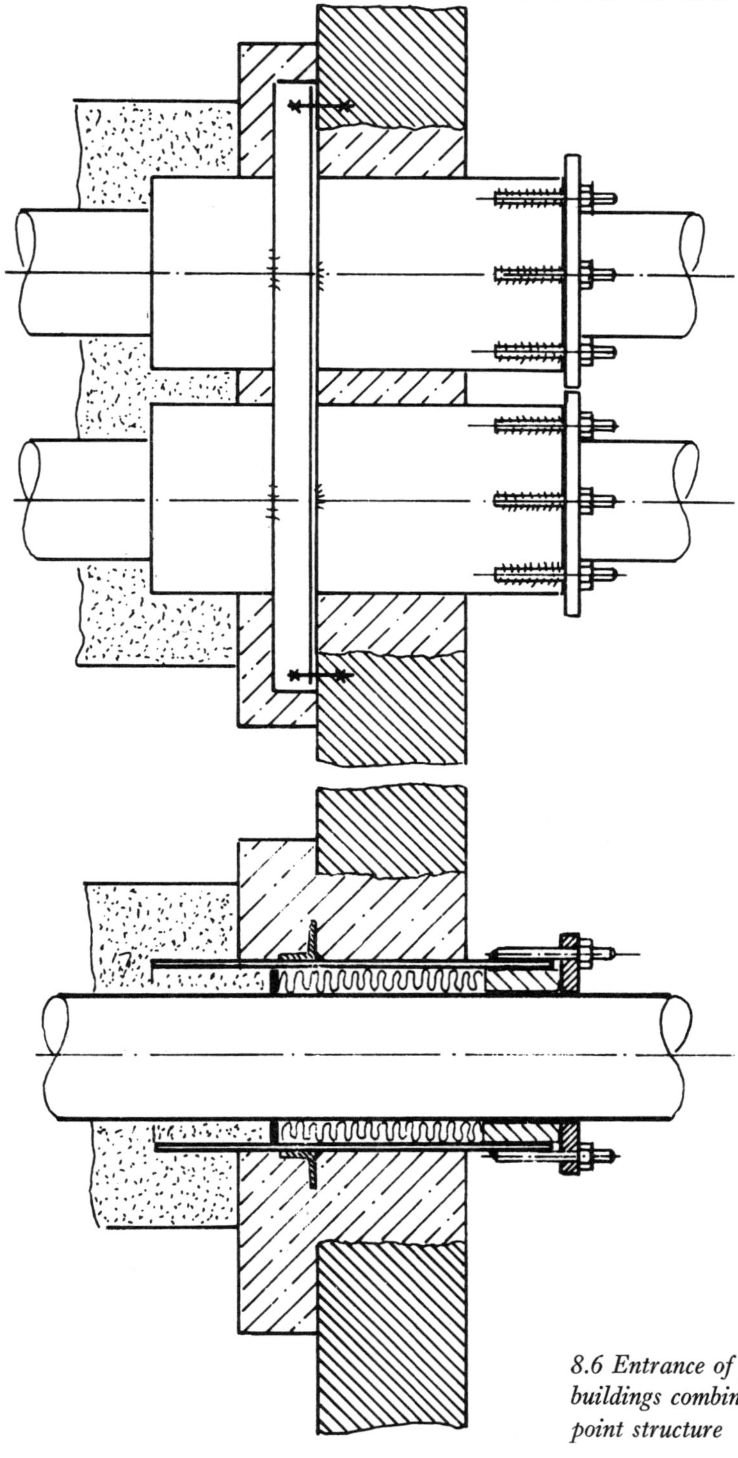

8.6 Entrance of pipelines into buildings combined with fixed point structure

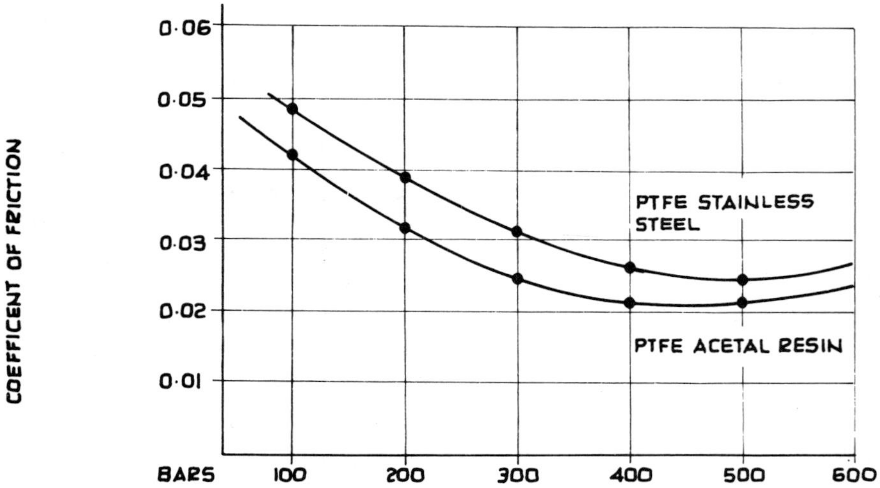

8.7 Coefficients of friction of sliding bearers

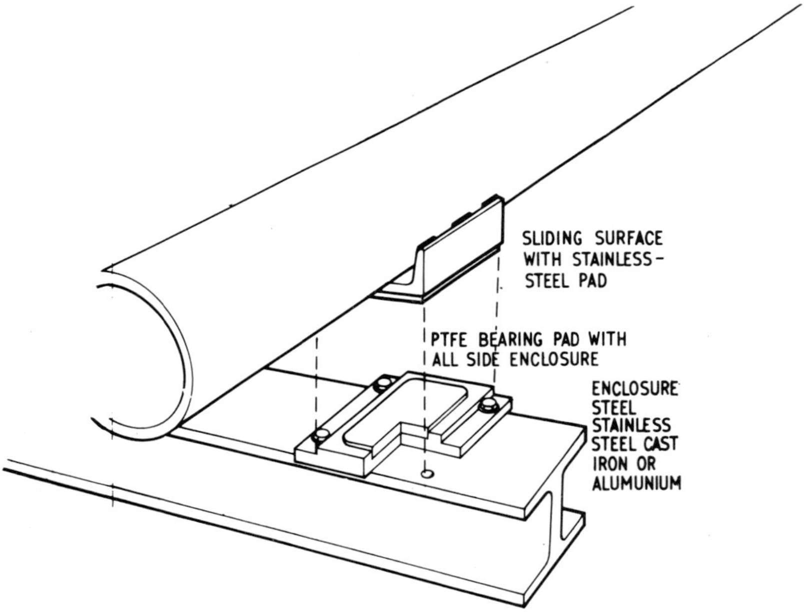

8.8 Sliding support using a PTFE bearing surface

sections placed inside manholes which can also serve as drainage positions, inspection chambers and junction boxes (**plate 18** and **8.9**).

The spacing of the compensators depends on the manner in which the pipelines are anchored and on the maximum temperature of the water being

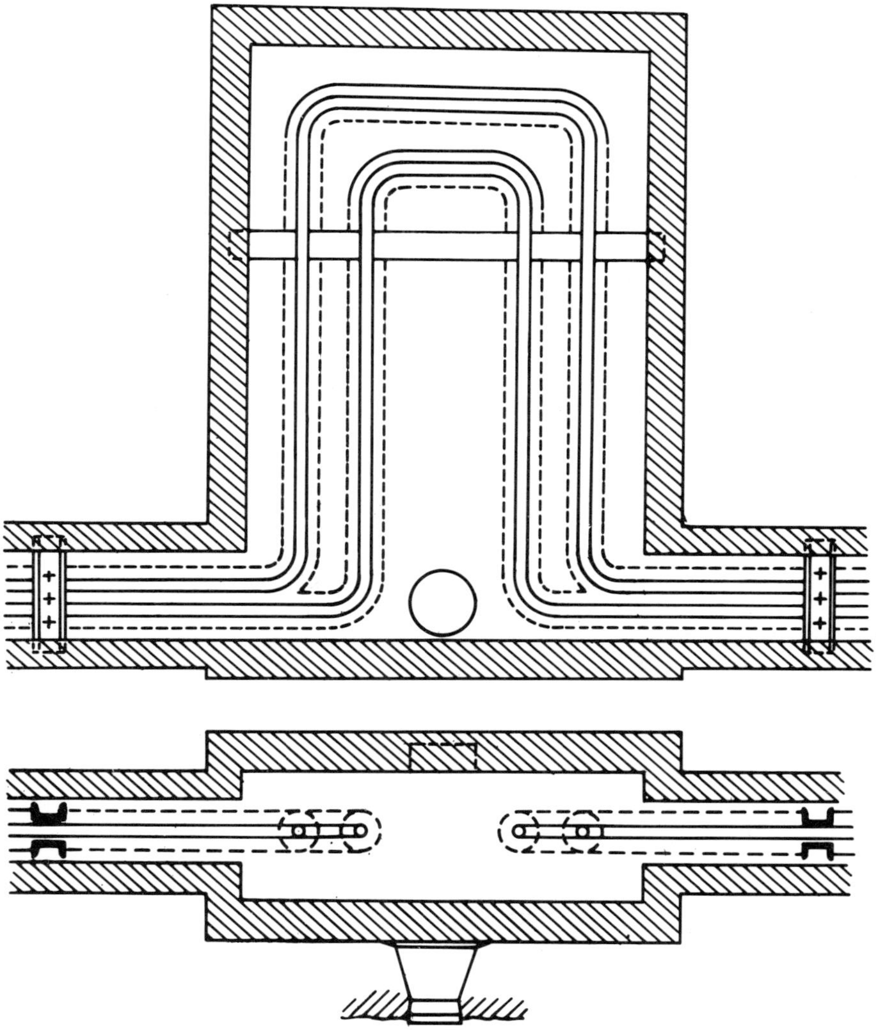

8.9 Expansion chamber with U-shaped welded compensators and drainage duct

conveyed. In the Rotterdam district heating system, for example, where the maximum water temperature is 140°C (284°F) the inspection and compensator chambers are spaced 40 m apart. In Utrecht where the water temperature is up to 100°C (212°F) the spacing between compensator chambers is up to 60 m. Each of these two pipelines is installed with a single fixed point at the centre of the line. However, if two fixed points are used with an expansion unit in between, the spacings must be halved. Having two fixed points for each pipe reduces the shear stresses between the thermal insulation and the pipe.

Ball joint compensators

Ball joint compensators are rigid steel or malleable iron assemblies which can flex angularly through between 15 and 40° and also swivel through 360° if so required. Ball joints can withstand high temperatures and pressures without leakage. The type of gaskets used may be either asbestos based, PTFE or metal. Ball joints are generally more costly to install than plain U-loops or Z-compensators but they can accommodate very considerable dimensional movements in pipelines and their expansion chambers can be much smaller. Ball joints compete directly with bellows units and sliding expansion joints (see **8.10**).

Bellows expansion joints

Bellows type expansion joints are extensively used for the compensation of thermal pipe movements. Their main advantage lies in completely sealed construction: their expansion movement is absorbed by the flexing of a carefully selected and accurately engineered convoluted section of high grade metal of thinner section than that of the pipe. In favourable circum-

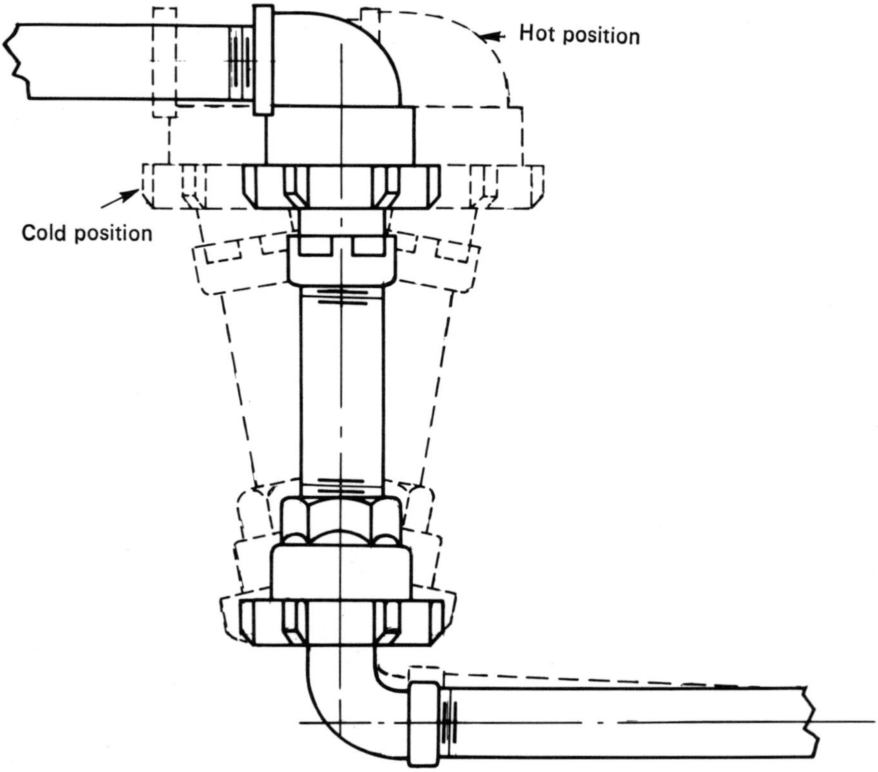

8.10 Ball joints as expansion units

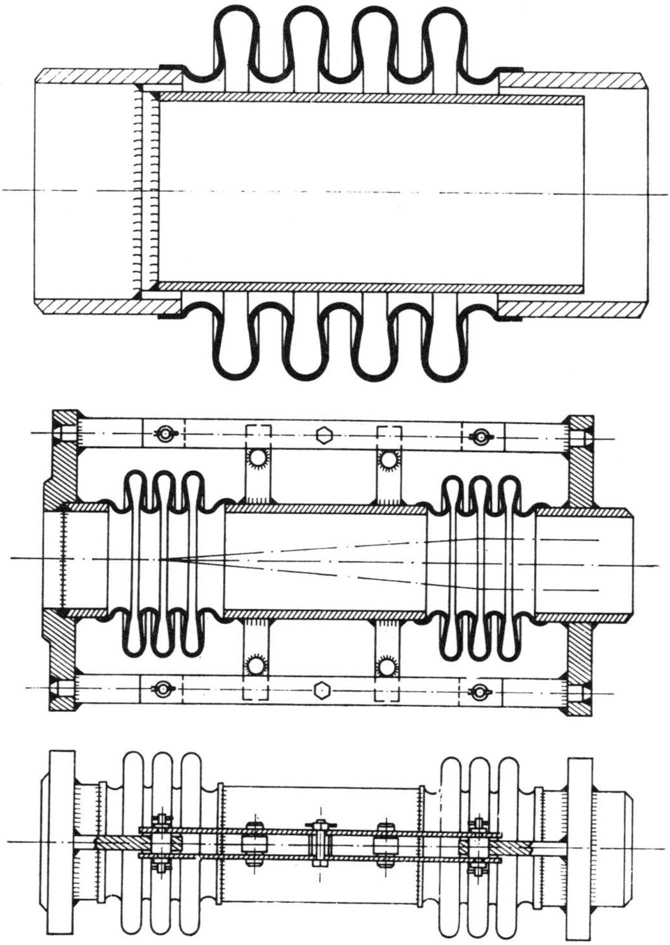

8.11 Various types of axial compensators

stances the life of the bellows should equal that of the pipe system into which it is fitted.

Two basically different bellows systems are available, the axial type and the articulated and angular type. The axial type is a plain bellows used in compression and extension along its axis in the conventional manner. It is effective where relatively small amounts of movement have to be absorbed and where pipes are installed in a confined space in close proximity to other pipes or in trenches or ducts.

When using the axial type the pipe must be anchored firmly and guided effectively. The action of the bellows as a spring in combination with the thrust on the pipe due to internal pressure tending to open it out lengthwise can impose considerable forces on the pipe and its anchors. Because of

compressive forces which acts on the pipe run, the tendency for the pipe to bow at any point must be prevented. If this is to be effective:
(a) a pipe which includes an axial type bellows must have a firm anchor on each side of the bellows
(b) the pipe must be straight between anchor points, in plan and in elevation
(c) the pipe guides must be able to resist vertical and horizontal forces at right-angles to the pipe run while allowing free movement along the pipe axis.
Because of these requirements a pipe with axial type bellows must never be supported on hangers.

The articulated and angular type of bellows is very different to the axial type and works on an entirely different principle. It is designed to bend, absorbing expansion at right angles to its axis but restrained lengthwise by a hinge. The angular unit is a single hinged bellow used in sets of two or three, similar to a two-pin or three-pin hinged structure. The articulated unit is a package comprising two bellows within a hinged framework (**plates 22** and **28**).

Whatever pressure is applied to such a compensated system, the internal pressure tending to open out the bellows lengthwise is held by the tie bars and hinges and restrained from acting as a force on the pipe or on the pipe anchors. The system is therefore truly pressure-balanced (unlike the plain pipe loop) and is particularly suitable for medium and high pressure applications. The movement to be compensated depends only on the angularity taken up by the bellows and the distance between them: very large movements can be accommodated, 500 to 800 mm compensation at one point is not uncommon. In view of the hinged arrangement, which achieves a pressure-balanced system, and the very low thrust required to bend the bellows (rather than compress and extend it) a system with articulated or angular bellows can in fact be treated as though the units did not exist as far as forces are concerned. It will compensate effectively so long as the pipe can move along suitable supports and guides.

The advantages given by the various types of bellows expansion joints have earned them a very prominent place in district heating networks. However such bellows must be selected correctly by the pipeline designer, preferably jointly with the manufacturer of the bellows. The completed installation must be most thoroughly checked on site before heat is applied to the compensators so as to obviate the risk of uncontrolled or twisting movement which might cause serious damage.

Certain types of bellows are suitable for suppressing or smoothing out vibration of pumps, compressors and the like. **Plates 17, 19, 20, 21, 23, 24** and **25** show various interesting bellows applications.

Packed expansion joints

Packed expansion joints are mainly employed to compensate for the move-

20 years in district heating, 500 schemes and 250,000 compensators.

Operating conditions don't come more demanding than they do in Nordic district heating.

Neither do customers.

That's why we're so proud of the fact that Teddington Bellows expansion compensators are the ones they keep specifying. In 20 years we've supplied more than 250,000 compensators for more than 500 cities and towns in four countries.

From Stockholm to Helsinki to Esbjerg they know what's good for them.

Do you?

Teddington

A MEMBER OF THE UGI GROUP OF COMPANIES

Teddington Bellows Limited

Expansion comes naturally

Teddington Bellows Ltd.,
Pontardulais, Swansea SA4 1RP, West Glamorgan, U.K.
Telephone 0792 882591. Telex 48155.

are you resourceful?

APPLIED SOLAR ENERGY

A guide to the design, installation and maintenance
of heating and hot water services

David Kut & Gerard Hare

- a complete introduction to solar energy systems
- practical guidance on installation and maintenance of solar heating
- comprehensive coverage of solar energy for hot water supply, space heating, swimming pools systems and new applications
- fully illustrated with photographs and line drawings
- invaluable as a manual for services engineers, related building professionals—and advanced DIY enthusiasts alike

160pp A5 Illustrated cloth £7.95

The Architectural Press
9 Queen Anne's Gate, London SW1H 9BY

ment in steam mains (**8.12**). They are most popular in the USA where steam systems are much more widely adopted than in Europe. A joint of this kind consists of a sliding tube made from crack-free hard chromium plated steel which slips into the unit itself. Self-lubricating stuffing boxes are provided to achieve a good seal. The US company Advanced Thermal Systems Inc manufacture standard units for steam pressures up to about 20 bars and temperatures of up to 320°C (579°F) (**plate 26**). Units for even higher pressures and temperatures are manufactured as specials (**plate 27**). The joints are suitable for internal pipeline diameters of between 40 mm and 600 mm, with traverses per slip in modules of 100 mm up to a maximum of 900 mm. The most common designs have slips of up to 30 mm.

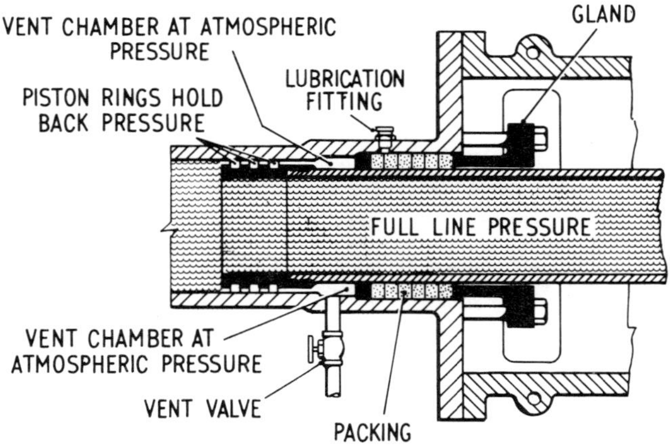

8.12 Piston ring expansion joint unit

Pipelines are usually butt-welded or flange-jointed to expansion joints of this kind. It is absolutely essential that the correct guides are fitted to pipelines to ensure that the expansion joints are not subjected to sideways pressures. Single and double joints complete with anchor base are available with service (branch) connections, as are expansion joints with an anchored 45° or 90° elbow.

The traverse chamber and stuffing box of these joints are made from seamless steel A53 Grade B or equivalent coated externally with aluminium enamel. The slip is machined from similar steel pipe, coated with twin layers of 0.025 mm thick chromium to comply with ASTM standard B 499. The stuffing box is factory-packed with self-lubricating PTFE-asbestos semi-plastic packing and ring packing. The maximum frictional resistance is given by the following table:

TABLE 8.1 PACKED EXPANSION JOINTS

Pipe size (mm)	Expansion joint thrust area (mm^2)	Frictional resistance of joint (kg)
38.1	1751	409
76.2	9940	1090
203.2	36 515	2179
304.8	79 844	3269
406.4	125 738	4358
508	196 454	5448
609.6	282 863	6538

The frictional resistance given in the above table is that required to initiate motion in a new joint. Once pipe movement has commenced the required force drops by about half.

Systems without compensators

Proprietary expansion units can represent a costly element in district heating networks and in certain circumstances it may be possible to omit these. One such technique is the Nocomp system developed by the Swedish Rorkonsult for Fjarrvarme i Orebro company. In the application of this system the pipeline is heated after positioning to a temperature about midway between the cold pipe laying temperature and the operational temperature and then rigidly attached at fixed points. As a result the pipeline is in tension when in the cold state and in compression when at the operating temperature, both of these stresses being within the elastic limit of the pipeline system.

Such a compensatorless system is said to be between 10 and 20 per cent cheaper than a corresponding system with compensators but the following conditions must be met to make such a system applicable:

(a) the soil must be able to exert a pressure upon the pipe at every point of its length: a system which has an air gap between pipeline and insulation is not suitable

(b) the pipes must be firmly jointed to the pipe insulation: the joint between the pipeline and the insulation must be capable of withstanding the resultant maximum frictional stresses

(c) the compressive strength and elasticity of the thermal insulation must be capable of withstanding the resultant pressures

(d) the joints between the pipes and between the pipes and the associated valves must be as strong as the pipelines themselves.

The maximum permissible temperature changes with respect to different grades of steel are indicated in Table 8.2 in which T_{lim} is the tensile strength of steel (including safety factor of 1.7) within the elastic limit and Δt_{max} is

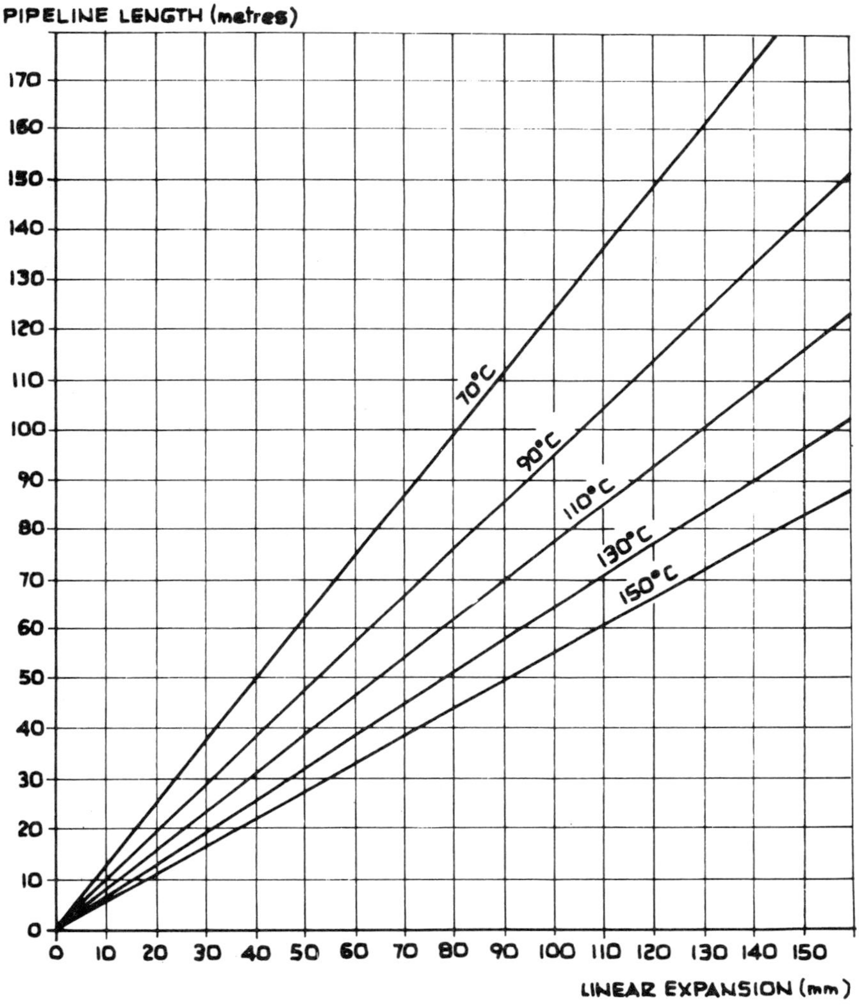

813 Expansion of mild steel pipelines with varying temperature differences

the maximum permissible change in temperature permitted with the appropriate pipeline system within the limit of elasticity of the steel.
Some allowance must be made for any eventual corrosion, the shape of the structure, welded joints etc which, taken together, yield welding factors by which the maximum permissible temperature changes are to be multiplied: these vary from 0.70 to about 0.85.

When the maximum amount of heating up or cooling down has been determined for a given pipeline, the system is heated to the so called 'fix-temperature'. This is set at a point which ensures that maximum and minimum operating temperatures do not lie outside the permitted range for.

TABLE 8.2 PIPELINES WITHOUT COMPENSATORS: MAXIMUM PERMISSIBLE TEMPERATURE CHANGES

T_{lim} (megapascals)	Δt_{max} °C	°F
104.44	42.4	108.3
115.72	47.0	116.6
130.43	53.0	127.4
138.3	56.0	132.8

contraction and expansion. When the pipeline has reached its fix-temperature it is held rigidly in position by tightening up the fixed points.

A recent innovation is the I. C. Moeller 'E-muff'. It is a component which functions only once during the first (start-up) operation of a pipe system. During the installation of an I. C. Moeller district heating pipe system, E-muffs are welded-in at calculated intervals in accordance with the expansion requirement of the pipeline. The system in the ground is then back-filled, with the exception of the E-muffs, and the heating plant is started up. The pipes will expand as calculated and the bellows of the E-muffs will be compressed until the internal steel pipe ends meet. The E-muffs are then locked with circumferential welds of the casings. They are then protected and insulated with long I. C. Moeller joint muffs in the usual way for this type of pipe system. The system is now locked and the natural flexibility of the pipe materials will accept the variations in length that will occur.

The manufacturers of the above system state that the E-system has been checked and tested by them in detail during the last three years in co-operation with the Technical University of Denmark, both in the laboratory and on test sites. The claimed advantages are:

The E-technique simplifies the installation of district heating systems and improves its operational reliability.

Bellows compensators and expansion bends are not required to operate in the pipe systems, and a large number of anchors can be omitted.

A cost reduction over more conventional district heating pipe systems.

AIR LEAKS AND GAS BUBBLES

The advent of plastic district heating pipes has brought about an increased tendency to air leakage because many plastics, unlike mild steel, possess a certain degree of permeability to air.

The solubility of air in water is directly proportional to the pressure and falls with increasing temperature. Table 8.3 gives figures for the solubility of air in water.

When water flows through certain sections of the heating system there is a

TABLE 8.3 SOLUBILITY OF AIR IN WATER AT DIFFERENT TEMPERATURES AT PRESSURE OF 1 BAR

Temperature °C	°F	Solubility of air (parts per million)
60	140	15.5
70	158	14.7
80	176	14.3
90	194	13.3
100	212	12.5
110	230	11.7
120	248	11.0
130	266	10.3
140	284	9.6

sudden release of pressure which causes air to come out of solution. Assuming that there is a pressure loss of about 0.05 bar and that the water was saturated with air, a noticeable release of air occurs: eg at 80°C (176°F) the air released would be about 0.7 parts per million (by weight); or for each cubic metre of water 0.7 grammes of air could be released equivalent to:

$$\frac{22.4 \times 353 \times 1000 \times 0.7}{29 \times 273 \times 8} = 87.4 \text{ millilitres}$$

at a temperature of 80°C (176°F) and at a pressure of 8 bars.

Efficient air bleeding arrangements are essential. Bleed valves should always be positioned at locations where pressure inside pipelines is at a minimum due to sudden increase in flow rates.

Air release valves will generally require to be of the automatic type with a bleed pipe taken to a suitable location for moisture release. Such valves require periodic routine inspection to ensure that they release air and do not dribble water. Manual air vents require to be fitted with air collection reservoirs and air vent valves located in weather-tight accessible positions.

Dissolved nitrogen and other gases cause some nuisance because they form air cushions on parts of the radiator system and interfere with the proper working of pumps, valves and recording systems. The presence of oxygen in the system is however very harmful as it causes corrosion. Oxygen must be eliminated by the addition of hydrazine or other techniques of deoxygenation as described in chapter 10 below.

CORROSION

General considerations

Corrosion damage is in fact one of the major troubles which has beset buried district heating mains, particularly in the United Kingdom where the ground is likelv to be wet (**8.14**). Several systems have been initiated with

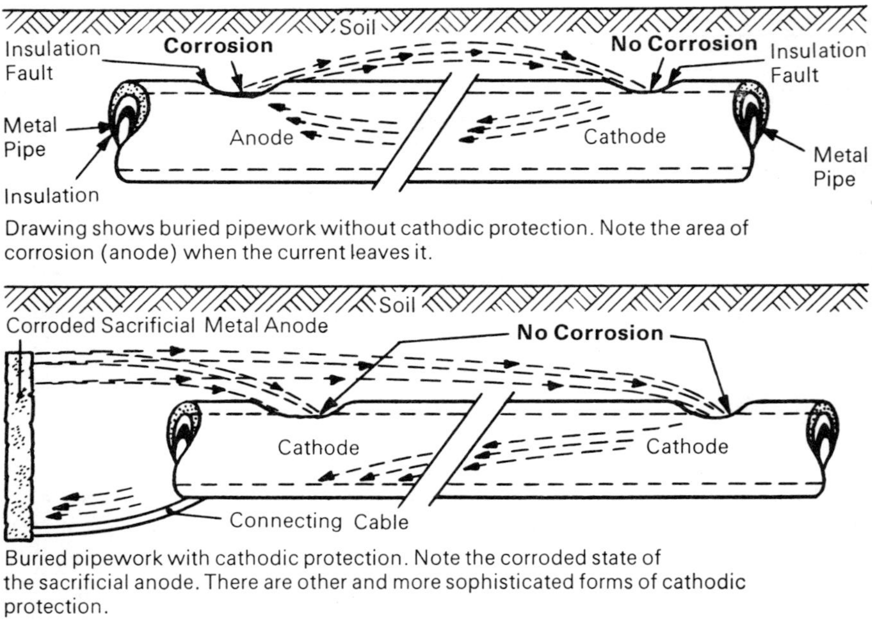

8.14 Corrosion effects with buried pipelines

great hopes but have quickly failed due to corrosion. As with all other chemical reactions the rate of corrosion reaction roughly doubles with each increase in temperature of 10°C (18°F) so that it takes place 128 (2^7) times faster at 90°C (194°F) than at 20°C (68°F). Insulated pipelines in which air is excluded from virtually all the network are particularly badly attacked at positions where water can penetrate to the steel pipeline itself.

Although pipelines of plastic, copper and stainless steel have been introduced for specific purposes most large district heating pipe networks are made of mild steel and will continue to be made of this material because steel offers major advantages over other systems in strength, durability, resistance to ageing and cost.

The internal surfaces of steel pipelines can be kept free from corrosion if the oxygen content of the hot water remains close to zero and if, in addition, the water is always conveyed at a pH value in excess of 9 (as the alkaline water converts the surface layer of iron to magnetite Fe_3O_4 which effectively protects the adjoining layers of steel).

The single most important factor in preventing external corrosion is to stop moisture reaching the surface of the steel pipeline.

When pipelines are laid in gas concrete a certain degree of protection is afforded by the alkaline nature of this material. It does however possess a strong positive capillarity: water must therefore penetrate to the surface of the pipe if the gas concrete gets wet. To prevent corrosion (a) as much dampness as possible must be excluded by effectively draining the duct and by protecting the upper surface of the gas concrete; and (b) any water that does penetrate must be allowed to evaporate through the open pores of the gas concrete assisted by the warmth inside the thermal insulation. It is essential for gas concrete and other hygroscopic materials to be able to breathe.

Mineral wool excludes water by virtue of its negative capillarity caused by its greasy nature which results in an angle of capillary contact greater than 90°. The inherent dangers with mineral wool and similar materials lurk at joints or at imperfections of the insulation. If water can penetrate, even only into a small section of the insulation, corrosion is rapid and exceedingly destructive. Pipelines insulated with mineral wool are recommended only for use in very dry conduits or in enclosed spaces where there can be certainty that no water will ever penetrate to the surface of the mineral wool insulation.

Loose fill powders are all extremely water repellant and can hold water away from pipeline surfaces by capillary repulsion. This safeguard should be augmented by corrosion proofing of the pipelines. Troubles arise when loose-fill insulation materials are not applied properly or where there is excessive ground movement or settlement. For example trenches may have badly constructed bottoms and sides so that parts of the powder charge may fall away leaving a section of pipe exposed to ground water; tamping may not have been carried out thoroughly and in consequence a small section of the pipe is left uncovered; the depth of loose-fill cover may be inadequate; the powder may have been mixed with earth or rubble so as to eliminate its resistance to water; layers of powder may have been moved by underground streams or land slips. In all such cases severe corrosion of pipelines is inevitable.

Prefabricated moisture-exclusive pipelines offer the best solution when contact with ground water or rainwater is expected. These systems are machine-made under close technical supervision and it is unlikely that actual pipeline sections will develop corrosion leaks. Mantle tubes are either rigid plastics, asbestos cement or well-protected steel. All these materials should be completely moisture proof. External factors may however cause penetration of the protective mantle pipe during operation: for example it may be punctured by a sharp stone pushed down by a heavy truck or damaged by workmen excavating in the vicinity. If the external mantle pipe is penetrated moisture can reach the insulation and hence the carrier pipe. By and large however the reported failure rate experienced with pipe in a pipe systems has been significantly less than with the more traditional pipelines. The most likely danger is that water will penetrate because of bad workmanship at the joints. Various techniques have been devised to ensure that corrosion does not occur from this cause. Many prefabricated pipeline systems include a moisture detecting system such as that developed by I. C. Moller of Denmark (described later in this chapter).

Electrical corrosion

Where there is a danger of stray electrical currents from direct-current conductors (electric trains and the like), or as a supplementary precaution against any such possible danger, pipelines may be protected by using sacrificial anodes (**8.15**). These are large lumps of steel or other metals connected to the pipeline to serve as an anode. Corrosion subsequently occurs on this sacrificial anode and not on the pipe. Such anodes are buried in a very corrosive environment using coke breeze surroundings as backfill. Table 8.4 lists the most common ones.

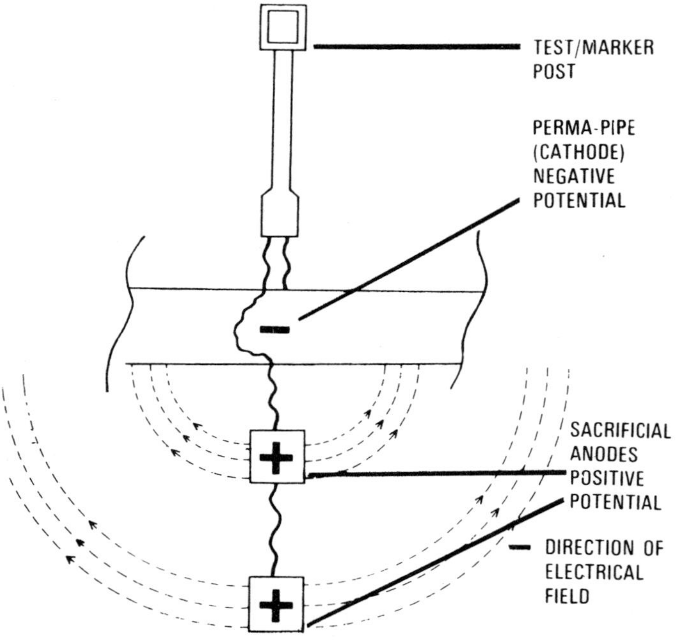

8.15 Sacrificial anode used for pipeline protection

Electrical drainage can be provided by connecting the pipeline system to a source which has an electromotive force (emf) exactly counterbalancing the emf induced by any anticipated stray current. Automatic control devices adjust to this emf immediately the stray currents fluctuate. Electrical drainage is usually applied at the expansion and inspection chambers of the pipeline.

Fretting corrosion

The bearer units of district heating networks may often be affected by fretting corrosion. This occurs when there is wear and tear between two

TABLE 8.4 SACRIFICIAL ANODES

Material	Loss (kg/year)	Maximum working capacity (A/m^2)
Pig iron	35	5
Scrap steel	45	5
Graphite	4	20
Silicon iron	4	40
Aluminium	20	20
Platinised titanium	zero	10 000
Magnesium	35	Depends on nature of backfill
Zinc	55	Depends on nature of backfill
Lead alloys (sea water only)	almost zero	200

pieces of steel so that the surfaces tear and the particles oxidise in the presence of air and under the action of heat. Fretting corrosion is 15 per cent less when air is excluded than when air is present. It has also been established that fretting damage is least when the relative humidity is 50 per cent and that it increases when this is either very low or very high. At 50 per cent relative humidity soft iron hydroxide is formed which provides a lubricating action.

Frettage has often been a major cause of fatigue because it causes cracks to develop: these can be prevented by lubricating surfaces with oil or with thin plastic or rubber films. Copper, tin and cadmium can also be applied to prevent fretting.

Cavitation

Irregular water flow in a district heating system causes damage, partly mechanical and partly chemical, by cavitation. Mechanical damage done in this way facilitates easy access for corrosive liquids which then destroy the metal chemically. Cavitation is best prevented by designing pipeline systems so as to avoid radical pressure differences. Where this cannot be done it is prudent to design the parts most affected by water hammer to be the most cathodic, employing either sacrificial anodes or electrical drainage.

Jointing of dissimilar metals

Iron and steel have the property of being anodic to almost all other metals except zinc. Thus when steel is connected electrically to, say, copper or aluminium, it corrodes far more quickly than otherwise. It is of course well known that steel and other metals should not be jointed together in a corrosive environment, although at times this may be unavoidable. Corrosion acts most rapidly when the surface area of the anode (ie the steel) is relatively small, while the surface area of the cathode (ie copper, aluminium, stainless steel etc) is relatively very large. To slow the corrosion process in such circumstances the surfaces of the steel pipe must be maintained as dry as possible but unprotected. The other metal in contact with the steel should however be surface-coated effectively with plastic, bitumen or rubber. Application of this coating should be particularly thick at joints between the two different metals and it should overlap to not less than 100 mm on the steel pipe so as to ensure that the cathodic effect of the other metal is reduced to almost zero. Any corrosion that does occur will then spread along a wide area of the steel pipe, instead of being concentrated (as would be the case if the pipe were painted and the paint damaged by eg a small pinhole).

The best way of protecting joints between dissimilar metals from corrosion is to insulate them from each other electrically by inserting a length of plastic pipe between them.

SINGLE-PIPE SYSTEMS

The Russian system

An important recent advance in the design of district heating pipeline networks is the single-pipe system developed in the USSR (**8.15, 8.16** and **8.17**). In this system very hot demineralised, filtered and mechanically de-aerated pressurised water is supplied to the district network at a temperature of between 150 and 200°C (302 and 392°F). This is injected into the consumers' heating system which operates at the customary 60 to 70°C (140 to 158°F). It then flows at somewhat lower temperature into domestic hot water uses and thence to waste. Ideally the volume of hot water consumed should equal that of the very hot water supplied from the heat generating station. However, even if the volume of the latter is greater it has proved cheaper to waste the excess water at the low temperature of about 50 to 60°C (122 to 140°F) rather than incur heavy expenditure on constructing return mains for recycling the water to the station.

The development of the single-pipe system in the Soviet Union has had a considerable effect upon the economics of constructing district heating installations. It must however be appreciated that unless space heating and domestic hot water use are in near-balance the use of the single-pipe system can involve the discharge to waste of large quantities of water. Such an arrangement is therefore unlikely to appeal to those water supply utility companies which have to draw on scarce or costly water sources.

The typical method of operating a single-pipe system in towns is as follows:

(a) Superheated hot water is fed into the heating network either from a combined heating and electricity generating plant (CHP) or from a central

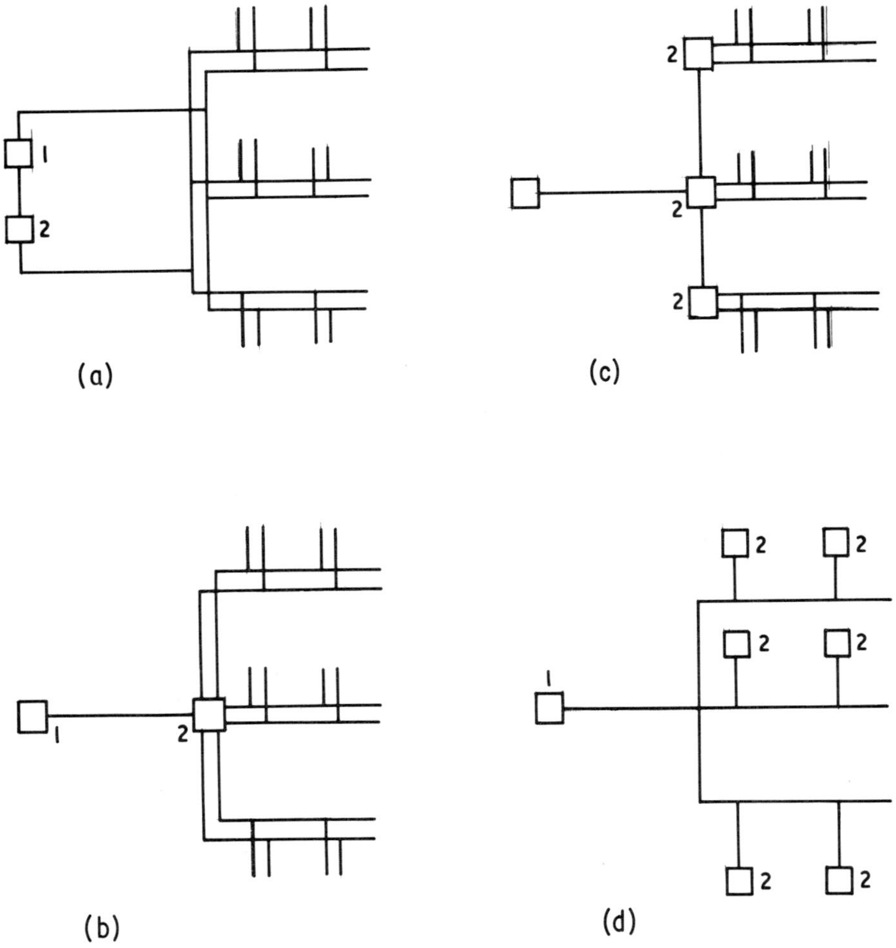

8.16 Russian single-pipe system: various connections

boiler plant and arrives at the consumer after passing through a gate valve, a sediment trap, a water meter/thermometer and the temperature and mixing control of the heating and hot water supply systems.

(b) The water then enters the radiators and is circulated by pumps at the customary circulation temperature. Heat can be supplied to the radiator system via a heat exchanger or by direct mixing. The injected quantity of superheated water is very small in comparison with the amount of circulating water.

(c) The cooler water discharged from the space heating system then flows into the domestic hot water supply mixing tank, or through an automatic valve set at minimum room temperature. Hot water in excess of the con-

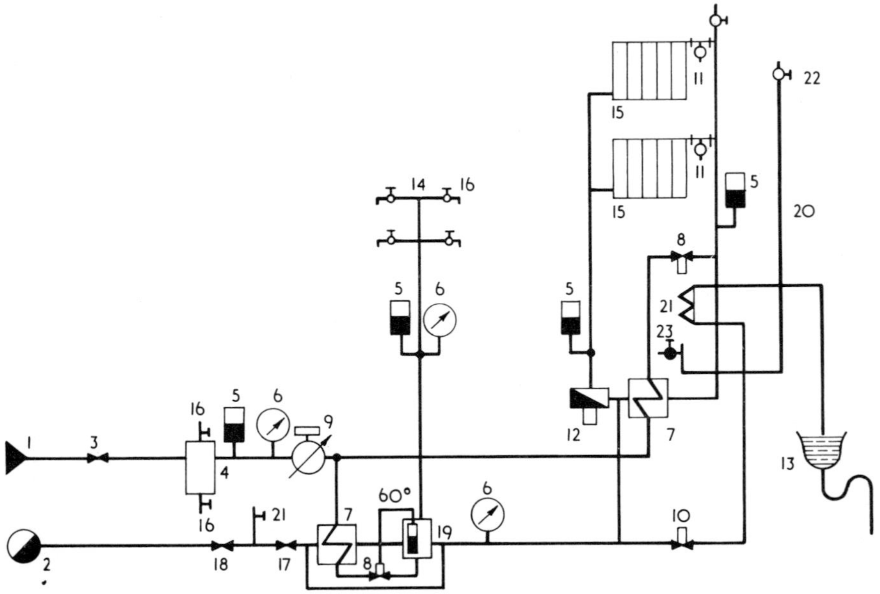

8.17 Layout of Russian single-pipe system at consumer's circuit
1 *Heating network*, **2** *Water main*, **3** *Valve*, **4** *Sediment tank*, **5** *Thermometer*, **6** *Pressure gauge*, **7** *Water cooler, eg from 200° to 150°C (m^2 per 1000 inhabitants)*, **8** *Regulating valve*, **9** *Calorimeter*, **10** *Hydrostatic head valve*, **11** *Automatic room temperature regulator*, **12** *Low pressure pump*, **13** *Drain line*, **14** *Hot water supply system*, **15** *Radiators*, **16** *Taps*, **17** *Return valve*, **18** *Valve in the water main*, **19** *Gas cylinder with boiling liquid (butane)*, **20** *Oil-filled pipe*, **21** *Feedback*, **22** *Oil filling valve*, **23** *Valve for release of oil and for compressed air*

sumer's needs is drained off to waste.
(d) When the pressure in the system has been reduced cold water from the town's water main enters the hot water supply mixer through a regulator. The static head of the heating system is maintained by an automatic valve installed in the water overflow line. The central regulator for controlling the volume of water entering the heating system is in the form of an oil-filled pipe 12 mm in diameter which passes vertically through all control departments. On being heated the oil expands and compresses the air in the bellows of the automatic regulator. The feedback in the overflow line warms the arm ahead of the bellows and maintains the valve closed until the temperature in the control compartments has fallen by about 1°C (1.8°F). The main link in the automatic process of this system is a simple room thermostat which uses a bimetallic spring and valve.

The equipment installed in Moscow has maintained average exit temperatures of 40°C (104°F) during the heating season with an inlet temperature to the system at 95°C (203°F). Such low exit temperatures have materially reduced the run-off heat losses. The most economical supply temperature of hot water in a single-pipe system has been established at 200°C (392°F).

It has been found that the water run-off from the heating system can be used for domestic hot water consumption for 18 out of 24 hours even during very cold periods. During the day buildings can be overheated to some extent and overnight, when the domestic hot water demands are lower, some of the hot water supply can be shut off. It has been established that even in the extreme Russian climate the return of water from the heating system is wholly uneconomic and that a single-pipe system therefore represents a great advance. It ought to be even more advantageous than in Russia for countries which have milder climates and which consequently require a smaller proportion of heat for space heating than for hot water consumption. During summer when space heating is at a minimum nearly all the heat is used for domestic hot water in any case and the supply temperature can therefore be lowered drastically.

The advantages of single-pipe networks are:

(a) the warm water, which is of limited use at the power station but can be used by the consumer, is consumed locally instead of being returned to the station

(b) the rate of pipeline construction is speeded up because single-pipe networks can be laid faster than double-pipe networks

(c) capital costs on pipelines, excavations and insulation are greatly reduced

(d) return pipes tend to corrode more easily than flow pipes (being at a lower temperature) and moisture can therefore reach the surface of the pipe more easily. With the very high temperature flow mains used in the single-pipe system external pipe corrosion is reduced to an absolute minimum

(e) as the water in the single-pipe system is de-aerated before being passed into the pipeline, corrosion inside the pipeline is reduced to a minimum. In a conventional hot water supply system, where a heat exchanger is used to heat up aerated town water heavy corrosion often takes place

(f) twin pipelines, pumping water at lower temperatures with higher viscosities, have much higher pumping costs than a single-pipeline system using very hot water.

The only material difference between laying single hot water pipes and laying conventional cold water mains is that in the former case it becomes necessary to install expansion compensators and to use thick thermal insulation materials which are poured into the trenches over the pipelines. The average thickness of such thermal insulation is 200 to 300 mm. One of the most widely used insulation materials for this application is a mixture of porous clay filling heated to a temperature of about 1000°C (1832°F) and waterproof roasted peat (obtained by drying peat at about 300°C (572°F)).

The high temperature of the circulating water maintains the insulation in a dry condition so that little or no corrosion has arisen.

The quantity of the heat supplied to the consumer is measured by a special calorimeter (developed by Yakimov) which differs from usual heat meters in that only the temperature of the consumer's heat supply pipe is measured.

Water purification

Two other features of the single-pipe system are that a town's water supply pipes can be smaller and that the municipal water works is relieved of the need to purify all the water for consumption. In the single-pipe system water is drawn from a river and purified at very high temperatures by coagulation, filtration and removal of cations by ion exchange processes; it is de-oxygenated and the CO_2 is removed before it is fed into the pipe network. At the prevailing high temperatures bacteria are eliminated and there is therefore no need to ozonise the water, as would be necessary with conventional water treatment processes.

CHP plants and the single-pipe system

Combined heat and power plants have hitherto generally been sited in the centre of towns. But they have usually been of small capacity and have therefore tended to be far more expensive (per megawatt output) to construct than large plants. The new super power stations (electrical output of 2000 megawatts upwards) must, on the other hand, for many reasons be sited some distance away from towns: they must be near adequate water supplies; they must have relatively free access for fuel transport; they must not release soot and sulphur dioxide close to densely inhabited areas; they occupy, with their cooling towers, switch gear, transformers etc, a lot of ground and may be regarded as environmental eyesores. The consequence is that such large power stations require far longer district heating mains than has previously been considered feasible. The single-pipe system, which is relatively inexpensive and operates virtually under corrosion-free conditions, is therefore almost ideal for use in conjunction with modern giant power stations.

A 300 MW capacity power station with combined heat and power generation supplies as much as 330 MW of heat which, under Russian conditions, would require a town of at least 100 000 heat consumers. It is often necessary therefore to build a combined heat and power station to supply several towns in conjunction with very long heating networks.

Single-pipe networks using heat from combined heat and power stations are best run in conjunction with a twin pipe system from town boiler plants to obviate excessive overflow of water into sewers at times when the space heating requirements are well in excess of the domestic hot water consumption. Combined heating and power stations should supply the base load of the heating system and the smaller local boiler plants the peak heat load. In practice while the pipelines are designed for hot water temperatures up to 200°C (392°F), the average temperature of the water on the outlet side is 102°C (216°F) during the summer and up to about 180°C (356°F) during the coldest period in winter (**8.16**). At times of peak demand the various auxiliary boiler plants are activated and these are commonly situated much closer to the consumer town.

Single-pipe heating is currently being adopted in many Russian cities and towns and a number of alternative layouts are being employed. Under Russian conditions it has proved economic to pump hot water through immense distances (up to 224 km from a 500 MW heat and power station) from combined heating and power stations. Although cost considerations are bound to differ in other countries, there is no doubt that much longer pipe runs are feasible with single-pipe systems than are conventionally accepted for twin pipe systems (**8.17**).

Single-pipe working implies that large quantities of water must be supplied to the boilers of CHP stations and other heating centrales in a condition which avoids scaling-up the heating element. A novel method of achieving this is by magnetic water softening. The hard water is passed through a permanent magnetic field and the precipitation of calcium salts is induced by providing conditions for multiple precipitation nuclei. The precipitated calcium carbonate is in the form of a loose sludge which can easily be filtered off leaving the water free of scale-producing salts.

Single-pipe working permits the use of newly developed types of boilers, which operate some 20 per cent more economically than conventional ones. The emerging flue gases pass into a recuperator section and are washed by a water spray dosed with calcium hydroxide. This cools the flue gases to about 32 to 37°C (90 to 100°F) and extracts the CO_2 and SO_2 gases which react with the calcium hydroxide in the water spray to form calcium carbonate and calcium sulphite. The hot impure water is then passed through a magnetic field where the impurities are precipitated and filtered off. The purified water, containing both the sensible heat of the flue gases and the latent heat of the water vapour driven off during the combustion of the fuel, is then fed to the boilers. The flue gases which finally emerge and are exhausted to atmosphere are cold and virtually free from sulphur dioxide, grit and soot.

DETECTION AND REPAIR OF FAULTS

District heating mains systems are apt to develop faults caused by leakage of the hot water medium from the pipeline into the insulation layer via imperfectly formed joints, cracked welds etc or by water penetrating the insulation layer from the outside through its surrounding envelope due to a manufacturing fault or damage from earth movement.

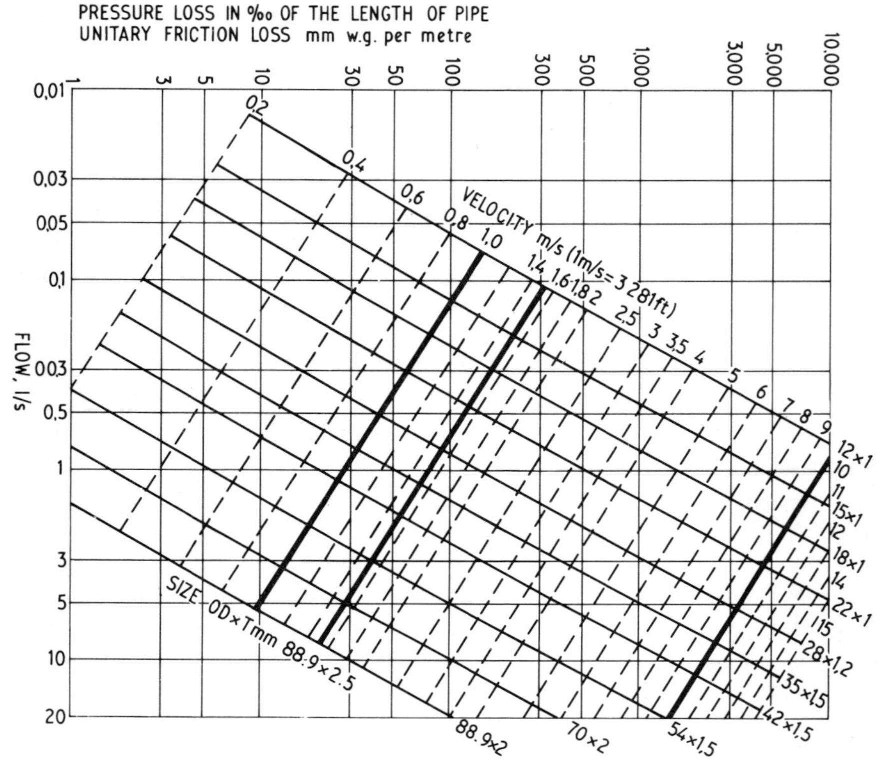

8.18 Nomogram for hot water flow

I. C. Moller of Denmark have developed an electronic alarm system which can be fitted to all types of pipelines and is designed to detect the presence of moisture or leaks in any type of installed pipe system. It can be used both for testing the installation before the pipes are buried in the ground and for detecting leaks in operating installations before serious corrosion damage has developed.

The system consists of two copper wires surrounded by a water absorbing felt inserted into the insulation layer of the pipeline during manufacture. In the case of prefabricated pipelines insulated with foamed polyurethane the detecting wires are inserted prior to the foaming process. The wires are joined together by special connectors to provide a continuous circuit. Should the felt between the two copper wires get wet, the electrical conductivity of the system rises. Audible and visual alarms are available. An associated digital read-out system has been developed which relies on an impulse reflectometer technique: this indicates that there is a fault and shows its position from a given point with an accuracy of plus or minus one metre.

Major heat leaks can also occur from faulty insulation alone without any leakage or water penetration. A method of infra-red photography (developed by the Danish firm of Bjeld and Lauridsen and handled in the UK by Tylin CAE Ltd) is of considerable assistance in detecting heat loss of any kind. An infra-red camera is employed to produce a heat picture of the entire district heating system in a large number of 35 mm exposures. Positions of excessive heat loss show up much higher exposure rates than areas where heat losses are slight. This system can be used to check upon the thermal insulation properties of pipelines and of buildings. In Denmark it is common for district heating systems to be surveyed annually by this thermographic method with a view to detecting any faults which may have arisen before much damage is caused. Pipeline water leaks show up as white patches on the thermographic negative.

The following other methods are available for the determination of leaks in district heating pipeline networks:

Observation of pressure and flow data: by deploying a number of flow meters and pressure gauges on each section of the network it becomes possible to find out where irregularities due to hot water or steam losses occur. Such a method is however none too accurate and only gives a rough guide at the time when leakages are considerable.

Sound measurements: leakage of the medium from pipelines is often accompanied by sound waves and the location of leaks can therefore be observed by means of a sound pressure meter. Electro-acoustic sound measurements measure the sound generated by leakage and transmitted along the pipeline. This sound reduces as the distance between the source of the leak and the electro-acoustic probe is increased, so that it is then possible to assess the distance of the leak from the measuring position. This method is unsuitable for systems which incorporate compensators or valves in the pipe lines as such components greatly reduce the energy of sound emitted. Another difficulty inherent with this method is the need to expose part of the pipeline. It is used with exposed pipelines which traverse cellars or tunnels.

Use of chemicals: a certain quantity of a chemical, such as sodium phosphate, is injected into the district heating network and the quantity of this chemical present in the return line is then analysed by the use of ammonium molybdate in a spectrophotometer. The analysis provides a clear indication of the amount of extraneous water which has leaked into the district heating network but it does not indicate the losses from the system to the surroundings.

Freon gas: the leaking pipeline network is drained and blanked off at both ends and the system is then filled with Freon gas under considerable pressure. The Freon will seek out even the smallest holes, and the location of the leak can then be detected readily by means of a highly sensitive electronic halogen detection device. This method can also be used for the detection of leaks in steel mantle pipeline systems: the Freon gas is pumped into the interstices between the conveying pipe and the mantle pipe thus making it possible to detect leaks in the mantle tube.

Radioactive isotopes: although these have been used successfully with cold water pipes there are some difficulties in their use with the insulated pipeline systems used in district heating. Highly active radio-isotopes emitting hard gamma radiation must be used and these can only be employed within areas to which there is no public access. Such methods have therefore not been widely employed with district heating systems but efforts have been made to find suitable isotopes and measuring instruments.

Surface thermography: leaks can be located by measuring the surface temperature of the earth above the pipeline system by means of an infra-red recorder to an accuracy within 0.2°C (0.36°F). The intensity of infra-red radiation is measured 16 times each second and is recorded by a camera as a 'heat map'. Wherever the insulation layer has been saturated by leaking hot water the district heating network is outlined in white. An accurate assessment of the position of any leak can be achieved.

Continuous check: it has been found valuable to monitor district heating networks continually so as to obviate major shut-downs caused through major failures. Two monitoring systems are available (a) the use of twin copper wires in a hygroscopic cover able to detect immediately when the insulation becomes waterlogged at any point, and to indicate the exact position of the leak; and (b) sonic testing of the system by means of a continuously applied tone spectrum and its detection with an electronic echo-test device: any fractures in the pipeline or leakages into the insulation layer are then readily detected.

When leaks have been detected on the steel pipeline which carries the medium there are alternative approaches to the required repair:

(a) the leaking section can be isolated and this can be followed by welding, brazing or replacement of the affected section. Such repair is facilitated by the existence of a ring-main system where the overall heat supply is not interrupted;

(b) when it is impracticable to drain the pipeline a system such as the Furmanite technique can be used. In this the affected pipe section is covered over by a substantial steel shuttering device which is sealed hermetically to the pipe and positioned tightly in place by means of U-bolts. A thermosetting plastic mass is injected via a valve in this shutter at a pressure approaching 400 bars. After setting, the shuttering is removed leaving a tight and permanent repair;

(c) for the smaller and medium size mains (say up to 250 mm) the pipe freezing technique can be employed. In this, a plug of ice is formed by the application of CO_2 freezer pads at either side of the leak, permitting repair work without the need to drain the pipe circuit.

CALCULATIONS RELATING TO PIPELINES

Insulating layers

When there are several insulation layers on a cylindrical pipeline (**8.19**) the following formula applies:

$$Q = \frac{(t_1 - t_2)}{r_o\left(\frac{\ln r_1/r_i}{1} + \frac{\ln r_2/r_1}{k_2} + \ldots \frac{\ln r_o/r_{n-1}}{k_n}\right)} \text{ W/m}^2$$

where

r_i and r_o are the inside and outside radii respectively (m)

r_1, r_2 etc designate the remaining radii

k_1, k_2 etc designate the k-values of the insulation materials (W/mK).

Example: Find the heat transferred through the insulation layer of a pipeline of internal diameter 200 mm, which is insulated with a 50 mm layer of insulation material 100 mm thick with k-value 0.095 W/mK, and a final layer 20 mm thick with k-value 0.68 W/mK. It is assumed that the internal surface temperature is 150°C (302°F); the external surface temperature is 25°C (77°F). (The thickness and reciprocal k-value of the pipe itself is generally ignored.)

$r_i = 0.1$ m
$r_1 = 0.15$ m
$r_2 = 0.25$ m
$r_o = 0.27$ m

Therefore:

$$Q = \frac{(150 - 25)}{0.27\left(\frac{\ln 0.15/0.1}{0.13} + \frac{\ln 0.25/0.15}{0.095} + \frac{\ln 0.27/0.25}{0.68}\right)} \text{ W/m}^2$$

$= 53.78$ W/m^2 surface of pipeline

Heat losses from buried pipelines

When computing the heat losses from a typical pipeline (**8.20** and **8.21**) it is possible to calculate the heat loss of the pipeline according to the following equations:

$$Q = \frac{(t_p - t_o)}{R_i + R_a + R_b + R_c + R_e}$$

where

Q is the heat loss per metre run of the pipe (watts/metre)

t_p is the pipeline temperature (°C)

t_o is the outside air temperature (°C)

R_i is the thermal resistance of the pipeline insulation (m K/W)

R_a is the thermal resistance of the pipeline insulation surface (m K/W)

R_b is the thermal resistance at the inside of the culvert channel (m K/W)

R_c is the thermal resistance of the culvert (m k/W)

R_e is the thermal resistance of the soil and the surface of the ground (m K/W).

When the whole culvert is filled with insulation material then $R_a = R_b = 0$. The heat transfer through culvert and soil is often the same and it is therefore possible to amalgamate R_c and R_e.

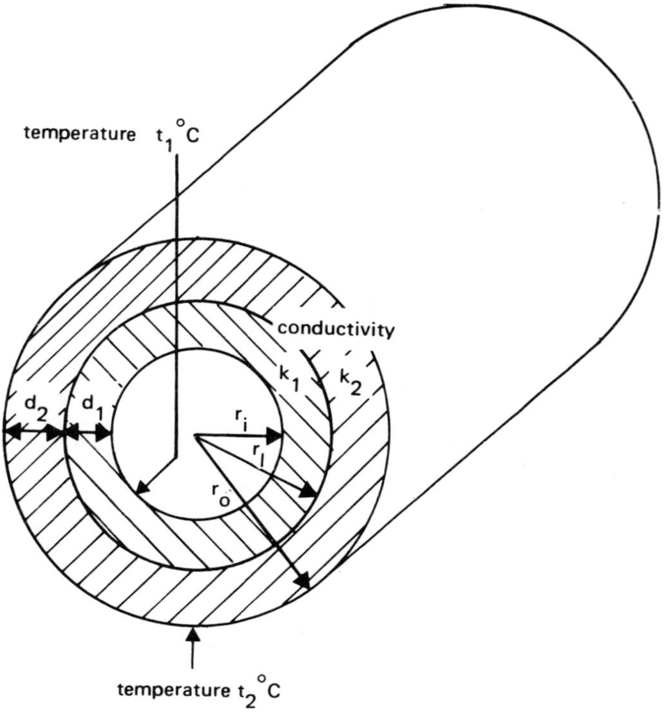

8.19 Pipelines with multiple insulation layers

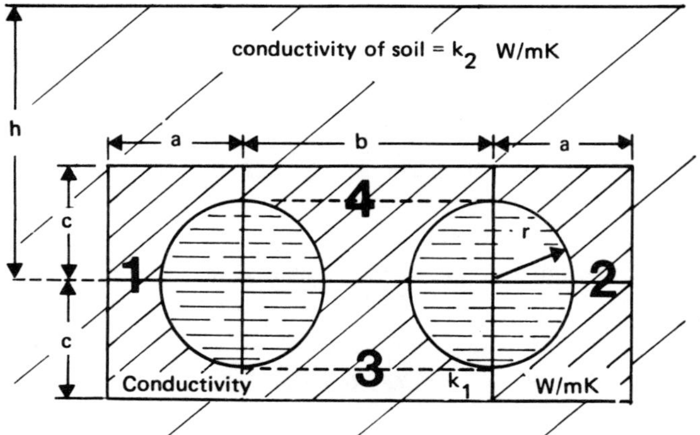

8.20 Evaluation of heat loss from buried pipeline

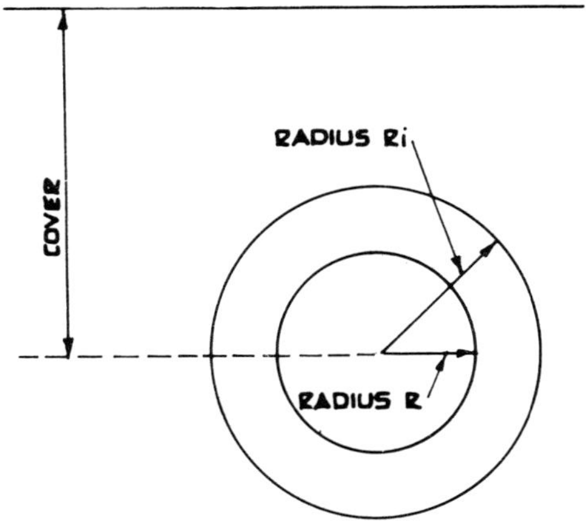

8.21 Heat losses from buried pipeline

The difference in temperature between the flow and return lines in a district heating system is usually not great (about 30 to 50°C (50 to 90°F)) and it is common to use the average temperature of the two lines as t_p.

The average external temperature over the year can be computed from degree-day or from meterological data: eg for London (UK) the degree-day figure is 2557 on a base of 18°C (64.5°F); so the average external temperature over the year may be computed as:

$$18 - \frac{2557}{365} = 11.0°\text{C } (52°\text{F})$$

Thermal resistance of insulation (R_i)

When there is a pipe with a concentric insulation, the following equation applies:

$$R_i = \frac{1n\,(r_i/r)}{2\,\pi\,k}\ \text{m K/W, where}$$

r_i is the radius of the pipeline insulation (m)
r is the radius of the metal pipe (m)
k is the thermal conductivity of the insulation material (W/m K).

In most cases of district heating networks there are twin pipelines and the insulation is not in a circular section. An approximation can then be made by dividing a typical pipeline (see **8.19**) into four parts: parts 1 and 2 can be amalgamated to resemble a single tube embedded in thermal insulation. The average radius of the insulation is thus:

$$\frac{a + c + 2\,(a^2 + c^2)^{\frac{1}{2}}}{4} = r_i$$

Example: in a pipeline of 400 mm diameter (r = 200 mm), a = 350 mm,

X = 450 mm, k = 0.09 W/m K: then r_i can be evaluated as:

$$\frac{0.35 + 0.45 + 0.2\,(3.5^2 - 4.5^2)^{\frac{1}{2}}}{4} = 0.342 \text{ metres}$$

$R_{i,2}$ is therefore:

$$\frac{1n\ 0.342/0.2}{2 \times 0.09} = 0.949 \text{ m K/W}$$

Parts (3) and (4) can be considered as simple areas and the thermal resistance is evaluated as such.

$R_3 = R_4$ therefore:

$$\frac{(c - r)}{b\,k} \text{ mK/W}$$

If b is 700 mm, then R_3 is

$$\frac{0.45 - 0.2}{0.7 \times 0.09} \text{ mK/V} = 0.992 \text{ mK/W}$$

R_i can be evaluated from:

$$\frac{1}{R_i} = \frac{1}{R_{1,2}} + \frac{1}{R_3} + \frac{1}{R_4}$$

Inserting the above data:

$1/r_i = 1/0.949 + 1/0.992 + 1/0.992$ W/m K, or

$R_i = 0.326$ m K/W

Thermal resistance of soil round a buried pipe

The standard applicable expression is the well-known Louden equation:

$$R_e = \frac{1}{2\,\pi\,k_e}\ 1n\,(h/r\,(1 + (1 - (r/h)^2)^{\frac{1}{2}}), \text{ where}$$

h is the vertical distance between the centre of the pipe and the surface of the soil (m)

k_e is the thermal conductivity of the soil (W/m K)

Assume that h equals 1.6 m and the conductivity of the soil (clay with about 30% moisture content) is 4.55 W/m K.

R_e is therefore:

$$\frac{1}{2 \times 4.55}\ 1n\left(\frac{1.6}{0.2}\,(1 + (1 - (0.2/1.6)^2)^{\frac{1}{2}}\right)$$

$= 0.304$ mK/W, therefore

$R_i + R_e = 0.326 + 0.304 = 0.630$ m K/W.

The heat loss of the district heating network can now be computed when the flow and return temperatures are known. It is assumed that the water flows out at 110°C (230°F) and returns at 65°C (419°F); an average temperature (t_p) of 87.5°C (189°F).

The average annual external temperature (t_o) has already been determined above at 11.0°C (52°F).
Therefore the heat loss (Q) is:

$$\frac{87.5 - 11.0}{0.630} \text{ W/m}$$

or 121.4 watts per metre run of the system.

Pressure losses in hot water pipelines

The pressure losses in pipelines (**8.22**) are evaluated from the following general formula:

$$\triangle P = \frac{4 f v^2 D}{d} \text{ pascals/metre, where}$$

f is the friction factor (dimensionless)
v is the flow velocity (m/s)
D is the density of the fluid (kg/m³)
d is the pipeline diameter (m)
The value of f is determined from the Reynold's number, which is a dimensionless function expressed by N_{Re}, where

$$N_{Re} = \frac{v d D}{\mu}$$

where μ is the viscosity of the fluid (kg/m s).

Streamline flow

If the Reynold's number is less than 2000, then the value of f is simply $8/N_{Re}$.

This condition is met only very rarely with district heating pipelines as it can apply only to small-bore pipes and sluggish fluid velocities.

Turbulent flow

When the Reynold's number is above 2000, the flow is turbulent. Over the range of N_{Re} = 3000 to 3000 000, the Koo equation is used for smooth copper, lead, glass and brass pipes:

$f = 0.00140 + 0.125/(N_{Re})^{0.32}$

The more accurate von Karman formula is as follows:

$1/f^{\frac{1}{2}} = 4.0 \log_{10} (N_{Re} f^{\frac{1}{2}}) - 0.40$

For rougher pipelines, such as the mild steel drawn pipes used in district heating practice, the von Karman formula is modified to:

$1/f^{\frac{1}{2}} = 3.2 \log_{10} (N_{Re} f^{\frac{1}{2}}) + 1.2$

The von Karman formulae must be solved by trial and error methods or by a computer programme.

Worked example: Water at 100°C (212°F) flows through a district heating main of 100 mm internal diameter at a velocity of 1 metre/second. Assuming that the viscosity of water at 100°C (212°F) is 0.000282 kg/m s and that the pipeline is completely smooth, determine the pressure loss per metre run.

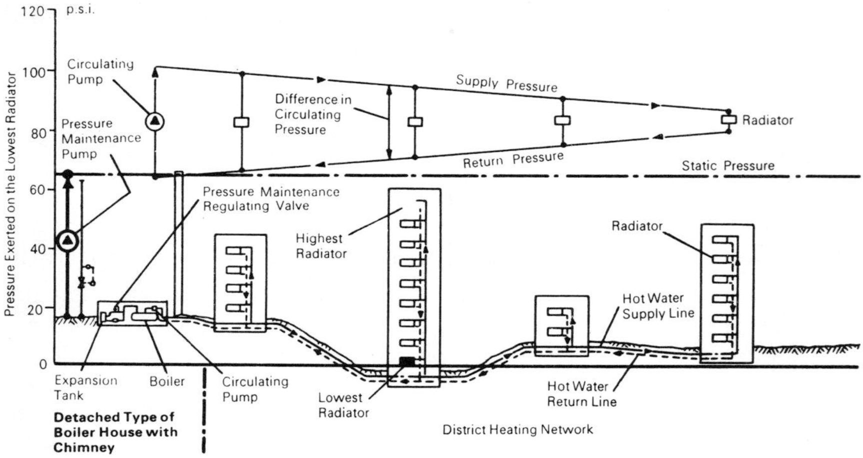

Above: a diagram of a 'closed' heating plant design. Static pressure in this system is produced by pressure maintenance pump, with the expansion tank sited within the boilerhouse.

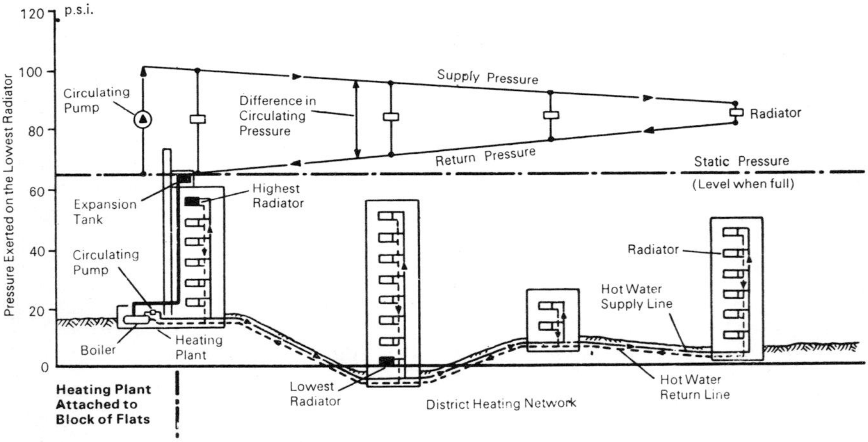

Above: a diagram of an 'open' heating plant design. It can be seen that the heating plant and its expansion tank should be sited close to, or incorporated in, the building containing the highest radiator in the system.

8.22 There is a constant reduction in pressure head when laying a district or group heating circuit

The density of water at 100°C (212°F) is 958.4 kg/m³. Reynold's number is then:

$$\frac{1 \times 0.1 \times 958.4}{0.000282} = 339\ 858$$

Using the Koo equation for perfectly smooth pipelines:

$f = 0.00140 + 0.125/(339\ 858)^{0.32}$

$= 0.00352$

Using the modified Von Karman equation f has to be evaluated by a number of trial and error steps. With the given Reynold's number an f value for mild steel pipelines of 0.00415 is obtained.

The head loss per metre run of pipeline is therefore

For smooth copper pipes:

$$\frac{4 \times 0.00352 \times 1 \times 958.4}{0.1} = 134.9 \text{ pascals/metre}$$

For mild steel pipes (using the friction factor obtained from the von Karman equation):

$$\frac{4 \times 0.004372 \times 1 \times 958.4}{0.1} = 167.6 \text{ pascals/metre}$$

It is inconvenient to evaluate the von Karman form of friction factor. If the Reynold's number of flow is above about 100 000 the following approximation may be adopted: the smooth pipe friction factor f is evaluated by the Koo equation, multiplied by 1.18, to obtain the friction factor applicable for mild steel pipelines. At low Reynold's values, such a method is not recommended, as the effect of surface roughness is far less, and the head losses of the perfectly smooth and the rougher pipes tend to be closer.

Head losses of pipeline components

All bends, constrictions, T-joints etc which form part of a pipeline can be evaluated (or converted) as equivalent lengths of straight pipelines. It is very difficult to establish these theoretically: the equivalent lengths are therefore best determined experimentally.

To evaluate the head loss of a given pipeline system one has to add the length of the straight pipeline to the sum of the equivalent lengths of such bends, expansion units, constrictions, etc to obtain the total effective length of the pipe system.

The velocity of flow (v) is determined by dividing the rate of volume flow (G m^3/s) by the cross-sectional area of the pipe (m^2).

If the entries and the discharge positions of the pipes are at different levels the required pump head is supplemented by 9.8 $\triangle H$ pascals for a discharge position higher than the entry. If the discharge is below the entry, 9.8 $\triangle H$ is subtracted from the required pump head. $\triangle H$ is the difference in height (m) between the entry and discharge location.

Computing for steam mains and district networks

Steam is a compressible fluid: hence there is a continuous change in density as it traverses the pipeline. In consequence calculations related to steam mains are more complicated than those related to hot water mains and the calculations for major networks are most conveniently progressed and processed by computer (see the computer program in Appendix E). For simple networks, it is reasonable to assume the arithmetic mean of the inlet and outlet pressures.

The two Kirchhoff laws are employed for the calculation of networks: At each junction of a network the heat added equals the heat lost. In each network component the sum of the pressure losses (friction in pipelines fitting and valves) equals the sum of the pressure gains (achieved by means of pumps etc).

In addition, use is made of the pressure loss equation:

$\Delta p = C\,Q^2$, where

Δp is the pressure loss (pascals or $kg/s^2\,m$)

C is the pipeline resistance (1/kg m)

Q is the flow rate (kg/s)

This can be expressed in differential form as:

$$dp/dx = \frac{-f\,w^2}{2\,V\,d}, \text{ where}$$

dp/dx is the fall in pressure per unit length

f is the friction factor of the pipe (dimensionless)

w is the flow rate (m/s)

V is the specific volume of the liquid (m^3/kg)

d is the internal diameter of the pipeline (m).

When hot water is the heat fluid the volume flow rate and the dynamic viscosity remain constant for the entire network calculation.

When steam is the heat fluid the calculation becomes much more complicated:

If the pipelines are well insulated, any changes are regarded to be adiabatic and pv^n constant. Then,

$t = (t_o + 273.15)\,(p/p_o)^{n-1/n} - 273.15$, and

$v = v_o(p_o/p)\;1/n$, where

n equals 1.31 for steam

v_o is the specific volume at the steam entry to the network (m^3/kg).

p_o is the initial pressure (pascals)

t_o is the initial temperature (°C)

v, p and t denote (in similar units) the final specific volume, pressure and temperature respectively.

If Reynold's number for the flow is above 2320 the Prandtl-Colebrook equation for the friction number can be applied as follows:

$1/f^{\frac{1}{2}} = -2\log\,(k/3.71\,d + 2.51/N_{Re}\,f^{\frac{1}{2}})$

f can be determined from this by an iterative technique.

Head losses in steam lines

Unlike water, steam is a compressible medium.

Let P_1 = initial steam pressure in bars

P_2 = final steam pressure in bars.

The head loss can then be calculated to amount to:

$$P = \frac{4\ f\ v^2\ D}{d}\ \frac{(1.617P_i)}{(P_i + P_2)} \qquad \text{pascals/metre}$$

Computer treatment

For computer treatment each section of the pipeline between junctions is entered on the programme card and the various data (pipeline diameter, equivalent lengths of pipeline bends etc) are stated. The network data are then arranged under the three headings: 'resistance data', 'line data' and 'junction data'. Using a suitable computer programme such as NERO (by Siemens AG of Vienna), the pressure and flow rate at any part of a large urban system can be accurately established. Stadtwerke Munich have been using this system for many years.

TABLE 8.5 VALUES OF Z FOR USE WITH TABLE 8.6 (P = steam pressure in bars absolute)

P	Z	P	Z	P	Z
1.1	1.2	2.2	4.58	4.5	18.2
1.2	1.42	2.4	5.41	5.0	22.3
1.3	1.66	2.6	6.32	5.5	26.8
1.4	1.91	2.8	7.29	6.0	31.7
1.5	2.19	3.0	8.32	6.5	37.0
1.6	2.48	3.2	9.43	7.0	42.7
1.7	2.78	3.4	10.60	7.5	48.8
1.8	3.11	3.6	11.83	8.0	55.2
1.9	3.45	3.8	13.13	9.0	69.3
2.0	3.81	4.0	14.5	10.0	84.9

TABLE 8.6 FLOW OF SATURATED STEAM IN STEEL PIPES TO BS 1387 (heavy grade in grammes per second (g/h))

Z_1 and Z_2 = initial and final pressure factors from Table 8.5. L = length of pipe in metres. P = absolute pressure in bar corresponding to 30 m/s.

Z_1–Z_2 L	mm: in:	15 ½	20 ¾	25 1	32 1¼	40 1½	50 2	65 2½	90 3	100 4	125 5	150 6	
	P												P
0.00		1	2	4	8	12	23	46	72	146	262	425	
0.002		1	3	5	11	17	33	66	103	210	379	613	*2*
0.004		2	4	8	16	25	47	96	149	303	547	885	*3*
0.0063		2	5	9	21	31	60	122	190	386	695	1130	
0.008		3	6	11	23	36	68	381	215	438	789	1280	
0.0	*1*	3	7	12	26	40	76	156	242	492	888	1440	*5*
0.02		4	10	18	35	60	110	224	349	711	1280	2070	
0.04	*2*	6	14	25	55	84	139	324	514	1030	1850	2990	
0.063		8	17	32	70	106	202	412	641	1300	2350	3310	*02*
0.08	*3*	9	20	36	80	211	230	468	728	1481	2670	4320	
0.		10	22	41	89	136	238	526	819	1670	3000	4860	
0.2	*5*	14	32	59	128	196	373	760	1180	2140	4340	7020	
0.4		20	46	85	185	283	538	1100	1710	3470	6360	10 100	
0.63		26	59	108	236	360	684	1390	2170	4420	7960	12 900	
0.8	*11*	29	67	123	267	408	777	1580	2460	3020	9030	14 600	
.0		33	75	139	301	460	874	1780	2770	5640	10 200	16 500	
2.0		47	109	200	435	664	1260	2570	4000	8140	14 700	23 700	
4.0		68	157	289	628	958	1820	3710	5770	11 700	21 200	34 300	
6.3		87	200	367	799	1220	2320	4720	7340	14 900	26 900	43 600	
8.0		99	226	417	906	1380	2630	5350	8330	17 000	30 600	49 500	

Note: 1 The above is abstracted from the *IHVE Guide*. Mass flow rates are rounded to the nearest whole number

2 Equivalent length of single resistance in metres for K = 1 given to the *Guide* Table may be approximated as follows:

½ in.: EL = 0.5	1½ in.: EL = 2	4 in.: EL = 6
¾ in.: EL = 0.8	2 in.: EL = 3	5 in.: EL = 8
1 in.: EL = 0.1	2½ in.: EL = 4	6 in.: EL = 10
1¼ in.: EL = 1.5	3 in.: EL = 4.5	

3 The value of 30 m/s velocity has no quantitative significance. Quantities corresponding to other velocities may be obtained by simple proportion and, to other initial pressures, by interpolation.

TABLE 8.7 CONVENTIONAL STEAM VELOCITIES (SATURATED STEAM)

	m/s
Boiler outlet connections (LP)	5 to 10
Boiler outlet connections (HP)	15 to 20
Headers (LP and HP)	20 to 30
Steam mains (LP and HP)	30 to 50
Vacuum steam mains	100
Calorifier connections	25 to 30

Cross sectional area of piping

Steel to BS 1387 heavy weight and BS 806 (over 150 mm)

Nominal bore		Area
in.	mm	m^2
2	50	0.0020
2½	65	0.0036
3	80	0.0050
4	100	0.0085
5	125	0.014
6	150	0.018
8	200	0.032
10	260	0.05
12	300	0.08

CHECK LIST

1. Criteria for choice of construction line and design
2. Ground conditions
3. Traffic pressures
4. Presence of other services
5. Requirements for pipe spacing, fixed points and supports
6. Uses of pipeline guides

7 Uses of compensators to counteract thermal expansion and contraction of pipelines
8 Suitability of systems without compensators
9 Effects of air leaks into system
10 Need to prevent internal and external corrosion of pipelines
11 Suitability single-pipe district heating system
12 Methods of detecting faults in district heating mains
13 Methods of detecting heat and water leaks from networks
14 Basic calculations necessary to obtain proper insulation of pipelines
15 Method of calculating pressure losses in pipe networks
16 Use of computers to calculate heat distribution in networks

REFERENCES

Fernwärme International (*articles in German*)
G. Fauser, 'House connections', 2 (1973) 1, pp 10–12
W. Hollander, 'Domestic hot water production in district heating practice', 2 (1973) 6, pp 139–146
D. Bublitz, '50 years of district heating in Berlin', 5 (1976) 5, pp 173–174
K. Curth, 'Monitoring district heating systems', 8 (1979) 1, pp 5–9

Other sources
H. Brun, 'Filling of pipeline conduits for district heating with loose material', *Installation*, 91 (13), 1971, pp 357–358 (in German)
H. Brun, 'Problems involved in laying heating pipes into the soil', *Elektrizitätswirtschaft*, 70 (25), 1971, pp 718–724 (in German)
'District heating distribution', *Construction*, March 1979, pp 19–20
M. R. Dorn and others, *Efficient control of district heating schemes*, Samson Controls Ltd, London, 1976
A. A. Field, 'Embedded coil floor heating', *Heating-Piping*, 49, August 1977, p 85
M. J. Hannah, 'How to instal and service packed expansion joints', *Heating, Piping, Air Conditioning*, April 1973, pp 67–75
P. Margen, 'A study of low temperature district heating', *Proclim*, May 1975, pp 81–114 (in French)
P. Margen, 'New types of hot water distribution systems for low density heat areas', *Building Systems Design*, 76, December/January 1979, pp 23–44
D. J. Neil, 'Spiral heat exchangers', *Heating and Air Conditioning Journal*, May 1974, p 23
A. Paterson, 'General design and controls for district heating', *Energy Digest*, 4, May/June 1975, pp 9–10
'VDEW Technical guidelines for the building of district heating networks', VWEW, Frankfurt 70, 1966 (in German)
W. L. Viar, 'Inter-building heat energy distribution systems', *District Heating*, 60 (2), 1974, pp 16–23

CHAPTER 9

Pipeline systems

MODES OF CONVEYING THE HEATING MEDIUM

Heat can be transported from supply stations to the consumers in one of two modes: steam or hot water. The advantages of using steam are:

(a) it is far easier to convey steam uphill than hot water: indeed the difficulties of pumping hot water to (or across) considerable heights may make a given scheme uneconomic; this also applies when tall blocks of flats are to be supplied with district heat

(b) the volume and temperature of steam can be varied by increasing the boiler pressure so that the installation of expensive accumulators can be avoided

(c) the accurate metering of steam is easier and cheaper than of hot water: accurate quantity meters which determine the amount of steam condensed can be read to an accuracy of $\pm$ one per cent.

However there are the following important advantages in using hot water:

(a) low temperature heat energy is used; this permits higher efficiency in the operation of thermoelectric heat generating systems

(b) the piping arrangement enables return mains to be fitted which do not corrode

(c) although the heat stored per unit weight of hot water is far smaller than that of steam as it is present only as sensible heat, the amount of heat stored per unit volume may in fact be higher so that pipe dimensions can often be made smaller for given velocities of heat flow

(d) the amount of heat carried by hot water can be fairly easily related to the requirements of the consumer, without the need to increase the flow rate by increasing the temperature difference between the flow and the return pipes (with steam systems the heat flow is varied by throttling or raising the steam throughput but this is not always desirable).

When a given heating network is established and it is desired to add to it by connecting further adjoining districts, it is usually necessary either to lay an entirely new steam piping system whenever a section has become over-expanded or to oversize the piping network in the first place to allow for future expansions. It is therefore best to use hot water heating in concentrated urban areas, where no great pressure head differences exist.

Steam constitutes a much higher grade of energy than hot water. When it is desired to provide the largest heating capacity possible by the use of waste heat from power stations the provision of steam for district heating results in very much greater loss of electricity than when hot water is provided. Steam networks are therefore generally only adopted when it is particularly difficult to use hot water networks. In the USA steam systems are widespread because fuel is relatively cheap and because steam is more useful for refrigeration than hot water. In Europe fuel costs are high: steam is therefore generally limited to district heating systems which supply industrial heating or to areas where the terrain is mountainous.

The latent heat of steam is about 2245 kJ/kg at the temperature and pressure conditions at which it is commonly delivered to the consumer. If the temperature of the feedwater to the boiler is taken as 10°C (50°F) then the sensible heat in the condensate is 377 kJ/kg, about 16.8 per cent of the total heat contained in the saturated steam. (The sensible heat contained in superheated steam is almost negligible.) It is debatable whether it is worth while to reclaim this sensible heat by the installation of condensate return pipes. In the USA where labour charges are high return mains are seldom provided. In Europe return lines are commonly fitted, mainly to reduce water treatment costs by conserving the condensate. Special filters are placed upstream into the return lines and hydrazine is added in small quantities to remove traces of oxygen which would corrode the return pipe lines by differential oxygenation. The condensate is then returned to the boiler as feed water. In some American systems condensate return lines are made from stainless steel because of corrosion difficulties.

PIPE SYSTEMS

Four different types of pipe systems are generally used for the supply of hot water or steam in district heating systems. These are: the single-pipe system, the two-pipe system, the three-pipe system and the four-pipe system.

The single-pipe system is used with most steam systems and no condensate return is provided. It is also used for certain schemes which distribute highly pressurised and superheated water pumped over extended distances. In geothermal district heating schemes (where the heat is virtually free of cost) the water is cooled down as far as possible in the system and then discharged to the public sewers.

The two-pipe system is most commonly used on the continent of Europe. Hot water is pumped from the thermal power station (or other heating centrale) to heat exchangers located in a sub-station near the housing estate served, heating the secondary (consumers') circuit which is fed with treated town's water. The district heating (primary) water is returned to the heating centrale via the return line. The secondary heat may pass directly into the hot water tank connected in series with the heating radiators. Alternatively a bypass scheme can be employed whereby about only a third of the hot water passes through the radiator systems.

The three-pipe system is more expensive than the two-pipe system but offers some considerable advantages over it (**plate 30**). Two mains are laid constituting the main supply pipe and the main return pipe. A much smaller main is added for the supply of the hot water and space heating demand in the summer. The common return line serves both supply pipes. With this method two streams of hot water are dispatched from the heating centrale. During summer the main heating line is isolated and only the smaller bore pipe is alive. In this way heat losses from the network are considerably reduced during the summer months. The largest system of this type is that located in West Berlin.

In the four-pipe system domestic hot water (treated) and space heating hot water (dosed) are circulated through two separate sets of pipes. This system suffers numerous disadvantages. Pipelines supplying domestic hot water must for example be suitably corrosion resistant. At the Urmston district heating scheme near Manchester (UK) this caused considerable difficulties. The district heating centrale was intended to supply 425 houses in a pilot scheme with hot water and central heating. The domestic hot water was carried in a copper network and the water for space heating in a steel network. Both sets of pipes were located in the same conduit and thermally insulated with magnesia: this became damp and galvanic (electrolytic) action arose between the pipes causing rapid corrosion of the steel pipes. The four-pipe district circuit is nowadays generally discouraged.

Most hot water district heating systems employ a two-pipe system although there are several very successful systems with a three-pipe distribution. Steam systems in the United States are commonly single-pipe. European steam systems adopt condensate return (steam pipe and condensate pipe).

PUMPING HOT WATER

The hot water pumped from a district heating mains station is always pressurised to about 15 bars (absolute) the pressure being maintained by a nitrogen blanket on top of the hot water accumulator. The return pressure of the water is about 2 bars (absolute). The velocity of water flow through the district heating mains is about 2 to 3 m/s; the usual head loss equations apply. Much higher velocities are likely to cause erosion particularly at changes of direction. In very long pipeline networks booster stations have to be used to provide an adequate head to supply the more distant consumers. In general pumping costs are not a major consideration in the distribution of district heat being of the order of only 2 to 3 per cent of the total costs, or about 7 to 8 per cent of the cost of the heat delivered. Nevertheless in the interest of energy conservation a careful balance must be established between the selection of pipe sizes and the related pumping costs.

ECONOMIC PIPE DIAMETER

The economic pipe diameter for a particular scheme depends on the following three variables:
(a) The cost of pumping the water. This varies roughly inversely with the fifth power of the pipe diameter. Pumping costs include fixed costs of the annual expenditure on depreciation, interest charges, maintenance and service of the pumps and energy (usually electricity) costs.
(b) The annual cost of the pipeline system. This includes interest on capital, depreciation of the system over its life, annual cost of the system over its life and annual cost of maintenance and repair.
(c) The annual heat loss of the pipeline system. Assuming that the thermal insulation is selected so that the external temperature of the lagging is the same in every kind of pipeline system, then the heat loss is directly proportional to the external surface area, roughly proportional to the pipe diameter.

Such elementary economic considerations apply only to a simple pipeline system from the station directly to a cluster of consumers (a number of blocks of flats or houses).

In more complex systems other factors often influence the calculations. Consider a network system where two branches join the main pipeline. One of the branches extends to a high-lying district some distance away and the other serves some small houses close to the main pipe and at the same level as the pipe. If the two pipelines were of the same diameter either too much water would bypass the small houses or the pressure would not be enough to reach the high-lying estate. To avoid the installation of circuit booster pumps the supply pressure can be adjusted by making the pipelines to the high-lying district larger than those which serve the low-lying district. In some of the very large Continental pipeline schemes calculations of pipe diameters have been very exact to ensure that the head conditions at the different consumer outlet stations are similar.

THE IMPORTANCE OF INSULATION

Much investment is involved in the thermal insulation of pipelines. The choice of method used therefore requires great care because tragic errors can be made. Materials such as magnesia insulation have for example been found virtually useless: this material is strongly hygroscopic and during periods when there is a temperature drop within the pipeline system (eg in summer) the magnesia absorbs water which passes to the surface of the steel pipes by capillary action making rapid corrosion almost inevitable. It does not assist to wrap the insulation with a waterproof bituminous material. Even a pinhole or hair crack anywhere on this coating will cause water to enter, indeed bituminous coating prevents re-evaporation and promotes rapid localised corrosion.

The more important methods adopted in Europe for insulating municipal heating mains are considered below.

INSULATING SYSTEMS USING CONDUITS

Concrete ducts with gas concrete insulation

This is probably the most popular method of insulating pipelines used in the Scandinavian countries (**plate 29**). It has the advantage of cheapness and is generally reliable except during floods. Gas concrete by itself offers a good protective action upon a mild steel surface although pipes are sometimes first wrapped in corrugated cardboard.

A trench of rectangular section is constructed. Provided that the walls and the bottom of the trench are sufficiently firm, concrete can be cast directly against the soil. If not then a layer of lean concrete in the ratio of 1 : $2\frac{1}{2}$: $3\frac{1}{2}$, cement : sand : ballast: is first cast onto the bottom of the trench. Both sides and the bottom of the trench are reinforced with steel rods and stirrups. Side shuttering is positioned and concrete is cast, small seep drains being left at the base to facilitate the escape of small quantities of moisture. Blocks of steam-cured gas concrete are laid upon the concrete base slab and are contoured at the top to accommodate the pipes. After that the pipelines are positioned and butt welded electrically. A slurry of a mixture of finely ground silica, lime and aluminium powder is then poured into the trench and allowed to set. This forms round the pipelines a very porous structure (with many air cells) having a density of only 0.3 kg/m^3. The k-value of such a material is 0.09 W/mK under temperature conditions similar to those existing in a district heating duct. When the gas concrete has set the top surface of this soft material is scraped smooth and a precast heavy concrete reinforced lid, flat underneath and convex above, is placed on top of the pipe conduit. The top surface is overlaid with asphalt to protect the concrete against penetration by rainwater. Finally soil is filled back into the trench to bury the pipes. As soon as hot water flows through the pipes the gas concrete dries out. The water is distilled off and leaves the duct via the drain holes, which are provided either at the bottom of the duct or, more usually at the expansion chambers positioned at regular intervals along the pipeline. When the water has been driven off the gas concrete is kept dry by being vented to the external cold air while also protected by the lid against wetting by rainwater. The alkaline nature of the gas concrete protects the external surfaces of the mild steel pipe against most forms of corrosion particularly the very harmful bacterial action of sulphate-reducing organisms. This technique of pipeline insulation has proved eminently satisfactory.

During a visit to Odense in 1965 an inspection was made of some district heating pipelines which had been exposed for the first time since the middle 1920s. The external surfaces of the mild steel pipes were found to have been so well preserved that it was still possible to see the original machining marks.

Although a write-off period of about 35 years is being assumed for such pipeline networks in most continental municipal heating schemes it is likely that they will continue to provide completely satisfactory service over a much longer period of time. In most European district heating systems using such pipelines, the maintenance costs represent, on average, $\frac{1}{2}\%$ of the installation costs per annum.

When pipelines have to be laid in damp soil the design of conduits must be modified (**9.1** and **9.2**). If the ground water level is below the top of the conduit it is practicable to use a twin-layer concrete duct with plastic foil between the two layers or a similar duct with a layer of shingles or loose stones which can be drained by a pumping system. In general under such

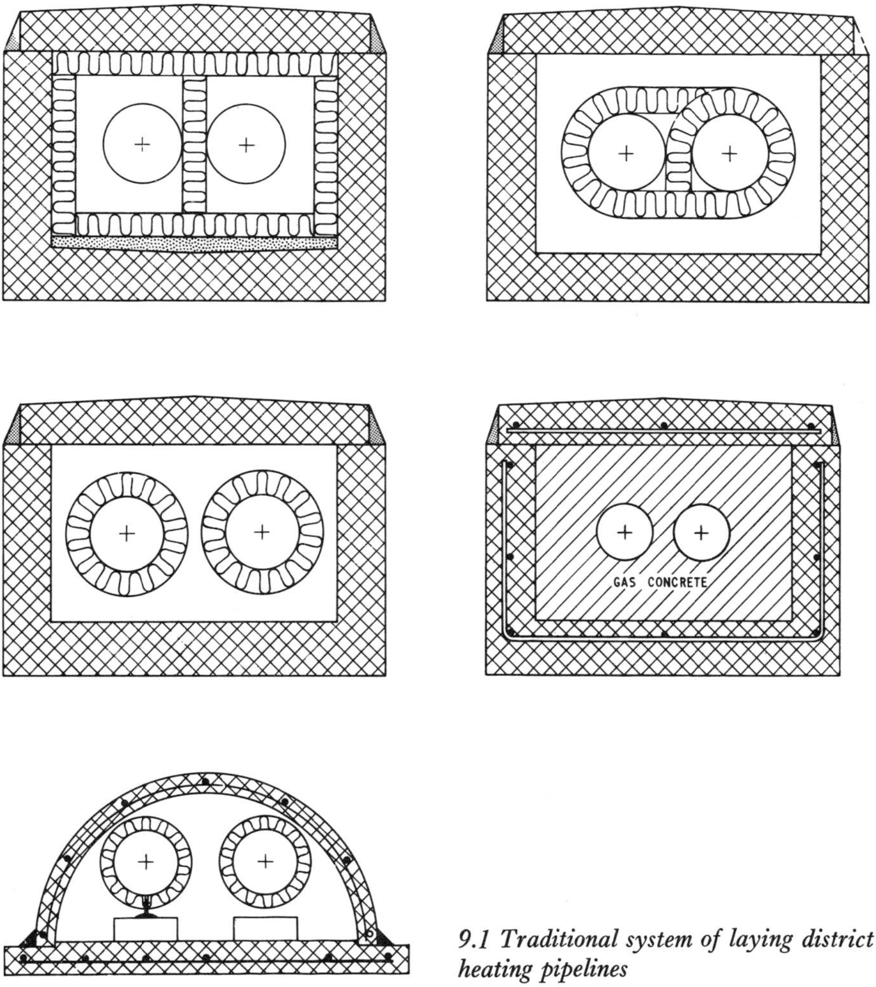

9.1 Traditional system of laying district heating pipelines

9.2 Types of conduits used for high water table conditions

conditions mineral wool as insulation is preferred to gas concrete. (The cost of constructing district heating networks is affected not only by pipeline diameters (which govern the dimensions of the concrete ducts) but also by local conditions: it is therefore essential that the latter should be accurately ascertained before pipelines are designed and installed.)

Concrete ducts can also be produced in the form of precast units usually between 2 and 3 metres long. Such ducts resist mechanical forces well but are more prone to the effects of settlement and other soil conditions than in-situ cast concrete conduits. It is more difficult to install extra pipe connections. Precast units are therefore commonly employed only when pipeline diameters are less than about 100 mm.

Concrete ducts with mineral wool insulation

In this popular method concrete ducts are constructed as already described and equipped either with drains at the bottom or, when the ground water level is above the level of the bottom of the conduit, with pump-drained channels. Mineral wool can be wrapped round each pipeline individually or around both pipelines at the same time using a thick pad between the pipes to prevent the transfer of heat from the hotter supply line to the colder return line (**plate 30**).

Mineral wool (or rock wool) is made from the mineral diabase (a volcanic deposit forming part of the earth's crust and found in Scandinavia and in other parts of the world). The rock is quarried, crushed and melted in cupolas with limestone and coke. The molten rock is then converted into fibres by a spinning method (the cascade process) in which a stream of liquid rock is poured onto a series of spinning discs and flung off the final disc as a fibre. The fibres are cooled rapidly from their melting temperature, about 1450°C (2642°F), to a little over room temperature. The size and quality of the fibres are controlled both by preliminary selection of the raw

1. The coal fired district heating station at Essen, W. Germany

2. *Aerial view of the group heating scheme, Rowletts Hill, Leicester*

3. *Aerial view of the district heating station complex at Essen-Innenstadt*

4. *Municipal swimming pool connected to group heating scheme*

5. *Billingham group heating scheme coal fired boiler house*

6. Group heating centrale in Denmark

7. Packaged coal fired steam boiler, with an output of 8.6 tonnes/h at 10.5 bars (gauge)

8. Small packaged heating plant ready for transport to site

9. A Brown-Bovri open-cycle gas turbine plant supplying heat and power

10. Rotary Kiln used by Volund with hydraulic friction drive

11. Volund Vest Incinerator of Copenhagen incorporating four units to give combined output of 1 000 tons/day

12. Volund Thisted incinerator plant incorporating one furnace line of 3 tonnes/h output (72 tonnes/day) with provision for one additional line

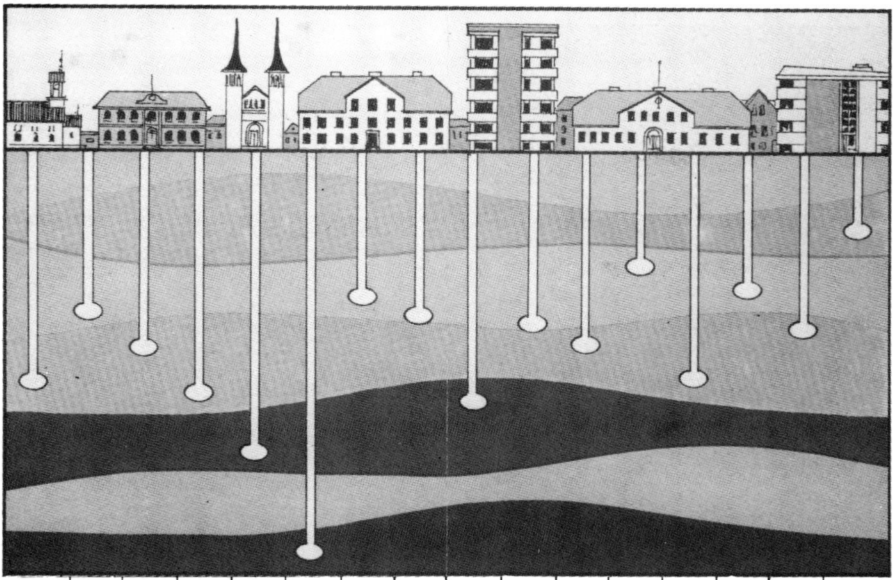

Depth	Water Supply	Temperature °C.
1266 m	36.8 l/sek	96.2°C
861	31.0	107.5°C
1077	40.3	100.5°C
1316	34.0	105.0°C
1613	23.2	94.0°C
2198	8.2	135.0°C
741	49.1	131.0°C
862	32.4	128.0°C
1306	15.1	130.0°C
928	32.8	130.0°C
1014	23.3	126.0°C
634	15.2	122.0°C
1239	39.1	128.0°C
764	47.0	129.0°C
978	44.2	128.0°C
403	14.7	115.0°C

13. Thermal energy distribution underneath Reykjavik

14. Prefabricated manhole in position (under construction)

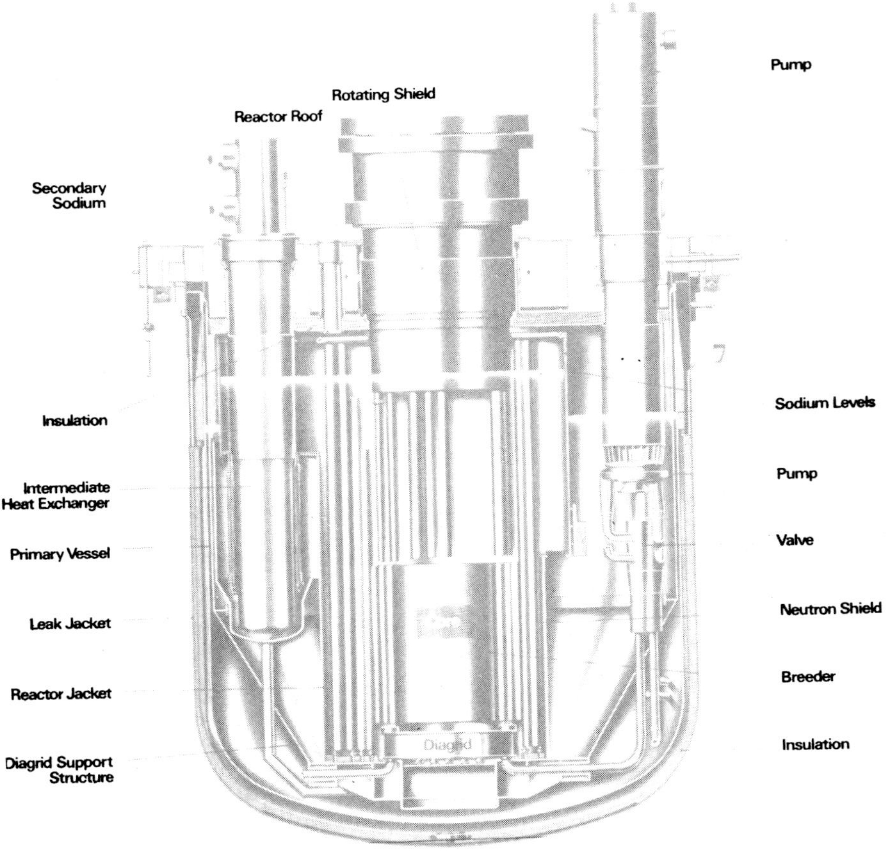

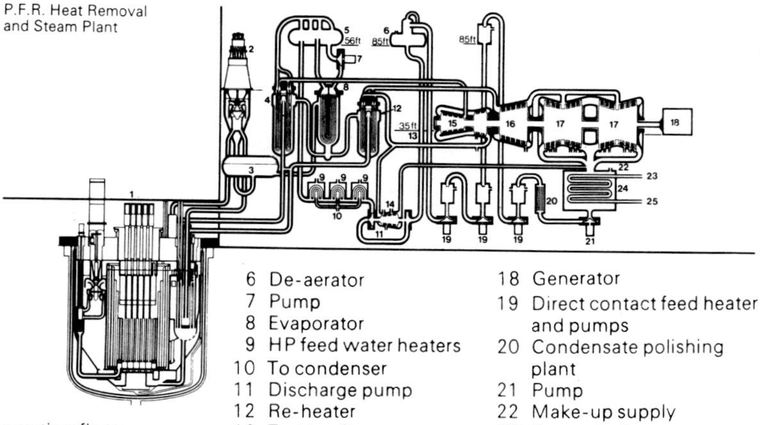

1 Operating floor
2 Secondary sodium pump
3 Sodium expansion tank
4 Super heater
5 Steam drum
6 De-aerator
7 Pump
8 Evaporator
9 HP feed water heaters
10 To condenser
11 Discharge pump
12 Re-heater
13 Turbine floor
14 Turbine
15 HP turbine
16 IP turbine
17 LP turbine
18 Generator
19 Direct contact feed heater and pumps
20 Condensate polishing plant
21 Pump
22 Make-up supply
23 Sea water in
24 Condenser
25 Sea water out

15. Design details of Dounreay fast breeder reactor

16. External view of Dounreay fast breeder reactor

17. A Tuboflex bellows compensator for a very large district heating main

18. Expansion loop (without use of compensators)

19. Installation of an IWKA bellows compensator in a German district heating network

20. Pipeline with anchors, bellows and supports

21. Vertical expansion loops with compensators

22. Use of axial bellows for angular movement

23. Expansion bellows for angular movement

24. Array of vertical pipe loops with supports and bellows

25. Pipe installation in duct with bellows
26. Large diameter single and double packed expansion joints for steam mains

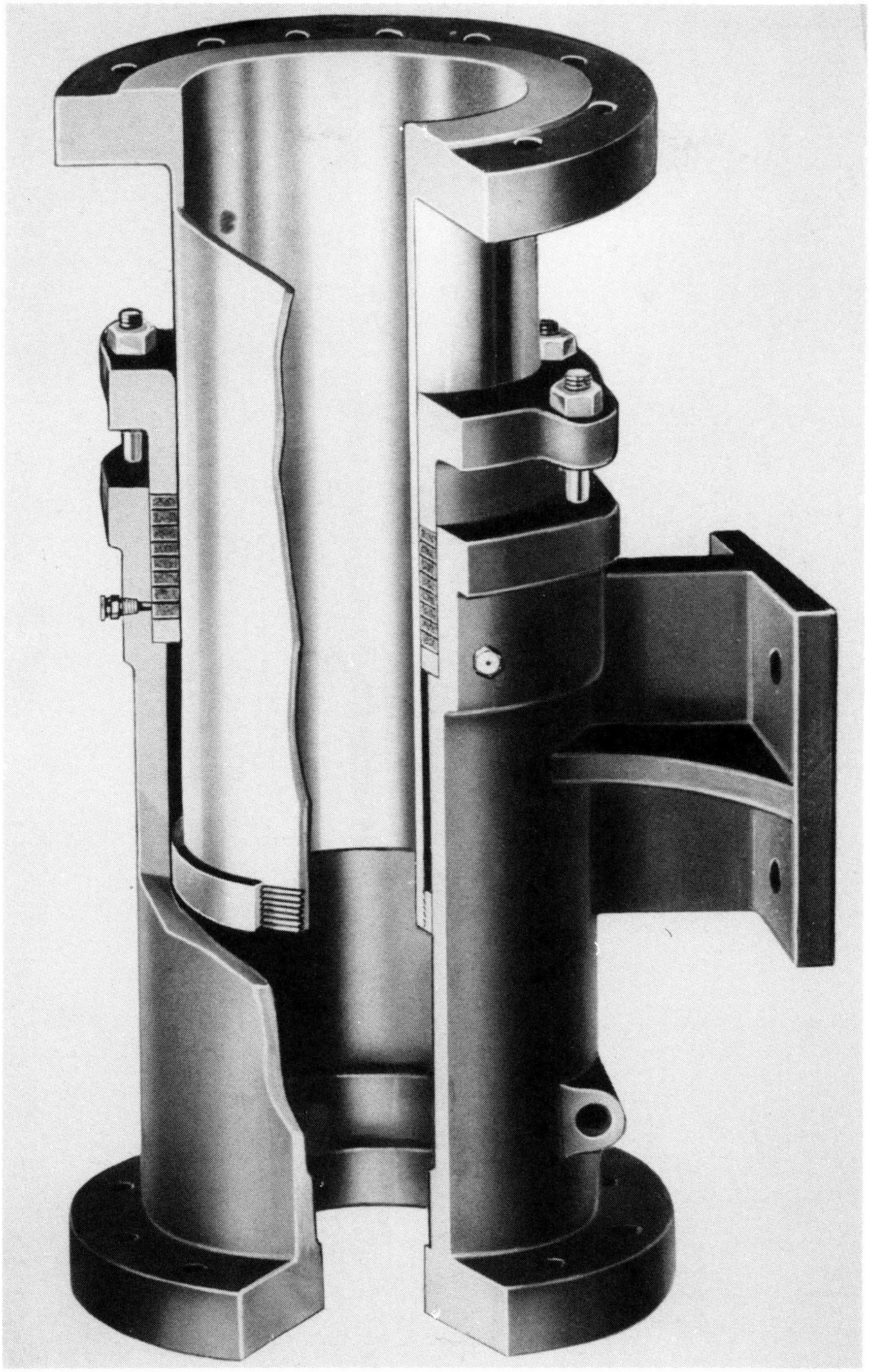

27. An Adsco expansion unit

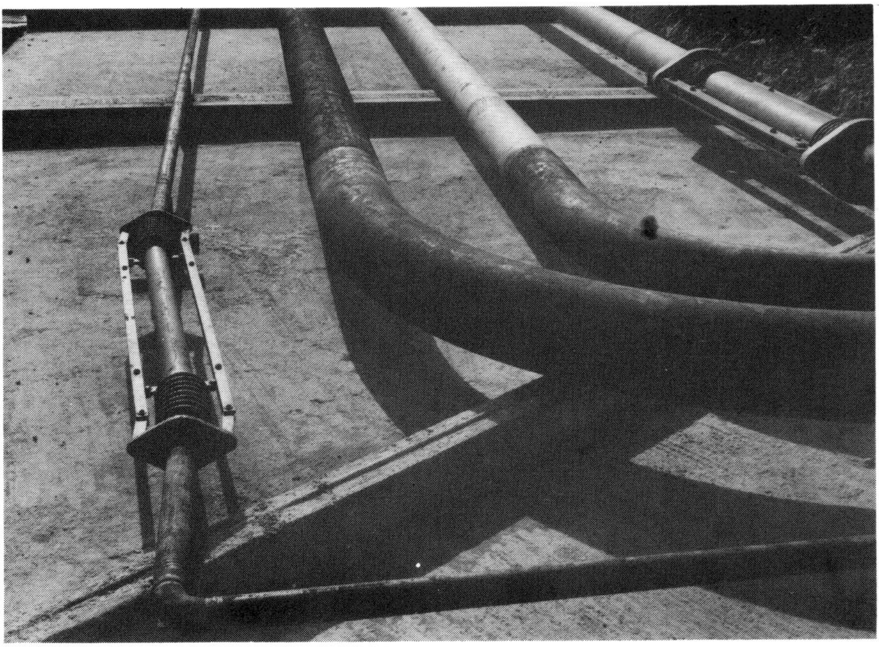

28. Close-up on pipetrack under construction of 75 mm articulated compensator, showing cold allowance

29. Poured gas concrete used for insulating pipelines

30. A 3-pipe line system consisting of steel pipes insulated in mineral wool and laid into a concrete conduit (West Berlin)

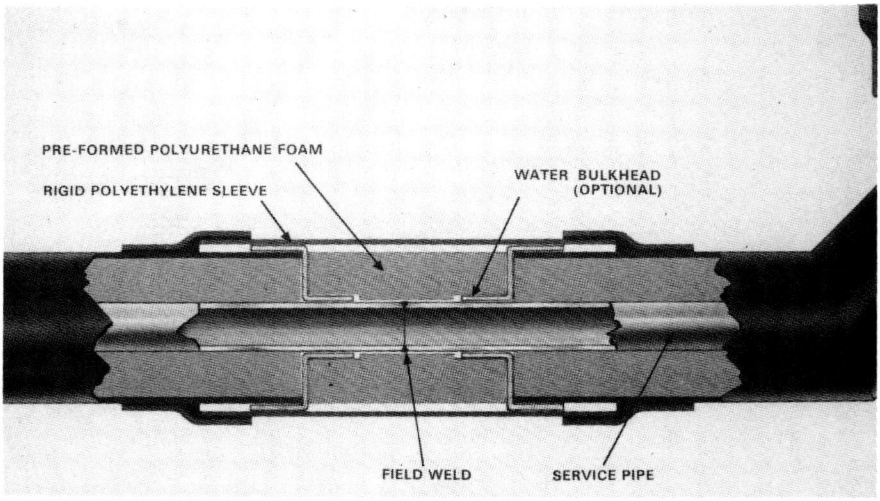

31. Thermo-pipe prefabricated mains system

32. Aquawarm pipeline system

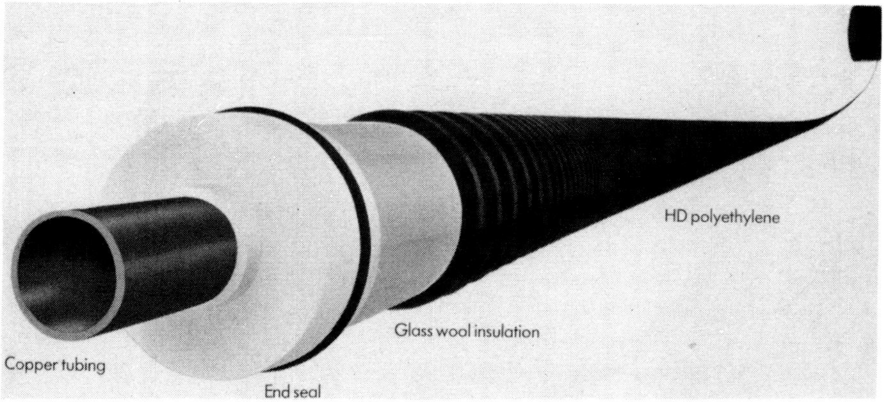

33. Installation of a 2 × 107 mm Permapipe high pressure hot water heating mains at Warwick University

34. Prefabricated pipeline being placed into trench

35. Flexwell district heating cable with leakage indicator

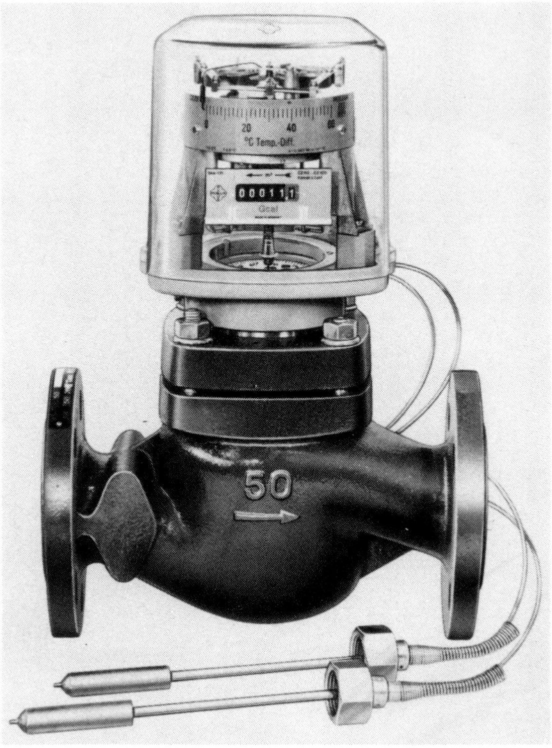

36. Spanner-Pollux mechanical heat meter

37. SVM electronic heat meter system. KEY: D *Differential signal unit*, AD *Analogue digital converter*, P *Counter*, T *Time clock*, Rk *Cold pipe temperature sensor*, Rw *Hot pipe temperature sensor*, Kt *Contact unit*, U *Emf*, I *Current*

Block diagram

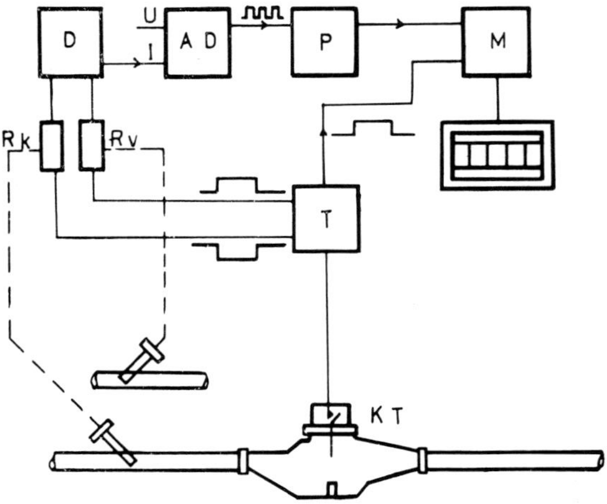

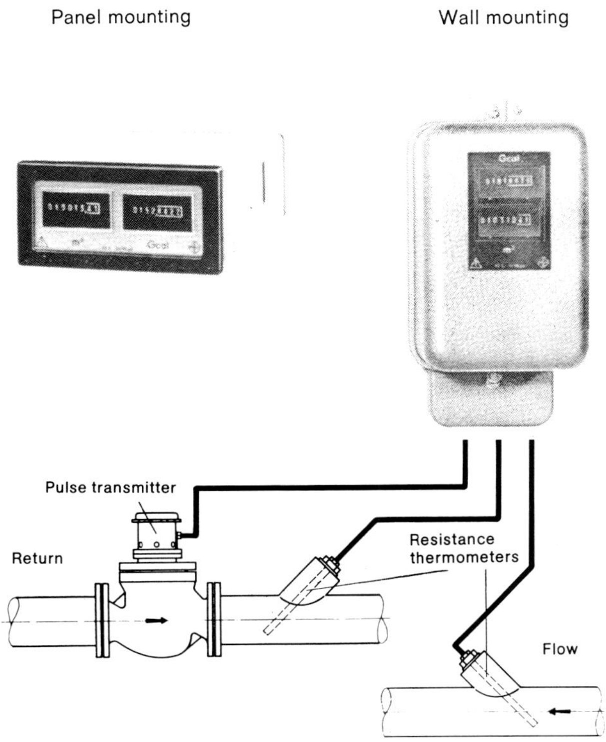

38. Pollux-Samson electronic heat meter

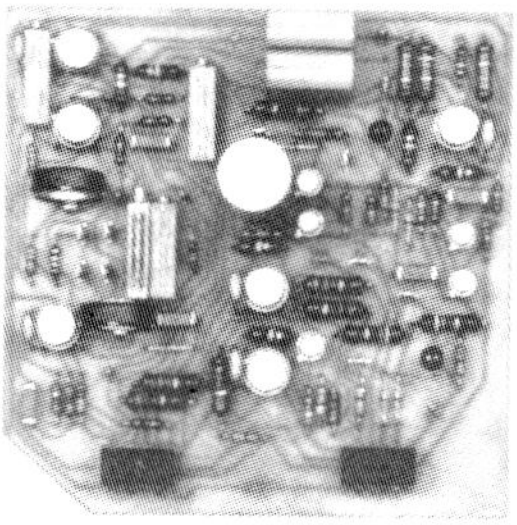

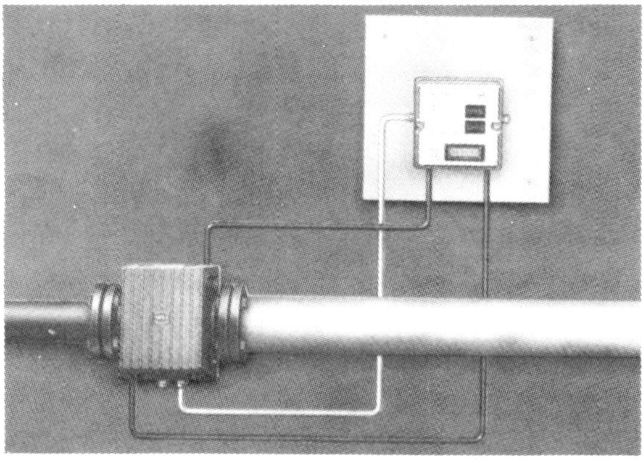

39. Components of Autarkon

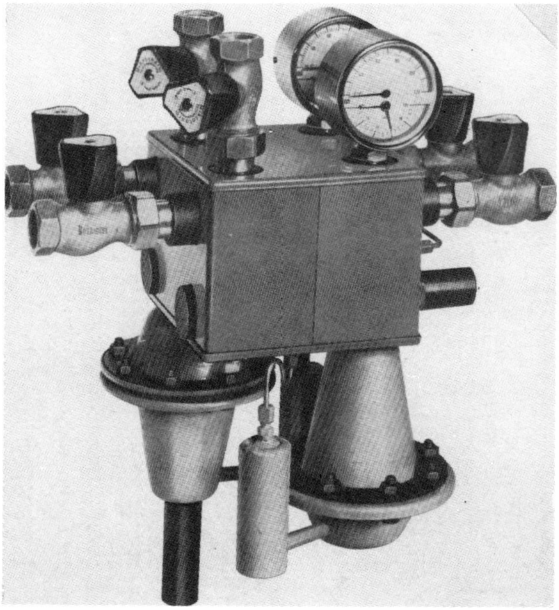

40. Packaged district heating sub-station for small house including pressure regulator, volume limiter, valves, filters, thermometers and manometers

41. Exploded view of thermostatically air controlled radiator valve

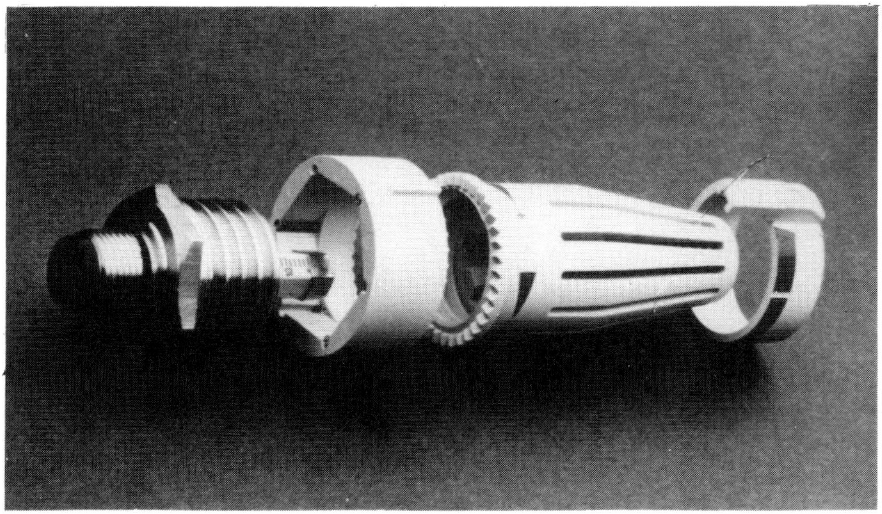

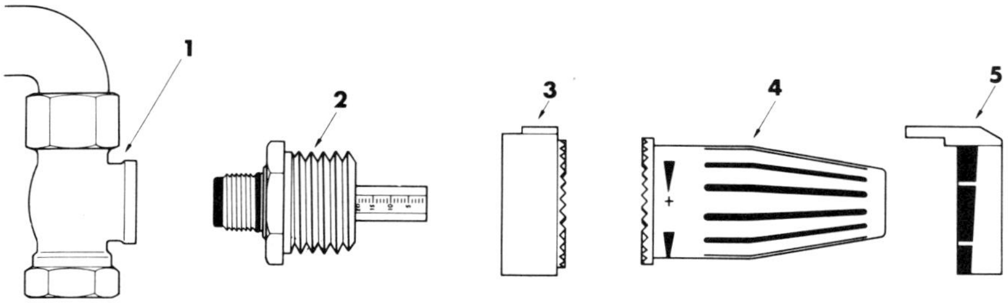

42. Danfoss RAVL radiator thermostat with remote sensor

43. Danfoss RAVL non-electric domestic hot water controller

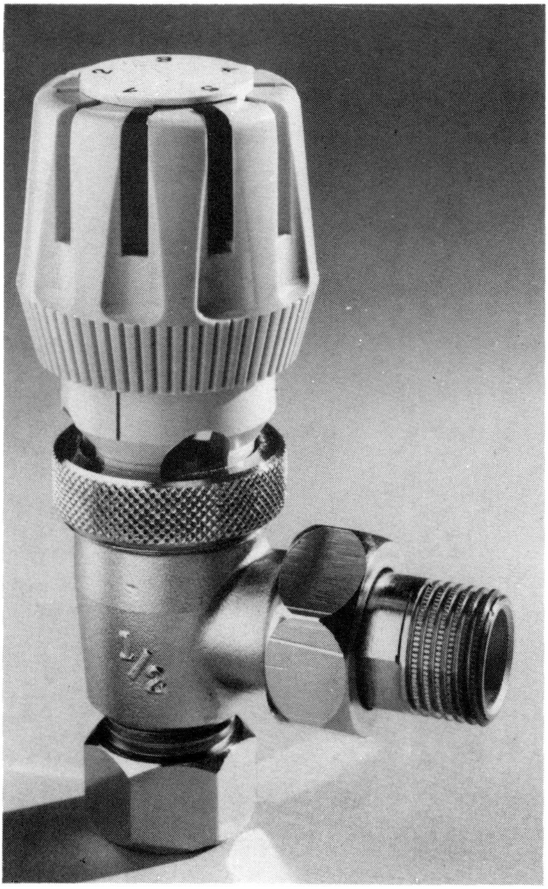

44. Danfoss RAVL radiator thermostat with built-in sensor

45. Satchwell self-contained thermostatic radiator valve

46. Building management system (BAS 00): control console

47. Autoscan supervisory control system: control panel

Satchwell
BAS 700

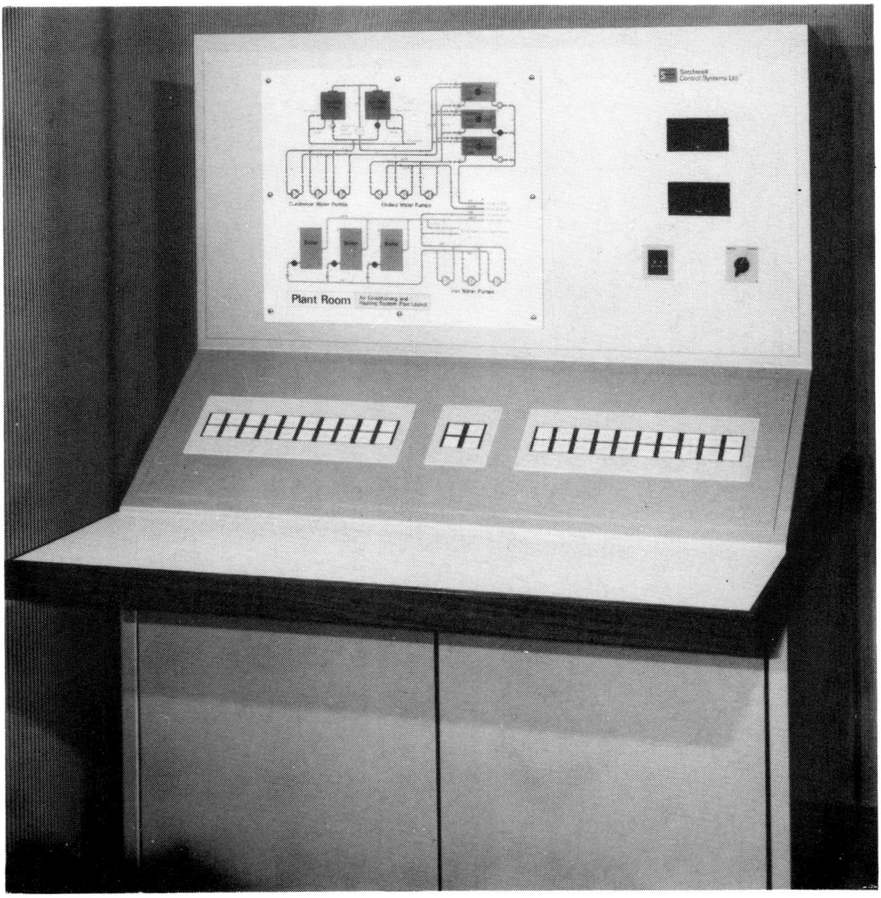
Plant Room

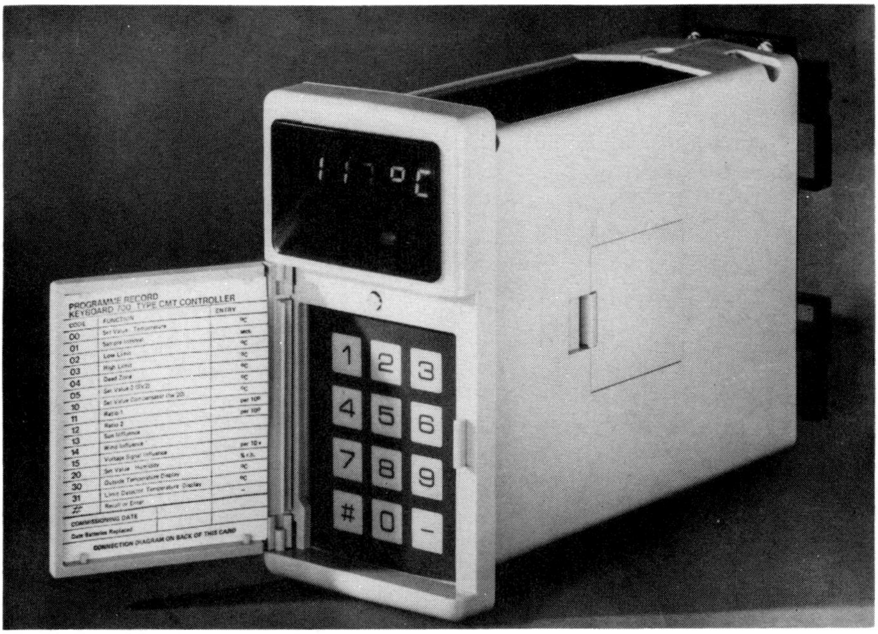

48. Micro-processor based keyboard (700 CMT) controller

49. Vertical small mild steel oil tank: note the vent pipe, fill pipe, fill terminal and oil catch pit

materials and during the manufacturing process. For pipeline insulation phenolic resin is added to bind the fibres together when slabs are made; mineral oil is also added when producing loose batt to be wound around the fibres. Pipe sections are rolled on a mandrel to the required thickness and then passed through the curing chamber. The rigid section is then trimmed and slit along its entire length. To provide mineral wool slabs with considerable compressive strength the fibres are arranged end-on and consolidated by the addition of resins. For some purposes it is best to produce the mineral wool in the form of wired mattresses when the organic content is kept low and the product is given added mechanical strength from a facing of wire netting.

Mineral wool itself cannot burn because it is inorganic. Even if no organic matter is present in rock wool a temperature limit of 1200°C (2192°F) is imposed because softening of the material can take place above this limit. The addition of organic materials does not increase the fire risk but would affect the temperature resistance. Mineral wool impregnated with phenolic resins is stable up to a temperature of around 250°C (482°F).

The water absorption of mineral wool is very low, amounting to only 0.02 per cent at a relative humidity of 95 per cent. The fibres do not absorb water at all when they have been impregnated with oil: internal convection is then very limited because air cannot easily leave the interstices so making it impossible for water to enter them. When mineral wool is used for insulation the main problems arise from water penetrating junctions between slabs or through cracks in slabs due to expansion and contraction stresses. If water does get through in this way rapid corrosion is caused through differential aeration.

For pipeline insulation standard sections of mineral wool 900 mm long or loose batts are used which are wound around and wired into place (**plate 30**). The outside surface of the mineral wool is then protected by one of the following substances:

Cocon: the trade name of a PVC solution which is sprayed onto the outside surface of mineral wool insulation to strengthen it and to make it impervious to water. This material can also be used in connection with glasswool. The PVC is sufficiently elastic to bridge small interstices or cracks in the mineral wool layer.

Impregnated paper: often wound around the mineral wool to prevent dusting and other damage. This technique is mainly used when the insulated pipelines are inside dry tunnels or pass through cellars.

Aluminium foil: used to improve the thermal insulation properties of the mineral wool by adding radiation insulation in the form of high reflectivity. It is essential that pipelines treated in this way never become damp as the aluminium foil constitutes an effective vapour barrier and prevents the mineral wool from drying out.

The thermal conductivity of mineral wool products is 0.037 W/m K at

10°C (50°F) when the density varies between 150 and 250 kg/m³, rising to 0.042 W/m K at 150°C (302°F). Table 9.1 indicates its supporting strength.

TABLE 9.1 SUPPORTING STRENGTH OF MINERAL WOOL

Density in kg/m³	Live load kg/m²
60	200
100	400
200	5000

Mineral wool does not possess equal anticorrosive properties to gas concrete when in contact with mild steel: it is therefore necessary to protect the surfaces of steel pipes prior to insulation by immersing them into a chromate bath and coating them with asphalt-impregnated felt.

In general mineral wool insulation is applied when heating pipes are routed through the down-town areas of cities where pipes often pass through cellars fronting the streets instead of being buried underneath them. It is also employed on pipelines running through rocky areas where the hot water pipes pass through blasted rock tunnels. Pipelines using mineral wool insulation are supported on steam-cured gas concrete blocks every one or two metres as with the gas concrete insulation techniques.

INSULATION SYSTEMS WITHOUT CONDUITS

Systems available

A number of different systems of insulating pipelines which claim to avoid the costly matter of construction of concrete conduits are on offer. A representative group of these is described below.

Loose fill insulation: when using this a trench is dug and the pipeline is laid into a powder with hydrophobic properties designed to exclude moisture. There are different systems and methods:

(a) The pipes are pre-insulated with a particular material and then laid into the soil. The insulation material is not heated or sintered in any way. Example: Manolith.

(b) The material is laid directly around the pipes without any pre-insulation but has to be sintered before use by passing the hot medium through the pipes. Example: Gilsulate.

(c) The material is not sintered and no pre-insulation is needed. Examples: Protexulate, Gilsogard, Gilsotherm, Trockengrun 20 etc.

Hard cover insulation: the pipeline is covered with insulation units on site and then lowered into the soil. The most important of these systems is the Foamglass technique but there are others which use plastic units instead of inorganic ones eg Extorbit and Getes.

Hard mantle insulation: systems are available with or without pre-insulation. Examples: Hydrobit, Thermocrete, Aerocrete, Hartmantel, Lebit, etc.

Loose fill insulation

In West Germany district heating has been employed for many years and much experience has been accumulated as to the most likely faults which happen to buried pipelines with loose fill insulation. W. Simon (1951) found that the causes of faults in pipelines which had been in the soil for 30 years were as follows:

Building faults			
Cracks leaks etc in the duct		19%	
Faulty pipes		11%	
Faulty insulation		10%	
Poor building connections		10%	
	Total	50%	
External water penetration			
Pipelines		13%	
Insulation		13%	
	Total	26%	
Other faults			
Pipelines		4%	
Insulation		5%	
Building connections		15%	
			24%
			100%

The most important circumstances causing faults were:

Ground water penetrated the insulation layer and caused subsequent corrosion of pipelines

Traffic loading resulted in compression of the insulation and its destruction

Land slip caused breaks in pipeline or insulation layer

Inadequate compensation for thermal expansion (linear or axial) damaged pipelines, auxiliaries and insulation

Expansion units, house connections, joints and bends were incorrectly designed

Stray electric currents combined with water penetration of insulation layers caused corrosion of pipelines

Incorrect falls of trenches led to build-up of rainwater and saturation of insulation layer

Inadequate drainage caused water in lagging
Frost penetration led to various damage to pipes and insulation
Inadequate construction of trenches caused their collapse
Poor welding of joints created leaks
Insufficient attention was paid to co-ordinated design and construction where other services, such as water pipes, sewers and power lines crossed
Contact between metals of different natural emf resulted in differential metal corrosion
Manholes were badly built
Fixed points and anchors were defectively constructed.

The thermal conductivity of insulation materials increases greatly when they become damp. When water replaces air in the pores the thermal conductivity rises drastically. Typical examples are set out in table 9.2.

TABLE 9.2 VARIATION OF THERMAL CONDUCTIVITY AT 10°C (50°F) WITH INCREASING MOISTURE CONTENT

Humidity (% **by volume**)	**Mineral wool** **(k in W/m K)**	**Loose perlite** **(k in W/m K)**
0	0.035	0.041
22	0.0465	0.052
50	0.37	0.41
70	0.465	0.56
98	0.605	—

At higher temperatures water has a greater tendency to penetrate insulation layers as its viscosity then reduces. This is counteracted by the combined action of capillary resistance and diffusion-proof layers. The surface tension of water against solid materials is about 0.072 N/m at 20°C (68°F).

When liquids wet surfaces of most building materials or soil, paper etc there are considerable capillary forces which drive the water through the material. Construction with materials prone to capillary action is therefore only admissible when there is a guarantee that no water will ever reach the surface of the insulation material. Covering the material with impermeable membranes only makes matters worse as it then becomes impossible for water which has penetrated into the insulation material to re-evaporate.

Manolith

Manolith is the trade name of a greyish powder made by the Mannesmann company (Germany). It is a by-product of the distillation of coal and is totally water repellent even if submerged under water for several years. It has the following physical properties:

Density: (loose) 900 kg/m^3
(consolidated) 1060 kg/m^3
Grain size: up to 1 mm diameter

Thermal conductivity when in position in consolidated state: 0.117 W/m K at 65°C (149°F)
Unaffected up to temperatures of 82°C (180°F)
Begins to sinter slightly at 100°C (212°F)
Granular structure affected at 118°C (244°F)
Appreciable contraction (8 per cent by volume) at 138°C (280°F).
These data show conclusively that water penetration into the material is negligible probably as a result of the powerful negative capillary effect exerted by the powder which has a greasy surface.

To place hot water pipes into Manolith a trench is dug with smooth vertical walls. The pipes can be laid even on very wet days and it does not matter whether they run above or below ground water level. The pipes are supported temporarily in the centre of the trench by bricks, pieces of timber or wires. The granular Manolith material is then spread underneath the pipelines and consolidated by a vibrator. No concrete ducting is required but the walls of the trench must be adequately strong to withstand the consolidation pressure of the Manolith. When shallow trenches are required or when the bearing strength of the soil is low it is necessary to use shuttering before positioning the fill. Pipelines are usually pre-insulated by the use of 25 mm thick layers of mineral wool, covered on the surface with polythene foil. Such methods of pipe laying offer likely difficulties at pipe junctions or where the insulation is not correctly applied. Problems with settlement of the powder have not been fully solved.

Gilsulate

Gilsulate is somewhat similar in its properties to Manolith but has the advantage of not dusting. When Gilsulate was used for pipeline insulation in Sweden it was found necessary to pass superheated water through the pipes at least annually to maintain the water repellency properties of the material (when heat is applied in this way Gilsulate melts closest to the pipes and forms an impervious corrosion resistant film on their surface).

Protexulate

Protexulate is a proprietary white powder of mineral origin offered as an underground loose fill or for use in ducted systems as protection against corrosion and for thermal insulation. The powder must be applied with expert care as there may be a tendency for moisture to enter the system via pipe supports. The cover thickness is also crucial to the long term success of this method of pipe insulation. Once moisture has entered it will channel along the pipes making it very difficult to trace the first leak.

Hard cover insulation

Foam glass

Foam glass is a completely inert and impermeable insulation material possessing high compressive strength (**9.3**). It is made by drawing out borosilicate glass into tubes, crushing these, mixing with carbon and then

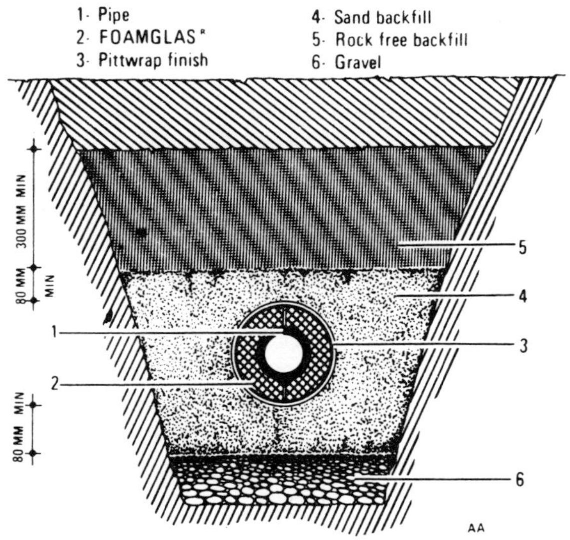

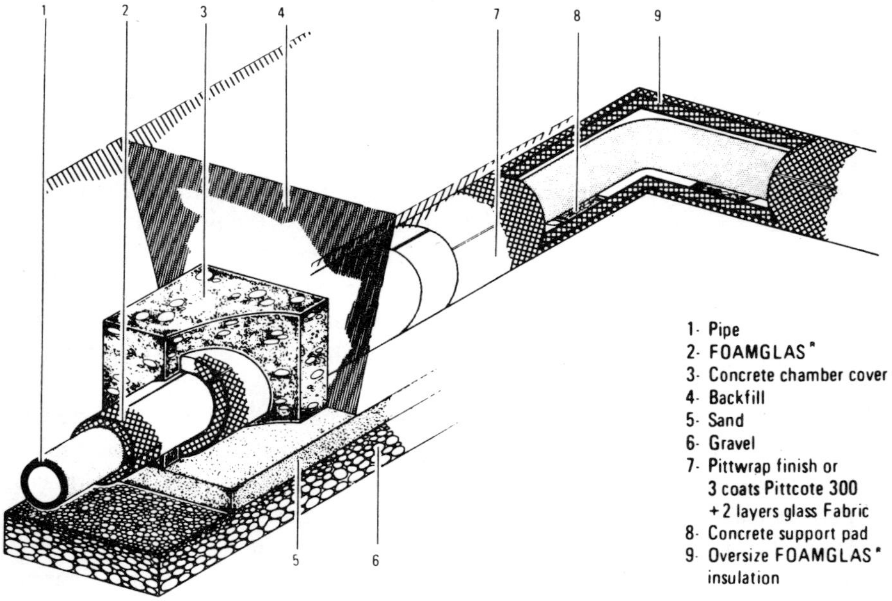

9.3 Use of foam glass insulation on pipelines

heating. As the carbon is burnt to carbon dioxide the glass expands into a mass containing innumerable non-interconnecting air cells. The blocks of this porous glass are then annealed and cut into preformed semi-annular shapes to fit the various pipe sizes.

For straight pipe runs pipe coverings of foam glass are applied on site with butt joints staggered and tightly butted. Each section of foam glass is held in place by two wraps of PC tape 25 with a 50 per cent circumferential overlap per wrap. The longitudinal and butt joints are left dry. The insulation is finished with a prefabricated laminate (such as pitt wrap), containing a 16 mesh/cm^2 glass fabric and a 0.04 mm thick aluminium foil sandwiched between three layers of bituminous mastic. The material is coated with a protective plastic film and the inner surface with special release paper. T-joints and bends are insulated with specially shaped pieces.

Foam glass is virtually pure glass containing no binder and is stable between 260°C and 430°C (500°F and 806°F). It has a density of 140 kg/m^3 and a thermal conductivity at 20°C of 0.055 W/m K (only slightly higher than that of polyurethane foam and expanded polystyrene which have conductivities between 0.034 and 0.038 W/m K when dry). Although the material contains a high percentage of voids it is remarkably strong but tends to be rather brittle. Thermal movement is compensated in expansion chambers; an anti-abrasive compound is used to overcome any abrasion due to thermal movements between the pipe and the foam glass covering. The coefficient of thermal expansion of foam glass is 8.3×10^{-6}/deg C.

The most important property of the material, when applied in conditions where the water table rises above the level of pipelines, is its resistance to the absorption and transmission of moisture. It is completely impermeable to water except for some temporary retention on the surface where damaged air cells can be filled with moisture. As it is entirely inorganic it cannot burn and it also has perfect dimensional stability.

Hard mantle insulation

Hydrobit

Hydrobit is obtained as a by-product of the distillation of oil. Before it is used a trapezium-shaped trench is formed and an unreinforced concrete sole slab, cast either in situ or made from precast units, is placed in position. The pipes are next encased in glass wool and positioned in the centre of a light steel sheet shuttering using temporary supports. The Hydrobit is then heated and poured in the molten state causing moisture in the insulation to evaporate. As the mass cools down it contracts very slightly and a topping-up further quantity of molten Hydrobit is added on the next day. After a further 24 hours the steel plate shuttering can be removed and the Hydrobit will be sufficiently hard to allow workmen to walk on it. The solidity of the material is such that mechanical equipment can be used to push the earth back without causing any damage.

Thermocrete

This system employs a modified gas concrete together with preinsulation to provide a duct-free insulation system which is claimed to be cheaper to construct and better from the point of view of heat losses than competing systems. The Thermocrete system is recommended for urban areas but not for new districts where the level of the water table is uncertain and where the ground conditions are likely to affect the insulation.

The system uses twin insulation: pre-insulation with glass fibre mats and external insulation with gas concrete. A firm and dry trench must be provided (with a sole of lean concrete if the floor is soft). The soil should be rainwater permeable. If it is not, pumping facilities must be provided during the construction process because Thermocrete takes 10 to 14 days to achieve impermeability. The pipes are laid on to special thrust bearings, welded, and guide points are installed. The thickness of the pre-insulation depends on the temperature for which the pipes are intended. If the fluid temperature is below 130°C (266°F) fibreglass mats are used alone. For higher temperatures a 5 mm thick layer of expanded polystyrene is applied first and is then followed by fibreglass mats: the expanded polystyrene then collapses to coat the pipe and produce an annular space between the pipe and the fibreglass; plastic foil is then wrapped around to exclude moisture. The advantage of using pre-insulation is to achieve good thermal insulation and to ensure that pipes subject to considerable axial expansion can move readily without breaking up the much harder external insulation.

When pre-insulation is complete external insulation is applied. For this purpose lightweight moulds are positioned in the trench around the pipes, the floor of the trench is covered with plastic sheeting and the Thermocrete mixture correctly blended with water is poured into the trench. After tamping it is covered by a 50 mm thickness of sand. Two days after filling the moulds can be removed and the trench filled.

Table 9.3 gives the distances at which pipes have to be supported in Thermocrete (centre to centre).

TABLE 9.3 SUPPORT DISTANCE FOR PIPELINES SUPPORTED IN THERMOCRETE

Pipe diameter (mm)	Support spacings (m)
32	2.40
40	2.60
50	3.00
65	3.70
80	4.00
100	4.50
150	5.50
200	6.00
250	6.50
300	7.00

When Thermocrete is employed, special techniques are used for the thermal and moisture insulation of expansion units, bends, junctions and entry sections into buildings.

PREFABRICATED PIPELINES

Advantages

Escalating labour costs have caused a swing away from relatively expensive in situ concrete conduits and in situ insulated pipelines (except for very large diameter mains). In general no less than ten stages are involved in the installation of conventional pipeline systems:

Excavation of trench
Consolidation and laying of concrete on base of trench
Concreting of walls
Positioning of pipelines on their supports
Insulation of pipelines
Making the concrete ducts waterproof and sealing them
Providing drainage of conduits
Partial filling with shingles
Back-filling of trench
Transportation to tip of the residual soil.

When pre-insulated prefabricated pipelines are used there are only six such stages and these are simpler and cheaper. They are:

Excavation of trench (in general the trench need have only 60 per cent of the cross sectional area of conventional trenches for equal pipeline diameters)
Assembly of pre-insulated pipelines on supports
In situ insulation of joints (this only need be done on one twentieth of the pipeline run)
Laying of sand around the pipes
Back-filling
Transportation of residual soil to tip (generally about half that involved in traditional construction).

The new prefabricated pipeline systems are mostly cheaper and better than the traditional methods as regards resistance to ground water, most of them can in fact be totally immersed in water without becoming waterlogged. Many manufacturers have entered this growth field, marketing various types of prefabricated pipelines. Some of these systems are described below – there is no space to be fully comprehensive.

Pan-Isovit system

This West German system is used for district heating pipelines of between 20 mm and 600 mm internal diameter (**9.4**). The insulation consists of hard polyurethane foam of a density 80 to 90 kg/m^3 and a thermal conductivity of 0.0256 W/m K at 50°C (122°F) which is foamed into the space between the service and the protecting pipes. The protecting pipe is of high density

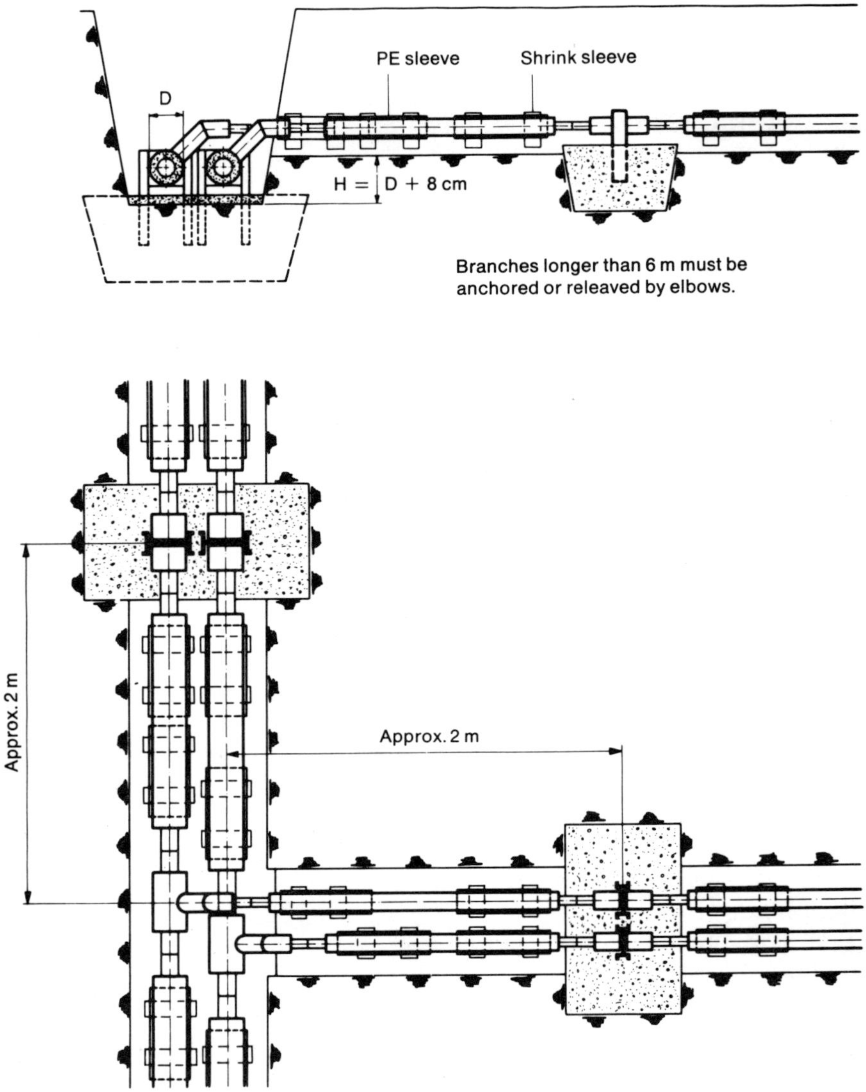

9.4 Laying of Pan-Isovit system

polyethylene which retains its elastic properties down to −50°C (−58°F). It has the following relevant properties:

Density:	955 kg/m³
Melting point:	127 to 131°C (261 to 268°F)
Coefficient of thermal expansion between 20 and 90°C (68 and 194°F):	2×10^{-4} °C^{-1}
Thermal conductivity:	0.041 W/m K at 50°C (122°F)

Table 9.4 below gives the recommended dimensions used for the different internal diameter pipelines. Heat loss is determined from the equation:
$q(t_1 - t_2)$ W/ metre run, where
t_1 is the average operating temperature (°C)
t_2 is the average soil temperature surrounding the pipes (°C)
L is the distance between fixed points or compensators (metres).
The insulated pipelines are supplied in lengths of 6 m for up to 200 mm internal diameter and of 12 metres for 200 to 600 mm internal diameter.

TABLE 9.4 DESIGN DATA FOR PAN-ISOVIT PREFABRICATED PIPELINES

Internal diameter (mm)	External diameter (mm)	Insulation thickness (mm)	Weight (kg/m)	Heat loss (q) (W/m K)	Assembly length (L) (m)
20	90	28	2.7	0.144	40
40	110	31	4.4	0.211	60
65	140	28	7.6	0.287	90
80	160	32	9.5	0.295	100
100	200	39	14.2	0.311	110
150	266	44	22.5	0.379	130
200	334	47	30.1	0.411	130
300	450	57	50.1	0.537	150
400	560	70	78.7	0.547	170
500	670	72	107.7	0.641	200
600	800	84	161.0	0.651	250

Intermediate internal pipeline diameters are also made: the above have been selected as examples.

Prefabricated insulated systems are available in copper (used for domestic hot water, condensate and/or chilled water). These are made in internal diameter sizes from 15 to 70 mm or as duplex systems with twin pipes with different diameters.

Pipeline joints are butt-welded and insulated separately. Special units are made for T-junctions, bends, fixed points and other attachments. The system can be used for hot water or steam up to a maximum temperature of +130°C (266°F) and a maximum pressure of 20 bars.

Kelit pipelines

These pipelines are made in Austria and by 1975 150 kilometres had been laid. Kelit units may be either smooth seamless mild steel pipelines with internal diameters between 16 and 352 mm featuring plain ends suitable for butt welding or galvanised steel pipes with threaded ends. The insulation consists of hard polyurethane foam with the following properties:

Density:	Minimum of 80 kg/m^3
Closed pores:	Minimum of 90 to 95%
k-value:	0.023 to 0.024 W/m K at 50°C (122°F)
Crushing strength:	400 to 600 k Pa

Two materials are used for the external protecting cover. For pipelines with external diameter below 250 mm, impact-resistant hard PVC 100 is used which resists chemicals and flames. For diameters in excess of 250 mm isotatic bright-grey RAL 7032 polyolefin is employed which is light, resistant to ultraviolet rays and chemically inert.

Pipelines are delivered to site in lengths of 6 metres with 150 mm of uninsulated sections at both ends, so that the length of insulation and protecting jacket is 5.7 metres. The company also supplies the following:

Socket pipe connectors: plastic sleeves with an internal diameter of about 3 mm more than the external pipe diameters to be joined. The sleeve is slid upon the end of one of the pipes to be joined and the pipeline ends are butt-welded, using the usual X-ray or radiographic checks. The sleeve is then pushed over the joint and polyurethane foam is pumped in through the valve onto the sleeve under considerable pressure so as to glue the sleeves to the covers, making these rigid and watertight. When polyurethane foam is pumped under pressure into a space 2 to 3 mm wide, it becomes almost completely compressed

Elbows with bends of 15°, 30°, 45° and 90°

Branched units of various types

Reduction sleeves used to enable pipelines of different external diameters to be joined together

Fixed point units: a coated steel plate is welded concentrically to the internal steel pipe and the mantle is attached to the steel plate by silicone to exclude penetration. After assembly the fixed point is cast into concrete.

Wall connections are made by means of plastic sleeves with a diameter of 300 mm (other diameters are also available) with twin grooves into which synthetic elastomer rings are inserted. The insulated pipe is then pushed through. The rings make a good and watertight seal.

The pipelines resist a maximum direct pressure upon them of about 200 kPa (kN/m^2). The materials neither age nor are they attacked chemically by any external agents with which the pipes may come into contact.

When laying Kelit pipelines a trench of at least twice the external width of the pipeline is dug of sufficient depth to provide a minimum of earth cover of 800 mm. Wooden supports with 100 mm $\times$ 100 mm cross section are positioned every 3 to 4 metres and the compensator shafts are built. A sleeve is slipped over each pipeline and the units together with bends, junctions etc are laid into the trenches. Butt-welding and painting of the welded ends with corrosion resistant paint is carried out. The junctions are completed with polyurethane foaming using a special machine. Sand is placed to provide a 100 mm cover all around the pipes. Fixed points and wall entries are concreted and the excavation is back filled.

Numerous firms throughout the world market systems of prefabricated pipelines consisting respectively, of steel service pipes, polyurethane foam insulation and a mantle pipe made from either rigid polyethylene, PVC or other plastic copolymers. Such systems are becoming increasingly popular as they combine cheapness with reliability. Some of them are dealt with below.

The G and H system

This system is marketed by Grunzweig and Hartmann AG of West Germany. It employs pre-insulated steel pipelines in 6 metres lengths which are supplied together with fixed points, junctions, reduction units etc. The system differs from others available in that there is a patented heat-resistant gliding layer on the surface of the steel pipes. The steel pipes are held concentrically within hard PVC protecting pipes and the space in between is foamed with polyurethane foam. The gliding layer ensures that the insulation layer is not damaged as the pipelines expand and contract. At bends an elastic foam is used to allow for thermal expansion and contraction. The protecting PVC pipes are completely watertight and no water penetration is possible at the joints. The polyurethane is foamed with Freon 11S and has completely closed pores. The thickness of polyurethane is between 26.1 and 37.6 mm. It is advocated that changes in direction should be carried out as far as possible at angles of 90° although other angles are also made. Standard U-compensators are supplied with fixed points at 67 metre intervals using twin pipelines suitable for operating temperatures below 130°C (266°F). The plastic cover at the ends of the pipelines must be protected against damage during the welding operation. The joints are insulated with twin semi-annular Moltopreen insulation pieces positioned round the joint of the steel pipes, made watertight and consolidated by bitumenised rubber foil which is glued around the joint. A further cover is then shrunk on in situ. Alternatively a similar technique can be used to the one employed by the Kelit company. The density of the polyurethane foam used is 50 kg/m^3; it has a conductivity of 0.029 W/m K at an average temperature of 20°C (68°F). The external unplasticised PVC is stable up to a maximum temperature of 80°C (176°F).

This system is suitable for hot water or steam at a maximum temperature of 130°C (266°F). The fixed points are pipeline units on to which the PVC protecting pipes join and they are intended to be concreted in. Pipeline reducing pieces are made in standard lengths of 1.2 and 1.5 metres.

Thermo-pipe

This system is marketed by the British firm of Pipe Conduits Ltd. It consists of steel service pipes which are covered by polyurethane hard foam insulation and a high density polyethylene casing. The polyurethane foam has a density of 80 to 90 kg/m^3, a thermal conductivity of 0.026 W/m K at 20°C (68°F) and a compressive strength of about 830 kPa. The cell structure is about 91 to 94 per cent of non-connecting pores. The high density polyethylene has a density of 955 kg/m^3, and a tensile strength of 21.6 to 26.5 MPa. Jointing is carried out by means of the technique shown in **plate 31** using a high density polyethylene sleeve with polyethylene shrink sleeves.

Wanit

This is another West German system. It employs seamless steel pipes coated with a layer of protecting paint and inserted on site into the asbestos cement core pipe/insulation/mantle pipe unit so that there remains an air gap between 4 and 15 mm width. The width of the gap between the core pipe and the external mantle pipe depends upon whether the line is a flow or a return main. The space between the two asbestos cement pipes is filled with polyurethane foam of density 60 kg/m^3. The system is stable for use with operational temperatures of up to 130°C (266°F) and for short period temperatures of up to 150°C (302°F).

The pipeline units are made in lengths of 4 metres and are joined by means of threaded asbestos cement connectors with rubber sealing rings. A special asbestos cement support ring is inserted at each end of the core pipe to support the steel pipe inside. Fixed points are made of one metre long asbestos cement piping shaped in such a way that it can readily be embedded in concrete. The steel pipes are supplied with welded flanges which hold them firmly in position against this asbestos cement unit. A very similar technique is used at entries to buildings. These fixed points have to be insulated separately and are made watertight on site.

The asbestos cement used for the core pipe and mantle pipe is made from Portland cement and has very considerable strength. Its thermal conductivity is 0.33 W/m K. The manufacturers recommend the use of axial compensators.

This system is particularly suitable when installed underneath roads with heavy traffic conditions. Mantle pipes are made in diameters of up to 700 mm and should not be decked in with an earth cover exceeding 6 metres. The asbestos cement is normally painted on the outside with protective paint but is wholly resistant to water and other penetrations.

Various supplementaries are available:

Pipeline units with oval core pipes are used close to bends and U-type compensators to permit the steel pipe to deflect without touching the core pipe
Junction units to connect pipelines of differing cross section
Standardised 90° bends
Straight run fixed points
Fixed points at bends
Prefabricated manholes.

After installation the system is tested with compressed air.

Eternit

This is a Swedish system of prefabricated pipelines and it also employs asbestos cement as an external mantle pipe (**9.5** and **9.6**). The density of the pipes is between 1900 and 2000 kg/m³. They are completely resistant to

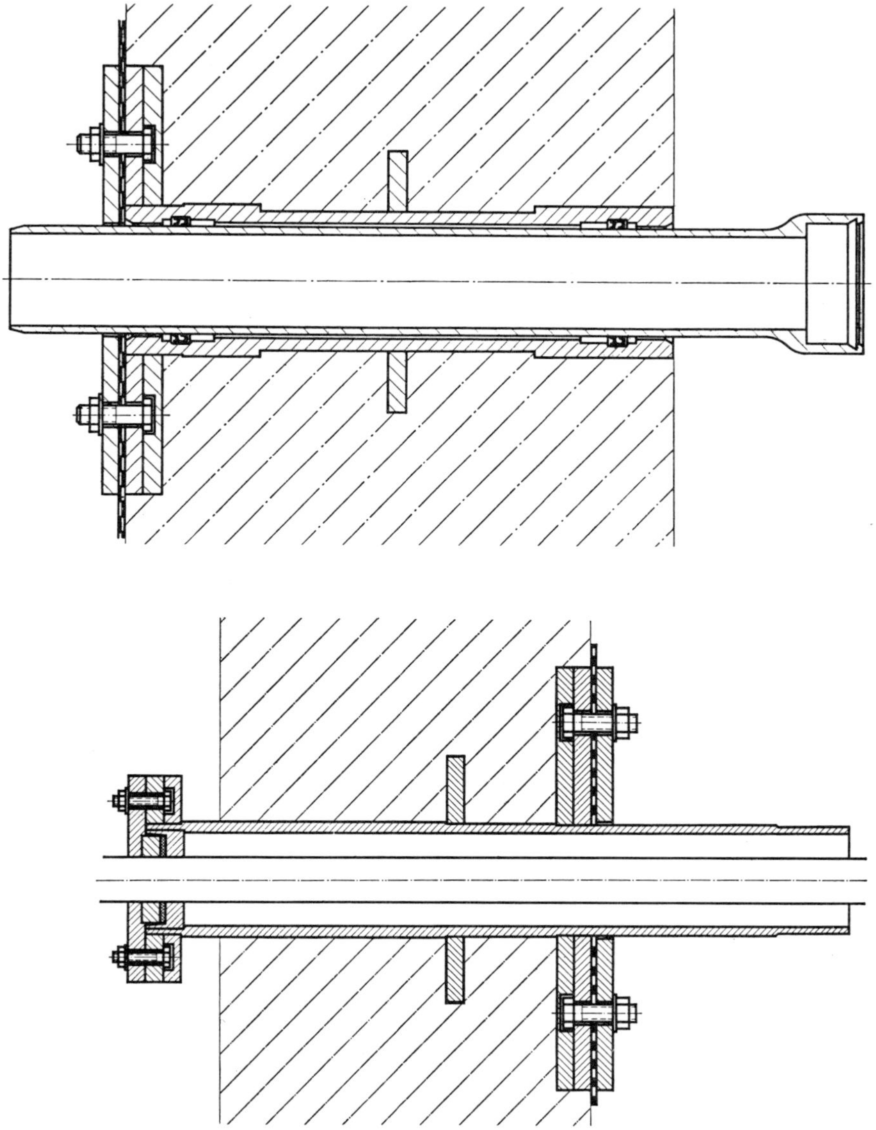

9.5 Combined house connections and fixed point with Eternit systems

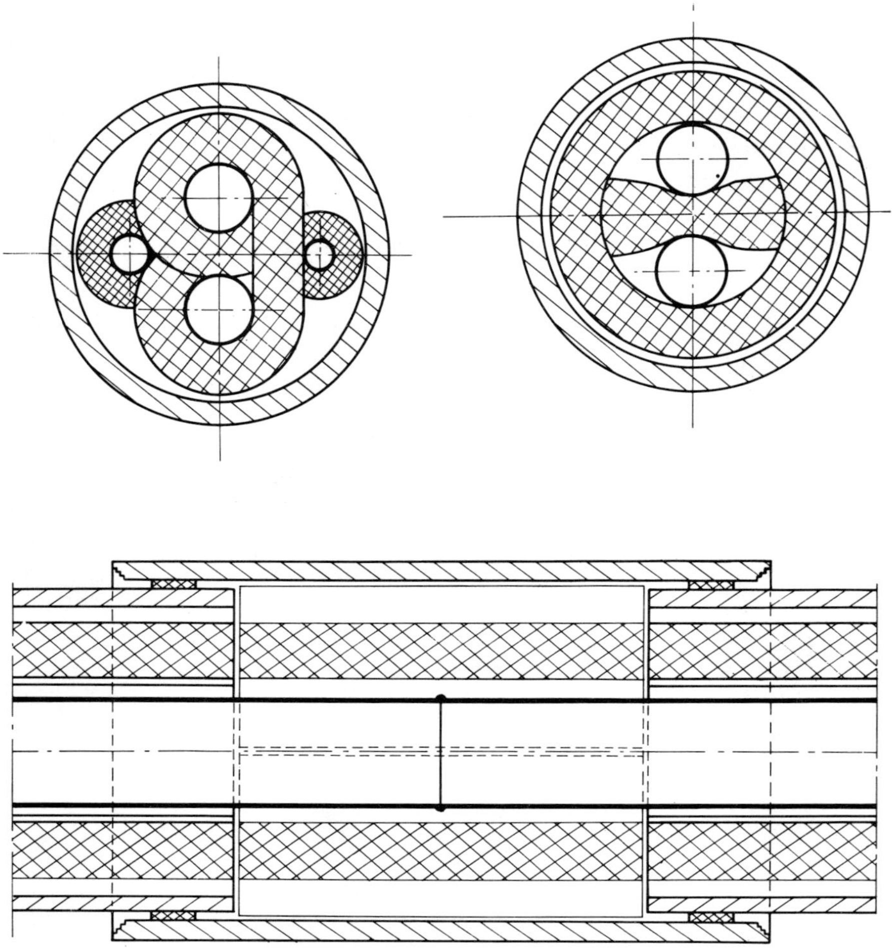

9.6 Sections through Eternit pipeline systems

neutral and alkaline waters, have a thermal conductivity of 0.465 W/K and are completely stable within the temperature range of $-50°C$ to $+300°C$ (-58 to 572°F). The mantle pipes are made in two grades for normal and for heavy traffic respectively. The normal type used for district heating is capable of withstanding static axle pressure of 14 tonnes + 40 per cent additional dynamic pressure when buried at a depth of 800 mm. The pipes are made in lengths of 5 metres. For district heating the strong external asbestos cement mantle pipe is combined with a thin-walled internal asbestos cement pipe and the space in between is filled with foamed polyurethane. Alternatively pipelines insulated with mineral wool are laid directly into the mantle pipes. The system can be used with single pipelines, twin lines or multiple lines. Steel pipeline supports or asbestos cement separators are used to locate the pipes within the mantle tubes.

The pipelines are butt-welded in the usual way naturally incorporating the various expansion units. The asbestos cement mantle pipes can be joined flexibly by means of a specially designed rolling coupling which employs an outer asbestos cement tube together with two heat and degradation-resistant rubber rings. A more rigid connection is the Gibault connection, which consists of a 600 mm long asbestos cement pipe, two rubber annular packings and steel flanges with threaded rod connectors. Modified connection units are used when pipelines of differing diameters have to be joined.

The Eternit pipeline system is laid into a well-drained trench, upon a layer of shingle about 150 mm deep, followed by coarse-grained sand up to a depth of about 150 mm above the level of the pipe. The top of the pipeline should be about 800 mm below the road surface.

Permapipe

This is a British system which used to be marketed by Haden-Carrier Ltd (**9.7 and plate 33**). It consists of single or multiple internal pipes designed to comply with specification and application requirements. All interior joints are tested to at least one and a half times the operating pressure. The internal pipes are insulated by means of rigid, pre-formed calcium silicate sections which have a k-value of 0.05 W/mk at an average temperature of 95°C (203°F), high compressive strength and good resistance to deformation. The calcium silicate withstands repeated boiling out and drying cycles extremely well. It is unaffected by water and cannot warp, shrink or

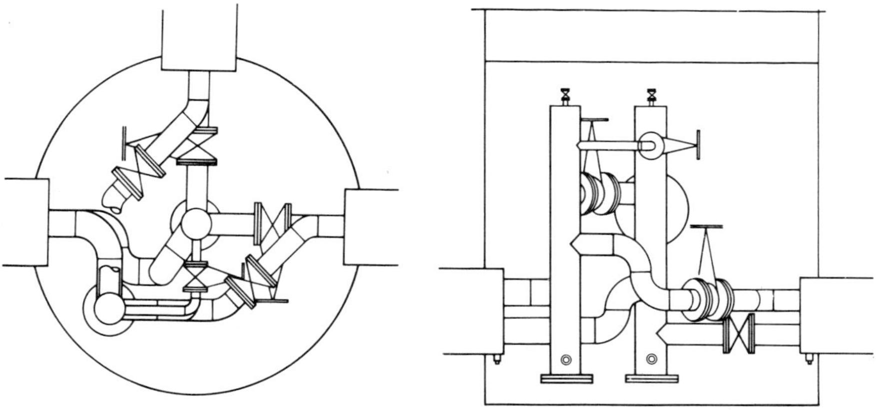

9.7 Permapipe prefabricated manholes: plan view (left) and elevation (right)

crack. It is temperature stable to 650°C (1200°F) and can dry out readily even if soaked through with water.

The insulation units are supported at intervals not exceeding 3 metres by special spacer supports which isolate the inner service pipes thermally and electrically. Abrasive action between the spacer support, inner piping and casing is eliminated. A nodal slip ring provides non-restricted drainage and air flow and spreads the loading over the lower 180° of the casing. The steel lagging cover which is banded tightly around the insulation slides easily inside the nodal ring with thermal movement. Slip ring arrestors prevent movement beyond the load bearing lagging plate. There is a 25 mm wide air gap between the inner conduit wall and the outer insulation surface which provides a thermal barrier as well as a passage for ventilation and drainage.

The casing for the Permapipe system is of spirally welded steel never less than 3.2 mm thick. Most installations are given a second line of defence against corrosion by cathodic protection. It is possible to enclose one, two, three or even more service pipes within the outer casing which may be either steel or copper. Steel casing is suitably coated and wrapped with impregnated reinforced glassfibre tissue.

The Permapipe system is individually designed to suit each specific application. Special components are provided:

A field closure kit to enable pipeline sections to be joined together
Twin piece expansion loops
Elbows
Tee-junctions
End fittings
Anchors
Combination anchor and end fitting
Combination anchors and elbows
Gland seals
Eccentrics.

The prefabricated sections which already include expansion loops, elbows and tees are transported to the site ready for assembly. Nylon or canvas slings are used during handling to prevent damage to pipeline coatings and all service pipes and conduit ends are braced and protected during transit. For welding purposes the interior service pipes extend 150 mm beyond each end.

The Permapipe system is bedded on sand or pea gravel, compacted to a depth of 100 mm (**Plate 33**). After testing the service pipes, insulation is applied to the field joints and the conduit closure sleeves are welded into position. The conduit is then tested by means of compressed air and the field coating is applied to the sleeve joints. The complete conduit is 'Holiday' spark tested at 7000 volts and the excavated soil is back filled. Under grass or other soft surfaces a minimum earth cover of 450 mm should be provided above the conduit; underneath concrete, pavements or roads there should be a minimum cover of 100 mm of sand and at least 600 mm of earth.

Table 9.5 gives the average heat losses experienced with various twin pipe Permapipe systems.

TABLE 9.5 HEAT LOSSES FROM TWIN PIPE PERMAPIPE SYSTEMS (watts per metre run)

Internal pipeline diameter (mm)	External conduit diameter (mm)	Average temp 80°C (176°F)	Average temp 100°C (212°F)	Average temp 120°C (248°F)
15	244	33	41	50
25	244	43	55	66
40	273	53	67	82
50	324	62	79	95
65	355	71	90	108
80	355	77	98	119
100	486	91	116	140
150	559	93	117	142
200	711	112	142	172
250	813	132	187	202

Nitrogen can be used within the conduit in lieu of air. It should be pressurised to 0.33 bar (gauge) and monitored at various sections of the distribution network. It offers the following advantages:

Nitrogen is an inert gas with low moisture content: it reduces the risk of corrosion

Uninsulated lines do not sweat due to reduction of moisture in the system

A pressure of $\frac{1}{3}$ bar is adequate to exclude ground water from the system should a leak occur

Instant alarm is given when the pressure changes. An increase in pressure indicates a leak in the service pipe while a reduction in pressure gives warning of a leak in the external conduit.

The prefabricated manholes are of welded steel construction and contain valves, gauges, future connections, steam trap assemblies, high and low vents, a sump and other equipment. Complete off-site assembly minimises site labour and installation time. Each prefabricated manhole is air pressure tested and adequately corrosion proofed.

The company also offered prefabricated walk-through tunnel systems in lengths of 12 metres with access for inspection, maintenance or modification without requiring further excavation.

Pass STM

This system is West German and consists of steel internal pipes insulated with calcium silicate in full, half or segmented sections. The system is temperature resistant up to 700°C (1242°F). The mantle pipes consist of longitudinal or spirally welded steel protected on the outside by glassfibre reinforced bitumen 6 to 8 mm thick. Units are made in lengths of 12 m with various accessories such as compensators, bends and T-junctions already built in at the factory. After installation the space between the mantle pipe and the internal pipe is filled with an inert gas.

Lebit and Perrubit (bituminised insulation)

The two German methods Lebit and Perrubit employ a special patented type of bituminous material which is thermoplastic, corrosion-free, resistant to aging and made in three grades to withstand operating temperatures of 110°C, 150°C and 180°C (230°F, 302°F and 356°F) respectively (**9.8**). The physical properties of the Lebit material are as follows:
Density: 880 kg/m³ ± 3%
Compressive strength (according to DIN 1996): 2.746 MPa
Tensile strength (according to DIN 1996): 0.539 MPa
Thermal conductivity at 100°C (212°F): 0.105 W/m K

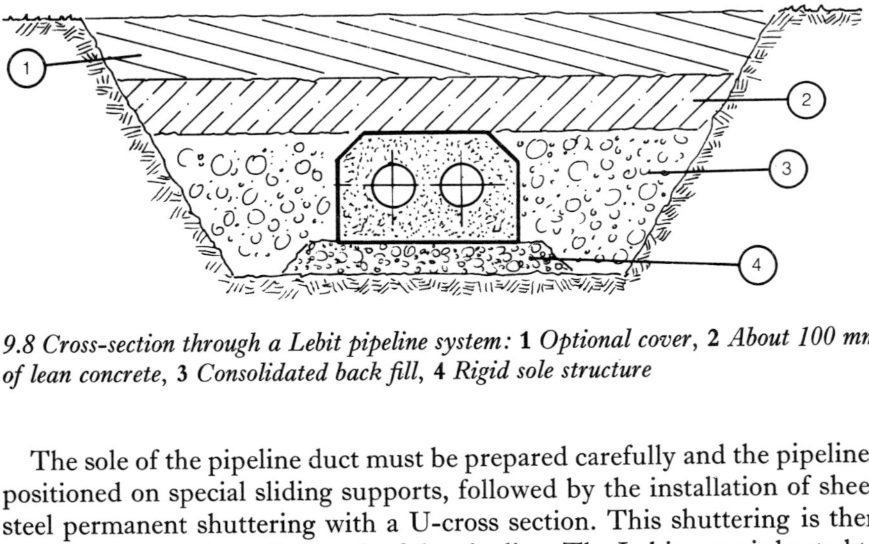

9.8 Cross-section through a Lebit pipeline system: **1** *Optional cover,* **2** *About 100 mm of lean concrete,* **3** *Consolidated back fill,* **4** *Rigid sole structure*

The sole of the pipeline duct must be prepared carefully and the pipelines positioned on special sliding supports, followed by the installation of sheet steel permanent shuttering with a U-cross section. This shuttering is then riveted together along the length of the pipeline. The Lebit mass is heated to between 170 and 210°C (338 to 410°F), depending on grade, and poured into the permanent shuttering. After about 48 hours the mass has solidified into a dense highly resistant (but still elastic) insulation block. The sheet steel shuttering is not removed but remains in the soil.

This system can also be used for renovating existing district heating pipelines where the insulation has become defective.

Perrubit pipelines use materials for the insulation of prefabricated pipelines which resemble Lebit. Two types are made for maximum operating temperatures of 140°C (284°F) and 200°C (392°F) respectively. The pipelines are made in lengths of 6 metres with 5.5 metres of insulation leaving the joints to be insulated in situ after butt-welding. The insulation comprises easily fusible bitumen which enables the pipe to expand freely independent of the insulation layer. The Perrubit layer follows: the outer cover consists of compressed cardboard impregnated with bitumen and suitably coated on the outside.

The advantages claimed for this system are:

Good thermal insulation

The sliding area close to the steel pipe prolongs the life of the insulation indefinitely (**9.9** and **9.10**)

Low weight permits easy assembly

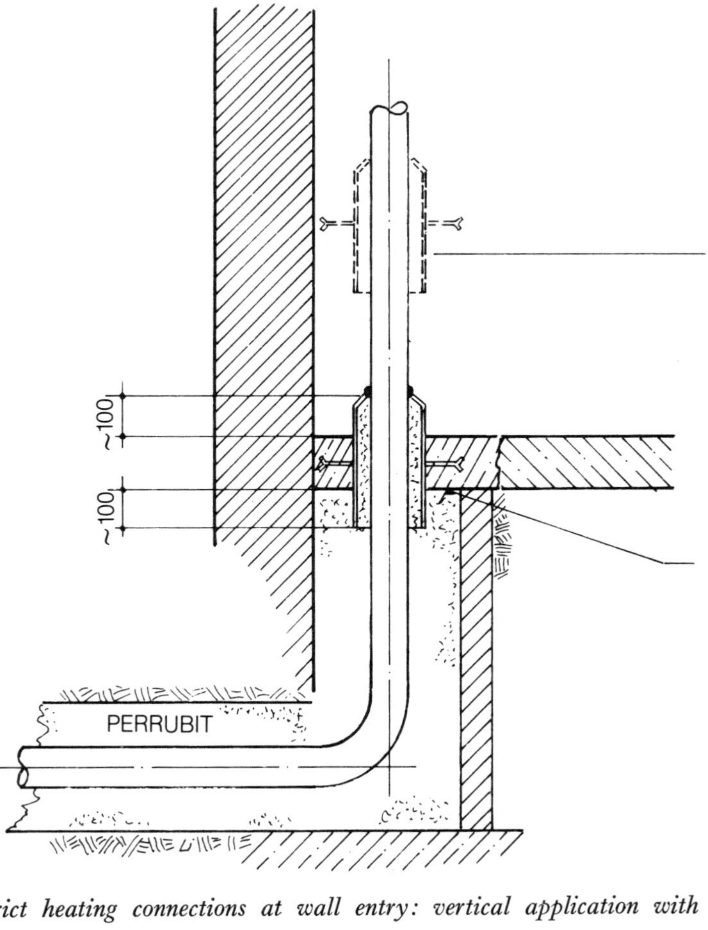

9.9 District heating connections at wall entry: vertical application with sliding connection

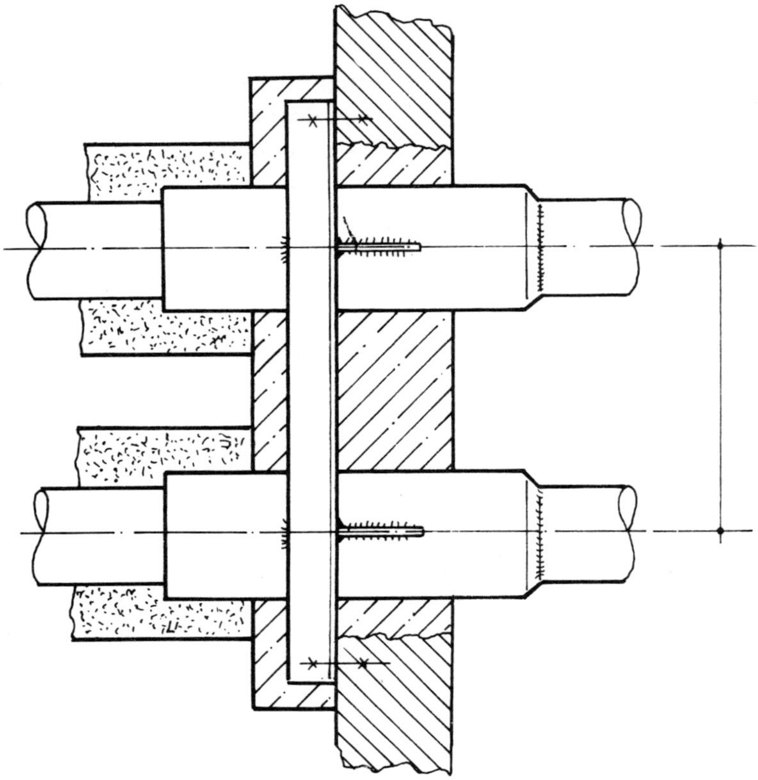

9.10 District heating connections at wall entry: horizontal application with fixed point

The system can easily be assembled as other components such as tees, bends etc are all made to fit in with the main pipeline units
It is possible to insulate the joints by the use of a two-component mix without the need for an asphalt heater
The system is completely water and corrosion proof under all circumstances.

Ric-Wil
This system is American in origin. Ric-Wil FRP consists of an inner pipe of fibreglass reinforced plastic. The internal pipe is provided with a PVC mantle pipe encapsulated at both ends and the space in between is filled with polyurethane foam. Field joints are not insulated because of the low thermal conductivity and high corrosion resistance of the carrier pipe. The system has the considerable advantage that the service pipe cannot corrode even if ground water penetrates to its surface. As the coefficient of thermal expansion is only 1.5 mm per 100 m/°C, expansion units are not required. The maximum operating temperature is normally 120°C (248°F) but brief

periods of operation at up to 150°C (302°F) can be tolerated. The jointing technique employs Buna N 3.2 mm thick full-faced gaskets which only have a working pressure resistance of between 10 and 3 bars depending upon pipeline diameter. Filament-wound flanges are designed to a working pressure of 10 bars throughout. Alternatively the pipes can be glued together using two-pot adhesive and the strength of such a joint is equal to that of the standard pipe. The Ric-Wil company also markets other more traditional systems of prefabricated pipelines, manholes etc (**9.11**).

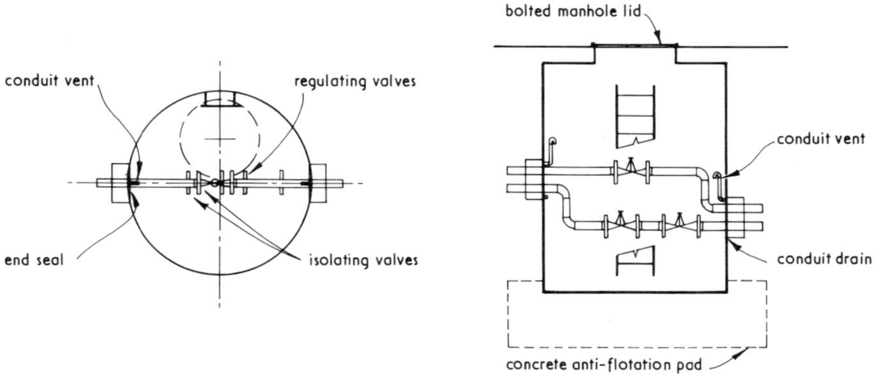

9.11 Ric-Wil prefabricated manhole

Kabelmetal

This German company markets two systems. The first of these is a steel encased pipe-within-a-pipe system employing internal service pipes with effective diameters of between 25 mm and 500 mm. These are insulated with calcium silicate and enclosed in a steel mantle pipe which is protected by tropical quality bitumastic on glass weave. Cathodic protection and fault monitoring wiring is incorporated. The units are supplied in prefabricated sections of 14 metres in length.

The second system available (Flexwell) consists of prefabricated piping with internal diameters up to 125 mm coiled onto a drum ready for transport (like an electric cable) with a maximum single length of 600 metres. The corrugated inner service pipe consists of copper or stainless steel and is insulated by means of flexible polyurethane foam positioned between the service pipe and the corrugated steel outer protection casing. The protection pipe, in its turn, is encased in a double layer of flexible bitumen rubber compound. Extruded polyethylene up to 5 mm thick is used as the external sheathing. Monitoring wire is inserted inside the polyurethane insulation layer adjacent to the service pipe for connection to a central fault warning and locating apparatus. Any damage to the service pipe can be detected immediately by a visual and acoustic warning system and can be localised within one metre of the position of damage (**plate 35**). The internal pipe is designed to ensure streamline flow for water. Copper is used for hot water for central heating purposes while stainless steel pipes are used for domestic hot water supply. The polyurethane employed as insulation inside the pipe has a k-value of 0.023 W/m K at 20°C (68°F). The maximum operational temperature is 150°C (266°F) and the maximum operational pressure 16 bars. Flexwell cables require no expansion joints and should be so used that joints are not needed on runs between buildings. Connections to supply or to heating systems can be made inside buildings by means of flanges. Special wall entries and connecting pieces are available. Flexwell cables can be combined if required with the standard steel mantle pipe. Due to the corrugated nature of the service pipe pressure losses are roughly twice those of traditional smooth surface pipes.

Aquawarm

This Swedish technique uses smooth walled copper pipes as the carrier pipe (**plate 32**). Internal diameters range between 10 and 84 mm and are designed to accommodate water flow velocities of 10 m/s at a maximum pressure of 16 bars and at a maximum temperature of 120°C (248°F). The copper pipe is insulated with glass wool of k-value 0.03 W/m K at 50°C (122°F) and the external sheathing is of corrugated high density polyethylene. End seals are provided to prevent water from seeping through to adjacent pipe sections should the sheathing of one section be damaged. The units are supplied in coils varying in length between 8 metres for the largest bore up to 25 metres for the narrowest bore. All sizes can be handled readily by two men without lifting tackle. Although copper has a higher coefficient of thermal expansion than steel the units are intended to be used without expansion joints. They are laid in trenches in a continuous sinosoidal pattern. A special jointing technique is used: the sheathing and insulation are cut away about 100 mm on both ends and the shrink-on sleeve is slipped over one of the pipeline sections; a copper sleeve is then placed over the adjoining butt ends of the copper pipes and a secure joint is made by brazing; after pressure testing the joint is insulated with semi-annular sections of glass fibre and a jointing sleeve of corrugated polythene is fitted and secured with tape; the shrink-on sleeve is next slid into position and carefully placed centrally around the joint; it is finally heat-shrunk by gas flame until the mastic coating inside the sleeve has molten and extruded at the ends of the sleeve. The resulting joint is fully watertight and has the same mechanical strength as the straight runs of pipelines. The usual supplementary units are supplied with the system.

The pipes can be laid into any soil without the need for drainage or fixed points but the back-fill to the pipelines should be surrounded by sand and should be to a depth of 0.3 metres when laid in parks or similar areas without traffic, 0.5 metres underneath pavements and 0.8 metres underneath side streets. The system appears to be unsuitable for very heavy traffic loading conditions.

The EBKO systems

The US firm of E. B. Kaiser of Glenville Illinois markets district heating pipeline systems based upon strong continuously spiral welded steel casings **(9.12)**. These are protected externally by a layer of high temperature coal tar enamel followed by a layer of glass fibre reinforcing mesh and another layer of coal tar enamel; a final wrapping of coal tar saturated glass fibre reinforced asbestos pipeline is applied under tension. The final coating thickness is a minimum of 3.2 mm with a very high impermeability rating. The casing can be insulated on the inside. The service pipes are carried inside. They may be left uninsulated, leaving an air gap between them and the inside surface of the insulation layer, or insulated by wrapping various types of pipeline insulating material around them. Metal casings are manufactured

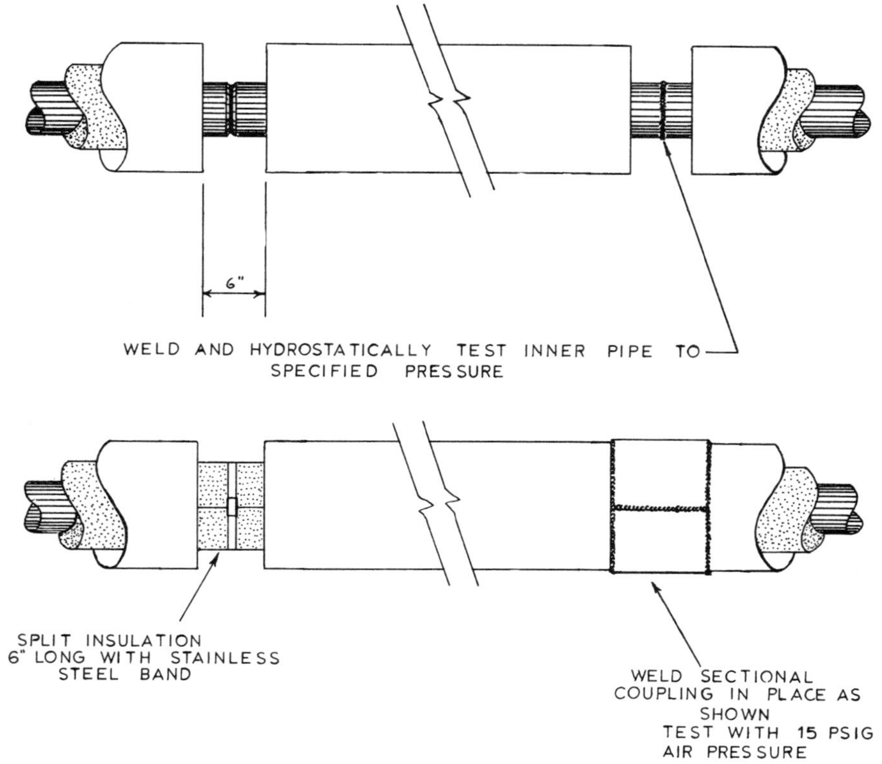

9.12 Laying an assembly of Kaiser pipeline unit

with a wall thickness of 3.42 mm to 6.35 mm. Plastic casings have thicknesses between 4.76 mm and 7.94 mm. All the units are made in various diameters with maximum sizes of 7.62 mm.

The Lubo system

This Swedish system which is operated by N. Lundberg AB of Fristad consists of steel inner tubes and phenolic resin mantle tubes (**9.13**). The space between these is filled with rigid polyurethane foam. A small drainage channel is left inside the insulation layer to enable quantities of water vapour to evaporate. Lengths of units are 5000 mm with internal pipeline lengths of 5500 mm, allowing the units to be butt-welded and insulated in situ. Joints are made from fibreglass reinforced polyester and a special rubber ring

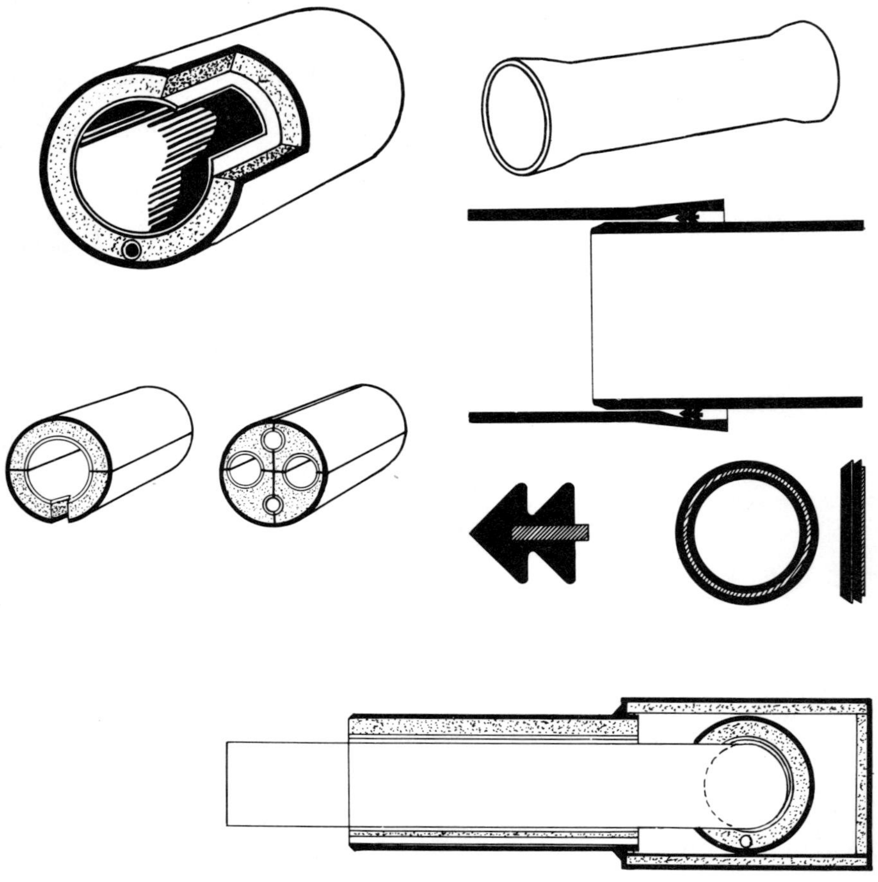

9.13 Lubo system using steel inner tubes, polyurethane insulation and phenolic resin mantle tubes

makes the joint between the jointing pipe and the protection sleeve. The compression of the nylon impregnated rubber ring is about 25 per cent.

CHEMICAL HEAT PIPES

The most expensive aspect of operating a district heating or group heating scheme is the transfer of thermal energy from the heat source to the point of use. Traditionally, this is achieved by conveying either water or steam as a heat exchange medium in pipes. Experiments have been carried out into the use of heat pipes to provide more efficient transfer of heat energy (**9.14** and **9.15**). These can transfer up to 1000 times more heat per given area than a solid rod of copper can transfer by conduction.

A heat pipe has no moving parts, requires no energy input (other than the heat it transfers) and in most cases does not even rely upon gravity for its operation. It is basically a container made of a metal which is compatible with the completely sealed working fluid. It is lined on the inside with a wick: a porous material which can be sintered metal powder combined with metal weave screens, fibrous glass cloth, felt, or similar materials. The operating liquid may be either water, ammonia, acetone, fluoro-carbon, alcohol or even liquid metal depending on the conditions under which the heat pipe has to operate. Only enough fluid to saturate the wick is introduced into the heat pipe. In the evaporator section (the part intended to absorb the heat) thermal energy is transferred by conduction through the container wall and vapourises the liquid inside the wick. This vapour then travels through the central section of the tube to the other end (the point of use) where it recondenses and gives its latent heat to that section of the tube. The liquid content of the wick at the evaporator section is replenished by capillary action.

A chemical heat pipe is constructed of two chemical reactors which are separated by twin pipelines which may be many kilometres long. At the heat absorption end a mixture of chemicals undergoes an endothermic chemical reaction (ie a reaction which absorbs heat energy). The resultant com-

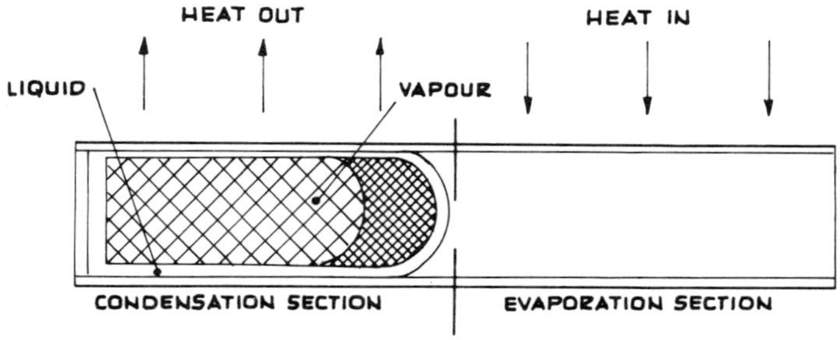

9.14 Evaporation heat pipe

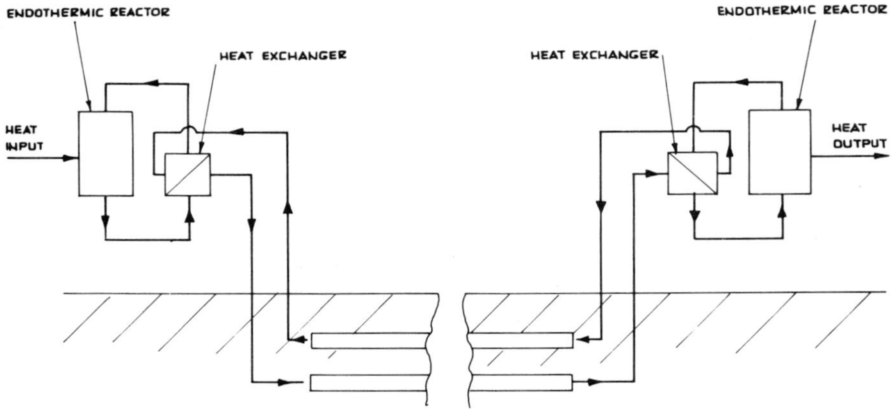

9.15 Flow sheet of chemical heat pipe system

pounds are then pumped along the pipeline to the point of use where their heat energy is released by an exothermic reaction (ie a reaction which releases heat) in a suitable chemical reactor. The de-energised chemicals are then pumped back again to the heat supply station and the process, a continuous one, re-commences. The recommended chemical reaction mixes at different temperatures are:
Above 700°C (1292°F):
$CH_4 + H_2O$ to form $3\ H_2 + CO$
275°C to 575°C (507°F to 947°F): hydrogenation/dehydrogenation reactions such as:
$C_6H_{12} \rightleftharpoons C_6H_6 + 3\ H_2$ are the most popular. Experiments have also been conducted with reactions of this type involving toluene and naphthalane. So far complete technical feasibility has not been proved: the right catalysts have still to be developed and there are also problems with regard to carbon deposition.

In the practical operation of a chemical heat pipe system, the temperature at the inlet must always be considerably higher than that at the outlet. Even if the claimed overall efficiency of heat transfer (between 80 and 90 per cent) can be achieved there is an indirect thermal loss as entropy increases with a lowering of energy grading. Taking the thermodynamic heat loss into account, transmission efficiencies are unlikely to exceed 75 per cent.

It is anticipated that chemical heat pipes may eventually become viable and useful for the transmission of heat over long distances from a nuclear reactor to an urban district heating centrale. There have for example been claims that it would be technically feasible to use a chemical heat pipe for the transmission of heat energy from a centrally located nuclear, coal or solar plant in a remote area of the country to industrial and urban areas up to 320 km distant from the heat source – much further than could be envisaged with pipelines carrying heated fluids (which would be subject to considerable sensible heat losses to the surroundings, however well insulated). There is no heat loss with chemical heat pipes because the chemicals are pumped at ambient temperature. Pressure losses involved in pumping compressed gases are low when using high temperature conversion plants and still only moderate when pumping liquids such as benzene and cyclohexane. Pipe diameters have to be very much smaller than those associated with conventional heat transfer media (for a net enthalpy flow of 1000 MW the required pumping volume per second is only 3.24 moles of the methane/carbon monoxide/hydrogen mix).

CALCULATING AN ECONOMIC THICKNESS FOR INSULATION

Finding the most economic thickness for thermal insulation is a subject which lends itself to accurate mathematical treatment: the equations derived by L. B. McMillan in 1926 are regarded as the main criteria.

Flat surfaces

Let the cost of heat loss per m^2 and year be £M. This can be equated as

$$M = \frac{h(t_1 - t_o)c}{d/k + R} \text{ £/m}^2\text{, where}$$

h is the number of seconds/year of operation
t_1 is the internal temperature (°C)
t_o is the external temperature (°C)
c is the cost of heat (£/J)
d is the thickness of insulation (m)
k is the thermal conductivity of the insulation (W/m K)
R is the combined boundary layer heat transfer resistance (m^2 K/W).
Let the cost of insulation per m^2/year be $C = xd + y$, where x is the cost of installed insulation per metre insulation thickness in £/m^3 and y is a constant.
The total cost per annum per m^2 and year is therefore
$M + C = C_{tot}$
Substituting, differentiating and rearranging, one arrives at the most economical thickness of insulation according to the McMillan equation:

$$d = \sqrt{\frac{h\,(t_1 - t_o)\,c\,k}{x}} - Rk \text{ metres}$$

Cylindrical insulation

The derivation of the McMillan equation for cylindrical insulation is much more complicated than that of an equation for rectilinear flow through a wall. The method of evaluating proceeds the same way, by first of all determining C_{tot} and differentiating with respect to d. In this case d is $(r_2 - r_1)$, the difference in insulated and uninsulated radii. On equating to zero the optimum degree of insulation can be found.
The final equation works out at:

$$\sqrt{\frac{h\,(t_1 - t_o)\,c\,k}{x}} = (r_2\,1n\,r_2/r_1 + Rk)\sqrt{\frac{2r_2 - r_1}{r_2 - Rk}}$$

where:

r_1 is the radius of the uninsulated pipe (m)

r_2 is the radius of the uninsulated pipe (m)

(1n) denotes Haperian logarithm of, or log to, the base *e*

and the other values bear the same significance as before.

When solving the problem it is possible to determine all the variables except r_2.

Example: Determine the optimum pipeline insulation if the following data apply:

$h = 250$ days/annum $= 250 \times 24 \times 3600 = 2.16 \times 10^7$ s/y

$t_1 = 120°C$

$t_o = 20°C$

$C = £2.00$ per GJ $= 2 \times 10^{-9}$ £/J

$k = 0.1$ W/m K

$x = £50/m^3$ Y

$r_1 = 0.15$ m

$R = 0.03\ m^2$ K/W

We can therefore write:

$$\frac{2.16 \times 10^7\,(120 - 20) \times 2 \times 10^{-9} \times 0.1}{50} = 0.09295$$

$$= (r_2\,1n\,r_2/0.15 + 0.03 \times 0.1)\sqrt{\frac{2\,r_2 - 0.15}{r_2 - 0.03 \times 0.1}}$$

$$\text{or } 0.09295 = (r_2\,1n\,\frac{r_2}{0.15} + 0.003)\sqrt{\frac{2\,r_2 - 0.15}{r_2 - 0.003}}$$

Solving this equation represents some difficulties. It is best to use a trial and error calculation, in that one guesses at various values for r_2 and keeps on substituting in the equation until the equation balances.

Without a scientific calculator, such an operation would be rather laborious. When using a good calculator, it does not take all that long. When many different such calculations are required a computer program can be arranged and the computer will provide the answers (for a sample programme see R. M. E. Diamant, *Insulation deskbook*, pp 51–54, Heating & Ventilating Publications, Croydon, 1977).

r_2 (metres)	$(r_2\,1n\,r_2/0.15 + 0.003)\sqrt{\frac{2\,r_2 - 0.15}{r_2 - 0.003}} - 0.09295$
0.2	−0.0248
0.25	+0.0626
0.21	−0.0088
0.215	−0.00055
0.216	+0.001129

By trial and error it would thus seem that the correct answer for r_2 equals 0.2153 metres or 215.3 mm, giving a thickness of insulation of 215.3 − 150 = 65.3 mm.

Economic thickness of pipeline lagging

The Patton equation is employed to evaluate the most economic thickness of insulation for the lagging of pipelines (**9.16**). One form of this is the following:

$$K = 0.08333/r_{unins.} \left(\frac{100\ Y\ c_h\ k\ (t_i - t_o)^{\frac{1}{2}}}{a\ C} - \frac{k}{h} \right)$$

where

$r_{unins.}$ is the radius of the uninsulated pipe (m)
($r_{ins.}$ being the radius of the insulated pipe (m))
Y is the period of operation per annum (seconds). (This is obtained by multiplying the annual number of hours of operation by 3600.)
c_h is the heating cost in £ per useful joule
k is the thermal conductivity of the lagging material (W/m K)
t_i is the temperature of the medium inside the pipe (°C)

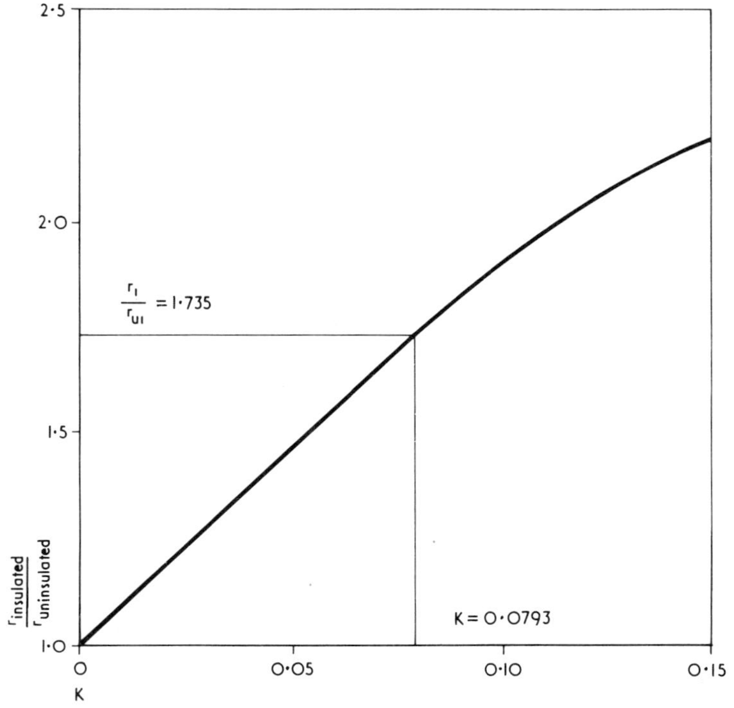

9.16 Relationship between Patton coefficient K and ratio of insulated and uninsulated pipeline radii

t_o is the external ambient temperature during the period of operation (average value) (°C)
a is the sum of the interest, depreciation and maintenance charges as a percentage of the capital spent on insulation
C is the cost of the material plus fixing charges evaluated as £/m³
h is the boundary layer coefficient (W/m² K)
Once a value of K has been obtained, this is related to the ratio

$\frac{r_{insulated}}{r_{uninsulated}}$ (from table 9.6 below).

The ratio of $R_{ins}/R_{unins.}$ can also be obtained by the following regression analysis equation:

$R_{ins.}/R_{unins.} = 0.999 + 12.0137\,K - 57.1957\,K^2$
$+ 389.374\,K^3 - 1533.4185\,K^4 + 2439.5\,K^5$

or if $R_{ins.}/R_{unins.}$ is expressed as R, then

$K = -0.01525 - 0.07615R + 0.1123075\,R^2$
$- 0.03792\,R^3 + 0.0068576\,R^4 - 0.000491\,R^5$

The economic thickness of pipeline insulation can thus be evaluated.

Example: Find the most economical thickness of pipeline insulation for a steam pipe held at 150°C (302°F) with an outside diameter of 100 mm. The lagging material costs £500 per m³ to purchase and to fit. Its thermal conductivity is 0.13 W/m K. The system is operated 3500 hours per annum and the average outside temperature during this time is 6°C (43°F). The cost of heat is £1.5 per useful gigajoule (10^9J) (or £1.5×10^{-9} per joule). The system is to be written off in 20 years and maintenance is estimated at 2% per annum. Bank interest amounts to 12% per annum.
a is therefore 12 + 5 + 2 = 19%
h under the given conditions is 9.65 W/m² K
Y is 3500 × 3600 seconds = 12.6×10^6 seconds
Substituting in our equation we get:

$$K = \frac{0.08333}{0.050}\left(\frac{100 \times 12.6 \times 10^6 \times 1.5 \times 10^{-9} \times 0.13\,(150-6)^{\frac{1}{2}}}{19 \times 500} - \frac{0.13}{9.65}\right)$$

$= 1.6667\,(0.06103 - 0.01347) = 0.0793$

Using the regression analysis equation it can be calculated that the value of

$K = 0.0793$ corresponds to a value for $\frac{r_{ins}}{r_{unins}}$. of 1.735

The insulated diameter of the pipe is thus 0.050 × 1.735 = 0.08675 m or 86.75 mm. The optimum amount of insulation is therefore a layer (87 − 50) = 37 mm in thickness.

CHECK LIST

1 Advantages/disadvantages of steam as heat conveying medium
2 Advantages/disadvantages of hot water as heat conveying medium

TABLE 9.6 TABLE OF PATTON CONSTANTS AGAINST r_{ins}/r_{unins} RATIOS

K	r_{ins}/r_{unins}	K	r_{ins}/r_{unins}
0.00	1.000	0.11	1.961
0.01	1.112	0.12	2.032
0.02	1.218	0.13	2.108
0.03	1.322	0.14	2.170
0.04	1.407	0.15	2.232
0.05	1.498	0.16	2.303
0.06	1.578	0.17	2.369
0.07	1.662	0.18	2.431
0.08	1.737	0.19	2.493
0.09	1.816	0.20	2.556
0.10	1.888		

3 Types of pipe systems available and their preferred use
4 Costs of pumping hot water
5 Ways of determining an economic pipe diameter
6 Principal methods available for constructing municipal heating mains
7 Thermal insulation systems available for use on heating mains
8 Methods of protecting pipelines with hard covers
9 Prefabricated pipeline systems available
10 Use of thermal plastic bitumen for insulating pipelines
11 Proprietary systems available for installing pipe mains
12 Use of heat pipes within the district heating field
13 Economic thickness of thermal insulation.

REFERENCES

Fernwärme International (*articles in German*)
K. H. Suttor, 'Long distance heat transmission', 3 (1974) 4, pp 115–118
W. Schmidt, 'Conduit free district heating systems insulated with foamed polyurethane', 4 (1975) 5, pp 172–178
W. Kopp, 'Supply of heat to district heating customers', 4 (1975) 6, pp L5–L13
J. Lilljequist, 'Installation problems in district heating', 4 (1975) 6, pp 206–210
K. Kirvela and R. Seppala, 'Planning nuclear district heating networks in Finland', 5 (1976) 2, pp 36–42
J. Valasec, 'Optimum thickness of insulation of district heating', 5 (1976) 3, pp 74–81
H. Zuhlke, 'Heat losses in district heating networks', 5 (1976) 6, pp 193–198
D. Bartsch, 'Searching for leaks in district heating systems', 8 (1979) 1, pp 2–5

Other sources

W. J. Brown, 'Thermographic surveying of heated pipelines pipes', *Pipelines International*, 23rd December 1978, pp 14–16

H. Brun, 'Comparative construction techniques of district heating pipelines', *Energie*, 17 (11), November 1965, pp 3–7

R. G. Courtney and J. Maccadam, 'An economic assessment of long distance heat transport and the utilisation of power station reject heat for district heating', BRE Paper 46/47, 1977

G. M. Hoyle, 'Hot water distribution in district heating', *Heating and Ventilation Engineer*, November 1974, pp 158–166

J. Lind, 'District heating pipelines', *Building Services Engineer*, 43, September 1975, p A 23

J. S. Quigg, 'Underground distribution systems', *Building Services Engineer*, 44, September 1976, p A 50

N. Rueth, 'The pipeline with a difference', *District Heating*, 61 (3), 1976, pp 24–29

CHAPTER 10

Water treatment

PRINCIPLES

The water temperatures in district heating networks are relatively high and for reasons of cost mild steel pipelines are used throughout. Corrosion anywhere in the system can be serious and expensive. It should be prevented by the adoption of thorough methods of water treatment.

It is most important that there be an almost total absence of traces of oxygen in the circulating hot water. This is usually achieved by the addition of hydrazine to the district heating water. The method is indeed most effective but has been found to lead to the formation of ammonia in the water with a consequent attack on the components of the system. To avoid this problem it is now more common to reduce the initial oxygen content by means of thermal techniques which then make it possible to use much smaller quantities of expensive hydrazine.

Maintenance of the correct pH value of the water is also most important. At very high pH values water reacts with steel surfaces to form top layers of magnetite (Fe_3O_4) which protect the rest of the steel against further corrosion. Most district heating systems are therefore operated at such high pH values, usually between 9.5 and 10. This high alkalinity is achieved by adding pure sodium hydroxide to the circulating water. It has however been found that aluminium radiators suffer severe damage through the very alkaline nature of the circulating water. They should therefore only be used in cases where there is an indirect system of district heating in which the very alkaline primary flow circuit heats up a secondary circuit with a far lower free alkalinity to which the aluminium radiators are connected.

To avoid scaling (or furring) in the system a level of 1 to 2 g of phosphate concentration P_2O_5 per cubic metre of circulating water should be maintained; this will avoid the build-up of hardness salts in the water caused by external sources of untreated water penetrating into the network.

TREATMENT OF CRUDE WATER PRIOR TO USE

When crude feed water contains insoluble materials these have to be removed by filters so as to avoid blocking valves and other parts of the system (**10.1**). Such damage will be particularly serious where the district heat is supplied directly and where sensitive equipment (eg Woltmann velocity meters, electronic heat integrators etc) is fitted. Ion exchange resins used in ion exchange equipment can for example be ruined by suspended

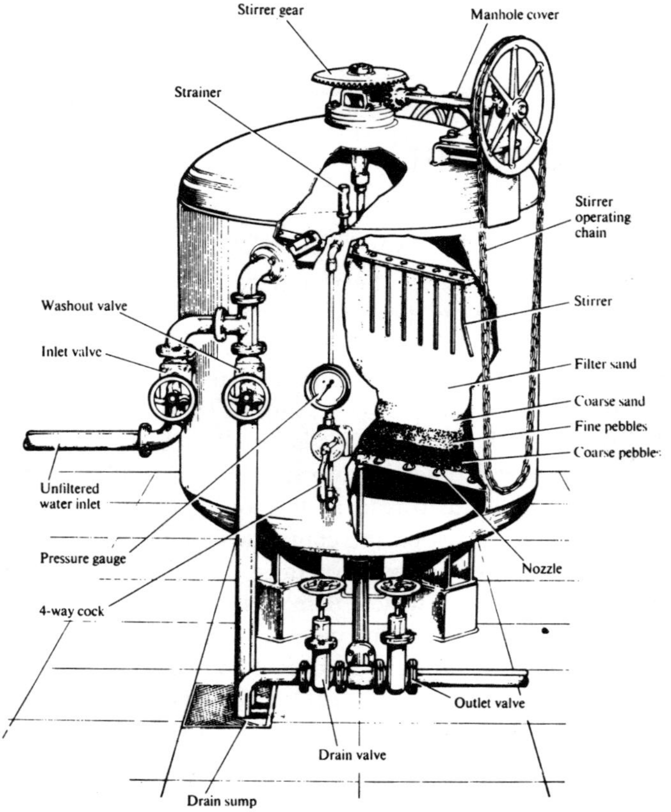

10.1 Cut away view of industrial sand filter

particles of iron oxide which act as oxidation catalysts.

Sand filters are most commonly employed to remove insoluble materials although carbon filters are more effective. The filtering medium in the latter is a mixture of ground anthracite and hydroanthracite; as a filter mass it retains iron oxide better than sand filters and, in addition, it lasts longer between backwashes. But carbon filter masses must be replaced more frequently than sand because they tend to adsorb some of the impurities and on balance their disadvantages therefore outweigh their advantages.

REMOVAL OF OXYGEN FROM CIRCULATING WATER

Mechanical deaeration methods

The common methods for distilling water permit oxygen concentration to be reduced to 0.1 ppm but better results can be achieved using modern vapour purification plant which can achieve oxygen concentrations down to 0.01 ppm.

Oxygen and other dissolved gases can be removed from steam by various 'scrubbing' techniques. Some plants using such techniques are capable of reducing oxygen content to below 0.007 ppm, the target figure in British power station practice. With less efficient plant a final reduction of oxygen concentration is carried out by the use of hydrazine.

In the desorption process water is mixed with completely oxygen-free nitrogen in a jet injector at a pressure exceeding 3 bars. This reduces oxygen concentration to about 0.04 ppm and final reduction (to virtually zero) is then made with hydrazine.

In flash-type deaeration pressurised water is heated to a temperature above its boiling point (**10.2**). It is then sprayed into an evacuated chamber where it

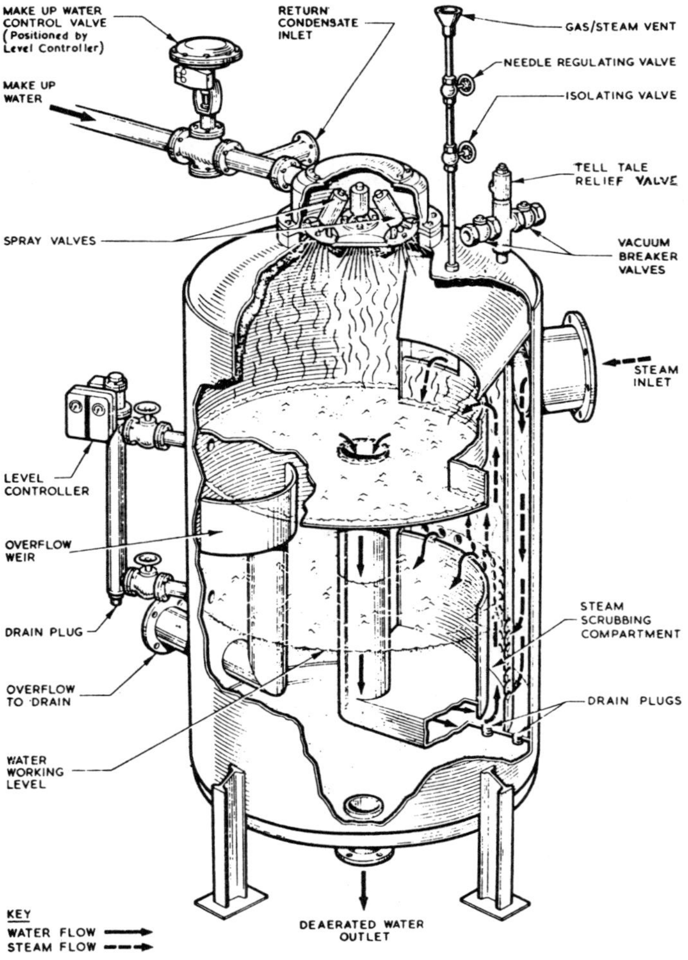

10.2 Section through an industrial water deaerator

boils. Most of the incondensible gases can then be pumped off. If the feed water comes from a turbine exhaust it is possible to reduce the oxygen content to below 0.07 ppm.

Use of hydrazine

The oxygen content of water can be reduced to about 2 ppm by storing it for about 30 minutes at a temperature just below its boiling point. When other methods are inadequate in reducing the oxygen concentration to less than 0.005 to 0.007 ppm (required for most systems), hydrazine treatment is best adopted. Hydrazine reacts with oxygen as follows:

$$N_2H_4 + O_2 \rightarrow N_2 + 2H_2O$$

The liberated nitrogen serves the useful function of driving the oxygen out of solution so that it can be bled off. Some of the hydrazine tends to decompose under the action of heat as follows:

$$3N_2H_4 \rightarrow 4NH_3 + N_2$$

For this reason it is generally necessary to double the stoichiometric quantity of hydrazine to eliminate oxygen. As the molecular weight of hydrazine is identical to that of oxygen (ie 32) two parts by weight of hydrazine are required to eliminate one part by weight of oxygen. The nitrogen produced by this process is harmful in district heating networks and must be removed separately.

Hydrazine is far more expensive than sodium sulphite (see below) so that no more of the material than strictly necessary should be used. The hydrazine-oxygen reaction is accelerated by the presence of activated carbon; hence filters using this material can be usefully provided between the point where hydrazine is added and the pump. Hydrazine reacts with the free ferric oxide which is present in the water; it is therefore commonly necessary to add a considerable excess of hydrazine to the system when starting up.

Hydrazine offers the advantage that no harmful dissolved solids are formed and the ammonia which is liberated in the process can reduce the concentration of free CO_2.

Use of sodium sulphite

Sodium sulphite reacts readily with oxygen as follows:

$$2Na_2SO_3 + O_2 \rightarrow 2Na_2SO_4$$

About eight parts by weight of anhydrous sodium sulphite (or 16 parts by weight of hydrated sodium sulphite) is required to remove one part by weight of oxygen. The reaction is catalysed by copper and cobalt salts and occurs most readily at elevated temperatures between pH values of 5 and 8. Organic compounds such as glycerol or sugars tend to inhibit the reaction. Commercial sodium sulphite used for deoxygenation of boiler water usually contains 0.25 per cent of cobalt sulphate. At high temperatures sodium sulphite splits up into sodium hydroxide and sulphur dioxide. Sulphur dioxide produces sulphurous acid with water (extremely corrosive in its own right). The advantages of using sodium sulphite are that it is cheap and that the sodium sulphate produced during the deoxygenation reaction is often needed to prevent caustic cracking. The disadvantages are that quantities of sulphates are introduced which may cause scaling in the presence of free calcium ions.

Use of ion exchange

A number of different ion exchange resins are available which effectively reduce oxygen concentration in boiler water to very low values. These are: anion resins which have reduced copper or silver, ferrous hydroxide, manganous hydroxide, or other reducing agents co-ordinated with them. The oxygen concentration of water pumped through such resins can be reduced to less than 0.01 ppm. Metals are regenerated by sodium hyposulphite and ferrous or manganous compounds with a mixture of ferrous sulphate and sodium hydroxide solutions but this method is not currently as widely used as the other available techniques.

CORROSION AND OTHER OBSTACLES TO EFFICIENCY

Additives to high pressure boiler water

Octadecylamine (chemical formula $C_{18}H_{37}NH_2$) can reduce the amount of corrosion in pipelines by over 90 per cent and is stable up to temperatures in excess of 500°C (932°F). However it possesses the unwelcome tendency of detaching existing rust deposits with likely deleterious effects because the loose rust can block up steam traps, valves etc. The octadecylamine is usually fed at a rate to maintain about 1 to 2 ppm in the water; at this concentration the required corrosion-resistant surface film on metals is not unduly disturbed.

Cyclohexylamine or morpholine can also be used as additives. For water free of carbon dioxide (CO_2) a concentration of either 1 ppm of cyclohexylamine or of 4 ppm of morpholine will achieve the desired pH of 9. These substances are preferred to hydrazine as they do not attack the copper fittings in the system (**10.3**).

Nitrogen in circulating water

Nitrogen is an inert gas and does not aid corrosion (unlike oxygen); nevertheless its presence in district heating pipelines is most undesirable.

Nitrogen is usually formed during the reaction of hydrazine with oxygen. It has a tendency towards coming out of solution in high-lying sections of the district heating network, forming nitrogen cushions inside radiators and pipelines which then interrupt the heat supply and have to be frequently vented from the system. To avoid this nuisance district heating water must pass through a degassing plant following the addition of hydrazine. The degasser would be positioned at the hottest part of the pipeline and the gas extracted by subjecting the water to agitation and low pressure.

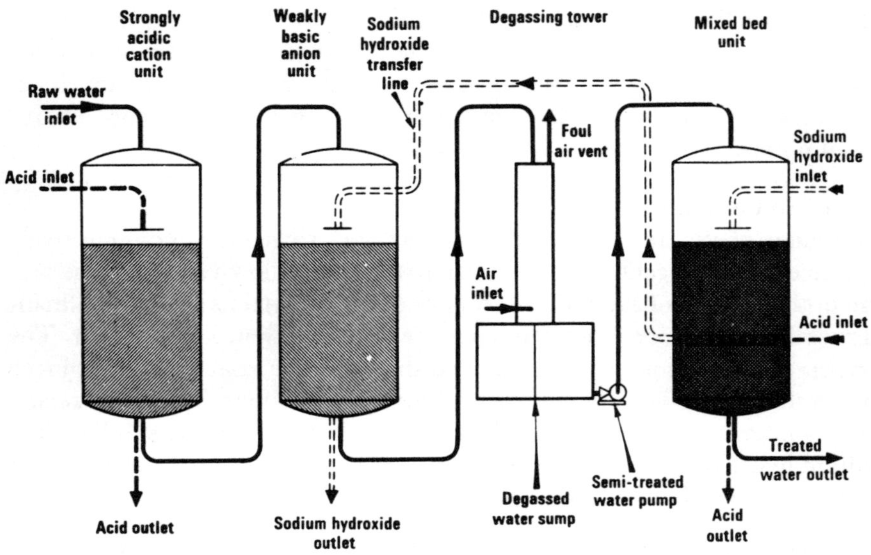

10.3 Feed water treatment for high pressure boiler plants

Sulphides in circulating water

When the circulating water is alkaline (as is more usual) sulphides exist in the form of sodium sulphide (Na_2S). This is hydrolised to NaHS and H_2S to such an extent that the distinctive odour of H_2S is noticeable whenever the sulphide concentration exceeds 0.1 mg/litre. There is little doubt that hydrogen sulphide in district heating pipelines originates from the sulphur content in the steel of the pipelines. In some European pipelines the concentration of hydrogen sulphide in the district heating water is permitted to rise to as much as 0.5 mg/litre, at which concentrations hydrogen sulphide acts quite corrosively, particularly upon pump components, valves and other auxiliary equipment.

The hydrogen sulphide content in district heating water can be reduced somewhat by maintaining a high alkalinity of the water. In the main method used however the flow of the water is diverted for special chemical treatment in which a solution of chlorine in water is added. This oxidises the hydrogen sulphide to sulphuric acid as follows:

$$4\,Cl_2 + 4\,H_2O + H_2S \rightarrow H_2SO_4 + 8\,HCl$$

The acids are neutralised by the natural alkalinity of the water. After treatment the water passes through a degassing plant for removal of any dissolved gases; additional quantities of sodium hydroxide are added to bring the pH of the water back to the value of 9.5 to 10 required for district heating networks.

Scale and sludge formation

Scale within heat exchanger tubes or on other inside surfaces subjected to heat greatly reduces the heat transfer coefficient of a heat exchanger; it therefore adversely affects the output of the plant. Excessive scaling can lead to tube failure due to local overheating of the surface covered by the scale. Once formed scale is seldom easily removed, requiring mechanical or chemical descaling which disrupts plant operation and can be costly.

Sludge is soft and does not generally adhere strongly to internal boiler surfaces, tending rather to accumulate at the low points of the system such as in the bottom of boilers, cylinders etc from whence it cannot in all cases be easily dislodged, particularly where it has settled below the drain-off connection(s). Sludge can be engendered during scale formation and can itself cause blockages in valves and in pipes. It is commonly removed by blow-down: the water is agitated and drawn off the equipment and in most cases should then carry the sludge with it to waste.

REMOVAL OF HARDNESS

The lime/soda and associated processes

The bases of the lime/soda process are the following reactions:

$Ca(HCO_3)_2 + Ca(OH)_{22} \rightarrow 2\,CaCO_3 + 2\,H_2O$

$CaSO_4 + Na_2CO_3 \rightarrow CaCO_3 + Na_2SO_4$

These reactions are typical of a whole series of reactions which have the object of removing the alkaline (temporary) hardness and the permanent hardness from the system. The lime/soda process has to be strictly controlled by accurate analytical techniques:

The number of equivalents of lime required are the sum of the equivalents of CO_3^{2-}, H^+, HCO_3^-, and Mg^{2+} present in the water plus the number of equivalents required by any coagulant such as aluminium sulphate; alum or sodium aluminate are added to produce a precipitate which is easy to filter.

The number of equivalents of soda required are the sum of equivalents of H^+, Ca^{2+}, Mg^{2+} and coagulant less the number of equivalents of bicarbonate present, after addition of the lime.

Although the process is still termed the 'lime/soda process' a variety of other chemicals which are more effective in removing hardness salts from the water are increasingly being used, such as a mixture of trisodium phosphate and sodium hydroxide, or of sodium aluminate in conjunction with lime and sodium carbonate.

Flocculating agents (or coagulants)

In addition to the solubility of the calcium and magnesium components precipitated in the water treatment, the rate at which the precipitate is formed is of equal importance in determining the effectiveness of a water softening process. Coagulant chemicals are therefore added in controlled quantities to increase the rate of precipitation to that which is required. Sodium aluminate reacts with water as follows:

$Na_2Al_2O_4 + 4\,H_2O \rightarrow 2\,NaOH + 2\,Al(OH)_3$

Aluminium hydroxide which has a gelatinous nature is an extremely good flocculation agent. Activated sodium silicate solutions are also added to aid flocculation. Activation is carried out by treating the silicates with acids, sodium bicarbonate and chlorine thereby producing hydrated silica. A typical reaction is the following:

$Na_2SiO_3 + 2\,NaHCO_3 \rightarrow SiO_2.H_2O + 2\,Na_2CO_3$

Activation is carried out just before addition to the water. The addition of such flocculating agents much improves the rate and completeness of the chemical reactions which take place during the lime/soda process because the precipitating particles are larger and therefore travel far faster through the solution carrying the more minute particles with them.

Magnetic water treatment

This method of softening water is widely used in marine engineering and in district heating in the Soviet Union (**10.4**). The hard water flows at high velocity through a set of permanent magnets and the magnetic field then varies the nature of crystallisation of the hardness producing salts. The main result is the creation of numerous nuclei within the solution so that crystallisation produces a sludge (instead of a hard scale) which can then be removed easily by simple blow-down methods. In use, the make-up water is first filtered and then passed through the permanent magnet device. The magnetic method works equally well with hot and with cold water. It has

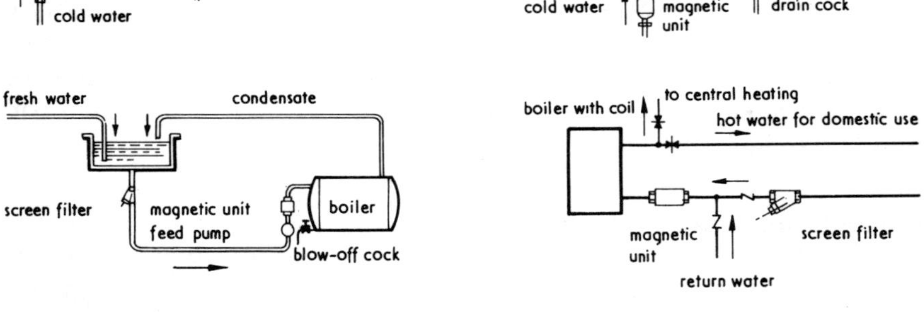

10.4 The use of magnetic water treatment units

been found that even when deposits are formed they are inevitably softer than they would be without magnetic water treatment so that they can be removed from the heat exchanger surfaces with a stiff bristle brush instead of requiring much harsher treatment.

ION EXCHANGE PROCESSES

Principles of ion exchange

To obviate corrosion and scaling it is necessary to subject the feed-water for a district heating system to an ion exchange process. When salts dissolve in water they separate into positively charged cations and negatively charged anions. The cations comprise metals as well as some non-metallic groups such as the ammonium ion and the amine group. The anions are acidic groups, such as sulphates and carbonates and also include some metallic ions. Ion exchange materials therefore comprise distinct anions and cations, but characteristically only one of these components is mobile while the other is static. When exchange of ions occurs the mobile ions in the exchange material are substituted for the appropriate ions in the solution. The solid material stays unchanged. If a solution in contact with the ion exchange resin contains the same sign of ions as the moveable part of the ion exchange, a reversible reaction takes place. This proceeds until an equilibrium is established between the ions contained in the solution and those in the framework of the ion exchange materials.

Ion exchange resins are currently based in the main upon polymer structures of styrene, acrylates and a number of copolymers of these and other plastics which form the skeleton of the exchange material to which the various ionic groups concerned are attached. Such resins are subdivided into the following categories:

(a) water softeners which are materials able to substitute sodium ions for other metal ions such as calcium, magnesium and iron.

(b) cation removers which substitute hydrogen ions for metallic ions

(c) anion removers which substitute hydroxyl ions for anions.

Cation removers

These materials serve to replace metallic ions inside solutions by hydrogen ions leaving the anions in the form of dilute acids.

Weak acid exchangers are only capable of removing metallic ions which are bound to weak acids such as the various carbonates, bicarbonates, etc. They only operate efficiently with salts which have a pH value above 7. They possess a very high efficiency of regeneration but have difficulty in removing metallic ions in the presence of strong acidic ions such as chlorides, sulphates and the like. Weak acid ion exchange resins are best used for dealkalisation. They are almost ideal for such purposes because they have a high exchange capacity with very low acid regenerant requirements which barely exceed stoichiometric levels. Weak acid cation exchangers are usually in the form of irregular granules of various kinds of condensation resins or as bead-shaped polyacrylics.

Strongly acidic cation removers are nearly always produced by copolymerising styrene with materials such as divinyl benzene which acts as a crosslinking agent producing tough and temperature resistant beads. Concentrated sulphuric acid is then used to produce sulphuric acid groups which are attached to the polymer skeleton. Strongly acidic exchange resins are used for the conversion of salts into the appropriate acids. This forms the first stage of the demineralisation process in which all the dissolved salt impurities are being removed. Regeneration takes place by means of sulphuric acid. But hydrochloric acid is used for regeneration in some countries with the advantage that the waste products obtained, such as calcium chloride and magnesium chloride are much more soluble than the equivalent sulphates; stronger acid solutions can be used resulting in a speedier regeneration process and in a greater degree of completion than is possible with sulphuric acid. The disadvantage of hydrochloric acid is mainly one of higher cost so that this must be balanced against its better efficiency as a cation exchange resin regenerator. Strong acid exchange resins such as polystyrene resins containing free-SO_3H groups are capable of splitting salts along the entire pH range, removing all the basic ions and leaving the anions behind in the form of a mixture of acids. Strong acid ion exchange materials can operate at any pH value but their capacity is limited and they need a large excess of strong acid for regeneration.

Weak acid ion exchange resins may be used in conjunction with strong acid exchange resins either by placing the two units one behind the other or by using a mixture of the two resins. Such a combination can materially reduce the cost of regeneration.

The bulk density of standard 8 per cent crosslinked sulphonated polystyrene beads with a diameter of about 100 to 250 μm is about 850 kg/m^3. The moisture content as delivered averages 45 per cent. As metallic ions are being replaced by hydrogen ions during the regeneration process the beads swell by about 5 per cent. The resins can be used at any pH and at any temperature up to a maximum of around 120°C (248°F).

Plants should be designed for a liquid flow rate through the ion bed, during both the loading and the regeneration cycle of 5 litres per second (average) across an area of 1 m^2 at 25°C (77°F). The pressure loss under such circumstances is about 20 kilopascals (kPa) per metre depth of ion exchanger bed. Larger industrial plants however often operate at flow rates of up to 35 litres/second/m^2.

Macroreticular sulphonated polystyrene

These types of ion exchange resins are made from polystyrene containing about 20 per cent or more divinyl benzene and they have a far more open structure than normal sulphonated polystyrene resins; yet their strength and resistance to abrasion remain high. They are resistant to fouling by organic materials and usually have a diameter of 500 microns against the normal bead diameters of between 100 and 250 microns. Hence when they are used in ion exchange plant the pressure drops average half those experienced with standard sulphonated polystyrene beads. Macroreticular sulphonated polystyrene beads are widely used in mixed bed demineralisers. The disadvantages compared to normal sulphonated polystyrene beads are higher cost and lower exchange capacity.

Anion removers

These are ion exchange resins capable of removing acidic ions from water and replacing them by hydroxyl ions. When used in conjunction with cation exchange resins they are capable of demineralising water ie removing all dissolved salts in the water and replacing them with water. Like the cation exchange resins the anion exchangers also occur in two modifications: weakly basic and strongly basic resins.

Weakly basic resins are made mainly from cross-linked polystyrene which contains $-NH_2$ and $-NH.CH_3$ groups. In addition some resins made by condensation reactions are also used but these are only capable of operating in acid solutions, tending to hydrolyse when the pH value exceeds 7. Like their opposite, the weak acid exchangers, they have little salt splitting capacity but can remove strong acids much more easily than weak ones; indeed their capacity for removing such strong acids is much higher than that of the strong base ion exchangers. These materials use only about half the amount of NaOH for regeneration needed by strong base ion exchange resins with the same capacity. Neutral salts produced during regeneration of weakly acidic ion exchange resins do not significantly affect the regeneration reaction (as is also the case with weakly basic resins). Thus the effluent from a strongly basic anion exchange unit can be used to regenerate a weakly basic unit. Weakly basic ion exchange systems are used mainly in conjunction with two-bed demineralising systems to remove the mineral acids produced in the cation exchange unit. Weakly basic ion exchange units are incapable of removing weak acid ions, such as SiO_3^{2-} and CO_3^{2-}.

Strongly basic anion resins are also made from cross-linked polystyrene but differ from weakly basic resins in the nature of the active groups which are attached to the framework. While the active groups with weakly basic ion exchange resins are mostly $-NH_2$ and $-NH.CH_3$ groups, $-N(CH_3)_2$ groups are used for anion exchange resins with medium strength alkalinity and $-N(CH_3)_3^+OH^-$ groups are used for strongly basic ion exchange resins. In practice the nature of anion exchange materials can be adjusted to any required value by varying the proportion of ammonia and primary, secondary and tertiary amines which are reacted with the basic polystyrene structure. Regeneration is carried out by means of sodium hydroxide but its efficiency is inevitably rather low. Strongly basic ion exchange resins are used for the removal of weak acids such as carbonic and silicic from solutions. This process can be combined with the removal of strong acids but care must be exercised to ensure that an 'overrun' is not caused by excessive water velocity through the plant. The strong anions are preferentially absorbed by the ion exchange medium. Strongly basic ion exchange resins are widely used in mixed beds.

Demineralisation

The required reactions for demineralisation (removal of all dissolved solids) are:

$$2\,HCl + \overline{Exch.}\,(OH)_2 \rightarrow \overline{Exch.}\,Cl_2 + 2\,H_2O$$
$$H_2CO_3 + \overline{Exch.}\,(OH)_2 \rightarrow \overline{Exch.}\,CO_3 + 2\,H_2O$$

Regeneration

The required reaction is:

$$\overline{Exch.}\,SO_4\,(CO_3) + 2\,NaOH \rightarrow \overline{Exch.}\,(OH)_2 + \begin{matrix} Na_2SO_4 \\ (Na_2CO_3) \end{matrix}$$

Average regenerant levels required are between 60 and 130 kg of sodium hydroxide per m^3 of ion exchange resin. Depending upon operating conditions it is possible to reduce the silicate content of water to between 0.01 and 0.05 parts per million.

The maximum theoretical capacity of a modern strongly acidic cation exchange material is about 90 to 100 kg of calcium carbonate per cubic metre; working to this maximum would require an uneconomically large quantity of regenerant.

Practical systems of ion exchange

To remove all the dissolved salts from water two different systems of ion exchange are adopted:

(a) duct the impure water through separate containers each filled with a different ion exchange resin and regenerate each of these resins in separate processes

(b) use mixed bed demineralisation and separate the resins from each other prior to regenerating each kind with its own regenerant.

The choice of ion exchange resins to be used depends upon the nature of the water to be purified.

Separate cation and anion removal

Cation demineralisers are regenerated with acids. It is therefore necessary to employ acid-proof construction. The most usual design consists of steel shells lined internally with 4 to 5 mm thick rubber sheeting. The ion exchange resins are usually supported on a layer of gravel or anthracite, 300 to 400 mm deep.

If the waters to be cleaned contain salts of weak acids only ie calcium and magnesium bicarbonates and similar substances, then the cation bed can consist of a weak carboxylic ion exchange resin which has a high capacity and low regenerant requirement. If the water consists mainly of permanently hard solutes such as chlorides and sulphates of heavy metals, iron and the like, it is necessary to use a strong acid ion exchange resin such as sulphonated polystyrene. If there is a mixture of temporary and permanent hardness, ion exchange resins with both sulphonate and carboxyl functional groups are best employed.

The cation which is most easily removed is Ca^{2+} followed by Mg^{2+}, Fe^{2+} and Fe^{3+}. In contrast alkali metal ions such as Na^{+} are far harder to separate out. In solutions where both heavy metal ions and sodium ions are present, sodium ions are likely to be removed only partially except when all the metal ions are in the form of bicarbonates. With solutions which have a high percentage of sodium ions an appreciable fraction tends to remain unremoved.

Anion exchange

The liquid which emerges from the cation exchanger contains all the anions present originally except for the carbonates and bicarbonates which have been partially split into carbon dioxide and water. Some of the sodium ions present have usually leached through.

Using a weakly basic ion exchange resin, nearly all the anions are removed but the water still contains some sodium salts, silicates and some other weak organic acids. Even under the most favourable conditions it is difficult to achieve chloride concentrations to below 0.3 ppm when using weak basic resins.

When strong basic resins are used, virtually all the anions are removed and any traces of sodium salts present are retained in the form of NaOH. If water from such an ion exchanger is then repassed through a further weak acidic ion exchange bed practially all the sodium is removed from the water.

Single bed demineraliser systems

Six different schemes of connecting different ion exchange beds to each other are commonly used in industry. These are:

(a) strong acid/weak base/degassing plant
(b) strong acid/degassing plant/strong base
(c) strong acid/weak base/degassing plant/strong base
(d) strong acid/degassing plant/strong base/weak acid/weak base
(e) strong acid/weak base/degassing plant/weak acid/strong base
(f) strong acid/degassing plant/weak base/weak acid/strong base.

Degassing equipment

During ion exchange processes the various carbonates and bicarbonates are converted into carbonic acid ie dissolved carbon dioxide. This carbon dioxide is extremely corrosive; it is therefore essential to remove it from the solution by a process of degassing. The following systems of degassing are commonly employed:

(a) by the release of fine bubbles of an inert gas such as nitrogen combined sometimes with mechanical agitation

(b) by a steam bleed into the de-aerator vessel which also releases any dissolved carbon dioxide and oxygen

(c) by applying a vacuum above the surface of the water combined with agitation to bring the dissolved gases out of solution.

System (a) can only be used if the silicates in the water do no harm. Systems which use three ion exchange beds in series are a good deal more efficient than those which use only one acid and one alkaline exchanger. The best systems by far are those which use two cation and two anion exchangers together with a vacuum de-aerator. These reduce the electrolyte content in the water to between 0.2 and 1 ppm, and the free silicates to below 0.02 ppm.

Mixed bed systems

As both acid and basic ion exchange materials are far more effective in the removal of their respective ions when the ions of the opposite sign have already been removed, the capacity of a resin can be stepped up enormously by using the mixed bed technique using cation and anion exchange resins simultaneously (**10.5**).

In this technique two suitable resins are selected which have differing densities and granule diameters. The water is first passed downwards through the mixed bed and the cations and anions are removed simultaneously at adjoining ion exchange granules. Provided such a mixed bed is not overloaded it is easy to achieve a conductivity of 50 nS/cm (in some cases even less) which compares well with the value of 37 nS/cm (the theoretical figure related to perfectly pure water at 18°C (65°F)). A mixed bed deioniser is thus capable of producing virtually pure water, far purer than one can hope to achieve by standard distillation methods. Dissolved silicates can be reduced to less than 0.002 ppm.

In most modern plants mixed bed deionisers are coupled to single units. The following basic schemes of using mixed bed ion exchangers are most commonly employed:

(a) Mixed bed alone
(b) Strong acid/vacuum deaerator/mixed bed
(c) Strong acid/weak base/vacuum deaerator/mixed bed
(d) Strong acid/vacuum deaerator/strong base/mixed bed
(e) Weak acid/mixed bed/vacuum deaerator

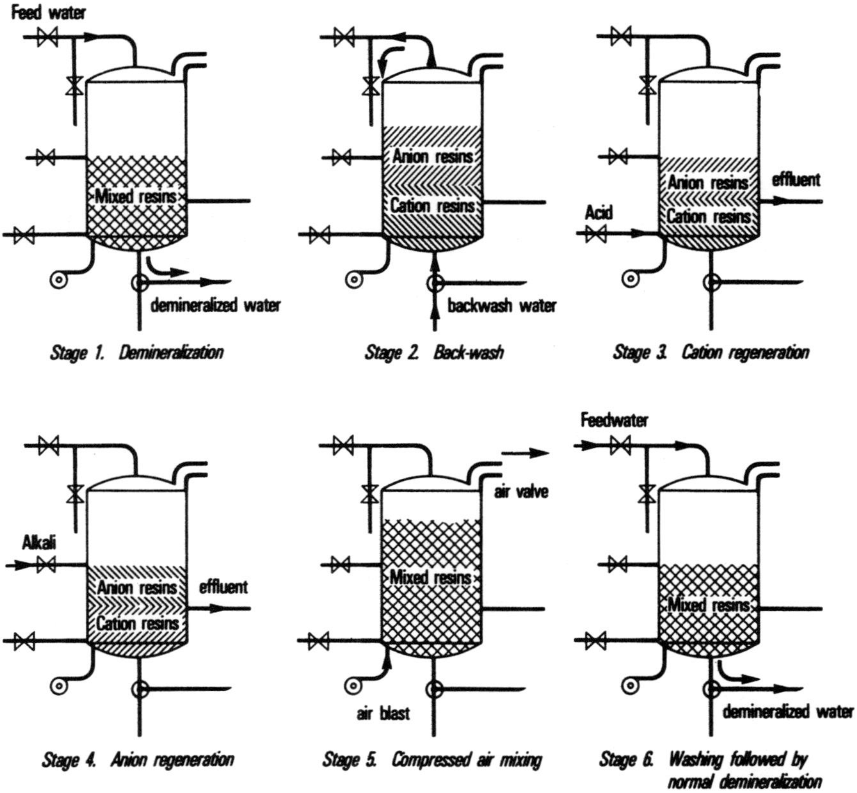

10.5 Mixed bed ion exchange

The purity of the water emerging from the different schemes is best quoted in terms of the electrical conductivity of the water produced.
Typical industrial figures are as follows:

TABLE 10.1 MIXED BED SYSTEMS: TYPICAL CONDUCTIVITIES OF WATER ACHIEVED IN INDUSTRIAL PLANTS

System	Conductivity at 20°C (68°F)
Two-bed	15 μScm^{-1}
Three-bed	5–10 μScm^{-1}
Four-bed	1000 $nScm^{-1}$
Mixed bed	500 $nScm^{-1}$
Two-bed and mixed bed	250 $nScm^{-1}$
Three-bed and mixed bed	100 $nScm^{-1}$
Note: (S = ohm^{-1})	

The conductivity of the emerging water is a measure of the quantities of ion impurities remaining in solution.

Regeneration of mixed bed ion exchangers

The density of cation resins is usually around 1.4 kg/dm³, while the anion resins have a density of about 1.1 kg/dm³. The cation resins are smaller and have a smoother surface. Once the mixed bed deioniser has become spent, water is pumped in from the bottom at the rate of about 2 litres/second per m² of bed cross section, causing the resin particles to become fluidised, separating the anionic particles (which are large and have a low density) from the cationic ones (which are small and have a relatively high density). The anion exchanger granules rise to the top; the cation exchanger granules remain at the bottom. The upper layer is then regenerated with a 4 per cent sodium hydroxide solution which is pumped in at the top and taken out at the centre, followed by several water washes. Next, a sulphuric acid solution about 4 per cent in strength is pumped in at about the centre of the tank to regenerate the cation exchanger resin. The waste acid is drawn off at the bottom followed by a water wash. All liquid is then drawn off and air is blown in from the bottom to mix the cation and anion exchanger resins. The plant is now ready for re-use.

Semi-continuous systems of ion exchange

A typical system of this type is shown in **figure 10.6.** During the production process valves A, B, C, D and Y are open. Valves X and Z are closed. Water is being pumped through the top section of the exchanger tower and its anions and cations are being exchanged by the resin particles. At the bottom of the exchanger tower the spent resin is being regenerated. At the end of each cycle valves A, B, C, D and Y are closed. Valves X and Z are opened. The piston pushes the resin globules along until some begin to rest on valve Y. Valves A, B, C, D and Y are now opened. Valves X and Z are closed. The ion exchange and regeneration processes restart. The spent resin falls on top

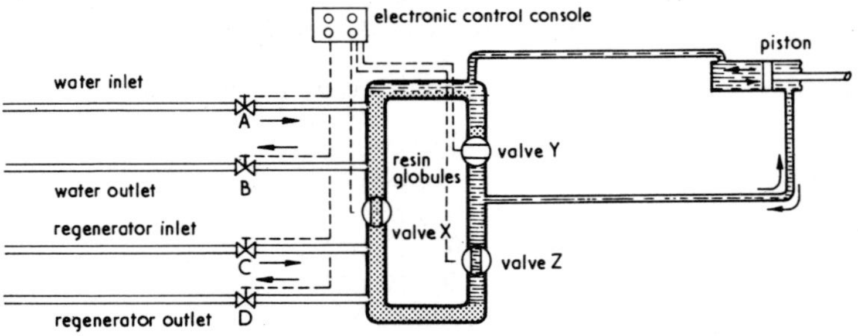

10.6 Sections through a continuous ion exchanger plant

of the now closed valve Z. All valve and piston movements are controlled automatically from an electronic control console.

Poisoning of ion exchange resins by organic substances

If the water used in the ion exchange contains appreciable quantities of dissolved organic materials the ion exchange resins may become 'poisoned' with the effect that the fine porous structure essential for an efficient exchange process becomes clogged up and therefore the capacity is reduced. It is essential that the water used should have a total chemical oxygen demand value below 1.5 ppm. The most common technique employed is to soak the resin in a mixture of sodium chloride and sodium hypochlorite for about 24 hours. An alternative technique is an overnight soak of the resin in a mixture of sodium chloride and sodium hydroxide.

Continuous ion exchange systems

Most ion exchange processes are batch systems in which the water treatment beds are used until they are exhausted. Regeneration then follows. It is practicable to modify mixed-bed ion exchange systems for continuous operation by this method. Raw water is passed through an ion exchange tower and resin is taken off continuously at the same end. It is then passed into a backwash tower where separation into anion and cation exchanger resins proceeds, followed by separate regeneration of the resins with sodium hydroxide and sulphuric acid respectively. The resins are mixed and the mixed regenerated ion exchanger is returned to the absorber vessel. All continuous ion exchange equipment is constructed on the principle of counterflow. Continuous ion exchange equipment possesses the major advantage over batch type exchangers that production need not be interrupted periodically for the regeneration of the spent acid.

Reverse osmosis

In this process, water is forced at a pressure of about 30 bars through a semi-permeable membrane in the opposite direction to normal osmotic diffusion (**10.7**). This removes the dissolved solids; undissolved materials such as high molecular weight substances are filtered off. The membrane used in reverse osmosis plants is either a hollow fibre polyamide or cellulose acetate supplied in spiral wound sheets. Reverse osmosis plants are built up in modules, each module containing a number of tubular bundles or permeators.

Reverse osmosis offers some advantages over conventional ion exchange systems: regeneration is not required and the plants which are very compact can be used continuously. The system is best used with brackish water with a high dissolved solids content. The plant is flushed out each month with detergent or acid and the effluent (a concentration of the solids originally present in the water) can be discharged into the public sewers. It is common

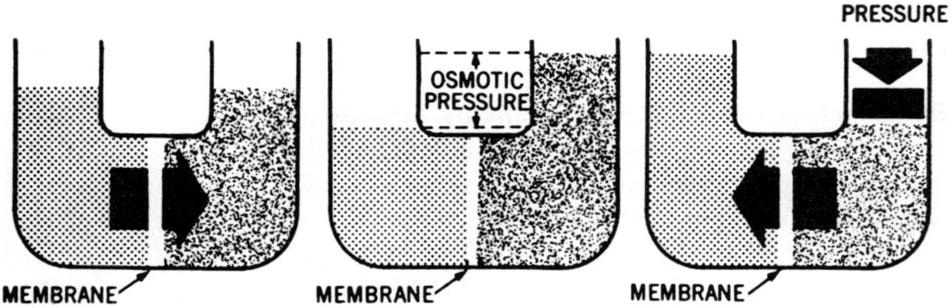

10.7 The principle of reverse osmosis

to employ reverse osmosis plants in conjunction with conventional base-exchange softeners and to use prefilters for the removal of many forms of suspended matter.

Practical aspects for group and district heating

The incoming make-up feed water is commonly filtered, demineralised, degassed and its pH adjusted to the correct level before the water is eventually fed into the network. Deionisation is required additionally to make good the water losses due to leakage and evaporation.

When steam is used for district heating, condensate return lines are seldom fitted and therefore very thorough methods of demineralisation are necessary to prevent scaling in the boilers and heat exchangers.

The water circulated in district cooling systems must be thoroughly treated. Scaling is a particular risk in such systems because of the low temperature at which the water is being pumped, leading to the precipitation of sparingly soluble constituents. District cooling systems are usually fitted with return lines to minimise the treatment costs. Cold water has a high solubility of oxygen; good systems of deoxygenation are therefore vital.

CHECK LIST

1. Principles of using water treatment in the district heating field and the objects it can achieve.
2. The various methods of treating crude water prior to use
3. The treatment required for high pressure boiler water
4. The effects of the presence of nitrogen in the circulatory water and the ways of treating these
5. The effect of the presence of sulphides in district heating water
6. The effects of scale and sludge formation and the ways in which this can be treated
7. Processes for the removal of hardness
8. Application of magnetic water treatment

9 Methods for the reduction of dissolved salt concentration
10 Methods of demineralisation
11 Ion exchange methods of treating water
12 Degassing equipment
13 Semi-continuous systems of ion exchange
14 Risk of poisoning of ion exchange resins and precautions which should be taken
15 Continuous ion exchange systems
16 Methods of treating water by reverse osmosis

REFERENCES

E. Beetz, 'Problems regarding the preparation of water for district heating purposes', *Fernwärme International*, 4 (1975) 1, pp 5–9 (in German)
G. A. Cappeline, 'Controlling corrosion in cooling water systems', *Plant Engineer*, 32, 9th November 1978 pp 162–165, 23rd November 1978 pp 129–132
G. V. James, *Water treatment*, Technical Press, Edinburgh, 1971
E. Nordell, *Water treatment*, van Nostrand-Reinhold, New York, 1961
D. S. Pybus, 'Problems associated with boiler feedwater treatment', *Plant Engineer*, August 1971, pp 41–46
B. A. Sard, 'Demineralising boiler feedwater', *Power and Works Engineering*, January/March 1967
A. I. J. Treanor, 'Treating water for low pressure boilers', *Industrial and Process Heating*, February/March 1967, pp 12–14
Water Treatment Handbook, Degremont Laing Ltd, Boreham Wood, Hertfordshire, UK, 1965

CHAPTER 11

Heat metering

EQUITABLE DISTRIBUTION OF AVAILABLE SUPPLY

The consumer who is connected into a district heating system has the heat supply immediately 'on tap'. The unrestricted use of such a facility leads to a number of undesirable consequences, for example:

(a) waste of heat through the careless consumption of excessive hot water

(b) waste of heat through excessive ventilation during the winter when valuable heat is dissipated through open windows, doors etc

(c) more than average use of heat through excessively long periods of heating (for good reasons or otherwise)

(d) more than average use of hot water in households with numerous children and/or occupants engaged in sports or 'dirty' occupations.

It is obvious, in the interests of fairness, energy conservation and economy, that some restrictions must be placed on the use of heat or that charges for heat supply must be closely related to the use of heat by the particular consumer (generally in a manner similar to that accepted for electricity and gas supplies). In comprehensive district heat or cooling systems there is likely to be a mix of different types and classes of consumers: these may include domestic households, shops, offices, warehouses, factories, public buildings, schools etc. It is self-evident that an equitable method of charging each individual consumer for the heat consumption under his control must be established and administered. The following sections consider various aspects of this need in relation to the subject of heat metering and describe the methods which are in use.

CHARGES FOR HEAT SUPPLIED

Flat charge

It is feasible to recoup the cost of heat supplied by some sort of flat charge per season or by including the estimated heating cost with rent or service charges. Such methods of charging offer no inducement towards economy in the consumption of heat. Uncaring consumers will leave heaters switched on with windows open, hot water will be wasted by washing-up under running taps and so on. In such a situation the overall flat rate charge has to be kept relatively high and many instances of unfairness (and vociferous complaints) are inevitable.

Charge based on metered consumption

Such methods meter or proportionately evaluate the heat supplied to each consumer and charge him accordingly. Usually a two-part tariff is employed: part of the bill is in the form of a fixed flat charge to recoup the common and capital costs and the other part a variable charge related to the actual heat consumed. There are variations within this framework: for example there may be a sliding scale in which quantities of heat in excess of a certain minimum become cheaper (similar to electricity and gas tariff structures) or there may be differential tariffs which favour one method of consumption over another (eg encouraging the use of off-peak summer heat).

In all recorded cases the heat consumed by each dwelling has fallen considerably following the introduction of heat metering: this has generally been due to the exercise of greater economy and care by consumers and has led to material reductions in the amount and cost of the energy supply.

The validity of the above statement can be illustrated by the results of an investigation carried out in Sweden concerning two identical blocks of flats occupied throughout by similar households and supplied with hot water for space and domestic heating from the central district heating station. One block had a flat charge levied; the other was fitted with meters and the consumer was charged by the amount of heat used. In the flat charge block the heat supply over the year was noted as follows:
Space heat per person/day: 12 820 KJ
Domestic hot water per person/day: 54 litres
The consumption data in the metered block was:
Heat per person/day: 6 240 KJ
Hot water per person/day: 26 litres
The startling conclusion: in the same period less than half as much hot water was used both for heating and for domestic supply in the metered block than in the 'flat charge' (un-metered) premises.

Similar experiments were carried out in Germany and Denmark with similar results. There was no evidence to indicate that the reduced heat consumption was caused by the maintenance of inadequate or uncomfortable temperatures within these dwellings or that hygiene suffered because smaller quantities of hot water were used. The savings were achieved essentially by the elimination of energy wastages.

METERING OF HEATING: STEAM

Steam consumption is easily metered either by the use of an integrating pressure flow meter (common in the United States) or by a standard condensate meter (common in most European countries).

Hallengren meter

This condensate meter consists of a rotating drum with three or four chambers and a geared counting train (**11.1**). The casing is made from a Silumin alloy which is fully resistant to water corrosion. As the steam

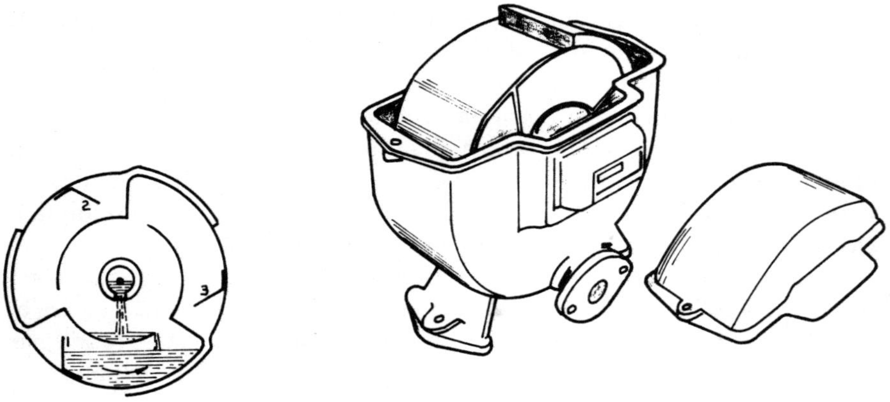

11.1 Condensate meter measuring supply of steam heat

condenses after passing through the heat exchanger the resultant water runs through the meter alternately filling and emptying the chambers of the rotating drum. The drum rotates on its axis as soon as the centre of gravity shifts to a certain position, dependent upon the degree of filling of the respective chambers. This technique measures liquid flow so accurately that even very small quantities of water are registered by the meter. The smallest of these instruments weighs 10.5 kg and can measure from zero flow to a maximum of 800 litres of water per hour. The instrument is very robust, of low cost and cheap to install. The amount of latent heat contained in fixed quantities of steam is known: it is almost independent of the degree to which the steam is superheated and pressurised. Hence such a condensate meter provides a reliable and accurate method of measuring the quantity of heat supply.

Cadillac meter

This steam meter is made in the USA and consists of a cylindrical copper metering drum revolving on roller bearings inside a cast iron case. The drum is divided into six scroll-shaped compartments. As steam condenses into water it is admitted axially by means of a spout at the centre of the drum. It then passes into the drum via a slot. Because of the peculiar scroll-shape of the compartments the centre of gravity is constantly moving towards one side as the water fills the compartment and this shift causes the drum to rotate. The motion of the drum is transmitted to an integrated clockwork counter which registers the total quantity of condensate that has passed through the meter. The counter is located within a separate housing and is calibrated in terms of the quantity of steam which has been condensed. The condensate must not be under-pressured and must be fully recoverable from the steam supply.

Two types of this steam meter are available. The first of these is the gravity meter which has a case open to atmospheric pressure and is intended for plants where discharge from the meter is either to a sewer or to a container which is open to the atmosphere. The second is the vacuum-type meter which is a modification of the gravity meter and is provided with a special air-tight counter housing and with gaskets at all joints to seal the casing; it is used on vacuum heating systems where it is desirable to meter the condensate on the suction side of the vacuum pump.

The mode of operation of the Cadillac meter can be seen from **figure 11.2.** Such meters are guaranteed to be accurate within 1 per cent at any rate of flow between zero and the full rated capacity. A meter must be installed at a point where it can receive all the condensate passing through radiators and/or other steam-using equipment. The return line must be free from leaks and

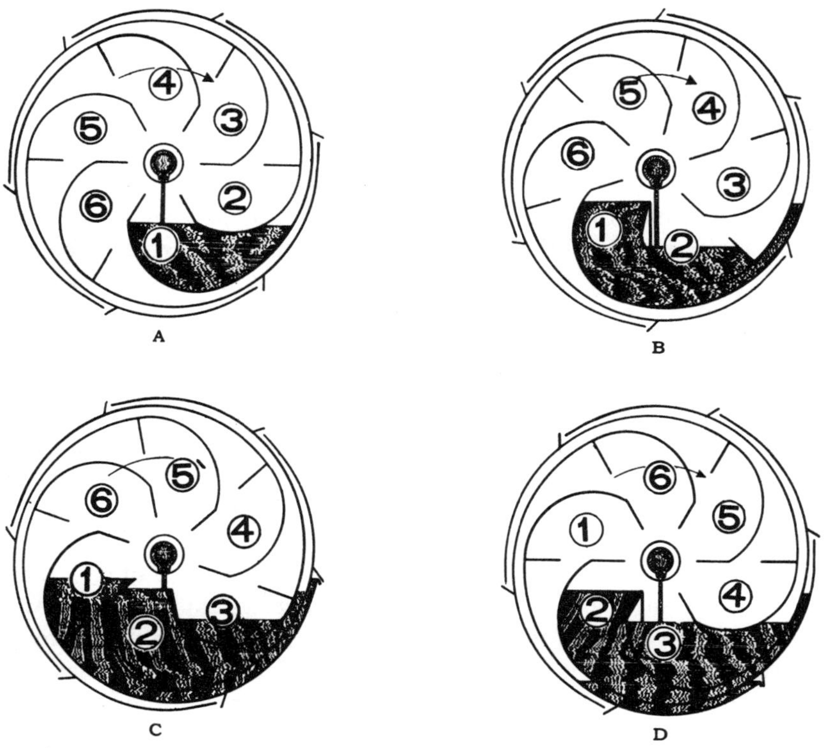

11.2 Illustration of operating principle of Cadillac steam meter
A *Compartment No 1 filling. Water extending to the right of center turns drum in direction of arrow.* **B** *Compartment No 1 filled and overflowing into Compartment No 2.* **C** *Compartment No 1 ready to empty. Compartment No 2 full and overflowing into Compartment No 3.* **D** *Compartment No 1 nearly empty. Compartment No 2 overflowing and Compartment No 3 nearly full*

properly graded and a suitable steam trap set with strainer must be installed upstream of the meter. The smallest available unit has a capacity of 113.5 litres per hour; the largest 5448 litres per hour (each litre of condensate represents one kg of steam condensed). The weights of the units vary between 18.2 kg and 147.5 kg.

For locations where it is difficult or inconvenient to read the Cadillac meter directly a special meter contactor is used which transforms the revolutions of the meter drum into electrical impulse type signals; these can then operate one of numerous different types of electrical devices (eg a remote reading output). It is even possible to use this electrical output for direct computer billing purposes thereby avoiding the use of personnel and possible human errors involved in reading meters.

Integrating steam flow meters

Integrating flow meters are used for metering steam consumption in large (exclusive) systems. The pressure differentials across orifice or Venturi meters are employed to determine the flow rate of the steam on the principle that the pressure exerted by a flowing liquid is directly proportional to the square of its velocity (Bernoulli's theorem). Meters such as the Republic meter integrate such pressure differential with the steam pressure and record the actual flow rate in appropriate units.

However some problems arise with such a simple system. In district heating practice, as in other industrial and commercial operations steam supply quantities vary widely over the year. Flow meters based upon the Venturi principle or orifice plate method do not provide acceptable accuracy when operating below 20 per cent of the rated meter capacity. In order to achieve acceptable accuracy at proportionally low flows several flow meters are installed in parallel and are arranged to turn the appropriate meters automatically on and off in a sequence determined by the nature of the steam load. This type of metering system is termed sequential metering; it is operated by either electrical or steam pressure control.

Electrical control of sequential metering is usually effected by a differential pressure switch which connects to the manometer taps of the different steam velocity meters. The sequential metering system is offered by the Bailey Company (USA): it generally operates with three different meters. Its advantages are that it involves low pressure losses and that it has a wide range of operation. Its disadvantage is its high cost so that it is mainly suitable only for users with loads in excess of about 10 tonnes of steam per hour.

Steam pressure is employed in the less expensive system of sequential metering offered by the Consolidated Edison Company of New York. In this a steam pressure reducing valve fitted in series with the base load meter is set to pass out steam at a pressure of between 0.3 and 0.7 bars above the pressure at the entry to the reducing valve which is placed in series with the high capacity meter. An orifice plate sized to provide critical flow at about one and a half times the meter rating is installed downstream of the base load meter: this limits the flow through the meter and provides the pressure drop required to open the pressure reducing valve in series with the second meter when the load exceeds the capacity of the base load meter.

Steam valves suitable for operation with pressure operated sequential meters are made by the C. E. Squires Company (USA); these have a globe valve body with a cylinder and piston at the top arranged to open and close the valve. The valve is so installed in the line that normal flow tends to close its seating down, while the differential pressure across the base load meter (applied to the cylinder and piston through tube connections) tends to open the valve up. The valve does not open until the latter force predominates. It closes again when the differential pressure drops and can no longer support the weight of the valve stem assembly. Metering systems embodying Squires valves use up to four different sequential meters.

The advantages claimed for the system are:

The opening and closing valve settings are independent of the service steam pressure

The valve is either fully open or fully closed: no steam can flow through the meter unregistered

No external source of power is required

Installation and maintenance costs are minimal.

Another system of steam metering used in the USA in conditions of very variable flow is the dual range type. This has two separate metering orifices and secondary instruments arranged in series. All the steam passes through the large metering orifice after which the sequence operates as follows: if the flow rate of steam is below the maximum capacity of the small orifice all the steam passes through it; if the flow rate exceeds the above maximum the excess quantity bypasses the system. The small pressure regulator is set about 0.35 bars above the larger pressure regulator and this differential keeps the latter closed during periods of low flow rates. When the flow rate comes up to the range of the large meter this is activated, and the small meter simultaneously deactivated, by a differential pressure switch placed across the small orifice. This action is reversed when the flow rate drops down to the range of the smaller meter.

Shuntflo meter

In this type of meter a shunt path (bypass) around the steam meter passes a fraction of the total flow driving a vaned wheel at a speed proportional to the steam flow rate through the meter and this is geared to indicate the total volume of steam which has passed through the meter. The top cover, which embodies the shunt path around the meter orifice, is the same for all meter sizes; in consequence the ratio of flow through the meter and through the shunt varies with different sizes of meters. A series of basic equations are used to determine the actual steam flow rates through the meter (see information by the BIF Organisation published in the 1962 NDHA Proceedings page 174).

METERING OF HEATING: HOT WATER

Combination of heat meters and quantity meters

It is far more difficult to measure the heat supplied in the form of hot water accurately and cheaply than to meter steam. For large consumers it is best to use a mechanical heat flow meter but these are rather expensive and difficult to install in the smaller dwellings. Heat meters are therefore frequently combined with simpler quantity meters which give an indication of the individual heat consumption in each dwelling. In such arrangements the mechanical heat meter may be placed either in the basement of a block of flats or at the sub-station supplying a housing estate and provides tolerably accurate figures for the total amount of heat supplied to all the consumers concerned. The individual share of heat received by each consumer is then determined by the use of smaller and less accurate meters installed in individual dwellings. The different meters available are discussed below.

Simple vane quantity meter

This system is commonly used in Holland. The total heat content of the water flowing into the system is evaluated by the main heat meters. Simple vane quantity water flow sub-meters installed in each dwelling then provide a reasonably accurate guide for fair subdivision of heat charges among individual consumers. Their advantages are that they are cheap and that it is easy to note and interpret readings. Their disadvantage is that the quantity of water which they measure is not integrated with its temperature: the position may therefore arise that some consumers have to subsidise others with different living habits since the temperature range of the hot water heating medium changes quite drastically according to the heating season.

In many blocks of flats pipe runs to radiators, hot water systems etc are not usually piped as separate pipe circulation for each flat but vertically from one storey to the next. This method of metering is inapplicable to such situations.

All vane quantity meters tend towards unduly low readings at small flow rates (when they favour the consumer).

Aquametro heat meter

This is a mechanical heat meter made by the Swiss Aquametro firm of Basle in various sizes, able to deal with maximum loads of between 0.5 m^3 and 340 m^3 of hot water per hour and to be fitted to pipes down to 12 mm in diameter (**11.3**). It consists of a water meter using a wheel with vanes which can turn along either a horizontal or a vertical axis. The temperatures of the

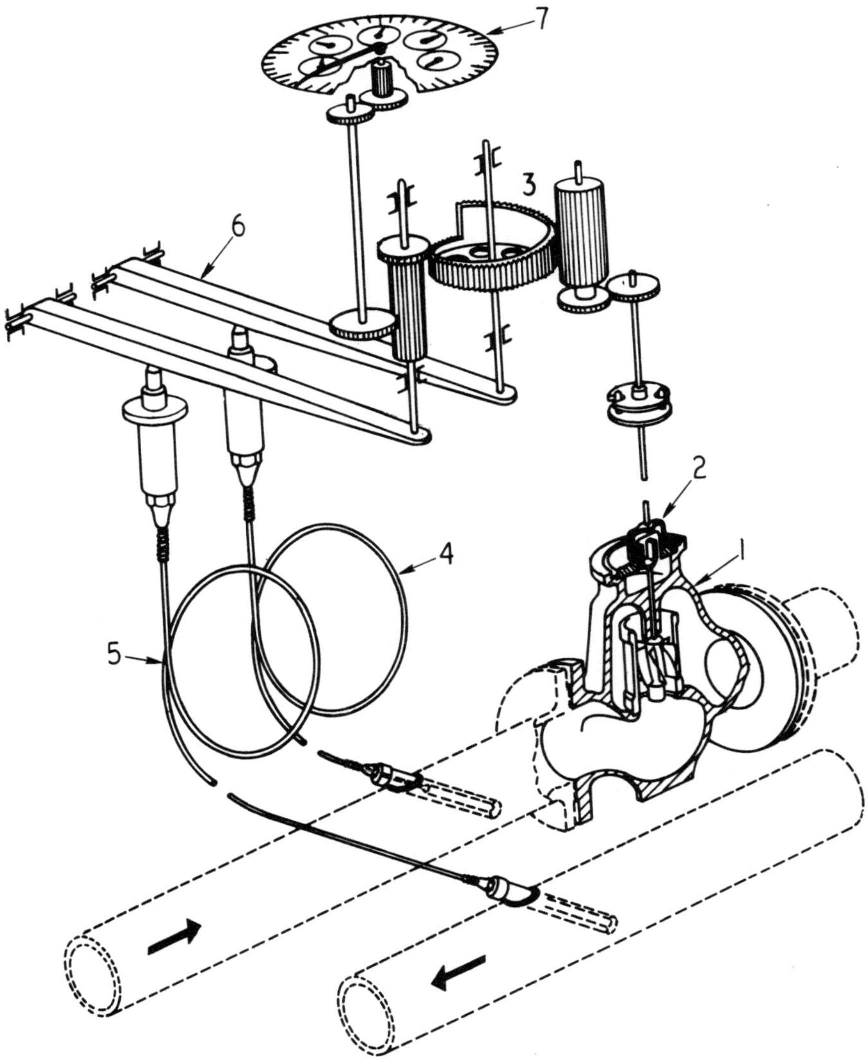

11.3 Aquametro heat meter

supply and return hot water lines are measured by means of liquid-in-steel thermometers which operate plungers that actuate the integration mechanism. The mechanism works in the following mode: the vaned wheel turns at a speed proportional to the amount of water flowing through the pipe per unit of time and is customarily located in the return pipe. The turning torque is transmitted to a truncated cogwheel with a sloping edge at the upper end. This truncated cogwheel is lifted or lowered depending upon the temperature of the water in the supply line. It may be engaged by a single cogwheel which is also lifted and lowered by a lever actuated by the oil in the steel thermometer of the return line. If the temperatures of the supply and return lines are the same the truncated cogwheel and the single cog do not engage: thus no reading is taken as no heat is abstracted. If there is a difference in temperature between the two lines, (ie heat is abstracted) the truncated cog and the single cog engage for a part of each revolution made by the truncated cog directly proportional to the difference in temperature. From the single cog the returning moment is then further transmitted via a gear train to a calibrated dial. The dial itself is in two parts: on the right-hand side the volume of water flowing through is measured directly; on the left-hand side the heat is measured by the dial which consists of five digits and a rotary pointer. The accuracy of the flow meter is ± 2 per cent within its measuring range, which runs between 2.5 and 100 per cent of the maximum loading figure; the accuracy of the temperature measurement and integration device is ± 1 per cent of the maximum temperature difference between supply and return line independent of the prevailing room temperature. The meter can be used for systems with pressures of up to 38 bars and temperatures up to about 180°C (356°F).

Special meters are available to measure the supply of cold water in 'district cooling' systems which feeds air conditioning plants (mainly used in the USA). The pressure losses suffered through these types of heat meters are quite considerable. The small meters with vaned wheels with a vertical axis lose a maximum of 6 kPa. Meters in which the vane is on a horizontal axis have far lower losses than that, only about 1.2 kPa maximum losses.

The various heat meters produced by Swiss Aquametro are suitable for measuring hot water for domestic and space heating purposes provided that there is a main common pipe through which the water enters the consumer's premises. Such meters are therefore more suitable for use in private houses, industrial and commercial premises as check meters for sub-stations etc. than in groups of dwellings where the pipe routes of the hot water to the radiators, kitchen and bathroom are vertical and not horizontal. The graduations are chosen in such a way that it is only necessary to read the meter once a year (or at other periods fixed to suit the tariff arrangement).

Bantam heat meter

The Aquametro meter is mainly suitable for metering supplies of hot water to public buildings, groups of dwellings and other larger users of heat. To meet the need for metering the amount of heat supplied to single dwellings from district heating systems, the smaller Bantam meter was developed by the same firm. Its operating principles are as follows: the quantity of heating water and the difference between the flow and return temperatures are continuously measured and mechanically integrated to give a dial indication of the extracted heat units. The quantity of water passing through the meter is established by the number of rotations on a vane wheel.

To ensure a long operating life a smooth rotation thrust on the vane wheel is obtained by means of multiple jets but only a single jet is fitted to the 10 mm model. The expansion of two precision capillary temperature sensors determines the temperatures in the flow and return lines. The integrator registers the heat used compensated for the average specific heat and density (heat factor 'k'). The dial mechanism incorporates two centre pointers and a five-position roller counter. The latter summates the heat consumption. The small pointer indicates the quantity of water for control purposes; the larger pointer gives fine reading of the heat quantity. The scale length is 230 mm divided into 100 divisions so that reading is made easy and accurate. The counter registers the number of rotations of the heat unit pointer.

The basic operating principle establishes that greater accuracy is obtained when the temperature differential approaches its designed maximum, whereas the water flow rate can be anywhere between the high and low flow limits. The selection of the meter must take this into account. The smaller units of the Bantam meters range can read down to a heat quantity of 40 kJ; the maximum roller counter registration is 400 GJ.

Pollux heat meter

In the Pollux heat meter two mercury-in-steel thermometers are provided, one in the inlet line and the other in the return line of the hot water supply. Each of these actuates a Bourdon spring which in turn actuates an arrangement of levers operating a cam indicating the temperature difference between inlet and return line. The temperature difference can be read off directly on a dial. A simple vane water meter is installed in the return line and gives a direct reading on a dial calibrated in units by volume of water; there is coordination with the temperature difference cam so that the water flow and the temperature are integrated and recorded as heat consumption on a third dial. The mechanism of the system of integration, which employs a hinged lever system with jockey wheel, can be seen from **figure 11.4** and **plate 36.**

Ferranti heat meter

This is a British made heat meter which can be operated with all known hot water systems including those which supply warm air heating. It measures the mass flow of water in the flow pipe by means of a spring-loaded flow sensor equipped with a disc which is pushed along the farther end the faster the water flows. The temperature differential between the flow and return pipes is measured by means of a piston assembly. Using a synchronous electric motor as well as systems of variable driving discs it is possible to integrate these values and to record them on the meter. Variations in flow from zero to 25 litres/minute can be accommodated by the selection of a suitable flow sensor.

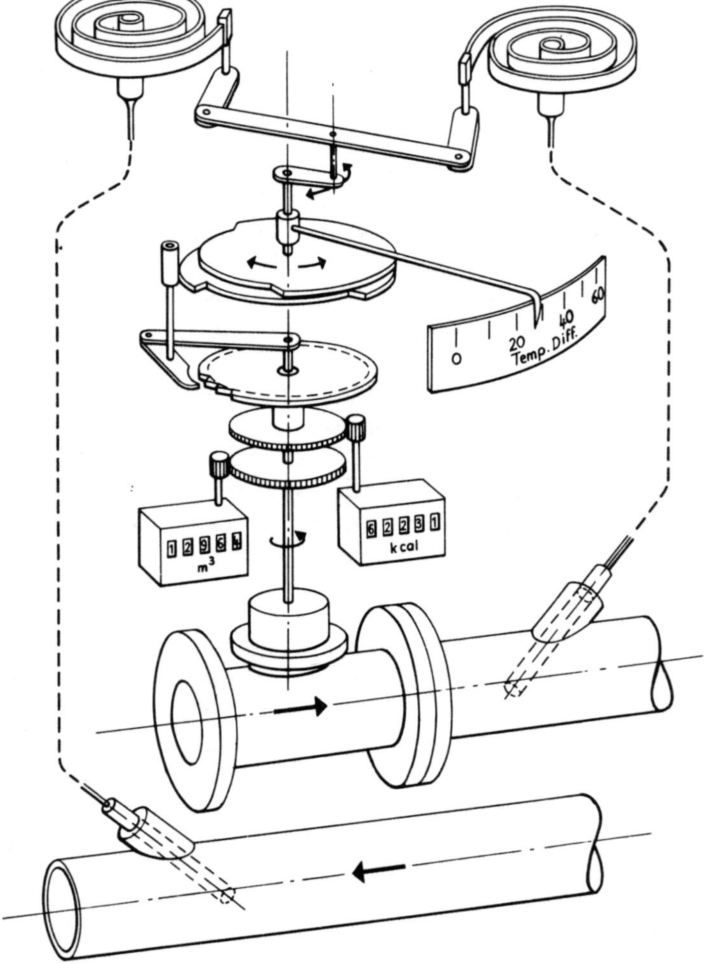

11.4 Pollux heat meter

The meter has the following characteristics: it can cater for temperature differentials between zero and 120°C (248°F) and can be operated at pressures of up to 15 bars. If needed the output can be in the form of electrical impulses for remote reading or direct computer billing. The read-out is similar to that of the commonly used domestic electricity meter; any desired unit can be employed.

Suitability of heat meters

A number of other mechanical heat meters are available such as the Dutch Pegus, the Danish Kamstrup, etc. The mechanisms vary but by and large mechanical heat meters while having the advantage of being accurate and easily read are rather expensive. Many authorities therefore consider that mechanical heat meters are too costly for fitting in private dwellings. Hence mechanical heat meters are mainly used for metering the heat used by large consumers such as offices, hotels, schools and other public buildings. For metering heat supplied to private customers mechanical heat meters of the type mentioned are commonly employed for measuring the heat supplied to groups of between 10 and 100 dwellings, the individual charge then being determined by the use of distribution (or relative) heat meters such as evaporation meters.

EVAPORATION METERS

Types available

Evaporation meters are of two types:
(a) those which measure the amount of space heat emitted by radiators (radiator evaporation meters)
(b) those which measure the amount of domestic hot water consumption (domestic hot water evaporation meters).

Radiator evaporation meters

The amount of heat supplied to each dwelling is measured by using a separate meter on each radiator in the dwelling. A heat conductor is attached to the radiator and makes contact with a glass tube containing a liquid having a high boiling point. This tube is supported in a plastic case so that only the scale and the glass tube are visible from the outside through a glass window. The meter tube itself is graduated, open at the top and 115 mm long. At the commencement of the heating season the fluid is levelled to a mark 30 mm below the top. The rate of evaporation of the metering liquid is determined by the vapour pressure of the fluid, the temperature of the liquid and the distance of fluid level from the top of the glass tube. Thus as the liquid evaporates the rate of evaporation becomes slower, a fact which has to be accommodated in the calibration of the scale. 51 different types of scales are produced each corresponding to a given size and make of radiator. Scales cover radiator surfaces up to 20 m^2. The meter casing which consists of the front and rear sections of the meter hinged together is secured by the use of a lead seal to prevent unauthorised interference. When the radiator is not being used, there is still an idling registration of about 18 per cent of the average registration and this is used in practice to operate a type of two-part tariff. Once a year all meter tubes are replaced by new ones. Different colours of liquid are used each year to ensure and demonstrate that the fluid has in fact been changed.

These meters are very cheap and if correctly serviced and read can provide a fair guide to heat consumption. Difficulties have arisen where the administration of this type of metering operation has been careless eg where the meters have not been matched to the radiators to which they are attached or where meters have been mixed up at the annual change-over. Meticulous identification of each meter is necessary.

Domestic hot water evaporation meters

Typical meters of this kind are manufactured by the Danish firms of Clorius and Brunata. The Clorius meter comprises two parts (a) a Venturi tee with a removable inserted plug member which is to be built into the hot water circuit and (b) the main meter housing which contains a heat accumulator (a closed meter glass partly filled with fluid and other components) **(11.5).**

The meter works on the following principle: when hot water is drawn off a shunt current amounting to about 0.4 per cent of the main current from the Venturi tee passes through the heat accumulator which is heated up to a temperature depending both upon the quantity of water passing and on the temperature of such water. The meter glass has the shape of an inverted U-tube. One limb of this tube which is partly filled with the meter liquid is in contact with the heat accumulator. The amount of liquid which is distilled over depends upon the temperature of the heat accumulator and also upon the time period during which it is at the given temperature. The cold limb of the U-tube is calibrated in cubic metres of water at a reference temperature of 60°C (140°F) and can be read to a reasonable accuracy at all times. Once each year the hot water meter is read, the glass is removed from the case, then inverted so that all the liquid is poured back into the hot limb, and finally reinserted into the case. The zero reading is adjusted by means of a setscrew. The meter is then sealed by means of a lead seal to avoid interference by unauthorised persons.

The capacity of the standard Clorius hot water meter is about 300 m^3 of water at 60°C (140°F). Special precautions are taken in the construction of the meter to enable both small and large quantities of water to be recorded accurately and to avoid idling recording (ie registering heat which would be conducted to the heat accumulator if the Venturi tee were situated too close to a constantly warm pipe line). The capacity of the heat accumulator is sufficient to cover for larger drawings of hot water than are usually met with in practice; the sensitivity is such that even quite small hot water draw-offs which follow large drawings of hot water are registered accurately. The metallic portion of the meter housing is attached at three points only to an outer insulated rear screen which is made from phenolic plastic and is mounted directly upon the flange of the Venturi tee. The inside of the meter is divided into two parts by a central screen. The measuring glass is placed into the front compartment so that the two limbs of the U-tube are at the same temperature when no water is drawn off. If the room temperature fluctuates both limbs are affected equally and no registration takes place. The heat accumulator is insulated and suspended in the rear chamber of the

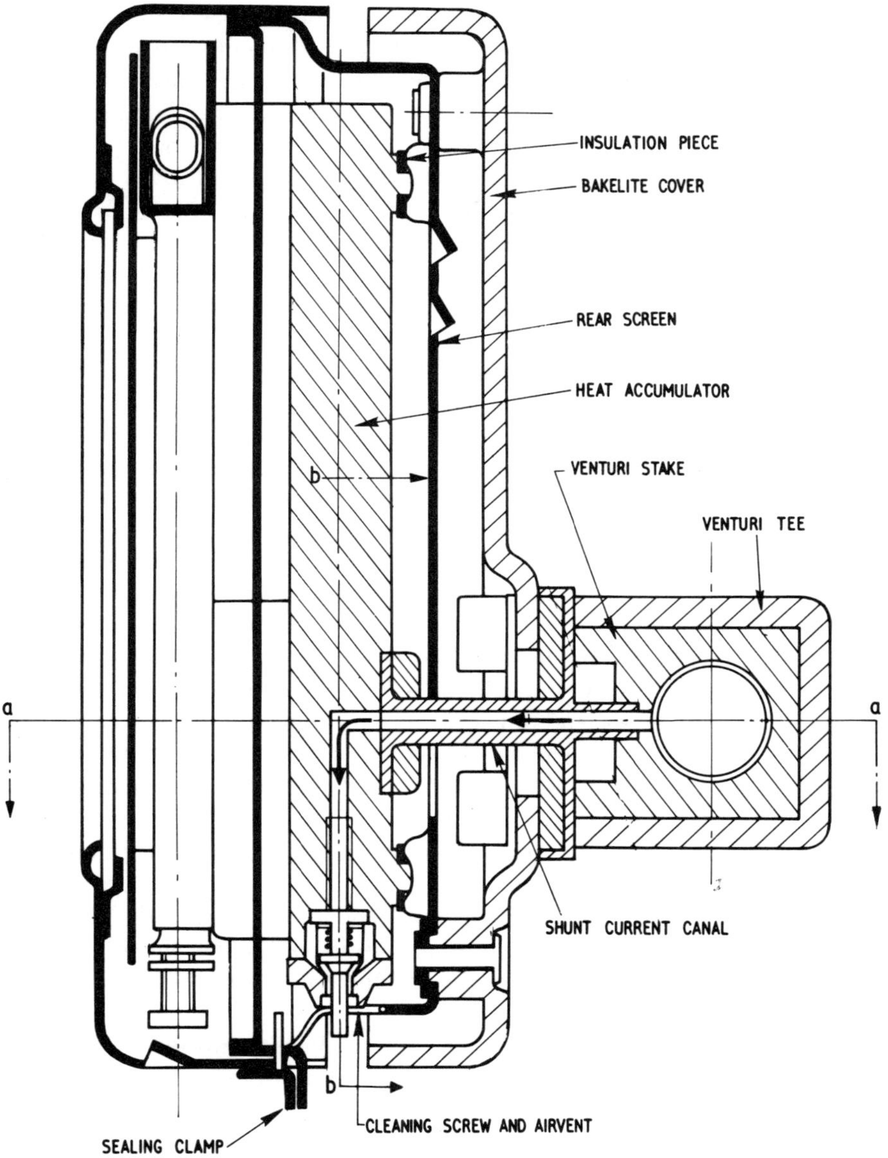

11.5 Clorius evaporation meter

meter housing. Contact between the heat accumulator and the evaporation limb of the U-tube is by means of a metal fin which leads from the accumulator and bears a V-profiled glass holder. The glass is pressed against this holder by a small spring attached to the scale so that good contact is made.

The connection between the Venturi tee and the heat accumulator is by means of a double nylon tube. This has the advantage of very low thermal conductivity and small cross-sectional area. Although the inner diameter of the nylon tube is quite large the length of the tube is short. Therefore, even if only small quantities of hot water are drawn off for very short periods of time, sufficient hot water reaches the heat accumulator to make a reading possible because only a very small quantity of cold water has to be driven away from the channel. The loss of heat in this way is quite negligible. The shunt current canal has a diameter of about 2.5 mm and is thus not easily clogged even if the water passing through the meter is somewhat dirty. In the event of clogging the cleaning screw is removed and the canal cleared. The removable plug member makes it possible to clean the Venturi tee effectively without the necessity of opening up the rest of the pipe system.

The Venturi tee is made from hot pressed brass and is about 50 mm in length. It is supplied with a square flange on which the meter casing is to be fastened. Its cover flange is tested to withstand a pressure of up to 10 bars. It is designed to be built into either vertical or horizontal pipes but cannot be built into pipes which run at an angle. Once the Venturi tee has been mounted the meter can be attached on top of it. The meter itself is made from an anodised aluminium alloy and has good corrosion resistant properties. The liquid moves from the heated limb to the cooled limb by a process of diffusion at a rate directly proportional to the difference in vapour pressure between the two limbs and inversely proportional to the resistance encountered on the way depending on the length and cross-sectional area of the gauge tube (which has to be manufactured carefully to standard specifications). However, since the vapour pressure of organic liquids is not directly proportional to their temperature, one is not really concerned with absolute temperatures, but rather with differences in temperature. Inaccuracies due to differences in the rate of increase of vapour pressure, with increasing temperature at different temperatures, are evened out so that the relationship between the difference in temperature and the difference in vapour pressure is almost linear.

Several problems occur with these meters when quantities of water are drawn off at frequent intervals as, under such circumstances, heat recordings may overlap and the amount of heat recorded is less than that actually used. Certain inaccuracies are also involved when the velocity of flow varies and when the temperature of the hot water fluctuates. The meters can be manufactured to a tolerance of about ± 3 per cent accuracy with regard to each other. It is not suggested that meters of this type are particularly accurate in use, especially when there are fluctuations in supply pressure and temperature, but they have been found to be sufficiently accurate to serve as a basis for charging consumers for the amount of hot water they have used during a heating period and they offer the advantage of low cost.

Uses of evaporation meters

The main purpose of employing evaporation meters is to subdivide (apportion) the cost of the heat supplied to a number of dwellings according to the amount of heat which has been supplied to each individual dwelling. The heat which is supplied to a group of 50 to 100 dwellings is measured accurately at a central location by a mechanical or electronic heat meter. Once each year a meter reader examines all the evaporation meters in a dwelling and notes down the number of divisions used over the year.

The share of heat apportioned to each individual dwelling is given as:

$$\frac{H\,n}{\Sigma\,n}, \text{ where}$$

H is the heat read off on the mechanical or electronic heat meter for the whole group of dwellings

n is the number of divisions read on evaporation meters in the individual dwelling in question

Σn is the total number of divisions read for the whole group of dwellings.

To best effect the entire billing operation should be carried out by a computer. The following precautions can be taken automatically by suitably pre-programming the computer:

(a) each radiator has a reference number: if this reference number is missing the results are thrown out

(b) the results are rejected for mechanical checking if the heat consumed by a dwelling appears to be outside a range of about ± 30 per cent of the previous year's readings

(c) the computer will reject the results of manual checking if the meter reading in respect of any one radiator gives a reading value which varies appreciably from the more usual (or standardised) values obtained for a radiator of this type.

Even if all the radiators in a dwelling have been turned off there is a certain small zero registration. This is considered to constitute a fixed charge to pay for the rental of the meter.

Where correctly fitted and administered evaporation meters have been found to be well-suited to their intended purpose because of their low cost and adequate accuracy and they are widely employed as the basis of computer billing. Their disadvantages are:

(a) it is necessary for the meter reader to visit every part of a dwelling. This involves negotiation for access, particularly difficult where all members of a household are at work. Some tenants are also concerned at the security aspect of permitting access to each heated room

(b) some tenants are nervous about the evaporation of organic liquids in their homes. Methylbenzoate has a very low toxicity but cyclohexanol has been considered by some authorities to be a suspected carcinogen

(c) it is not possible for individual tenants to read the meter themselves to

assess the likely size of the forthcoming fuel bill (in contrast to gas and electricity meters or other heat meters where readings can be taken with ease)
(d) meter reading and meter charging is labour intensive and errors can be made; these can be greatly minimised by suitably pre-programming the computer.

Until alternative systems of heat metering have been sufficiently developed to compete economically with evaporation meters such meters are likely to retain their popularity, if not with the public, then at least with the operating authorities. The only alternative to such metered apportionment is to have no meters at all and to make a flat charge for the heat but all experiences have shown that the heat consumption under such circumstances will exceed that in metered premises by anything between 20 to 50 per cent, thus increasing the cost to each consumer very appreciably.

THERMOELECTRIC HEAT METERS

This type of heat meter is more accurate than the evaporation meter. Its advantage for metering heat supplies to individual dwellings is that it can be read very quickly and easily at central positions outside the related dwelling if so desired. Its main disadvantage for that purpose is that it is twice as expensive and complex to instal as an evaporation meter.

A thermoelectric heat meter suitable for the determination of heat and hot water consumption supplied by a central source has been developed by the Danish Kemp and Lauritzen Company (**11.6** and **11.7**). It measures the heat consumed in a dwelling very accurately and can be installed at a central position adjacent to electricity and gas meters. If desired it can be placed outside the dwelling in a special meter cupboard on the landing, in a cellar or in a garage. In blocks of dwellings all the meters can be grouped together at a central position so that reading is a much easier process.

A thermoelement is fitted to the surface of each radiator consisting of a number of copper/constantan thermojunctions. A similar thermoelement is fitted to the wall adjacent to the radiator. In all cases the number of thermojunctions used is directly proportional to the surface area of the radiator, although allowance is also made for the emissivity of the radiator surface assuming that the difference in temperature between the heat emittor and the room is 45°C (81°F). Only slight inaccuracies have been noted due to the necessary approximation of the equation for radiation heat transfer. Co-operation with commercial hot water radiator firms has enabled the manufacturers to design the thermoelements in a way that reflects the capacity of each given radiator.

If a radiator is turned off its temperature will be the same as the temperature of the air in the room. Consequently no current will flow because the temperatures of the cold and hot junctions of the thermocouple system are

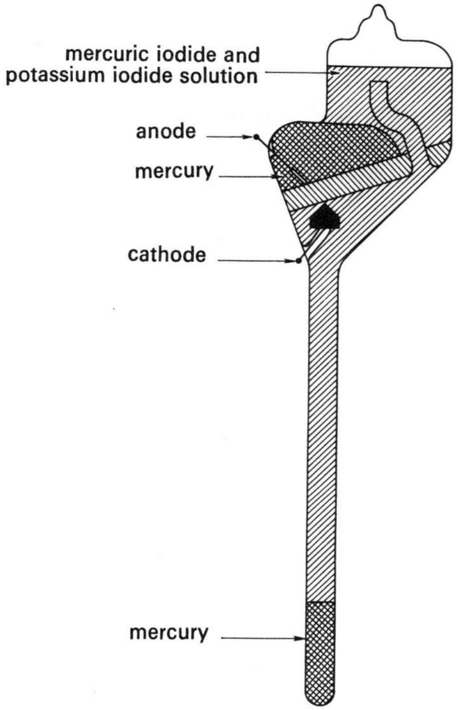

11.6 Kemp and Lauritzon thermoelectric heat meter

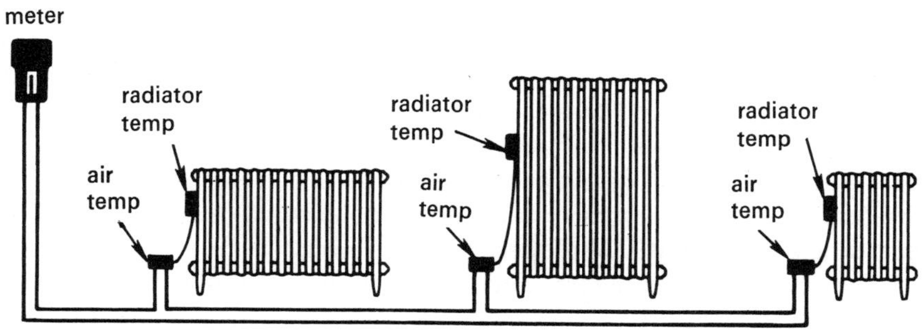

11.7 Application of Kemp and Lauritzon heat meter

the same. When the radiator is turned on its heat output can be taken as being directly proportional to its surface area multiplied by the temperature difference between its surface and the room air. The heat emitted by each radiator is expressed by the temperature difference between the two thermo-elements multiplied by the number of thermojunctions in each element and time. Within the range of use the current produced by the pair of thermo-elements is therefore directly proportional to the heat given off. If several radiators are employed the different thermocouples are connected in series.

The total current which is produced by all the thermocouples attached to the various radiators in the dwelling is measured by means of an electrolyte meter attached to the wall in a plastic box next to the electricity or gas meter or wherever desirable. The box in which the meter is housed measures 6 cm $\times$ 6 cm $\times$ 18 cm and contains a glass tube filled with a mixture of mercuric iodide and potassium iodide solutions, with a glass diaphragm in the top of the vessel and a funnel pointing upwards. The tube ends with a long extension of constant diameter. A bath of mercury on top of the glass diaphragm constitutes the anode; the cathode consists of a platinum plate below the diaphragm. When a current passes the mercuric iodide solution acts as the electrolyte in the following cell:

$Hg/Hg^{2+}/Hg$

Below the diaphragm the following reaction takes place:

$Hg^{2+} + 2e \rightarrow Hg$

while at the anode pool the reaction is as follows:

$Hg - 2e \rightarrow Hg^{2+}$

In this way the mercury is transferred electrolytically from the anode section of the tube to the cathode section dropping down in a fine stream to fill the sight glass which is calibrated in thermal units. After multiplying the reading by a single digit factor which is printed on each box a direct reading of the heat consumption is obtained. It is possible to check the amount of heat used at any time simply by reading the scale and multiplying by the factor: such a procedure should be well within the capabilities of the consumer.

During the operation of the metering device the meter is kept closed and vertical by the use of a lead seal to prevent unauthorised interference. After the meter has been read the lead seal is opened and the meter is inverted: this causes the mercury to flow from the cathode section (the sight glass) back into the anode pool. After the meter has been reset a new lead seal is fitted. The meter is designed so as to cover the usual heat consumption of an average flat or small house per heating season.

Connection to the radiator thermocouples is by means of twin core 0.5 mm diameter lead-covered cable which normally runs underneath the floor-boards. In existing houses this cable can be run on top of the skirting boards.

ELECTRONIC HEAT METERS

Types available

Much progress has recently been made in the design of electronic heat meters and it is likely that within a few years these will completely overtake other types of meters for determining the quantities of heat consumed in dwellings (**plates 37** and **38**). At the time of writing such meters are still rather expensive and are not as yet price-competitive with other metering systems. The cost of electronic equipment in general tends to tumble when quantity production is achieved and this may well happen in the case of electronic heat meters.

The following companies market electronic heat meters (at the time of writing):

Siemens GmbH of West Germany
Pollux-Elzol Company of West Germany (marketed in the United Kingdom by Samson Controls Ltd)
Aquametro AG of Switzerland (marketed in the United Kingdom by Ranger Instrument Ltd)
SVM AB of Sweden (marketed in the United Kingdom by Carlton Ltd)
Kamstrup-Metro A/S of Denmark (marketed in the United Kingdom by Clorius Ltd)
IWKA GmbH of West Germany.
Valmet OY of Finland
Bopp und Reuther GmbH of West Germany
Eduard Schinzel GmbH of Austria
Hallwachs GmbH of West Germany
Fischer und Porter GmbH of West Germany
Ludwig Krohne KG of West Germany
Eckard AG of West Germany
Mense KG of West Germany
Nordco GmbH of West Germany
Hartmann und Braun KG of West Germany
Sonceboz AG of Switzerland
Jernss AG of West Germany

Principles of design

There are considerable similarities in the design of these various makes of electronic heat meters which basically consist of three elements, the water flow meter section, the temperature sensors and the calculator and recorder.

(*a*) *The water flow meter section*

This consists of a vaned wheel or Woltmann counter which revolves at a rate proportional to the water flow. In some cases the Woltmann meter is positioned in a sidestream unit. Impulses are fed through a cable either magnetically or by the use of a light beam which impinges upon a photo-electric cell. The impulses are directly proportional to the flow rate of the water. It is of the utmost importance that the flow meter is always positioned in the return line and not in the flow line as its operation can be adversely affected by variations in the feed temperature of the water. Most systems are designed so that an impulse is sent out when a definite quantity of water has passed the meter. The water meter tends to be the most inaccurate part of the system. In spite of the fact that electronic heat meters incorporate a system of error adjustment to compensate for differences in temperature of the return water (which affects its viscosity) the limit of accuracy of such a meter is normally ±2 per cent when operating in its upper measuring range, and ±5 per cent when operating in its lower range.

The main difficulty found with all Woltmann meter systems is that after a certain time the vanes become dirtied by sediments in the circulating water and no longer operate as intended. All Woltmann meters therefore should either be operated with very clean water or frequently cleaned. Research work is proceeding with a view to improving the water flow measurement aspect of electronic and mechanical heat flow meters (**11.8**). In the Kamstrup-Metro system the Woltmann meter is completely separate from the impulse activator which is made from aluminium and can be fixed on top of any make of water meter. It operates by means of a rotating beam of light and a photoelectric cell. The electrical impulses from the water meter are then transmitted to the calculator by means of a cable with standardised electrical resistance. SVM market two rather similar impulse activator units capable of fitting virtually every type of Woltmann meter in common use in Europe. With some makes of meter it is possible to see the graduations of the original Woltmann flow meter and read off the quantity of water passing per unit of time through the transparent bottom of the impulse activator and thereby check whether there is a current failure if the heat meter does not appear to register. Some makes use an electromagnetic system instead of the Woltmann meter for measuring flow rates. There is at present little improvement in accuracy.

(*b*) *Temperature sensors*

Temperature sensors are high grade platinum resistance thermometers which are positioned in special junction pieces attached to the inlet and outlet pipelines (**plate 42**). The accuracy of the resistance thermometer system is far better than that of the water meter. The larger the difference between the flow and return line temperatures the more accurately is it possible to measure Δt, the difference in temperature. At a temperature difference of 20°C (68°F) the accuracy is ±1 per cent, at 40°C (104°F) ±0.6 per cent, and at 60°C (104°F) ±0.4 per cent. The linearity is better than 0.2 per cent and the stability at least 0.3 per cent over 10 000 operating hours. It is necessary for the cables between the resistance thermometers and the electronic instruments to be of standard length with standard resistances. If the distance between the thermometers and the Woltmann meter on the one hand and the electronic calculator on the other exceeds about 2 m a special junction box is required which is connected by a six-core cable to the calculator unit. It is essential for the cables to be made to fine tolerances so that their electric resistance per unit of length is constant. It can be calculated that an error in measured resistance of the cable of 1 ohm results in a temperature measurement error of 2.5°C (4.5°F). Resistance thermometers are usually made from nichrome which can be paired to register temperature differences of ±0.05°C.

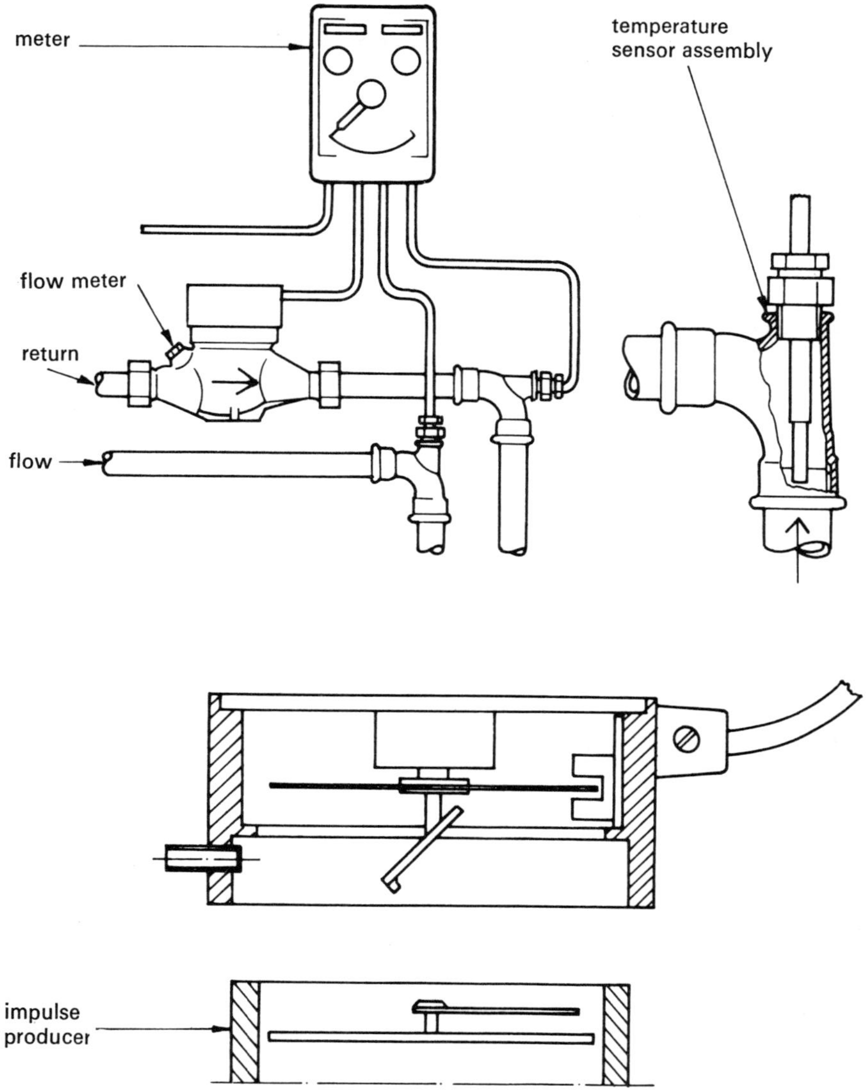

11.8 Kamstrup-Metro electronic heat meter

(c) The calculator and recorder

The six wires from the water meter and the two resistance thermometers are connected to the recorder which is fed with mains current, usually at 240 volts, to a transformer which is incorporated in the unit. An analogue frequency converter converts the temperature difference between the two resistance thermometers to electrical impulses which are then integrated with the electrical impulses given by the Woltmann water meter to constitute heat quantities. Various electrical resistances are incorporated in the system to compensate for the constants involved, eg conversion to G-calories, G-joules etc as required and differences in the lengths of connecting wires etc. In some instruments a special adjusting device is incorporated which is operated by the temperature of the return line and adjusts the Woltmann meter reading to compensate for differences in water viscosity and density. Using various types of electronic circuits, mostly in the form of 8-bit units, the product of the water meter impulses and the temperature difference impulses is recorded upon an electro-mechanical counting unit.

Adjustment

A special test instrument has to be used for the adjustment of electronic heat meters. Special variable resistances are operated with respect to:
(a) low temperature differences (3 to 10°C) (37.5 to 50°F)
(b) high temperature differences (30 to 100°C) (86 to 212°F)
(c) the integration time per water impulse
(d) the conversion factor to units in which the final data are given.
Electronic instruments must be superior in performance to mechanical integrators because the sophisticated adjustments to their operation which serve to correct enthalpy, density and specific heat changes in water cannot readily be carried out by mechanical integrators. Quite simple and inexpensive micro-chips can be added to electronic integrators to carry out additional refinement of the recorded heat quantities. All electronic recorders are nowadays made to read in four different types of units: kWh, MJ, Mcalories and therms.

The operating details vary with the different makes of electronic heat meters, but the most common applicable data are:
Smallest quantity of water which can be registered: 10 litres per impulse
Resistance of thermometers at 0°C (32°F): 100 ohms
Measuring current: 3 milliamps
Power consumption: 5 to 15 watts.
It is generally considered necessary to service such meters once every 3 to 5 years. Sediments have to be removed from the Woltmann meter and the temperature sensors but the electronic system itself should normally give no trouble whatsoever.

Future developments

Within such a rapidly expanding field as electronic heat calculators, information may date rather rapidly during the run of this book. At the time of writing such meters were still in their infancy and in consequence rather expensive. It is likely that further developments will progress along the following lines:
(a) more sophisticated measurement of water flow (currently the weakest and most uncertain aspect). Woltmann meters have considerable limitations and

even with electronic value adjustment are somewhat inaccurate
(b) direct computer billing (avoiding the need to physically read each meter) by sending out impulses from the recorder which are read and processed by a central billing computer.
Computer billing is a great advantage in that it saves considerably in labour costs and also eliminates human errors. It enables bills for district heating to be sent out in conjunction with those for electricity and gas: this is only an economic proposition where the same organisation is responsible for all three (as is indeed the case in most countries outside Great Britain).

METERING OF DISTRICT COOLING SUPPLY

Requirements

In district cooling systems water is usually supplied to the consumer at around 4°C (39°F) and returned at about atmospheric temperature. The return lines are never insulated. If the supply temperature of the water is kept constant there is no need to employ a heat meter: a simple vane quantity or Venturi meter will meet the requirements. Many different types of this sort of meter are made which record directly upon a mechanical reader or an electronic system. This equipment is far cheaper than using heat meters because it does not require temperature sensor and integrating devices. Measurement is also much more accurate because fluctuations in the density and viscosity of the flowing water do not have to be taken into consideration.

Venturi flow meter

A typical meter for measuring the flow of water in district cooling systems is the Venturi type made by Barco-Aeroquip (**11.9** and **11.10**). This consists of a Venturi constriction with twin pressure-sensing pipes one on the

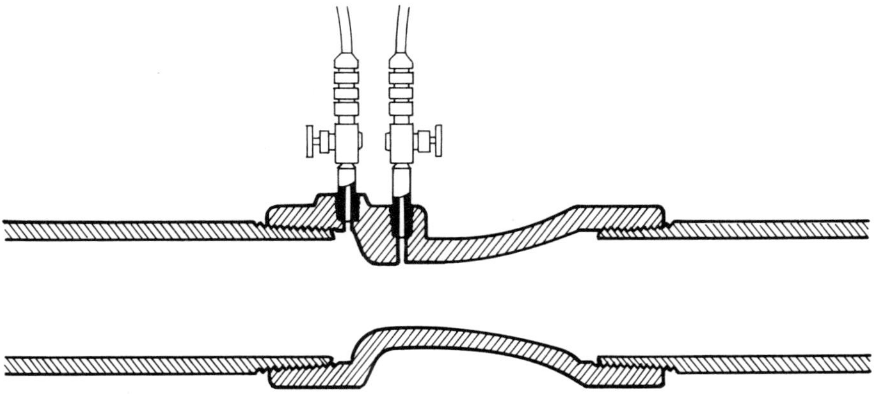

11.9 Barco-Aeroquip Venturi system for measuring water through velocities

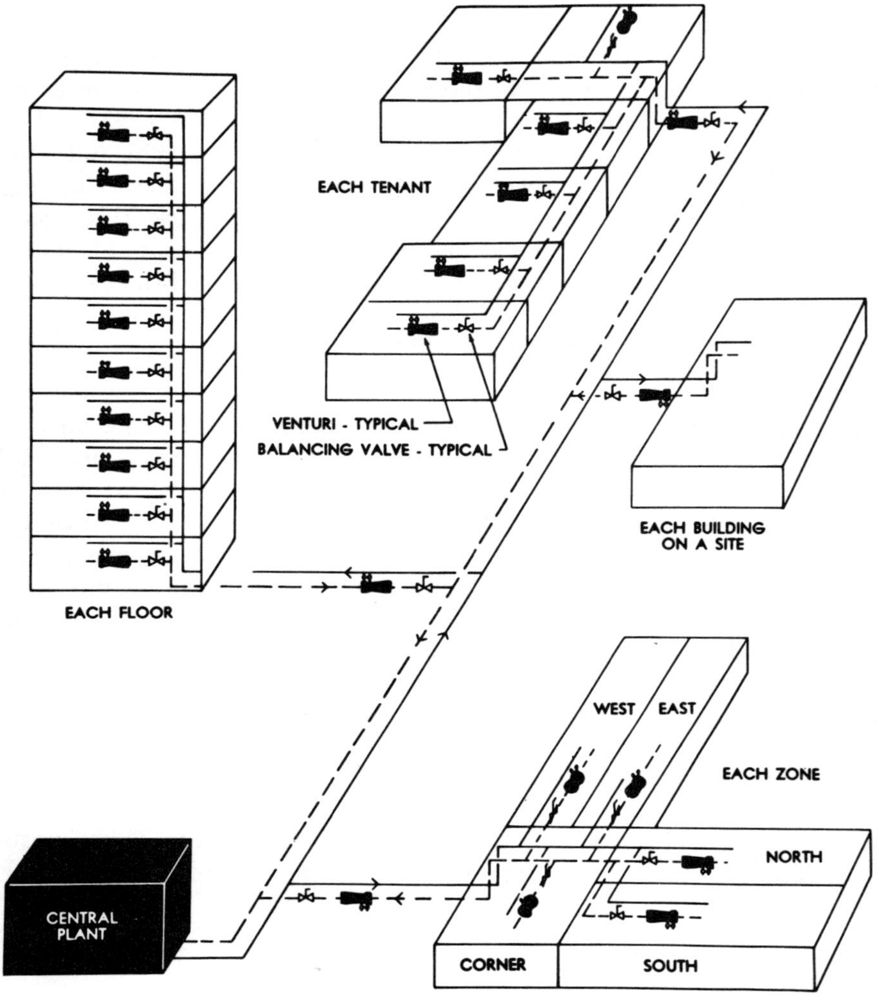

11.10 Meter system applied to a district cooling layout

constriction and one on the inlet of the pipe. The pressure difference can be read off directly using either a static or a portable manometer. The flow rate is directly proportional to the square root of the pressure difference. Such meters have the advantage of cheapness, sturdiness, low pressure head loss and reliability but they are not as accurate as more sophisticated meters.

Autarkon electronic heat meter

This meter has been developed by the West Germany company, IWKA (**plate 39**). It is unique among accurate heat meters in that it contains no moving parts whatsoever. It consists basically only of two units: the flow meter and the electronic system. The former operates on a pressure difference principle: after elimination of static pressure the difference in pressure between the flow line and the return line is proportional to the square of the fluid velocity. In consequence the electrical signals emitted from the flow meter equate to the square of the flow velocity. A special chip inside the electronic circuit takes the square root of this value so as to obtain the flow velocity directly from the pressure difference between the flow and return line. The actual flow rate is then equal to the flow velocity multiplied by the cross-sectional area of the measuring instrument. The electronic system integrates this flow rate by the temperature difference between the flow and return line and quotes it on a dial in terms of heat abstracted. At the same time the integrated readings of the flow rate are quoted in terms of m^3 of water consumed. The entire system is operated by a built-in thermogenerator which obtains electrical energy from the difference between the network temperature and the low temperature storage of the wall on which the instrument is mounted.

The instrument is capable of making allowances for differences in specific heat and can also be employed for different ranges. If the quantity of heat abstracted exceeds a given value, the instrument is switched automatically on to a less accurate scale. Two versions of the instrument are made. The technical specification is as follows:

Operation: either by means of 220 V 50 Hz mains current or by the built-in thermogenerator.

Maximum temperature: either 130°C (266°F) or 180°C (356°F) according to design

Accuracy of electronic calculation:	7 digits
Limit of accuracy above minimum value:	±2%
End of calculating section:	10 V and 14 V depending upon design
Current consumption:	2.5 mA
Maximum temperature difference:	120 deg C (216 deg F)
Maximum ambient air temperature in location of use:	50°C (122°F)

Thermogenerator: requires a minimum of 20 deg C (36 deg F) difference between supply temperature and wall temperature and is able to deliver 1.6 V and 0.5 A which can be stepped up by a transformer to the required level

The instrument can be used for all flow rates between 250 litres of water per hour to 1200 m^3 per hour.

CHECK LIST

1 Need for metering the heat consumption of district heating systems
2 Principles of charging for district heating consumption

3 Types of steam meters used for district heating
4 Uses of integrating flow meters
5 Types of hot water meters for district heating
6 Types of thermo-electric heat meters
7 Advantages/disadvantages of using evaporation meters
8 Types of electronic heat meters available for district heating systems
9 Methods of metering for district cooling systems.

REFERENCES

Fernwärme International (*articles in German*)

E. Dorn, 'Optimum subdivision of heat load at two positions of CHP stations', 2 (1973) 3, pp 41–54

E. Hartel and P. J. Hofbauer, 'Determination of heat distribution in district heating networks', 2 (1973) 4, pp 77–83

D. Magdeburg, 'Testing of heat meters', 2 (1973) 6, pp 147–151

D. Dommann, 'Methods of costing district heating to the consumer', 4 (1975) 1, pp 28–35

D. Dommann and W. Sdunzig, 'Costs of heat metering', 5 (1976) 3, pp 63–68

W. Hogreve, 'Measurement of heat quantity using magnetic induction flow meters', 7 (1978) 4, pp 137–142

Other sources

A Brown, 'Metered heat', *Building Services Engineer*, 43, September 1975, p A41

'Heat metering: a consumer view', Consumer Council, May 1978

R. Hromek, 'Subdivision of investment and production costs with combined heat and power generation', Brennstoff, Wärme, Kraft, 24 (9), September 1972 (in German)

E. Kraus, 'Use of heat meters in the city of Vienna', Conference on modern building, Berlin, 2nd September 1975 (in German)

H. Singh, 'Process controlled metering of district heating', *Building Services Engineer*, 43, February 1976, pp 234–239

'VDEW guide for the measurement and billing of heat', VWEW, Frankfurt, 70, 1966 (in German)

CHAPTER 12

Distribution of group or district heat

ADVANTAGES REQUIRED

District heating should offer the consumer a reliable heat supply which must also be cheap. Unless these two major advantages can be assured no consumer will abandon his private heating arrangements in favour of district heating and few authorities will initiate such schemes.

If district heating arrangements are organised correctly it should be possible to ensure a reliable heat supply to the consumer at a cost very much lower than current fuel price and still make a profit. The provision of district heating cannot be justified unless it can achieve this objective.

HEAT DEMAND

The amount of heat which is required by each consumer can vary greatly according to:

(a) Prevailing standard of thermal insulation
(b) Life style and personal habits of particular consumers
(c) Local climatic conditions
(d) System of temperature control
(e) Whether heat supply is metered or not
(f) Holiday pattern.

Given adequately insulated dwellings, the following maximum heat demand is likely to arise:

TABLE 12.1 DESIGN DEMAND FOR DISTRICT HEAT

Type of consumer	Maximum heat demand (W/m^3 of building volume)
Small house	80
Block of flats	45
School	60
Factory building	35
Offices	60

Each person in a housing estate draws off about 40 litres of hot water per day at an average temperature of 50°C (122°F). It is therefore necessary to budget for an annual hot water supply per individual consumer of 14.6m^3 at an average temperature of 50°C (122°F) or pro rata. This base load varies only slightly over the year.

University buildings and hospitals are heavy consumers of heat. Their maximum demand is likely to be of the following order:

Arts department (with workshops):	260 W/m^2 floor area
Science departments (including laboratories):	300 W/m^2 floor area
Hospitals:	400 W/m^2 floor area

These figures can only constitute guidelines (based on West German experience). Actual individual projects must be separately researched to establish a relevant design basis.

Some heating is necessary whenever the external temperature falls below about 16°C (61°F). Obviously the seasonal heating demand per dwelling in a particular year depends largely upon the climatic conditions that year. The average total annual space heating demand equals 1500 hours of the maximum heat demand in addition to the hot water requirements. Table 12.2 summarises the average heat demand for the major building types.

TABLE 12.2 AVERAGE ANNUAL HEAT DEMAND OF VARIOUS CONSUMERS

Type of consumer	Average annual heat demand (MJ/m^3 building volume)
Small house	430
Block of flats	245
School	325
Factory building	190
Office	325

The above figures are merely average figures based upon West German experience and must be modified to suit individual projects.

The average number of hours of space heating required per annum in various premises is as listed in table 12.3.

TABLE 12.3 AVERAGE NUMBER OF HOURS OF HEAT SUPPLY PER YEAR FOR SUNDRY PREMISES

Type of premises	Number of hours
Hospitals	2300 to 2800
Old age pensioners' homes	2000 to 2500
Swimming baths	2000 to 2500
Stores and shops	800 to 1400
Churches	300 to 600

If h is the number of hours of space heating required per annum and s is the specific heat demand (W/m^3) then the average annual heat demand of the particular premises is:

$$\frac{3.6\ h\ s}{1000}\ MJ/m^3.$$

HEAT SUPPLY

Economics of heat supply

It is obvious that district heating can only become viable and profitable if an adequate penetration of district heating in the consumer area is achieved. It is therefore generally necessary to include not only captive customers (eg local authority buildings, council house estates etc) but also private customers: these may be large users of heat such as factories, hotels, big offices and stores or smaller users such as small shops, offices and private householders. In every case it is necessary to promote and sell district heating aggressively, stressing the advantages and savings which will arise through the use of district heat. In the case of some consumers it may prove enticing to offer the installation of the entire hot water system complete with district heating connections as a package deal.

During the period when a district system is being installed less efficient methods of heat generation have to be employed (such as direct fired boilers or furnaces) instead of ITOC turbines and refuse incineration plant: the district heating will still only be sparsely used and the receipts will be relatively low. It may therefore be necessary to sell heat at a loss during the initial three or four years of operation so as not to frighten away potential customers by high charges. Such losses should be easily recouped in the future when most of the heat will be produced at a negligible cost using ITOC turbines and waste heat utilisation plant and income will be at a maximum.

Direct or indirect heat supply

It is always necessary to dimension district and group heating systems to ensure that the pressure difference between the furthest heat consumer and the station is about 8 bars for the flow line and 3.5 bars for the return line. There are two basic methods of supplying heat to the consumer, the direct method and the indirect method.

The direct heat supply method is illustrated in **figure 12.1.** The consumer is connected directly to the supply and return line and the district heating water or steam from the station flows through the radiator system. As most hot water type central heating systems are designed for operation at temperatures between 90°C (194°F) and 70°C (158°F) it is possible to run such systems efficiently with district heating supply at 110°C (230°F) and 60°C (140°F) return. Difficulties can arise when the circulating water is at a higher temperature as it is then difficult to co-ordinate direct supply from the central heating system without the use of bleed-in valves, by-passes etc which complicate matters and increase costs.

Direct consumer connection normally has the advantages that it is simpler to install and that it saves plant. The following disadvantages are however associated with it:

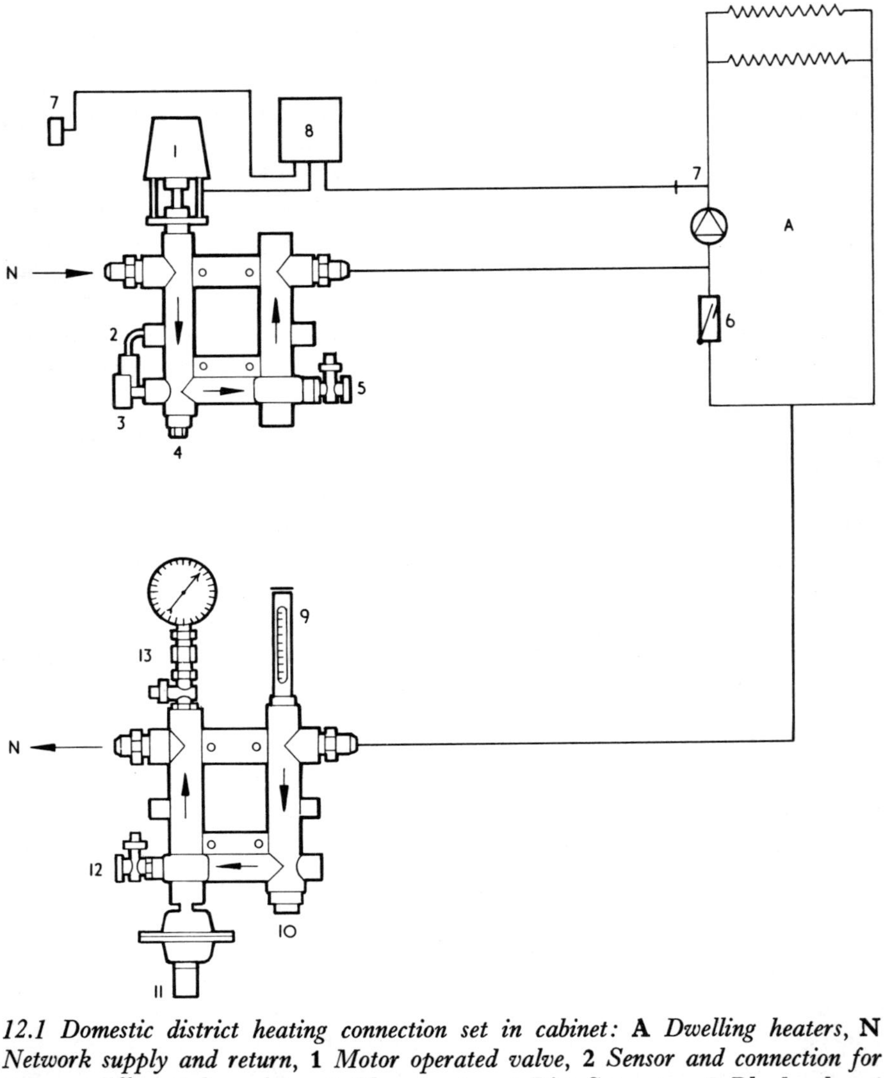

12.1 Domestic district heating connection set in cabinet: **A** *Dwelling heaters,* **N** *Network supply and return,* **1** *Motor operated valve,* **2** *Sensor and connection for pressure differential regulator,* **3** *Thermometer,* **4** *Dirt Separator,* **5** *Bleed valve,* **6** *Optional extra: non-return valve,* **7** *Sensor for external temperature regulator,* **8** *External temperature regulator,* **9** *Thermometer,* **10** *Dirt separator,* **11** *Pressure differential regulator,* **12** *Bleed valve,* **13** *Manometer*

(a) it is difficult to apply to tall blocks of flats where the head of the flow and return lines may be inadequate to serve the highest buildings
(b) district heating water has to be treated by the addition of large quantities of alkalis and deoxygenation materials such as hydrazine: a leakage of district heating water into the domestic circuit, or conversely a leakage of oxygen into the district system from the domestic circuit would be very harmful
(c) the system is less flexible during periods when repairs to the pipeline mains system are required
(d) if the district heating system operates with a higher supply temperature during periods of maximum demand, adjustment to this higher temperature must be made individually at each dwelling.

With an indirect heat supply system heat is delivered from the mains to sub-stations. Each of the latter abstracts heat via a heat exchanger to a separate circuit along which it is pumped to individual dwellings for space heating and hot water supply (**12.2** and **12.3**). The main disadvantage of the indirect system is that the sub-station, heat exchangers and other associated plant involve considerable extra cost. The major advantages are:
(a) if the district heating supply fails, the circulating water in the secondary circuit can be heated by means of an emergency gas or oil fired boiler (which may be provided with the scheme)
(b) the system can be operated with a smaller pumping head from the station
(c) by varying the speed of pumping (which can be controlled by a relay operated by a thermostat) it becomes possible to maintain the circulating water at a constant temperature however much the temperatures of the supply and return lines fluctuate
(d) it becomes possible to connect additional customers to an existing system without having to increase the pressure heads of the main flow and return lines.

Its numerous advantages have led to the adoption of the indirect supply method for most larger district heating systems. Direct water supply straight from the station is confined to the smaller group heating systems.

When group heating schemes are incorporated into the main district heating networks the group heating centrales are converted into sub-stations, the existing boilers being retained for emergency and peak supply purposes during periods when the central CHP turbines cannot cope with the demand.

Overall loading

The actual overall loading of a given district heating system varies enormously in terms of its maximum capacity according to types of consumers served, climate conditions and other factors. In general district heating systems are loaded at over 80 per cent of their design capacity for only 100 to 150 hours each year and at over 55 per cent of their capacity for only about 1000 hours each year. In terms of the overall economy of supply it is

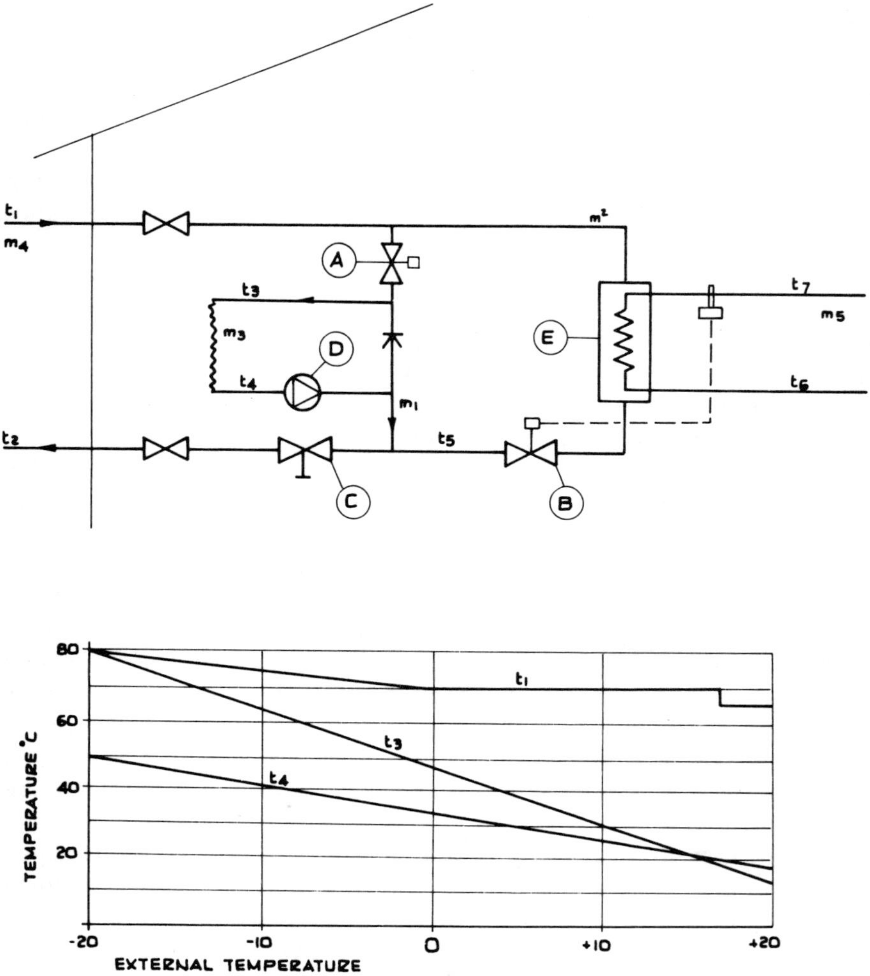

t_1 PRIMARY TEMP. INLET
t_2 PRIMARY TEMP. OUTLET
t_3 FLOW TEMP. TO HTG.
t_4 RET. TEMP. - HTG.
t_5 RET. TEMP. FROM HEAT EXCHANGER
t_6 HOT WATER FLOW TEMP.
t_7 COLD WATER RET. TEMP.

m_1 WATER FLOW TO HTG.
m_2 PRIMARY WATER FLOW
m_3 HTG. WATER FLOW
m_4 TOTAL PRIMARY FLOW
m_5 HOT WATER FLOW

A HEATING CONTROL
B HOT WATER CONTROL
C REDUCING VALVE
D CIRCULATING PUMP
E HEAT EXCHANGER

12.2 Layout of sub-station for small house

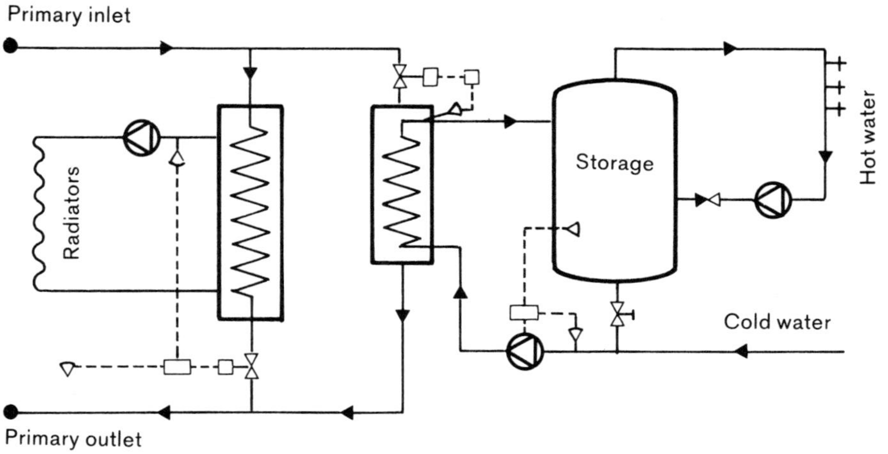

12.3 District heating sub-station using CTC heat exchangers

best to design a system in which low-cost base load stations (ITOC turbines, refuse incinerators etc) carry 50 per cent of the maximum load capacity with oil, coal or gas fired plant satisfying the peak heat demand. Some 85 to 90 per cent of the heat supplied to the system is then provided by plant with high capital but low running costs. The balance comes from peak supply stations with low capital but high running costs. The latter can also stand by effectively for emergency use in case of break-downs or interruptions at the main heat station.

House connections with direct heat supply

Most conventional radiator systems are constructed for a maximum operating pressure of 4 bars: district heating supply water is pressurised at the heat station to a pressure several times higher. For this reason it is necessary to build a throttle valve and various safety valves into the branch line from the main supply to protect the domestic system against excessive supply pressures: a positive pressure must always be maintained in the radiator system.

Most modern district heating networks are constructed on the ring main system so that two parallel rings of supply and return water are connected at the supply station (**12.4**). Thus when there is a blockage or leak anywhere in the system only a very small part of the network is directly affected.

DISTRIBUTION OF HEAT IN THE FORM OF HOT WATER

Twin pipeline systems

In a twin pipeline system heat is distributed via a mild steel pipe in the form of pressurised hot water at a temperature (t_f) and a pressure (p_f) with a

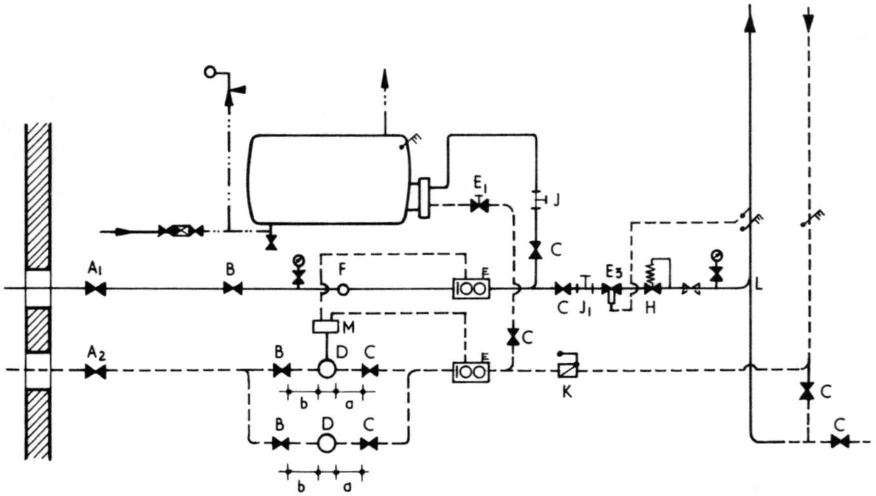

12.4 Diagram of heating centrale for district heating for circulating hot water: gravity system serving a small house. **Key**: **A_1, A_2** *District heating works' main shut-off valve,* **B** *building's main shut-off valve,* **C** *Shut-off valve,* **D** *Meters,* **E_1** *Thermostat without bulb,* **E_3** *Thermostat with bulb,* **F** *Strainer,* **H** *Shut-off valve with automatic pressure release,* **J** *Regulating tee or regulating cock,* **J_1** *Pressure reducing valve or as J,* **K** *Height loaded non-return flap,* a > *10×* *labelled dimension of meter,* b > *8×* *labelled dimension of meter,* **M** *Calorie meter head, pressure gauge scale 0–80 mwg*

return line system at a temperature (t_r) and a pressure (p_r) (**12.5** and **12.6**). Due to friction generated by the flow of fluids in the pipeline system the pressure in the flow main gradually decreases and the pressure in the return main gradually increases as the distance between supply station and consumer becomes larger. Thus the pressure head available to force the district heating water through the system becomes less the further the consumer is from the station. Pressure heads can of course be raised again by the installation of intermediate booster pumping stations which re-pressurise the feed water and lower the pressure of the return water.

Asymmetric pressure heads

If the diameters of the flow and return lines are the same, the pressure distribution in the pipe network is symmetrical and therefore limited to conditions where the pressure difference between the flow and return lines at the farthest point of the system is about 3.5 to 3.7 bars. The supply line pressure of such a system can be operated at a maximum of 7.5 to 8.0 bars from the station. Increasing the pressure of water in the flow line does not increase the flow rate through the pipeline system to any marked extent because this is governed by the pressure difference between the flow line and the return line (which is static at 3.5 to 3.7 bars). Better flexibility of opera-

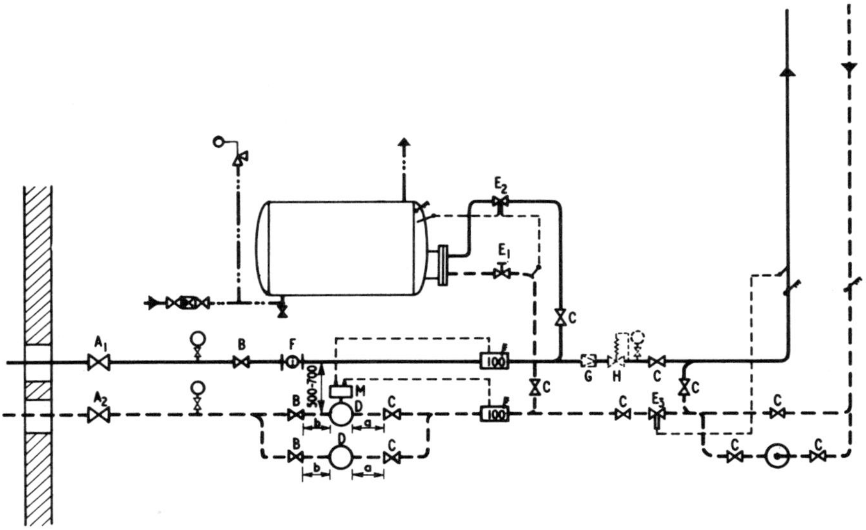

12.5 Diagram of heating centrale for district heating for circulating hot water: pumped system serving a tall block of flats. **Key**: **A_1, A_2** *District heating work's main shut-off valve,* **B** *Building's main shut-off valve,* **C** *Shut-off valve,* **D** *Meters,* **E_1** *Thermostat without bulbs,* **E_2** *Thermostat with two bulbs (only one of these required),* **E_3** *Thermostat with one bulb,* **F** *Strainer,* **G** *Reducing valve,* **H** *Shut-off valve with automatic pressure release (only if radiators cannot withstand 6.5 atm gauge),* **M** *Calorie meter head,* a > *10* × *labelled dimension of meter,* b > *8* × *labelled dimension of meter, pressure gauge scale 0–80 mwg*

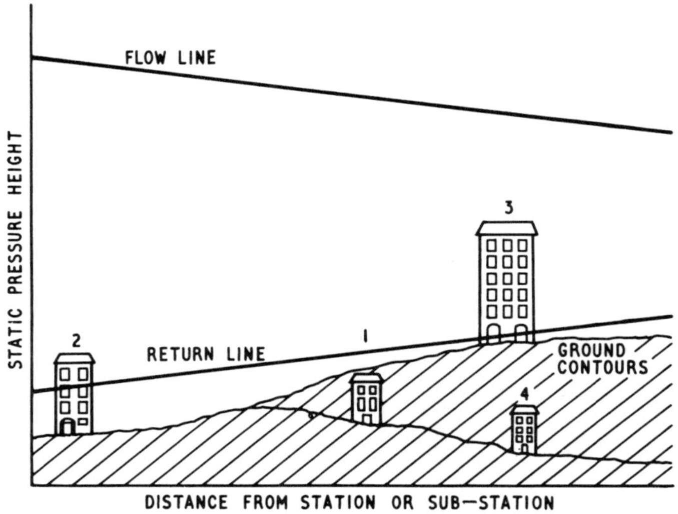

12.6 Supply conditions with buildings lying at different geographical heights

tion can be achieved if the return line is larger in diameter than the flow line. The pressure is then asymmetrically distributed, making it possible to increase the supply pressure at the head centrale when additional consumer areas have to be served or when it is necessary to cope with a rapid unplanned increase of hot water consumption. As asymmetric pipeline systems are more expensive than symmetric systems it is more usual to employ symmetric systems equipped with booster pump stations to provide flexibility.

Static head

In all district and group heating schemes it is necessary to give careful consideration to the degree of water pressure not only in the pipelines themselves but also in the radiators of individual dwellings served: it is essential to ensure that sufficient static head is available at all times for the radiators to remain filled with water (**12.7**).

The exact degree of static head required depends on the temperature of the water used and on the height of radiators above the station pump system serving them. Hot water exerts a vapour pressure and the barometric pressure in the pipelines has also to be considered. Where the pressure of the return line is maintained at a level higher than the static head of the system

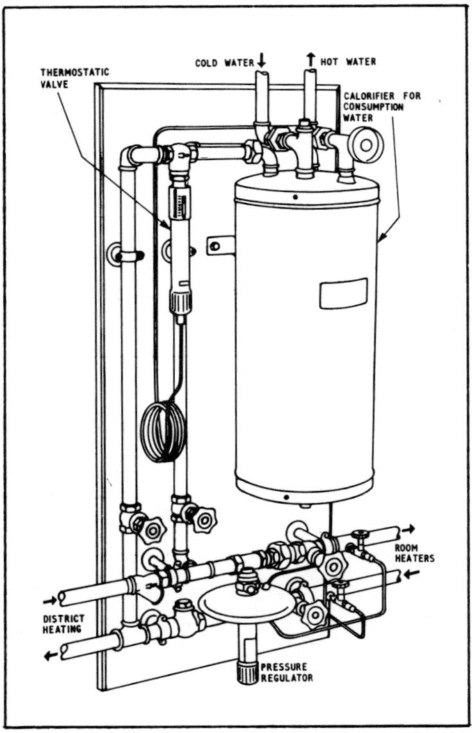

12.7 Diagram of heating centrale for steam district heating

there is no possibility that the system can ever operate at a vacuum or that flashing into steam can occur. With many systems however the return line operates at a pressure below the static head and it is necessary to provide check valves to protect the system from being emptied.

Sample calculation of static head: Determine the minimum water pressure necessary to prevent static head troubles in a district heating system operating at a supply temperature of 110°C (230°F) if the highest part of the building to be served is located 22 metres above the level of the supply station. At 110°C (230°F) the vapour pressure of water is 143.273 kPa. For each metre that the water has to rise at this temperature (when its density is approximately 950 kg/m^3) the vapour pressure increases by 9.316 kPa. The minimum degree of pressure to maintain the equilibrium value must therefore be:

$143.3 + 22 \times 9.32 = 348.3$ kPa or about $3\frac{1}{2}$ bars.

For safety's sake such a system should be maintained at an equilibrium pressure of not less than 4 bars. To reduce the overall pressurisation required in a network mainly serving low buildings at a similar level to that of the station it is often more economic to supply the upper floors of tall buildings indirectly by means of a supplementary secondary heat supply employing heat exchangers.

Supply water has to be maintained at sufficiently high pressure to give even the most distant consumer enough pressure head to force the water through his system (usually between 3.5 and 4.0 bars). This may cause the pressure difference available to other nearer consumers to be far too large or to fluctuate. It is therefore necessary to control the pressure differential applied at each sub-circuit of a district heating system at a constant value independently of the fluctuations in mains pressure. It is also necessary to limit the total quantity of water (and in consequence the amount of heat consumed) in each sub-circuit so as to balance the system.

Control of pressure differential and total pressure

The highest pressure differential exists close to the supply centrale or to booster pumping stations; this reduces along the network downstream from such points and is further reduced wherever a particular consumer abstracts a substantial amount of hot water. Differential pressure controllers are therefore fitted into the return main with impulse connections which sense the pressure at the start of the sub-circuit. These maintain a constant pressure differential under fluctuating mains pressure and/or flow rate. Close control of pressure differential at a constant head value permits the adoption of standard designs at all points in the district heating or group heating circuit whatever distances there may be between a sub-circuit and the main station or booster pumps. If the total pressure in the flow mains exceeds the pressure which the equipment in the sub-circuit can accept the pressure in the sub-circuit is reduced by pressure reducing valves. To safeguard the circuit against failure by any of the reducing valves a safety relief valve is fitted next to each of them.

Pumping systems

If water is pumped from the station so as to keep the pressure differential constant at the premises of the most distant consumer then there will be considerable variations in pressure differential at consumer connections closer to the station (**12.8**).

The pumping system can be arranged in one of the following three modes: (a) The pumps are placed in the flow line and produce in it a pressure sufficiently in excess of the static head acting on the return line to give enough pressure to supply the most distant consumer. If a single speed pump is used the pressure differential conditions near the station remain high but constant; at points more distant from the station they may vary greatly according to the heat load. When stage or variable speed pumps are

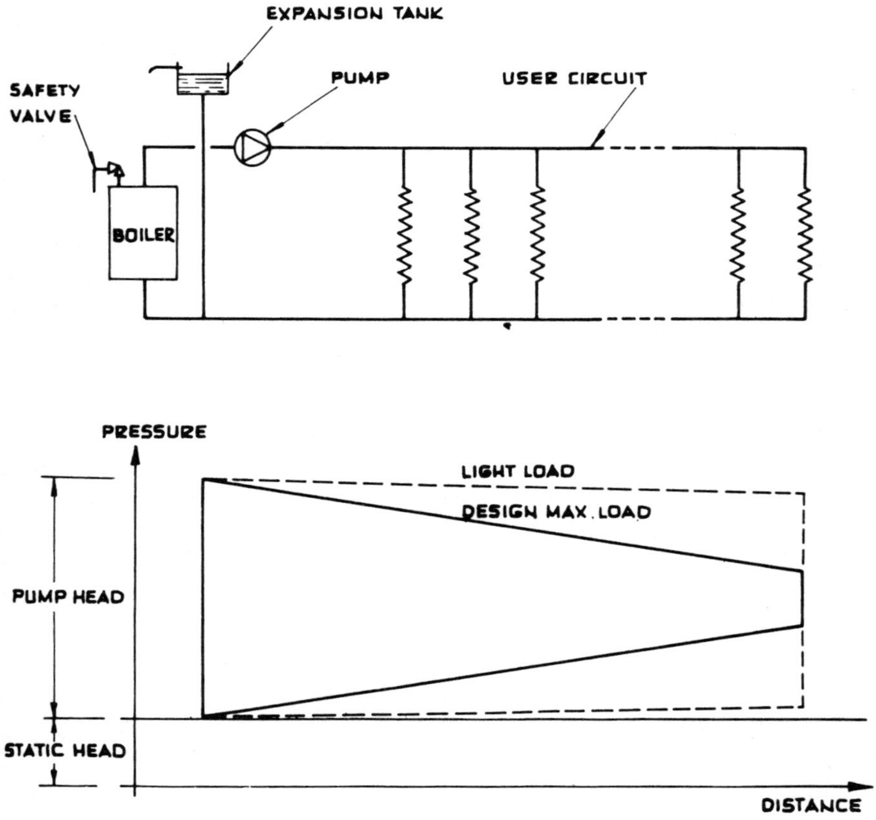

12.8 Pressure distance diagram for single fixed speed pump

used the pressure differential at the connections more distant from the station remains constant but varies close to the station. The pressure in the return line will always exceed the static head.

(b) The pump is placed in the return line to establish a static head which acts on the supply line system. Systems of this kind provide a lower total head than the system described in (a). The pressure in the return line is below that of the static head. Check valves must therefore be incorporated in the system.

(c) The pumps are placed both in the flow line and in the return line (giving a push-pull system). This arrangement gives a symmetrical pressure head with the static head being halfway between the flow and return pressure heads. It permits pressure reducing valves to be used to lower the pressure to the static head at any point in the system. It provides a lower total head than (a) for a given pump head and it is therefore possible to use equipment with a lower pressure resistance. The system offers advantages when installing district heating in older schemes. Cast iron valves can be used instead of the more expensive steel valves. Both variable and fixed speed pumps can be used with any of the above three schemes.

Pump controls with fixed speed pumping systems

When there is only one fixed speed and one standby pump available for pumping (often the case with small group heating schemes) the flow rate through the main pipeline system may vary extensively with changing heat demands. When the flow is very low the pump may suffer damage. Some boiler designs demand the maintenance of a specific minimum flow rate through the boilers. It is therefore useful to have a valved automatic pump by-pass which controls the pressure differential and opens the valve when ever the differential pressure exceeds a set value. The valve is usually adjusted to commence opening at a pump head which is slightly above the design operating point so that the circuit always maintains a constant pressure differential and the water flow through boilers and pumps remains constant at a set value. The best method of operating the pumping system uses several fixed speed pumps which are automatically switched on and off to cater for the varying circuit heat demands using a flow meter which measures the volumetric flow in the supply or return main.

Variable speed pumps

Variable speed pumps are economic in their use of electric energy. They are usually controlled by sensing the pressure differential at the index point. Given an adequate pressure differential at that point, the head must also be sufficient for sub-circuits closer to the heat station. The pressure differential at the index circuit is sensed by a pressure differential meter fitted with an electrical transmitter which signals the heat station to regulate the speed of the pumps as required (**12.9**).

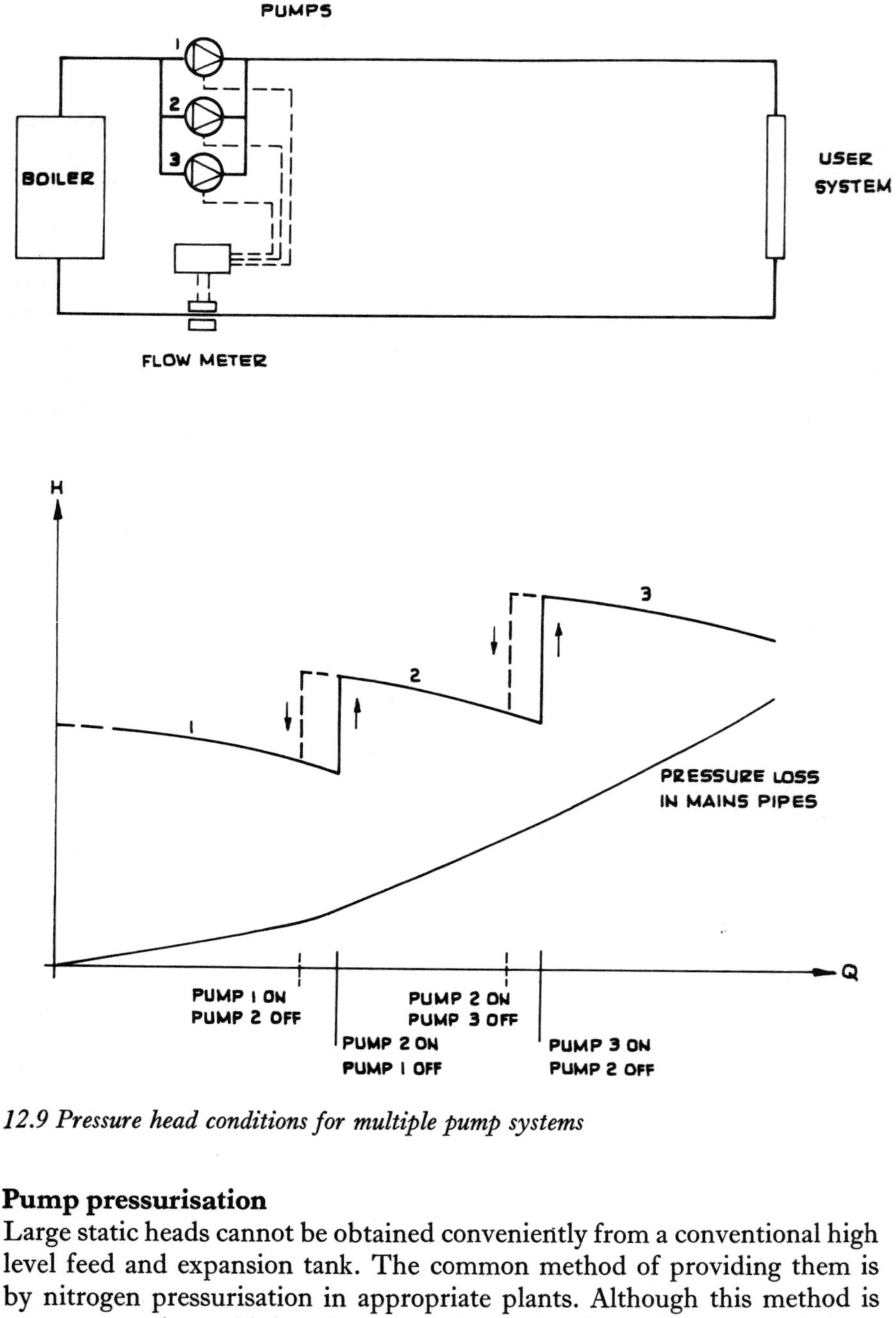

12.9 Pressure head conditions for multiple pump systems

Pump pressurisation

Large static heads cannot be obtained conveniently from a conventional high level feed and expansion tank. The common method of providing them is by nitrogen pressurisation in appropriate plants. Although this method is very commonly used it is rather expensive. A simpler and cheaper technique uses an expansion vessel open to the atmosphere in conjunction with a pump which superimposes a static pressure on the system. A pressure relief valve controls the pressure level in the system by allowing excess water to return to the expansion vessel. This equipment is usually dimensioned to have a capacity of between one and five per cent of the main circulating pumps and is provided with a standby pump, an excess pressure valve and with non-return valves on the outlet of each pump.

High buildings

A high static head is required to supply the upper floors of tall buildings with district heat and this causes the other sections of the system to become excessively pressurised. There are two common methods of overcoming this difficulty (**12.10**).

One method uses the supply and return water from the district heating centrale sufficiently pressurised to meet the basic requirements of most of the buildings connected to the system. At the entry to each relatively tall

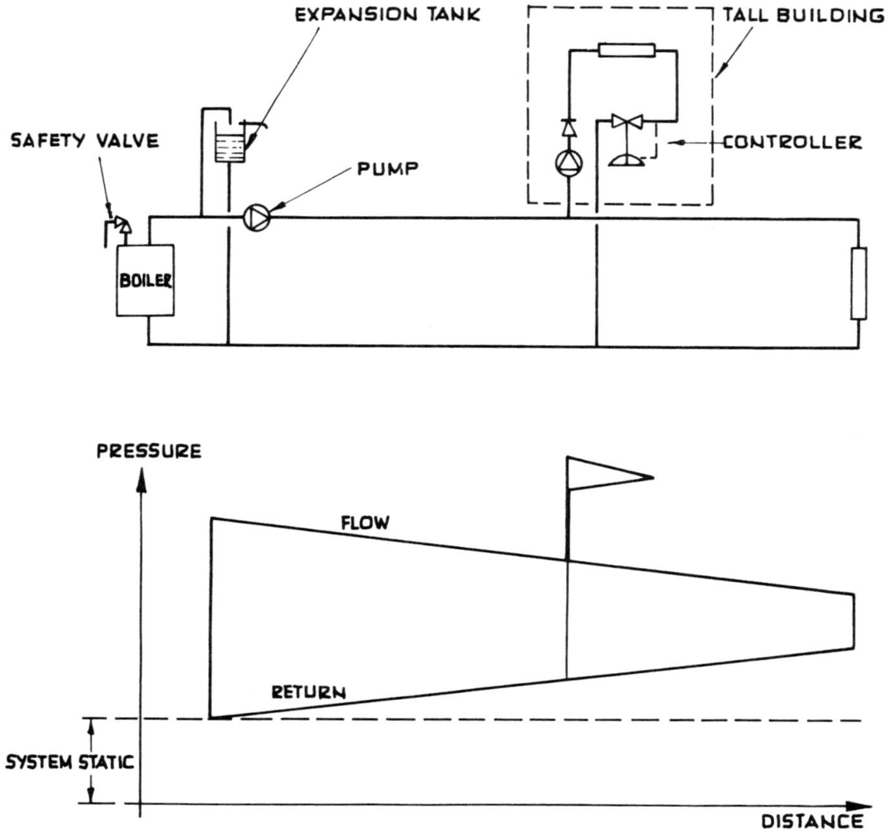

12.10 Pump pressurisation system for tall buildings

building the flow main pressure is raised by a pump; at the outlet an excess pressure valve maintains the return main pressure of the building above a set minimum value. It is necessary to fit a non-return valve near the pump so that water is retained in the circuit of the building even if the pump fails.

The other possibly more common method of supplying tall buildings uses an entirely separate secondary water circuit which is heated from the primary flow via a heat exchanger and is circulated by a separate pumping arrangement.

Auxiliary units in sub-circuit systems

It is necessary to provide several items of auxiliary equipment in each sub-circuit to ensure satisfactory and safe operation. These are:

Automatic by-pass valves

Automatic by-pass valves are fitted to ensure that the pump head in a secondary system containing two-port valves does not rise to an unacceptably high level. They prevent the total pressure from rising and also maintain good water circulation past the thermostat which senses the secondary temperature.

Temperature by-pass

The water in sections of a sub-circuit distant from the mains network can cool down if there are long periods during which no heat has been drawn; when there is a subsequent heat demand there will therefore be a time lag before hot water becomes available to the consumer. Such an undesirable situation can be avoided by fitting a thermostatically controlled by-pass in the distant sub-circuits. However the position can be kept in hand by providing adequate volume limiters in the system to ensure that the allocated water flow rate is not exceeded.

Strainers

The water used in a district heating system must be purified efficiently at the station and great care must also be taken to treat any supplies of make-up water fed into the system. Notwithstanding these precautions it is still necessary to provide pipeline strainers so as to remove any excessive build-up of sludge and/or other suspended solid particles.

Pressure gauges

The relatively high pressures involved in district heating make it desirable to use both direct-pressure gauges and pressure differential meters: these should always be installed at all the major distribution points. Provision should also be made for connecting portable test or check equipment of this type at the smaller stations. It is common practice to install into the network a metering orifice plate in each return pipe together with a permanently mounted flow rate indicator.

Control of group and district heating systems

The higher the temperature drop between the supply and return lines the

50. Borsari underground concrete oil storage tank under construction, part of the three-tank complex (two of 1 000 m³ and one of 800 m³)

51. Another view of the Borsari tanks in Plate 50 under construction

52. Two smaller sized Borsari tanks under construction: note the external finish and manhole detail

53. Interior of tile lined Borsari tank

54. Bulk tipper coal delivery truck

55. Pneumatic coal delivery truck

56. The brick-concrete boiler house at the centre of group heating scheme (Washington, UK): note the overhead bunkers and chimney to the left

57. Vertical steel silo for coal storage

58. Group of three internally flanged multi-flue mild steel clad chimneys (49 m high in each case)

59. Boiler house of the Nottingham, UK, district heating scheme: note the prominent concrete chimney

60. Three chimneys each extending 15 m above roof level, linked into a concrete structure for stability and to create an architectural feature

61. A striking modern coal fired boiler house with multiple free-standing chimneys (Chilwell Centre Ordnance Depot)

62. Allen Ygnis oil fired boiler plant serving district heating system (London, UK)

63. Vekos coal fired boiler plant (Billingham, UK, district heating scheme)

64. Pneumatic coal fired system with GWB boilers

65. Typical coal heating arrangement with screw elevator and horizontal trough conveyor

66. *Trianco gravity fed coal fired boiler installation*

67. *Checking and logging the plant performance*

68. National Coal Board district heating scheme at Edmonton, UK

69. Mobile heat service engineer

70. Evaluating combustion efficiency

71. Adjusting an oil burner

72. Routine inspection of warm air heater connected to a central plant

73. Maintenance of GWB oil fired boilers (Aviemore Leisure Centre, Scotland)

74. View of tidy boiler room; note control panel and tool rack

75. *Larger Mirrlees Blackstone diesel engine plant suitable for heat recovery*

76. *Larger Arkla rotary water chiller driven by low pressure steam*

77. Freon turbine driven turbo-refrigeration plant

78. Typical computer installation (ICL 1904A)

smaller the quantities of hot water which have to be transported around the system. Such a high drop therefore leads to savings such as smaller pipes, reduced thermal insulation and associated items. Where the hot water for group heating schemes is obtained from directly heated boilers it makes sense to use the highest possible supply temperature. This is also the case when gas turbines are the primary heat producers.

In district heating schemes which employ ITOC turbines to supply the heat it is however necessary to strike a balance between the thermodynamic loss of using a very high flow temperature (which is manifested in an excess loss of power generating efficiency) and the increased cost of operating the district heating system at a lower flow temperature. Such systems commonly operate with temperature drops of the order of 50 to 60°C (90 to 108°F). The temperature of the return line must be maintained at or below a certain design value. When the heat demand reduces, unwanted hot water must be prevented from entering the return line: two-port control valves are therefore used instead of three-port diverting valves because the latter would pass the unwanted flow water into the return main and thereby raise its temperature to above the designed value. The limitation of the return main temperature also reduces the heat losses (wastage) from these mains.

Volume controllers

It is necessary to ensure that no consumer can unbalance the design parameters of a given scheme by extracting an uncontrolled amount of heat from the system. Each consumer must therefore be allocated a maximum quantity of heat just sufficient to meet his maximum expected demand. This is done by installing volume controllers which guarantee that no consumer can take an undue amount of heat out of the system during periods of maximum demand and so deprive other consumers of heat.

Volume controllers are operated by pressure differential. A valve is used to restrict the flow passing through it to design level by sensing the pressure differential generated across an orifice plate or Venturi controller. If for example the design rate of flow is about 20 kPa, a controller can be adjusted to restrict the flow passing through it to that level immediately it is sensed by the orifice plate. This combination of orifice plate assembly and pressure differential controller is effective in limiting volume.

Small volume controllers can be used for groups of about 10 dwellings or for the smaller public buildings. These are made in sizes ranging from 12 to 25 mm and can be preset to any desired volume flow rate.

It is feasible to combine differential pressure and volume controls to actuate a single valve by adding a small computing relay. The valve then usually controls the pressure differential to a standard pre-set value. If the sub-circuit requires a greater flow rate the valve closes until the flow is limited to the maximum permissible allocated rate.

Temperature control of secondary circuits

When the temperature of the water maintained in the flow main of the district heating or group heating circuit is the same as that required in the consumer's sub-circuit no further temperature control is needed. More frequently however the temperature of the water in the supply line is higher than that required in the radiators of the consumer circuit. In such circumstances a mixing valve is employed to regulate the amount of water from the primary mains which is to be mixed into the secondary circuits. The control valve can either be self-operated or motorised. A return temperature limiter must be incorporated at the distribution station to ensure that the water in the return main is always at a fairly low temperature, thereby allowing the system to operate in the most efficient manner. This control should be set to operate at a slightly higher value than the design return temperature of the sub-circuit to prevent interaction between the return temperature limiter and the sub-circuit controls under normal operating conditions.

Temperature control of domestic radiators

A return temperature limiter is installed at the return outlet of each radiator. This has no control over the absolute flow rate but the water velocity through the radiator is adjusted in such a way that the exit temperature is always at or below the design value. As soon as there is any tendency for the temperature of the return water to rise, the return temperature limiter commences to close and thereby reduces the flow rate. A return temperature limiter of this kind is generally pre-set at the factory and can be locked to prevent unauthorised interference. The return temperature limiter has no effect upon room temperature. The heat emitted by the radiator can still be increased or decreased either manually by the use of a radiator valve or automatically by means of thermostatically operated valves. The return temperature limiter and the automatic thermostatically controlled radiator valve can be used together to control a group of radiators. For example it is possible to control all the radiators in a small house or in a flat by means of a single return temperature limiter and a single thermostatically operated radiator valve.

Volume flow limiters

These are used when the temperature of the water in the supply line is adjusted according to external temperature conditions. In such circumstances the central or sub-station automatically provides for an increase in the temperature of the supply water whenever the external temperature drops. The supply of hot water to any given dwelling is standardised on a volume flow rate basis. No return temperature limiters are fitted. Control of the internal temperatures is effected by a reduction in the volume flow rate to the radiators. It is usual to fit one such volume flow rate controller to each dwelling or group of dwellings.

RUSSIAN DISTRICT HEATING SYSTEMS

The large size of the Moscow district heating system and its widely varied types of consumers has led to the development and standardisation of special building connection schemes. In general hot water is used for heating most areas of the city of Moscow. Those on high-lying ground are however heated more economically from the steam network. The connection schemes for housing were evaluated to ensure that standardized components could be used throughout wherever possible so as to achieve efficiency in manufacturing them. However some individual designs had to differ because of variations in geographical height, density of population and rate of consumption. The following factors have been found relevant in choosing the most suitable connection scheme to adopt out of the alternatives available:

(a) difference in geographical height between station and consumer

(b) nature of district heating system (open or closed)

(c) required static and dynamic pressures at the junctions of the network

(d) thermal requirement of each single consumer at the minimum anticipated external temperature

(e) pressure drop in the network

(f) future operating data when the network is further expanded.

In towns and large housing estates open or closed twin-pipe systems have been preferred. Most of the connection systems incorporate injection systems and operate by gravity. Pumps are only used where the cheap types of injectors cannot be employed.

Hot water for domestic consumption is obtained from a twin-pipe system in one of two ways. In the case of open systems the town's water is heated directly by the supply and return water of the district heating system: this is the simplest form of heating used and is employed either with or without a hot water storage tank. In the case of closed systems a two-stage heating technique is employed: in the first stage the town's water is pre-heated by the return water of the heating system; in the second stage its temperature is raised to the level required in the heating supply network. This method has been found to be far more economical than heating the water in parallel (which requires three or four times as much hot water from the district heating network).

In general the supply of hot water for consumption purposes is not considered when calculating the dimensions of the district heating supply lines. In the event of a peak demand for the supply of hot water the temperature of the supply line water is raised. Usually the ratio of the heat required for hot water supply to that required for space heating is 1 : 10 (in Moscow). For special buildings, such as bath houses, hospitals, laundries etc where there is an abnormal requirement for hot water, parallel connections of heat exchanger and hot water tank are used.

SYSTEMS FOR SUPPLYING HOT WATER

In the Moscow district heating network the volume of water flowing through the pipes is kept constant but the temperature is varied with the external temperature: the lower the external temperature the higher the temperature of the water in the pipeline. The maximum supply temperature is commonly between 130°C (266°F) and 150°C (302°F). The minimum supply temperature is determined by the need to provide hot water for consumption purposes: it is not below 60°C (140°F) with open systems and not below 70°C (158°F) with closed systems even when the heating systems are not in operation. Because of the considerable pipe line heat losses which may then occur the hot water is pumped through only in batches at regular intervals of time. Hot water tanks at the consumer's premises provide the consumer with a 24 hours capacity storage of hot water.

The following control mechanisms are employed at the different supply stations:

Regulation of the air temperature in the rooms to be heated
Automatic pressure regulation of supply line
Automatic pressure regulation of return line
Automatic regulation of the temperature of the hot water which is heated in the calorifier.

The air temperature in the rooms is controlled by means of bimetal thermostats which turn the heat supply from the network on and off; these are used when hot water is not supplied by parallel connection. The room thermostats are not provided in buildings with a two-step hot water supply. At periods of maximum hot water consumption the temperature inside the room falls somewhat as the heat is diverted to the hot water draw-off. A constant supply of hot water is ensured by means of a throttle valve which is connected to the supply line and serves to maintain the pressure in the line to prevent emptying of the installation. When fan-assisted convectors are used for space heating the flow of the supply water is kept constant and temperature control is achieved by switching the fan on and off. The automatic throttle valve in the supply line is operated by the convector thermostat when the electric motor of the fan is started or stopped. To protect the consumer from being scalded by the consumption hot water a thermostat is provided to limit the temperature permanently to between 60°C (140°F) and 65°C (149°F) by regulating the flow of district heating supply water to the hot water calorifier. When there is a direct district heating supply, the regulation of the flow is controlled by means of a thermostat.

The following schemes of supply are in general use:

Case 1 where the consumer is situated below the level of the supply station: injectors can be used in most cases and no pressure regulators or return valves are then required

Case 2 where the highest part of the building served is above the return back

pressure but below the static pressure head of the system: it is necessary to fit an overflow regulator but check valves are not required
Case 3 where the highest point is above both the pressure in the main supply system and the static supply pressure and the pressure behind the injector is inadequate: when the system is operated on the short circuit (mixing lines) it is necessary to incorporate a pump; to ensure against the circuits being drained it is necessary to provide an overflow valve in the return line and a check valve in the supply line
Case 4 where the return pressure from the network upon the heating installation is higher than permissible: it is best to connect the consumer indirectly via backflow equipment so that the district heating network is fully separated from the local heating system.

The scheme shown in **figure 12.11** is used mainly when there is adequate pressure at the entrance to the dwelling: no control is then necessary. No domestic hot water supply is provided.

Key to figs 12.11–12.20: A *Radiators*, **B** *Water connection*, **C** *Room thermostats*, **D** *Electric contacts*, **E** *Town's water*, **F** *Hot water heater (calorifier)*, **G** *Circulating line*, **H** *Heat exchanger*, **I** *Water take-off*, **J** *1st stage*, **K** *2nd stage*, **L** *Accumulator*, **M** *Reversing flow equipment*, **N** *Temperature regulator*, **P** *Take-off for hot water*, **R** *Return line*, **Reg** *Regulating valve*, **S** *Flow line*, **1** *Dirt filter*, **2** *Water meter*, **3** *Injector*, **4** *Overflow valve*, **5** *Check valve*

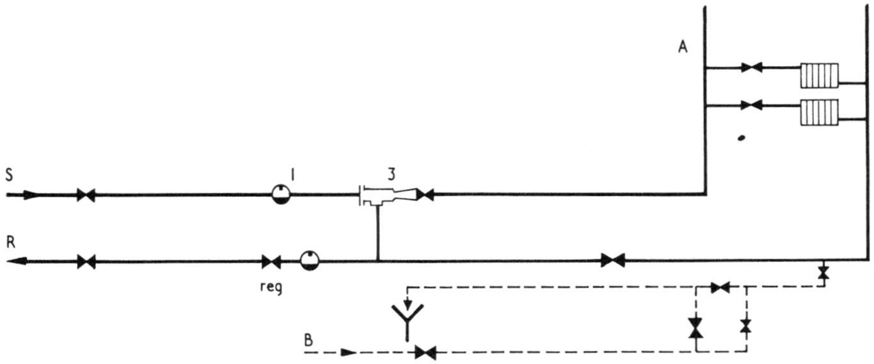

12.11 Connection at entrance to dwellings in domestic hot water supply provision

The scheme shown in **figure 12.12** is used with controlled connections when there is adequate pressure available and no domestic hot water supply is required. Constant water supply is achieved by the regulator (REG); control of the air temperature is by means of room thermostats which turn on or off the supply of heat to suit the required room temperature. The heating plant is protected against running empty by means of the overflow regulator and the return valve. The essential pressure in the heating system

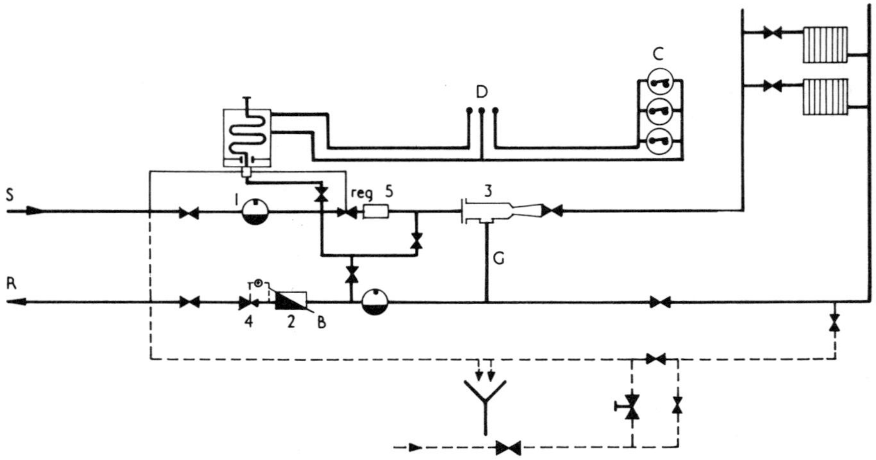

12.12 Consumer connection with constant heat supply control, room thermostats and overflow valve (see p. 311 for key)

is maintained by the overflow valve even if the return pressure in the network is inadequate. An electro-hydraulic relay checks on pressure losses in excess of 2 bars from the main supply line to the building supply point (point A) and on those below 2 bars between the main return line to the point B.

The scheme shown in **figure 12.13** is used when the available pressure is inadequate for the installation of injectors. The inlet temperature is regulated according to the external air temperature. The heating appliances are protected against running empty when the network pressure drops by means of an overflow valve and a check valve. The supply pressure is controlled when the return line pressure is inadequate.

The scheme shown in **figure 12.14** is used when under the dynamic or static pressure in the network the pressure in the heating circuit exceeds the maximum permitted pressure (in old systems: 5 bars). The temperature of the supply water is regulated relative to the external air temperature.

The scheme shown in **figure 12.15** is used when the supply of domestic hot water has priority over the supply of space heat (eg in hospitals). This scheme achieves a constant flow of hot water from the network to the local system. The circuit within the building is protected against emptying by means of the overflow valve and the check valve when the pressure inside the network is reduced. The essential pressure in the heating system is maintained, even when there is inadequate return line pressure in the network. The temperature of the domestic hot water is controlled beyond the heat exchangers by a thermostat and a flow regulator. The circulation pipe is connected to the centre of the hot water heater where the temperature is about 40°C (104°F) to 50°C (122°F). The impulse source is connected to

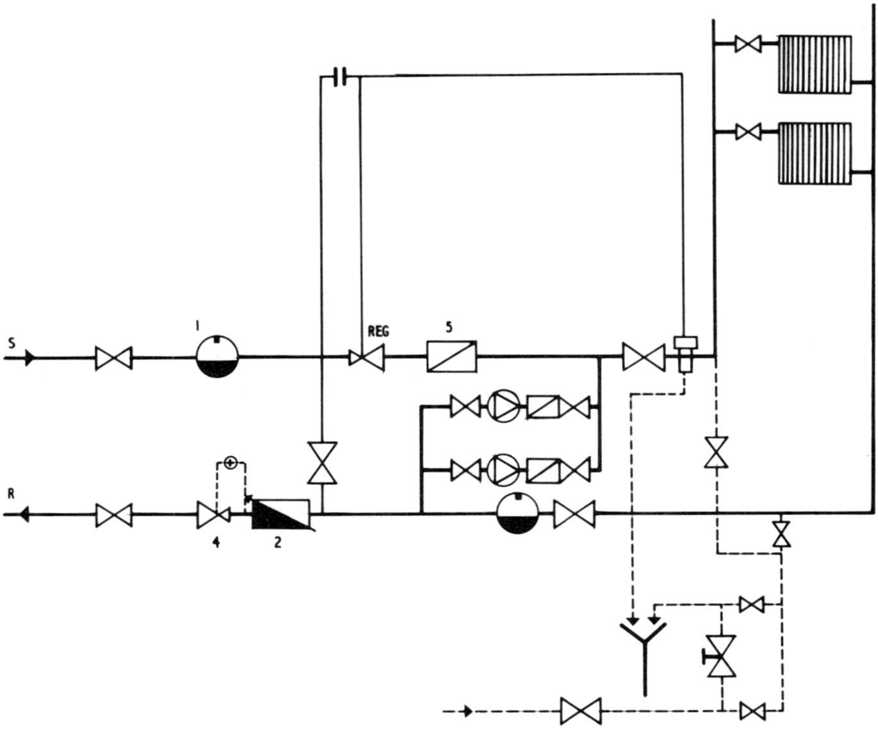

12.13 Consumer connection where inlet temperature is regulated according to the external air temperature: note protection devices (see p. 311 for key)

the water pipe where there is a pressure of 3 bars in both inlet and return line. The connection to the heating system is carried out in the manner indicated in **figure 12.16**

The scheme in **figure 12.17** is suitable for consumers with a ratio of hot water requirements to heating requirements of between 0.1 and 1. In this case the water flow in the supply line is kept constant by means of a flow regulator. The temperature of the domestic water is maintained constant by a temperature regulator. In the event of reduced network pressure the system is protected against running empty by an overflow valve in the return line and a check valve in the supply line. The necessary pressure in the heating system is maintained in the event of inadequate pressure in the network return line by means of an overflow pressure regulator. The pressure in the impulse source to the temperature must not be less than 3 bars: dependent on this, it is connected from the town's water supply to the supply or return line.

The scheme in **figure 12.18** is used only when the pressure in the heating system exceeds the permitted pressure (old systems: 5 bars; new systems: 5.5 bars) under the dynamic and static pressure conditions within the

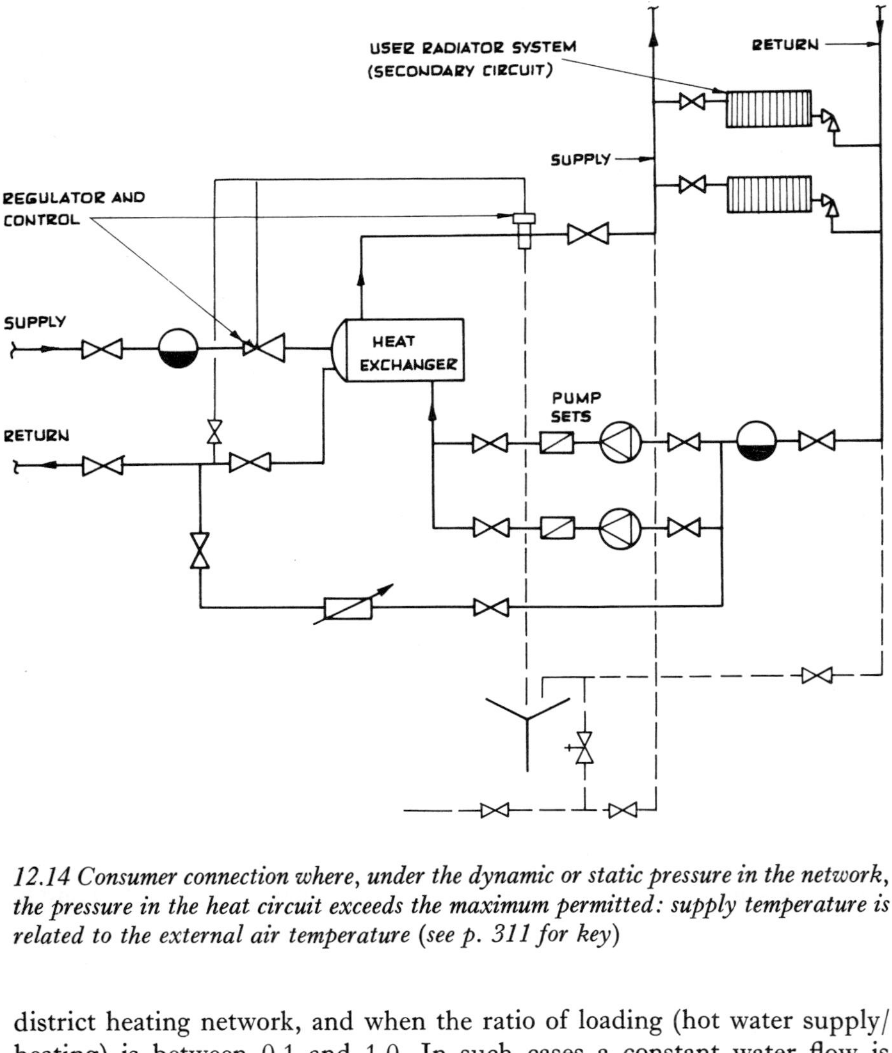

12.14 Consumer connection where, under the dynamic or static pressure in the network, the pressure in the heat circuit exceeds the maximum permitted: supply temperature is related to the external air temperature (see p. 311 for key)

district heating network, and when the ratio of loading (hot water supply/heating) is between 0.1 and 1.0. In such cases a constant water flow is maintained at the inlet by a flow regulator; the temperature of the domestic hot water is controlled beyond the calorifier by a temperature and a flow regulator. The network is protected against running empty (as above) and the necessary pressure within the heating system, under conditions of inadequate pressure in the return line, is controlled by a pressure regulator.

The scheme in **figure 12.19** is used for major consumers of hot water such as baths, laundries, hospitals etc. The control is somewhat similar to that adopted in 12.13. The circulation line is connected to the centre of the hot water calorifier where the temperature is about 40°C (104°F) to 50°C (122°F).

The scheme used with open systems where the main purpose is the

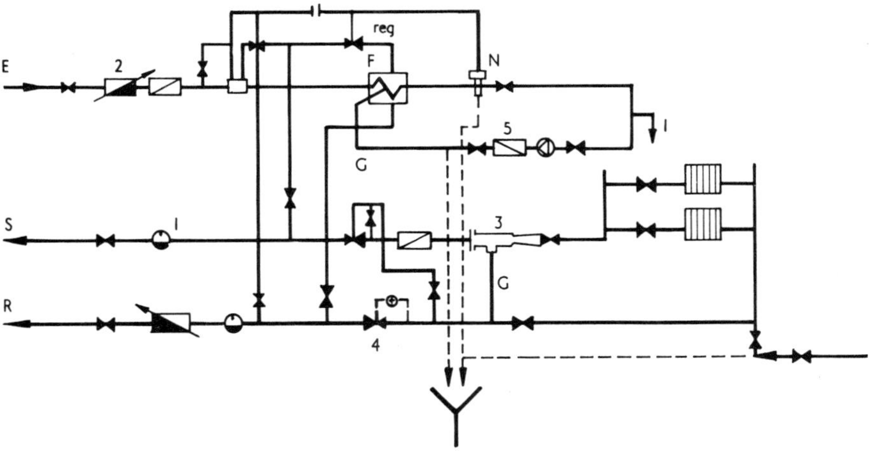

12.15 Consumer connection where supply of domestic hot water is the priority factor (see p. 311 for key)

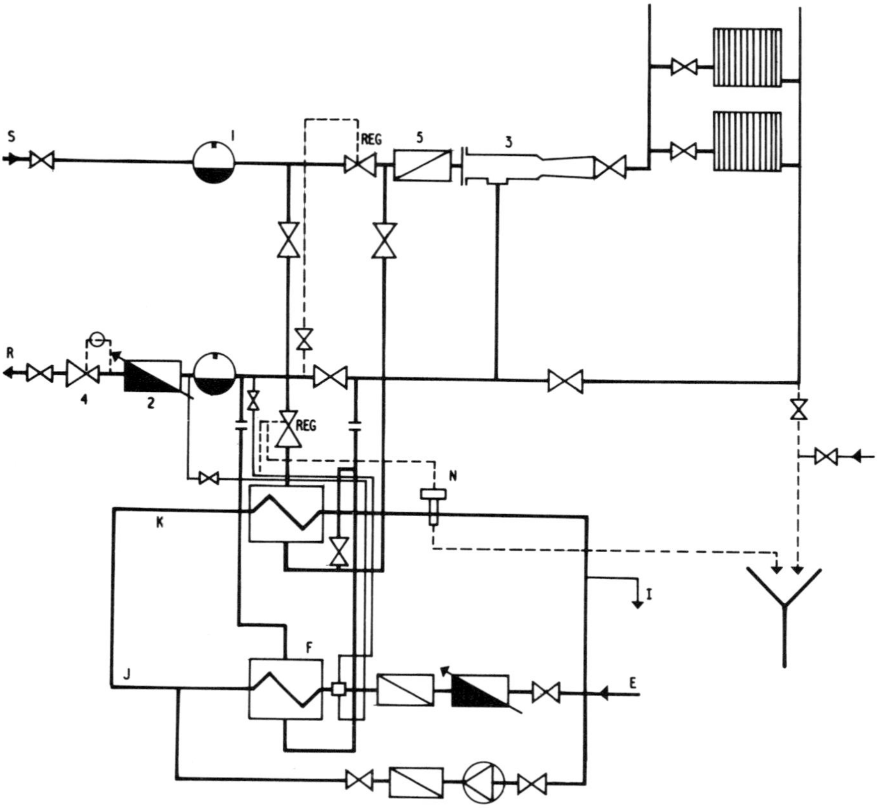

12.16 Consumer connection where the temperature of the domestic hot water is controlled beyond the heat exchanger by the thermostat and flow regulator (see p. 311 for key)

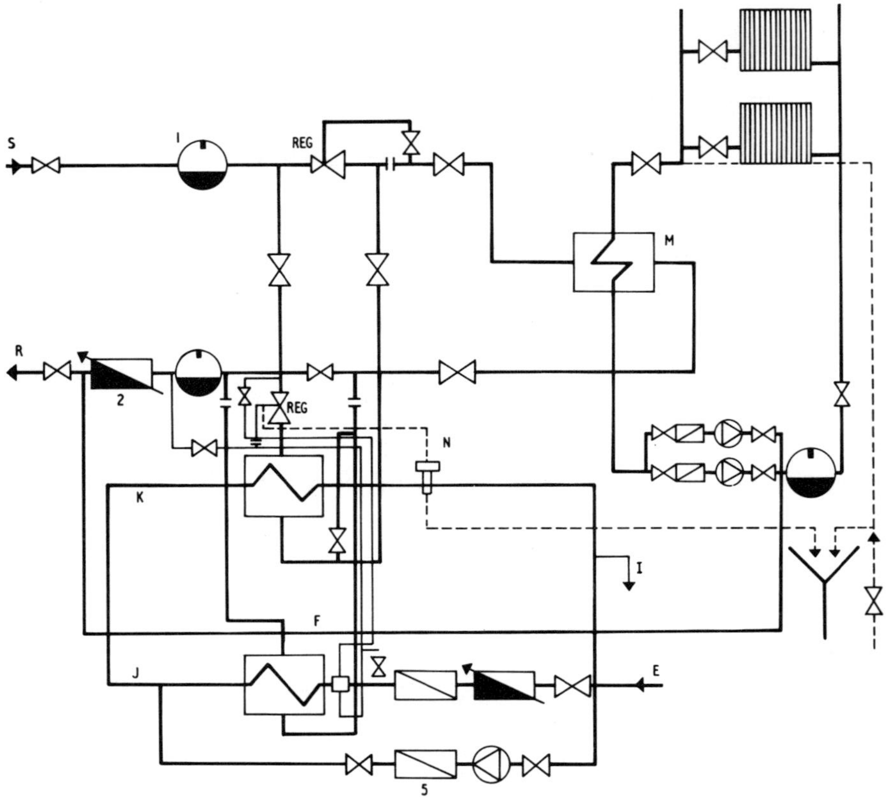

12.17 Consumer connection where the ratio of domestic hot water to space heating requirement is between 0.1 and 1: the water flow is kept constant by means of a flow water regulator and the domestic hot water by a temperature regulator (see p. 311 for key)

supply of domestic hot water is shown in **figure 12.20.** A flow regulator maintains a constant water flow rate in the heating system: a constant water temperature is maintained beyond the mixer of the hot water supply.

HEAT DISTRIBUTION EQUIPMENT IN DWELLINGS

Hot water radiators

Hot water radiators are by far the most widely used means of distributing heat inside a dwelling: there are numerous types and patterns eg single or multiple m.s. panels, cast iron classic or hospital column type (with feet or without feet), cast aluminium types and so on (**12.21**). Radiators can be fed with hot water from either a single- or twin-pipe system. There is a choice of connection: flow in at the top, return out at the bottom of the opposite side (TBOE); flow in at the top, return out at the bottom of the same side (TBSE); flow in at the bottom, return out at the bottom (BOE). It is un-

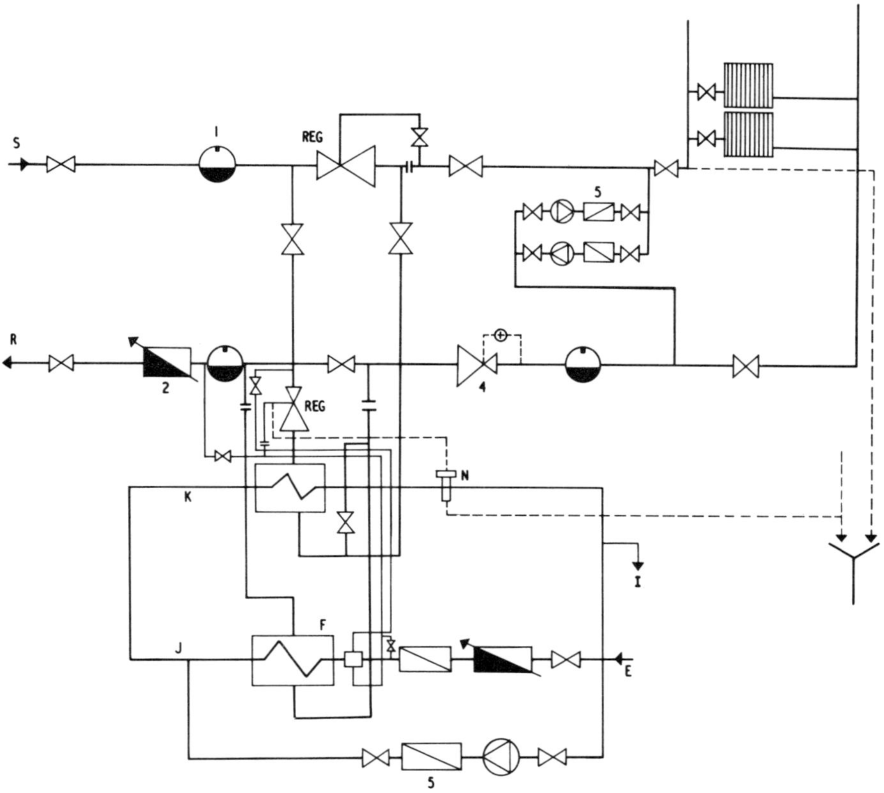

12.18 Consumer connection used only when the pressure in the heating system exceeds the permitted pressure (see p. 311 for key)

desirable for a radiator to operate at an excessively high temperature (which may cause scalding or drying of the air) and it is therefore necessary to control the inflow temperature to a maximum of about 70°C (158°F) by mixing return water with the inflow water. It is preferable for hot water radiators to be placed underneath windows. Radiators situated against internal walls tend to give unsatisfactory results due to a poor temperature gradient and lack of draught prevention. To avoid scaling and/or corrosion all types of radiators and convectors (and m.s. radiators in particular must operate on a closed hot water circuit.

Hot water underfloor heating

A modern version of this system is increasingly popular in West Germany (under the name *Bodenheizungssystem*). High density polyethylene or polypropylene pipes are clipped to the concrete reinforcement steel bars when casting the floor slabs of buildings. These pipes usually have internal

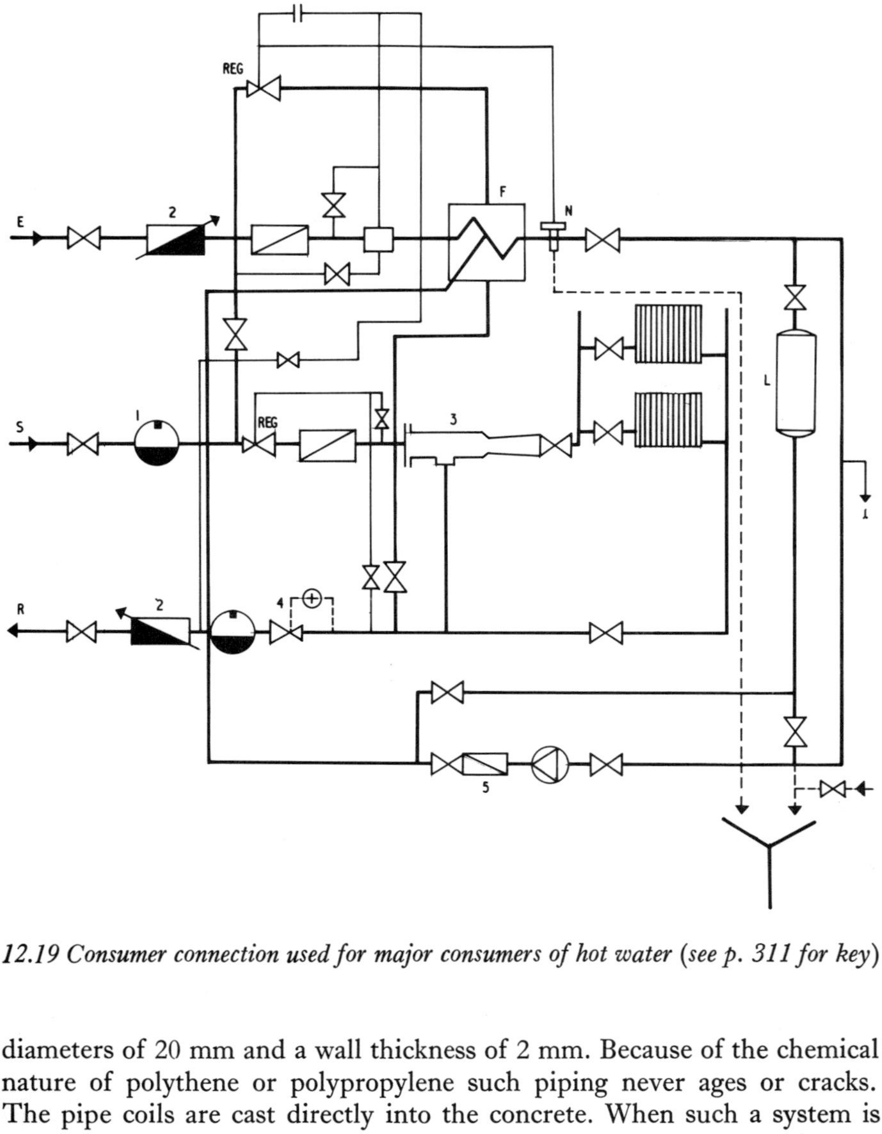

12.19 Consumer connection used for major consumers of hot water (*see p. 311 for key*)

diameters of 20 mm and a wall thickness of 2 mm. Because of the chemical nature of polythene or polypropylene such piping never ages or cracks. The pipe coils are cast directly into the concrete. When such a system is installed on the ground floor it is common to apply at least 100 mm thickness of high-void insulation in the form of mineral wool slabs, expanded polystyrene or similar material underneath to prevent the heat from being lost into the ground beneath. This degree of thickness in thermal insulation material is always necessary to counteract the insulating effect of floor coverings such as carpets.

The temperature at which hot water is fed through floor heating systems is usually about 40°C (104°F). They are therefore often connected only to the return system of the district heating network, thus using lower grade

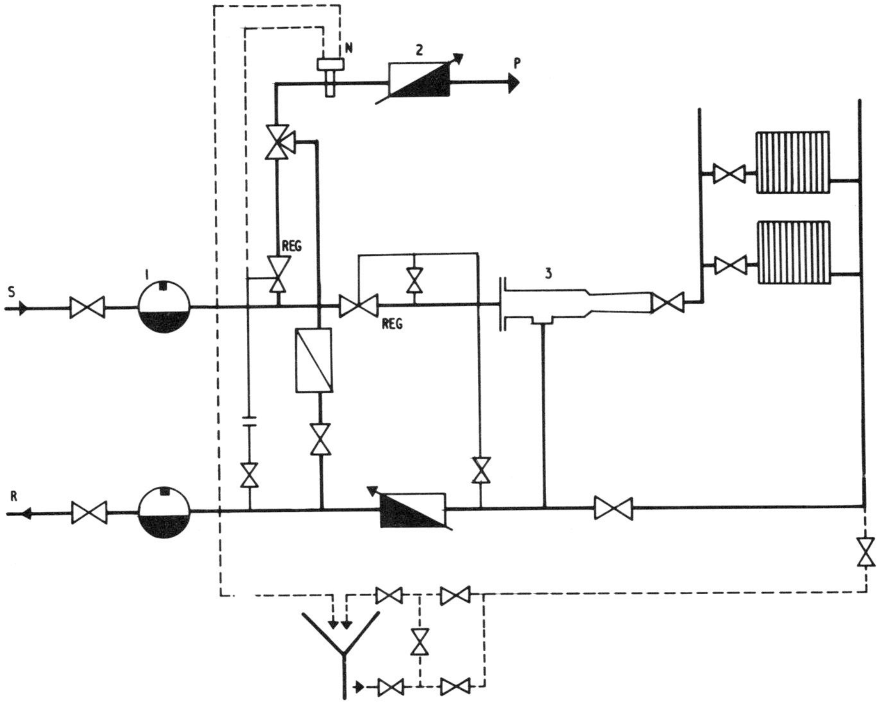

12.20 Consumer connection for use with open system: the main purpose is the supply of domestic hot water (see p. 311 for key)

energy (lower temperature hot water) than that supplied to conventional systems using steel or aluminium radiators. This is commonly recognised in West Germany by charging a lower tariff for floor heating. Although floor heating systems are rather more expensive to install than conventional radiator systems they have the two advantages that they are cheaper to run and that they provide a temperature gradient and distribution in a dwelling which is close to the optimum comfort condition.

Warm air systems

Warm air systems are widely used in group and district heating systems. They generally comprise centrally placed fan-assisted heater assemblies with or without distribution ductwork (**12.22** and **12.23**). Their advantages are their low cost and the ease with which heat can be metered by hours-run electric motors. Their disadvantages are (a) unless well-designed (and costly) ducted air distribution is used their heat distribution and temperature gradient is poor; (b) they may be noisy to operate; and (c) some authorities insist on the inclusion of expensive fire precautions in design specifications.

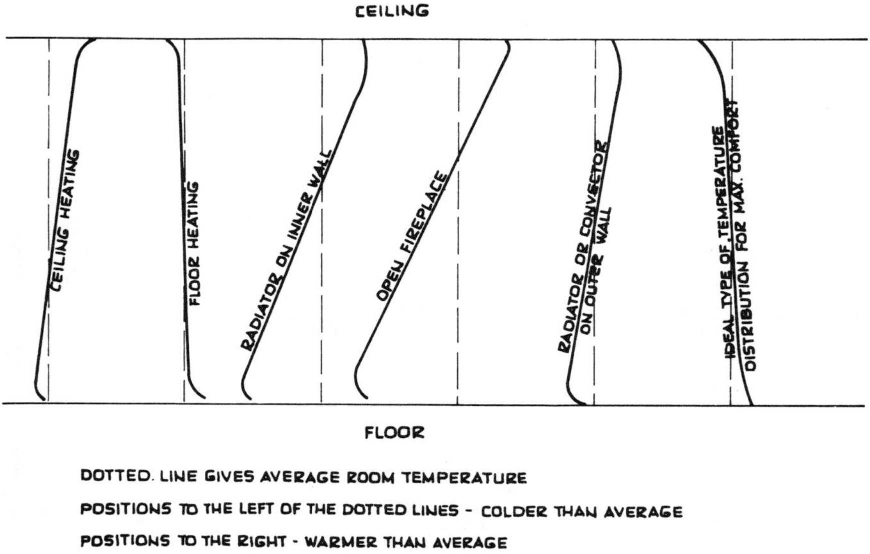

12.21 Vertical heat distribution (temperature gradient) curves for different types of heaters

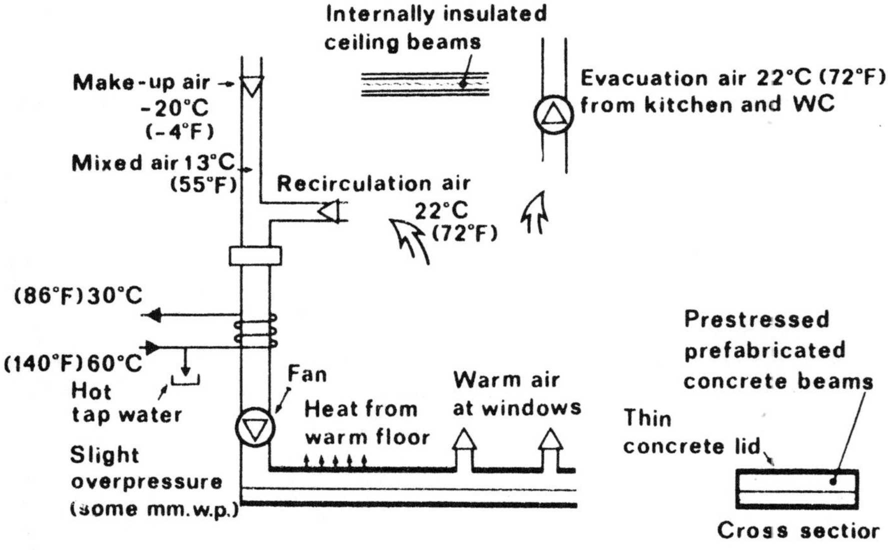

12.22 Schematic of hot air heating for small house

DISTRIBUTION OF HEAT IN THE FORM OF STEAM

Steam as a heat carrier

When steam is employed as the heat carrier the network pressure is de-

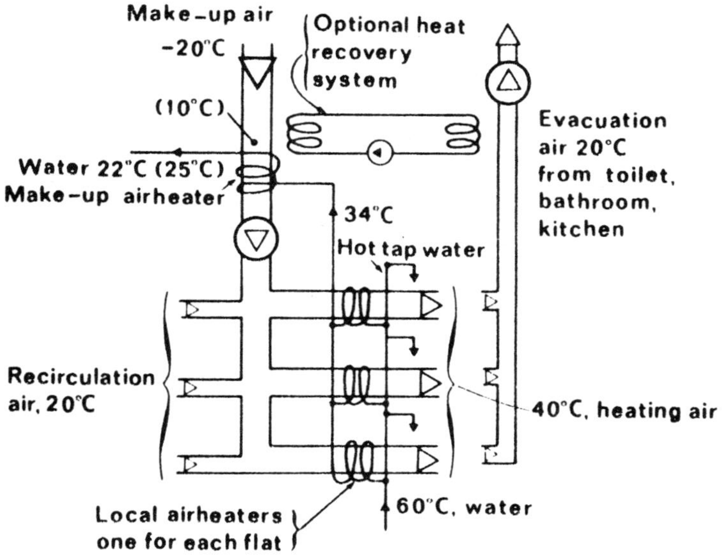

12.23 Warm air heating system for block of flats

termined by the requirements of the consumer and by the parameters of the district heating centrale and the size of the consumer network. The steam is dispatched into the network in the superheated state but this superheat is usually discounted and merely compensates for the heat losses in the pipeline network. The heat supplied to the consumer is assumed to be that of saturated steam (**12.24**).

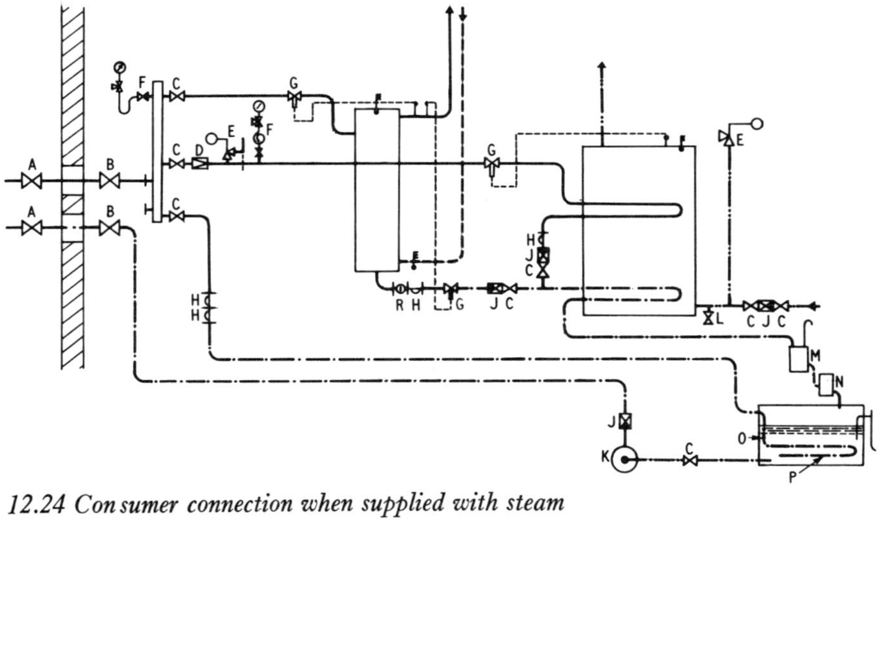

12.24 Consumer connection when supplied with steam

The heat in steam can be utilised either directly or indirectly. If the various radiators etc are supplied directly with steam, pressure governing devices have to be incorporated into the system to limit the steam temperature. Temperature control over the heat output can only marginally be affected by adjustment of steam pressure. Steam traps are required to prevent steam entering the condensate pipes. The surface temperature of the radiators must be high and tends to dry the ambient air excessively and so endanger the health of children and old people.

It is therefore more usual and more satisfactory to use steam indirectly via steam-to-water heat exchangers which heat up the circulating hot water in a closed system. The two systems must always be separated from each other because (a) the volume of condensate must be metered so that a separate charge can be made for heat supplied; (b) the condensate is pure water which should be returned to the steam plant and re-used with a minimum of further water treatment; and (c) the returning condensate tends to be highly corrosive to pipelines. In general it is advisable to use copper, stainless steel or other corrosion resistant metals for the return pipelines. In many cases the capital cost of a condensate return system cannot be justified economically in terms of heat recovered from the condensate and the saving of water treatment costs. In such cases the condensate is discharged into the sewers at the lowest achievable temperature.

Industrial consumers tend to purchase the steam directly without returning any condensate and it is then usual to employ simple flow meters coupled to pressure monitors as the basis for charging for the steam which they use. For domestic supply the steam is passed into a heat exchanger/condenser where it warms up circulating secondary hot water; in some cases, secondary steam is produced at the lower pressure which is required. The resultant condensate is metered and is either pumped back to the station or discharged to the sewers.

Steam can also be used directly without the need for heat exchangers (for heating vats, large baths etc). All steam lines have to be protected by safety valves and be efficiently insulated to conserve heat and to avoid scalding.

Cooling and return of condensate

The temperature of the returning condensate can be 100°C (212°F) or even higher. It is always advisable to cool it down to about 50°C (122°F). The consequent saving in sensible heat is almost 10 per cent of the total heat content of the steam as supplied. Corrosion of the return pipelines is also markedly reduced when condensate temperatures are lower. When using steam-to-water heat exchangers it is practicable if they are correctly dimensioned to ensure that the condensate is adequately cooled. It is often economical to cool down the condensate separately in additional condensate/warm water heat exchangers.

With open systems the condensate is collected and measured. Ideally it should flow by gravity towards the pumps, which return it to the station. Alternatively it can be collected in condensate receivers which are vented to atmosphere while pumps at the station produce a negative pressure of about 0.2 bars (absolute). If the condensate is at a temperature of about 70°C (158°F) it can be sucked to a height of up to 4 metres but it is better to keep the temperatures below this because the pumps operate better under low temperature conditions.

In a closed system the condensate can be returned to the station by means of steam pressure. In this method the condensate flows from the heat exchanger into a collection vessel and keeps the condensate under pressure. A float valve ensures that the condensate collection vessel never empties completely.

Supply limiters

Supply limiters must be fitted to steam lines to avoid any undue reduction of supply pressures. Each consumer is allocated a maximum steam ration which cannot be exceeded and the supply limiter valve is activated if attempts are made to exceed this allocation. Supply limiters can be operated either by the steam supply flow rate or by the condensate outflow.

USE OF HEAT EXCHANGERS

Steam-to-water heat exchangers are employed whenever an indirect means of domestic hot water supply or hot water space heating is required. Indirect systems of heat distribution offer considerable advantages for the supply of hilly areas or very tall blocks because it is far easier to control the pressure conditions in the primary circuits than in direct supply systems. On the other hand indirect systems are more costly and not quite as efficient in the abstraction of the primary heat as direct systems.

For the supply of domestic hot water for consumption purposes in kitchens and bathrooms an indirect system is always used. The transport of domestic hot water from the district heating centrale to the points of use is generally not economically feasible. It is therefore the usual practice to pipe cold town's water to the consumer's premises where a heat exchanger employs hot water or steam from the district heating circuit to raise the hot water to the required temperature.

The heat exchangers used in district heating usually incorporate batteries of seamless copper tubes with integral finned surfaces and operate with a head loss of about 2.5 m (w.g.) corresponding to 24.5 kPa when used as a water/water heat exchanger.

Lordan spiral heat exchangers are made in Israel: they consist of a tube bundle made from 10 mm outer diameter spirally wound tubes inside a conventional shell. The tubes are either copper, cupro-nickel or stainless steel. The main advantage claimed for spiral heat exchanger tubes is that the internal boundary layer heat transfer coefficient is higher than with equivalent straight tubes and gives improved heat transfer. Such heat exchangers may be installed vertically or horizontally.

It is essential to install the heat exchanger so that the fluid with the higher scaling factor flows in the shell and that only scale-free fluids are circulated within the spiral tube bundles. It is comparatively easy to de-scale the shell and the external surfaces of the spiral tube bundles but it is virtually impossible to de-scale the internal surfaces of the spirally wound heat exchanger tubes by mechanical means: suitable chemical cleaning techniques are available. When the spiral tube heat exchanger is used for steam-to-water heat exchange the steam is passed through the tube bundle.

CTC heat exchangers are of Swedish origin and are made in a wide variety of sizes and shapes for district and group heating applications. Where the risk of scaling is small, spiral finned tubes are used; where scaling is likely, straight pipes are used. Complete district heating connection sets for small houses are available to permit the connection of consumers to a district heating system at minimum cost. A typical unit is in the form of a cabinet measuring 600 mm × 600 mm × 1800 mm which contains the heat exchangers for radiators and the domestic hot water supply, as well as the hot water tank and the auxiliary components. Complete sub-stations for serving blocks of flats or groups of houses are also available.

TEMPERATURE CONTROL AND ENERGY CONSERVATION

Efficient temperature (thermostatic) control and thermal insulation measures must be taken to ensure that the district heating or cooling system offers comfort and optimum utilisation of energy. It is not possible in this book to describe such measures in great detail but there are a number of specialist books which do so. However within the context of district heating or cooling, the design concept must take into account the following essential requirements:

Choice of the most economic fuel

Selection of efficient plant at the heat station and at the sub-stations

Central supervisory and data logging facilities

Location of the heat station and sub-stations within the site to minimise the pipe runs and the associated heat losses

Routing of pipe mains to minimise lengths of runs and associated heat losses

Assessment and selection of optimum mains and pipe thermal insulation

Attainment of high standards of thermal insulation in buildings connected to the network including such matters as multiple glazing, loft insulation and cavity insulation

Selection of weather-compensation controls where appropriate

Selection of the most suitable arrangements for the temperature controls in dwellings and in other connected buildings (attention should be paid to means of preventing unauthorised interference with such controls)
Establishing appropriate operational programmes and optimised starting systems to ensure that the heat is provided only at such times as it is required
Effective and efficient methods of metering (or charging for) the supply of heat to encourage economic utilisation of heat
Good housekeeping and preventative maintenance procedures to maintain the whole system in continuing efficient operating condition.

Figure 12.25 and **Plates 40** to **48** show various related equipment.

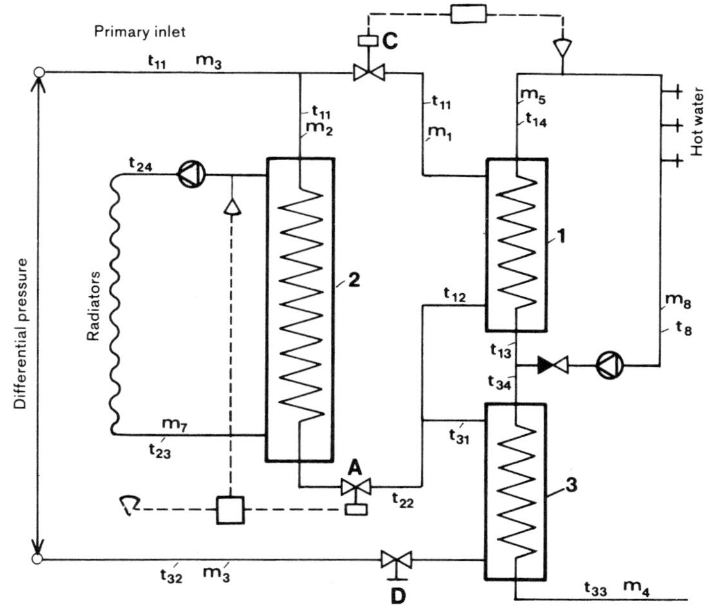

Temperatures:
t_{11}: Primary inlet
t_{32}: Primary outlet
t_{24}: Radiator temp.
t_{23}: Radiator temp.
t_{14}: Hot water temp.
t_8: t_{14}-2°C
t_{33}: 5°C

Numbers of apartments: 50
Heat demand/apartment: 5200 kcal/h
Differential pressure: 1 bar
2: Heatexchanger for radiators type SKR-84-1.5 R
1 and 3: Heatexchangers for hot water type SKR-126-0.5-2V

Control valve A: kv 8
Control valve C: kv 8
Valve D including other armatures and pipes: kv 7.2
Hot water circulating m_8: 0.1 l/s
Radiator flow m_7: 3.6 l/s
Fouling factor: 17500 kcal/m²h°C

12.25 District heat sub-station with 2-stage heat exchangers serving a block of flats

CHECK LIST

1 Major benefits given by district heating.
2 Extent of consumer demand for heat
3 Economics of heat supply

4 Advantages respectively of direct and indirect methods of heat supply
5 Overall loading of a district heating system
6 House connections
7 Pipeline pressure head required and provisions needed to maintain it
8 Control of pressure differential and total pressure
9 Pumping systems and controls
10 Fittings required in sub-circuits for control and safety
11 Standardised consumer connections used in Russia and methods used there for regulating heat supply
12 Heat distribution equipment used in dwellings
13 Suitability of steam as a carrier of heat
14 Cooling and return of heat in the condensate
15 Uses of supply limiters
16 Use of heat exchangers and types available
17 Use of temperature control for energy conservation.

REFERENCES

Fernwärme International (*articles in German*)

A. Karsch and others, 'House stations', 2 (1973) 2, pp 22–35

M. Proske, 'Density of heat connection', 4 (1975) 4, pp 120–122

U. Kaier, 'Distribution costs in district heating systems', 6 (1977) 2, pp 36–41

Other sources

J. Denbigh, 'Future British plans and prospects in district heating', *Energy Digest* 4, May/June 1975, pp 12–14

O. Faber and J. R. Kell, *Heating and air conditioning in buildings*, Architectural Press, London, 1974, pp 224–225

D. Fitzgerald, 'District heating radiators; return temperature differential', *Building Services Engineer* 44, September 1976, pp A58–59

Much of the remaining material for this chapter was obtained from technical handbooks issued by various industrial organisations active in the field of district heating.

CHAPTER 13

Boiler houses and fuel storage

CHOOSING A SUITABLE FUEL

Boiler plant for district and group heating schemes has to be suitably housed because it must have a large heat output capacity and occupy much space. The configuration of the buildings and the arrangement of the plant depend largely on the fuel which will be consumed by the boilers and on the heat output required.

The fuels in most general use are oil, natural gas and coal. There are also plants which burn peat and wood waste but such installations are inevitably special (one-off) ones and will not be considered in this chapter. The calorific values of the various common fuels are listed in table 13.1.

TABLE 13.1 TYPICAL CALORIFIC VALUES OF COMMON FUELS

Type of fuel	Calorific value
Gas oil (Class D) gross	3800 kJ/litre
Light fuel oil (Class E) gross	4400 kJ/litre
Medium fuel oil (Class F) gross	4100 kJ/litre
Heavy fuel oil (Class G) gross	4100 kJ/litre
Manufactured (Town) gas	15 to 16.5 MJ/m^3
Natural (North Sea) gas	50 MJ/m^3
Coals (size below 25 mm)	5800 to 39 000 kJ/kg
Ash content 6 to 20%	
Moisture content 5 to 15%	
Coals (size over 25 mm)	6000 to 5000 kJ/kg
Ash content 4 to 8%	
Moisture content 5 to 10%	
Gas coke	5500 to 4500 kJ/kg
Ash content 8 to 14%	
Moisture content 5 to 10%	
Wood waste (average for common timbers)	3000 kJ/kg

Note: The characteristics of a particular coal are related to the area at which it has been mined: hence the wide range of calorific values.

The choice of a particular type of fuel for a district or group heating scheme depends on the following major factors, all of which must be closely evaluated, separately and jointly:

Availability of fuel in the particular area
Relative fuel costs
Environmental and pollution standards applicable to the site
Fuel consumption and storage
Capital costs
Availability of operating staff
Space required for boiler plant and storage complex
Access to site for fuel supply vehicles
Pre-heating required
Height of chimneys required
Fire fighting arrangements and precautions
Ash handling and disposal (solid fuel)
Noise aspects.

In selecting the type of boiler plant the following additional factors must be analysed:

Magnitude of heat load
Whether steam or hot water is required as heat carrier and at what pressure
Degree of reliability required for plant and maintenance facilities available
Relationship of proposed boiler plant complex to the geography of the pipe distribution network
Nature of required chimney draught (natural, induced, forced or balanced)
Requirement for gas boosters (and associated noise attenuation)
Requirement for fuel supply meters
Arrangement for monitoring and verifying fuel supplies.

FUEL STORAGE CAPACITY

Boiler plants burning liquid or solid fuel must have fuel storage capacity adequate to suit the particular circumstances of a project relative to estimated fuel consumption at the peak of the heating season, ease with which fuel supplies can be obtained in the locality, availability of space for fuel storage, access for delivery vehicles and price reductions for quantity delivery. Where fuel supplies can be obtained without undue difficulty, it is usual to lay down a minimum store sufficient for 4 weeks' consumption at peak demand: greater storage capacity is necessary where there are specific fuel supply difficulties.

STORAGE OF FUEL OIL

Delivery

Oil is usually delivered to the consumer by road tanker vehicles usually having carrying capacities of 16 000 or 25 000 litres. Oil suppliers tend to favour the larger capacity vehicle, the smaller tankers being used only for special applications such as narrow or weak access roads. For users of large quantities of oil there may be a choice of delivery by road tanker, rail tanker, water-borne barge or by pipeline laid direct between refinery and customer's oil tanks.

Table 13.2 gives typical particulars of fuel oil road tankers (Esso).

TABLE 13.2 TYPICAL PARTICULARS OF ROAD TANKERS CARRYING DOMESTIC AND HEATING FUEL OILS

Type of vehicle	Capacity (litres)	Gross laden weight (kg)	Height (m)	Width (m)	Length (m)	Turning circle (m)
Rigid	15 500	24 000	3.12	2.44	9.14	21.00
Articulated	15 500	24 000	3.35	2.40	10.00	14.33
Articulated	25 000	30 000	3.05	2.44	13.30	16.50
New (proposals in hand)	32 000	To be advised	3.05	2.44	12.80	To be advised

One method of supplying oil to groups of houses or housing estates is by the CFD (Central Fuel Distribution) system whereby each dwelling is fitted with its own oil fired boiler but not with separate oil storage. All the oil storage requirements of the estate are met from an underground oil storage tank (or tanks) and a pump-assisted oil distribution system installed and usually operated under the direct control of the oil supplier. In the normal arrangement of a CFD system the oil-circulating pumps are provided in duplicate and located inside a pump chamber immediately above the underground oil tank. The pressure at which oil is distributed through the estate underground oil distribution network is between 172.3×10^3 N/m^2 and 344.8×10^3 N/m^2 according to the distances over which the oil is to be circulated. On large housing estates it is prudent to zone the oil distribution so as to confine a pipeline defect to a relatively small number of consumers and to fit each zone with appropriate isolating, drain and venting valves. The main underground distribution system is generally of 25 mm bore and the branch pipes to individual housing units of 12.5 mm. Each consumer is provided with a combined meter and control box located suitably above the oil-burner level. The meter box contains a pressure-reducing valve to reduce the high pressure in the oil main and (in the UK) a Board of Trade approved model oil meter. Below the meter box are located the local oil supply isolating valve, pipe fitting which reduces the oil line from the 12.5 mm branch off the main to the 6 mm local supply connection and, in addition, a phial-operated fire valve with phial positioned adjacent to the oil burner. The oil meter is read periodically by the oil suppliers and the consumer pays only for the oil actually consumed by his oil burner.

Oil tank construction

The oil tanks for district or group heating schemes will inevitably be relatively large. Above-ground tanks are generally constructed of mild steel of thicknesses related to the pressure and physical conditions under which the storage operates but mild steel tanks should not have a galvanised finish because the oil is inclined to dissolve the zinc coating. Fibreglass storage tanks have been used for above-ground storage in the USA petroleum industry.

Underground oil storage tanks may be constructed in concrete particularly where very large storage capacity is required. When the tank is built as a separate structure the concrete and its oil-proof lining are built as one unit capable of moving together in the event of ground settlement and the internal lining is commonly in the form of suitable tiles. Tanks holding up to 450 000 litres of oil in one single compartment have been successfully constructed in this manner. Where oil storage is built into the structure of a building, it is better for the lining itself to be fully oil-proof and capable of movement so as to compensate for ground settlement or damage to the concrete structure. Various oil-proof skins are available for this purpose including one made of fibreglass. The authors are however aware of numerous failures of such tank linings during service: extreme care must be exercised in selecting them and adapting them for storage in particular cases.

Plates 50, 51 and **52** show Borsari underground tile-lined concrete tanks under construction; **plate 53** shows the tile-lined tank interior.

Remote reading oil gauges

For the large oil storage tanks involved in district heating schemes, reliable remote reading oil gauges are required and these must have read-out facilities which are correctly calibrated and easy to read. The capillary pipe connections and electric cables between storage and gauge read-out must also be carefully installed and protected.

Various forms of remote reading gauges exist in which signals from the tank are transmitted either hydraulically or electrically to a remote indicator. With hydraulic transmission the dial and contents sensing device may be no more than a certain maximum distance apart (usually 8 to 11 m). In one model of this type the static pressure of the oil acts on a diaphragm or balance chamber installed inside the tank and indicates the contents as a reading on a dial gauge. The indicator and level-sensing device are connected by small bore copper (or plastic) tubing. Every care must be taken to ensure that all joints in the tubing and between tube and fittings are fully air-tight because indicator readings will be invalidated by air leakage into the system. It is therefore best to supply the connecting tube in a single unjointed length, leaving joints only at the actual connections between tubing and gauge equipment. Some hydrostatic gauge systems are supplied fully factory-sealed in respect of the complete assembly of indicator, level-sensing device and connecting tubing. This arrangement generally ensures a satisfactory and accurate system but it may sometimes be difficult to install or withdraw the system without breaking the factory seal. With electrical or electronic connection the distance between the level-sensing device in the tank and the remote indicator is not limited as it is with hydraulic transmission. Whatever the capacity of the oil storage tank the level indicator should be marked with two distinctive markings showing the minimum contents at which the tank should be re-filled and the upper limit at which filling must cease. Indicator dials must be calibrated for the particular grade of oil to be handled: different grades of oil have different specific gravity so that recalibration is necessary if the grade is changed.

Other essential fitments

Each oil storage tank must have the following fitments:
Fill pipe and over-filling alarm
Vent pipe
Drain valve or sludge withdrawal pipe
Access manhole
Oil pre-heater if required for grade of oil stored.

Fill pipes and alarms

Each oil storage tank must be provided with a suitable external filling arrangement. It may be practicable to fill very small tanks through a small inspection flap in the top of the tank by bringing the delivery hose pipe right up to the tank, but suitable external filling arrangements must be provided for all other tanks. The bore of the filling pipe should depend on the grade of oil, on the distance between fill point and oil tank and on the capacity of the tank. The most usual size of filling pipe for large tanks is 75 mm with hose coupling of the same size. However the size required should be checked with the oil suppliers before it is specified because not all oil supply tankers carry a mating hose coupling which suits the British Standards gauge recommended for fill points.

Care must be exercised in selecting a suitable location for the fill point. This should be within easy reach of the filling hose carried by the oil supply tanker and as close to the oil tank as practicable. Where it has to be located close to a public highway the filling connection should be installed inside a substantial vandal-proof locked box above ground or inside a hinged manhole below ground.

Where the fill point terminates below the top of the oil tank an isolating valve must be provided in the fill connection close to the fill point to prevent oil running back to the fill point when the filling hose is withdrawn. A gunmetal sealing cap with chain should be fitted at the fill point termination to prevent oil dripping from it and staining the surrounding ground or structure. Where there might be unauthorised interference the sealing cap should be secured with a padlock. With the heavier fuel oils a drain cock should be fitted to the fill point at the place where this terminates below the top of the oil tank: this will ensure that all oil is cleared from the filling line at the end of the filling operation and will avoid the possibility of the oil waxing (freezing) and obstructing the subsequent filling operation. A sandpit should be arranged under the valve to avoid staining the ground close to the fill point. It is an even better alternative to locate the oil fill point inside an oil-proof and watertight manhole.

Even the best oil-level indicators are likely to prove troublesome at some time or other. This may cause a larger quantity of oil to be ordered than the tank can hold and consequent overfilling with oil spillage in the vicinity of the fill point. Any oil spillage is a nuisance and stains from the heavier grades can cause severe damage. It is therefore prudent to incorporate a device which gives an audible alarm in the event of over-filling to warn the tanker attendant to cease filling. Over-filling alarms may be of the whistle or electrically operated type. In the former the air expelled from the tank causes a continuous whistle so long as the filling operation is proceeding correctly which ceases when the tank has reached the maximum safe level. The electrically operated device incorporates a float or diaphragm which transmits the oil level indication to an audible electrical alarm. This fitting must be provided with a test button so that periodic checks can be made to find out whether it is functioning correctly and with a cancelling button to cancel the alarm after over-filling has stopped the filling operation.

Vent pipes

A vent pipe is necessary to evacuate air which may cause pressurisation of oil tanks while they are being filled. If the tops of a range of tanks are constantly at the same level they may have one, but not more than one, common vent pipe. Usually however a separate vent pipe must be provided for each tank taken off its highest part. The cross-sectional area of each vent pipe connection (or of a common vent pipe where provided) must be big enough to ensure that there is no danger of excessive pressure if any tank is over-filled during the filling process. The vent pipe must be free from obstructions and it should therefore be without sharp bends and have a continuous rise from the oil tank to its termination in the open air. At that point it is likely to give off a nasty smell during filling. The outlet must therefore be kept clear of any area where this would be a nuisance eg below windows. The outlet should be turned down (generally by means of a U-bend) and fitted with an open mesh wire cage or special vent cap to exclude foreign matter from birds and other sources.

The height of the vent pipe is governed by the pressure which would be exerted by a column of oil standing to full height in the vent pipe in the event of over-filling. Where it is not practicable to limit this height to one within the safe design pressure head of the oil tank an unloading device should be provided which will automatically limit the pressure head caused by over-filling to a permissible level. Any spillage of oil from this device should be piped back to the catch pit or tank chamber. Its bore must not be less than that of the vent pipe it serves.

Drainage

A drain must be fitted at the lowest part of the underside of an above-ground oil storage tank for the periodic removal of sludge and emergency emptying. It should take the form of a screw-down gate valve or glanded cock constructed of gunmetal or cast steel. Cast iron valves are unsuitable because they are unlikely to remain oil-proof. The valve must be located at a point where a hose pipe can be fitted and its outlet should be fitted with a plug to prevent any discharge of oil if the valve is accidentally opened. Underground tanks must be provided with a sump (or depression) at their lowest internal area from which the sludge or oil can be withdrawn by suction through a standpipe.

It is good practice to specify diaphragm type valves for all valve applications to oil lines because they are not subject to gland leakage. The correct diaphragm must be specified for the particular grade of oil. For large tanks, the usual size of drain valve is 63 mm bore.

Manholes

All large oil tanks should be provided with manholes in accordance with British Standards recommendation. Manholes are usually circular in shape and not less than 0.5 m in diameter. They should be stiffened either by a flat-faced compensating ring welded or riveted above the opening or by a flanged neck and care should be taken to ensure that the holes for studs and set-screws do not penetrate the tank plate. Where tanks are located in the open air a flanged neck arrangement should always be provided because it will avoid the risk of water leaking into the tank via the manhole joint.

Wherever practicable manholes should be located on top of the tank and each should be provided with a securely fixed cover and a vapour-tight and oil-resistant joint made between the cover and the manhole. Vertical tanks over 3.5 m in height should be provided with an additional inspection manhole near the bottom of the base. Underground tanks located in areas of vehicle traffic must be provided with traffic-bearing manholes. Tanks of heights greater than 2.5 m may be provided to advantage with both external and internal access ladders to facilitate maintenance.

Oil pre-heating

Certain grades of oil must be maintained at a minimum temperature in storage and at the outflow, as follows:

Gas oil (maximum viscosity 45 sec)	No heating required
Light fuel oil (maximum viscosity 250 sec)	7.2°C (45°F) in storage and outflow

Heavy fuel (maximum viscosity 1000 sec)	18.3°C (65°F) in storage and 26.7°C (80°F) at outflow

Siting and maintenance

Storage tanks above ground should be located inside purpose-made tank chambers or within a catch pit. These receptacles should be lined with oil-resistant composition so that they will contain the oil in the event of leakage or tank fracture. The composition should reach high enough to accommodate 10 per cent more than the maximum contents of the oil tanks. To avoid corrosion in the area where the oil tank is in contact with its supporting piers and to help the tank to settle on its supports, a sheet of lead or a substantial damp-course material should be inserted between the top of the piers and the underside of the tank.

It is not permissible to provide a drain pipe from the catch pit so that dirt and rainwater tend to accumulate inside the catch pit area unless this is roofed over. If no roof can be constructed then it may be found convenient to lay the floor of the catch pit to fall in the direction of a sump from which water may be withdrawn by means of a semi-rotary hand pump or by an electrically operated pump which starts and stops under manual control.

Where a tank is underground it is best to locate it inside an accessible chamber in which the tank and its fittings may be periodically inspected and maintained. If such a chamber cannot be provided the proposed tank location and method of burying it must be agreed with the relevant authorities. Storage tanks should not in any case be buried direct in ground where the occurrence of an oil leak might lead to serious consequences such as the contamination of a water catchment area or the infiltration of the oil into the sewage system. Special care must be taken in locating, anchoring and externally protecting the tank in areas where the water table is likely to submerge the tank partly or fully and where the soil is acid bearing and corrosive to ferrous metal. **Plate 49** shows an externally located small vertical oil storage tank.

Control of oil supply to boiler rooms

Many authorities (certainly in the UK) do not permit a direct connection between the oil burners and the main oil storage which might lead to uncontrolled flooding of oil into the boiler room in the event of a control failure.

Two alternatives are generally acceptable:

(a) Provide a small service tank of the maximum capacity specified by the relevant codes or authorities (in the UK this is 900 litres). Such a tank may be filled automatically by level control. Once it is filled the automatic operation ceases and the pump must be restarted manually for the next filling operation. If automatic refilling were permitted much oil could flood into the boiler room before a malfunction of the control was noted or reported. On large installations the use of a daily service tank may be very inconvenient.

(b) Provide an oil-tight duct in which the oil pipes are run between the oil storage and the oil burners. This duct should be graded and should have a sump at its lowest point. A suitable float-operated sump pump should also be installed to keep the boiler room clear of water and oil spillages.

Fire precautions with oil fuel

It is essential to interrupt the firing of the oil burners in the event of a fire in the boiler house. An automatic shut-off valve actuated by the failure of fusible links above the burners is therefore provided at the entrance of the oil line to the boiler house. The linkage may be electrical or mechanical. The latter may however be difficult to provide for a large boiler plant with numerous oil burners.

A foam pipe installation is usually provided to large boiler houses. This consists of an arrangement of galvanised mild steel pipes terminating in a foam inlet box located in an accessible position outside the boiler room. The Fire Brigade can then pump foam into the boiler house from outside in the event of a serious fire.

Codes of practice

Detailed guidance on the requirements for oil fired installations is laid down in appropriate Codes, Standards and Building Regulations. The relevant British Standard is No 799 (Specification for Oil Burning Equipment).

STORAGE OF SOLID FUEL

Bulk storage delivery and handling

The storage of solid fuel in bulk must be co-ordinated with the means by which the fuel is mechanically handled from the store to the boilers. The supply of solid fuel, except at sites close to the pithead, is possibly somewhat less secure than that of fuel oil so that a minimum of 4 to 6 weeks' peak consumption of fuel should be held in stock.

When stored coal is exposed to atmosphere it is subject to attack by the oxygen in the air. The heat generated by this oxidisation may be only slight or it may be dissipated so rapidly as to be imperceptible. Under certain conditions however the coal may become noticeably warmer and eventually so hot that it decomposes or actually ignites. The latter phenomenon is known as spontaneous combustion. Three conditions must be satisfied before coal begins to heat up perceptibly in storage: the quantity of coal in store must be very large, the coal must be of a grade susceptible to spontaneous combustion and a high proportion of the coal must be in the form of

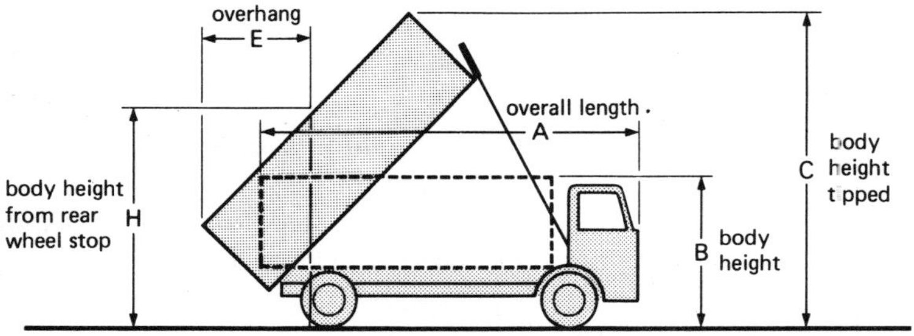

Types 'a' and 'b', 4 wheeler 2 axle, payload 8/10 tonne

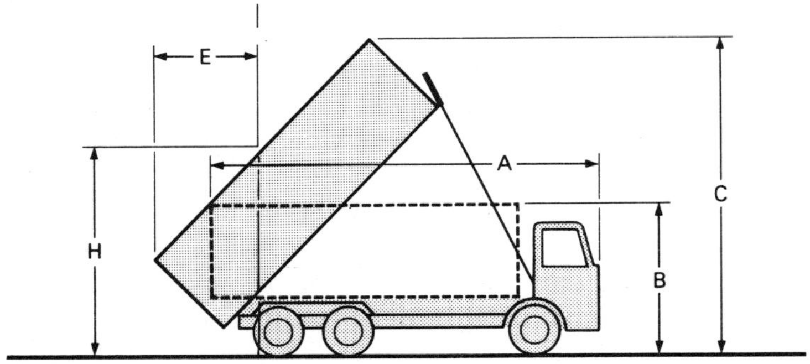

Types 'c' and 'd', 6 wheeler 3 axles, payload 12/13 tonne

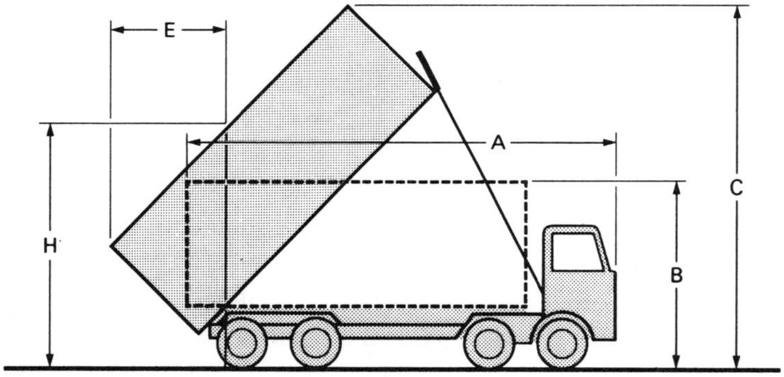

Types 'e' and 'f', 8 wheeler 4 axles, payload 14/20 tonne

13.1 Dimensions for bulk tipping vehicles:

Vehicle type		*a*	*b*	*c*	*d*	*e*	*f*
Gross vehicle weight	*tonne*	*13*	*16*	*20*	*22*	*24*	*30*
Approximate payload	*tonne*	*8*	*10*	*12*	*13*	*14*	*20*
Vehicle turning circle[1]	*m*	*14*	*17.5*	*17.5*	*18.5*	*19.5*	*25*
Dimension A	*m*	*5.5*	*6.4*	*7.4*	*8.0*	*8.0*	*9.5*
B	*m*	*2.6*	*2.8*	*2.6*	*2.8*	*2.8*	*3.2*
C	*m*	*4.2*	*5.2*	*5.4*	*6.3*	*6.3*	*6.8*
Maximum vehicle width D[2]	*m*	*2.6*	*2.6*	*2.6*	*2.6*	*2.6*	*2.6*
Vehicle over hang E	*m*	*1.9*	*1.9*	*2.2*	*2.2*	*2.2*	*2.2*
Clearance dimension H	*m*	*3.2*	*3.2*	*3.2*	*3.6*	*3.9*	*3.9*

fines (less than 1.6 mm in size). The first of these conditions is paramount: it is known that the heat due to oxidisation is dissipated so quickly in a small bulk store of coal that a high temperature cannot build up within it. Observance of the following general rules should ensure the safe storage of coal in bulk:

The coal should comprise nuts, cobbles or round coal and should exclude fines
The size of the store should be less than 200 tons
The height of the store should not exceed 2 to 2.5 m
Ventilation should be effectively suppressed.
The coal should have been mined at least 3 to 6 months previously.
The type of coal should be hard or dull, anthracite, anthracitic coal, splint, cannel.
The stack should be sited on clean firm ground.

Information concerning the properties of the solid fuels available in the area of the proposed plant should be readily provided by the supplier: this should facilitate decisions on sizing and other arrangements for storing solid fuel.

When arranging the stacking of coal the consumer should be aware of the following points:

(a) The angle of repose (angle assumed by the coal stack when at rest) lies between 45 and 55 degrees: this affects the allocation of the area to be occupied by a coal stack
(b) The moisture content of the coal should be ascertained (it may be as high as 15 per cent) because it affects bunker and stack volume and because drainage must be provided below the storage area to permit surplus moisture to drain away freely
(c) Fines (coal of size below 3 mm) may cause problems in open stacks and if moist can cause difficulties in handling and combustion
(d) The density of the coal must be known because it affects the provision required for storage: broadly the density ranges are:

Graded coal (over 12.5 mm)	1.4 m^3/tonne
Dry smalls (12.5 mm)	1.3 m^3/tonne
Washed Smalls (under 12.5 mm)	1.2 m^3/tonne

Wherever practicable coal should be stored in below-ground bunkers to permit simple delivery by tipper truck (**13.1, 13.2** and **plate 54**). In planning

Notes:
[1]*Vehicle turning circle: This is the distance between parallel kerbstones within which the vehicle can turn on maximum lock. However, to allow for overhang of the vehicle in advance of the front wheels, add 2 m to the turning circles shown in the table.*
[2]*Maximum vehicle width: This dimension is exclusive of exterior driving mirrors and indicator lights. Allow 200 mm for these items.*

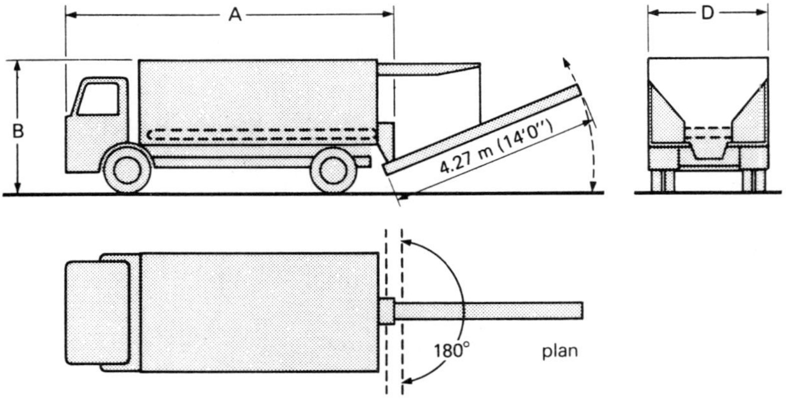

Conveyor delivery vehicle
Type 'a' 4 wheeler, 2 axles payload 6 tonne
Type 'd' 6 wheeler, 3 axles payload 15 tonne

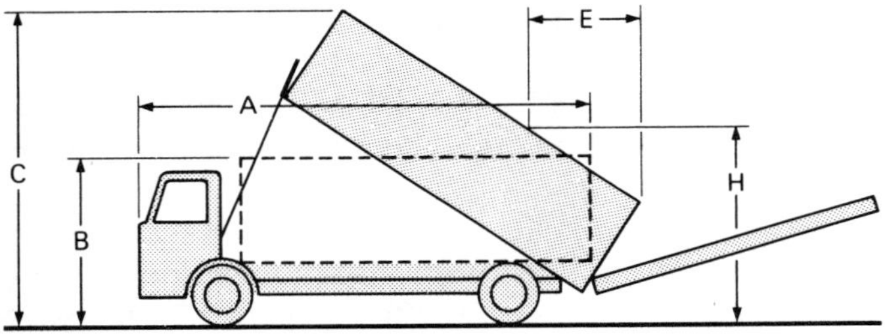

Tipper conveyor vehicle
Type 'b' 4 wheeler, 2 axles payload 8 tonne

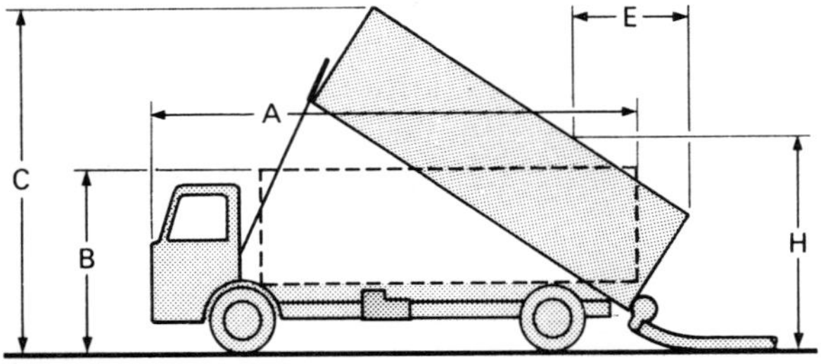

Tipper pneumatic vehicle
Type 'c' 4 wheeler, 2 axles payload 8 tonne
Type 'e' 6 wheeler, 3 axles payload 13 tonne

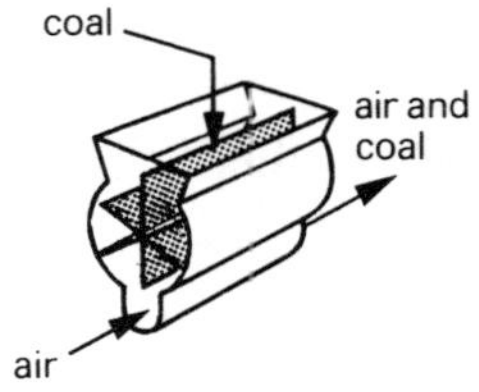

Detail of mixing valve (end removed to show rotary blades)

13.2 Dimensions for conveyor/pneymatic vehicles

such an arrangement all necessary provision must be made for easy vehicle access to the bunker and for safety in tipping (the tipping truck must be confined so that it cannot accidentally fall into the bunker). Underground bunkers are best used in locations where advantage can be taken of the slope of the land. The construction and lining of underground bunkers in other locations can be costly, particularly on wet sites. Delivery of coal by tipper truck tends to be the cheapest method of delivery.

Coal storage above ground can be most flexibly provisioned by pneumatic delivery; direct access by the transport vehicle to the coal store is then not required (**13.3** and **plate 55**). The delivery pipe (in the National Coal Board area 127 mm diameter) terminates with a coupling close to the point where the delivery truck is parked. When discharging coal into a bunker the displaced air must be vented via suitable filters (usually fibreglass) and the filter must be so placed that it cannot be coated with coal inside the bunker. The delivery tube should be as near straight as practicable with only a few bends. The final exit velocity of the coal into the store must be reduced by enlarging the delivery pipe (in the NCB area from 127 mm to 178 mm) in its final leg at a point about 2.2 m from the bunker entry. The bunkers must be fitted with access doors and hatches for safe operation and maintenance and to permit non-pneumatic deliveries to be made in exceptional circumstances. Delivery by pneumatic truck usually attracts a surcharge over delivery by tipper truck.

Coal may also be stored in freestanding mild steel silos (**plate 57**). These are usually provided with a watertight cover. They are available from specialist manufacturers and are generally assembled at site from mass-produced standard sections. Each silo will be supplied complete with access ladder(s), pneumatic fuel delivery fittings, access doors and corrosion protection. Such silos have a particular advantage on sites where ground storage area is restricted or undesirable because they offer a relatively high storage capacity per unit area of ground occupied.

All types of coal stores, silos and bunkers must incorporate drainage facilities so that the moisture in the coal can run away easily.

Vehicle type		*a*	*b*	*c*	*d*	*e*
Gross vehicle weight	*tonne*	*12*	*14*	*14*	*24*	*22*
Approximate payload	*tonne*	*6*	*8*	*8*	*15*	*13*
Vehicle turning circle	*m*	*15*	*18*	*18*	*21*	*19*
Dimension A	*m*	*7.1*	*7.4*	*7.4*	*9.5*	*8.1*
B	*m*	*2.9*	*2.6*	*2.6*	*3.1*	*2.8*
C	*m*	—	*5.1*	*5.1*	—	*6.8*
Maximum vehicle width D	*m*	*2.6*	*2.6*	*2.6*	*2.6*	*2.6*
Vehicle overhang E	*m*	—	*1.7*	*1.7*	—	*2.2*
Clearance dimension H	*m*	—	*2.9*	*2.9*	—	*2.5*

The parking area for the coal delivery truck should be selected and arranged bearing in mind both the vehicle's size and convenience to the consumer and to the public. Provision for receiving the coal should permit the discharge of a full load at each delivery. Various sizes of coal delivery trucks are in use with capacities from 6 to 20 tonnes. For obvious reasons coal deliveries by the larger trucks attract a lower delivery charge. Reception should therefore be arranged to permit delivery by the largest available truck but (sounding a note of caution) there is no advantage in making such provision if the truck has to approach the site via narrow roads or lanes which limit its size. This shows how important it is to study the total delivery situation before deciding what arrangements should be made for reception.

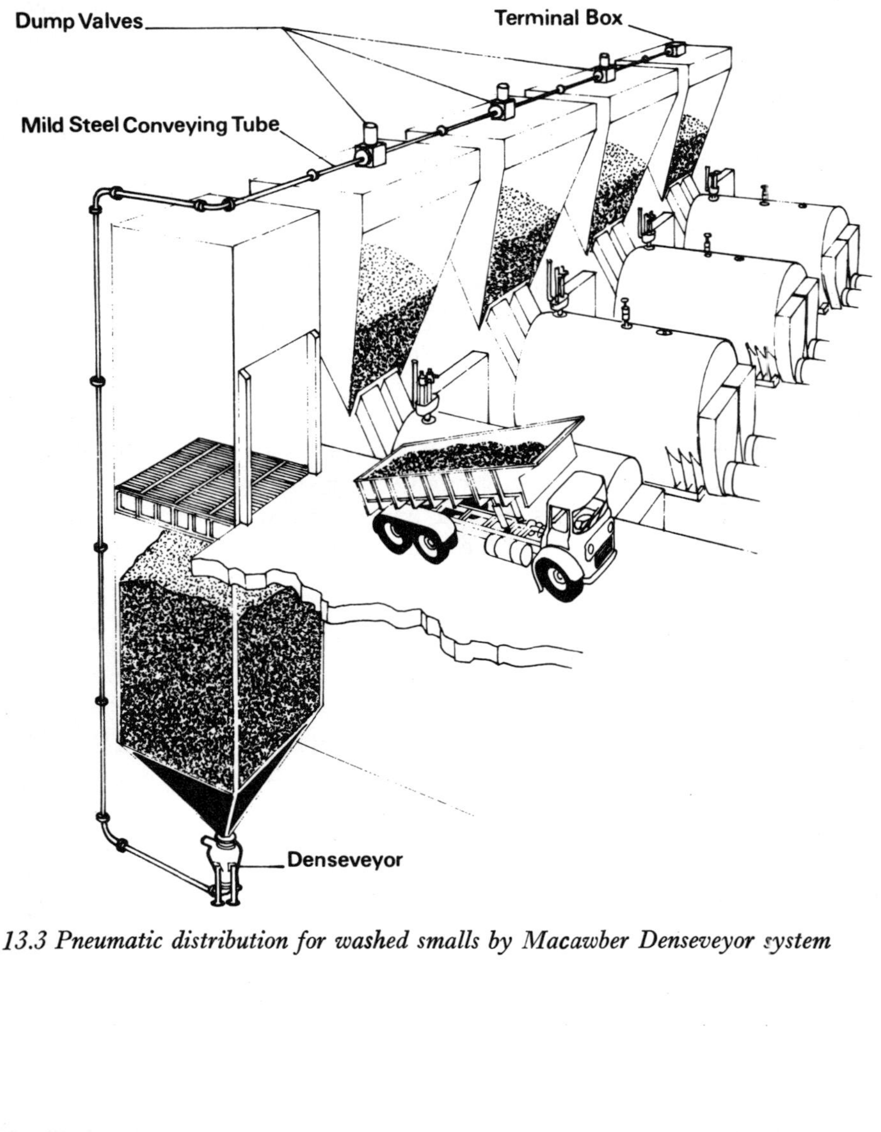

13.3 Pneumatic distribution for washed smalls by Macawber Denseveyor system

The most appropriate method of handling coal between the storage bunkers and the boilers depends on the size of boiler plant, how such plant is situated in relation to the coal store, the kind of coal which is being used and the cost involved. Each application must be considered with knowledge of all the relevant factors. A decision can then be taken whether to employ belt conveyor, mono-rail, gravity or bucket conveyor systems. Whichever system is adopted all components must be robust and suitable for handling coal (which tends to be abrasive); automatic or manual controls should be designed so as to be exactly suitable for the way in which they will be applied; all safety precautions must be taken; access provisions to enclosed components should be generous rather than sparse; and full operating instructions and training must be available to operatives.

Ash and its disposal

The combustion of coal leaves a residue of ash, which must be conveyed from the boilers into a store (or stores) and eventually taken off site. As a basis for planning, the ash storage and handling arrangements information must be available concerning the ash content of the coal (which may range between 5 and 16 per cent) and the coal consumption of the plant. The design of the equipment must be based on peak conditions and not on average consumption. Ash removal is as much part of the flow chart as coal delivery: a serious breakdown in the ashing facility will generally close down the boiler plant.

The average density of ash is of the following order:

Dry ash	640 kg/m^3
Wet ash	990 kg/m^3
Pulverised fuel (PF) ash, usually wet	1522 kg/m^3

Certain coals have a residue of ash and clinker, particularly those with a low ash fusion temperature. The presence of a proportion of solid clinker must therefore influence the ash handling provision and an allowance must also be included for the grit and flue dust which arises from boiler flue cleaning operations. Installations which produce less than about 4 tonne of ash per week may be conveniently serviced by individual containers of the dustbin type which allow ash to flow into them and can then periodically be removed, emptied and replaced.

Larger installations are better served by bulk ash storage arrangements. One popular system which uses individual containers comprises an overhead silo raised up sufficiently to permit the ash disposal vehicle to park under the outlet of the store. A skip hoist raises the bin containers along sloping guides and discharges them into the top of the silo. There must be a seal at the hopper entry to prevent ash being blown about during tipping. Such a system is suitable for handling cold ash with some clinker.

For the still larger boiler installations (about 1.5 MW and upwards), pneumatic ash and clinker handling systems are more suitable. The ash and clinker must first be cool. It is then wetted in an ash conditioner before being discharged into the ash disposal truck. Wetting for cooling purposes cannot be allowed at the entry to a pneumatic system because it would then interfere with the handling characteristics of the ash.

Other ash disposal systems use conveyors submerged in water. These include belt conveyors, vibratory conveyors and drag-link conveyors.

Grit collected by the grit arrestors must not be transported through the central system. The grit should be collected separately, bagged and dumped off site.

In certain places there is a commercial demand for ash. Where this is the case, sales can make a contribution towards the overall economy of the installation. Inquiries on this point should form part of early design and feasibility studies so as to permit the final scheme to include ash disposal facilities which suit potential customers.

GAS SUPPLIES

Gas is a convenience fuel piped into the site from some remote reservoir such as the North Sea (in respect of UK natural gas) or from a large central gas holder. After being metered the gas passes to the gas burners. If this plant is functioning correctly there is no ash and very little in the way of boiler or flueways fouling and maintenance and plant supervision is minimal.

The meter should be located inside a separate well ventilated meter room. Access should preferably be directly from the outside so that the meter reader need not traverse operating plant space.

Burners may be of the atmospheric or pressure jet type. In the case of district and group heating schemes, pressure jet plant will generally be used. Manufacturers specify the pressure at which their burners must operate for the specified full output (12.5 mb and upwards). At lower pressures the burner output is curtailed. The designer and operator must ensure that adequate pressure is available at the burner itself: if the mains gas supply pressure is below that required for full burner operation gas boosters must be employed between the meter and the burners. The latter tend to be noisy and vibratory and have therefore to be installed in a manner which attenuates noise and vibration.

A risk of gas leakage is associated with gas fired installations and it is therefore advisable to provide a gas leak detection system. This consists of detectors which cause a gas valve in the common gas pipe to shut off automatically if they sense a leak. A visual and/or audible alarm can be incorporated with the leak detection system.

Gas fired boiler rooms must be particularly well ventilated from, and to, the open air. Most gas utility companies stipulate minimum ventilation rates.

CHIMNEYS

The requirements for chimney height are far less stringent for gas than for oil and coal fired plants. The chimney must satisfy the following essential conditions:

(a) assist the combustion process by providing adequate draught

(b) ensure that the combustion gases are swept well clear of the stack and that temperature inversions do not take place in the chimney

(c) be suitably constructed and insulated to withstand the operating temperatures

(d) be adequately supported to retain stability in all likely wind and weather conditions

(e) be protected to resist both internal and external corrosion

(f) conform to national and local Codes Standards and bye-laws.

The detailed design guidance for chimneys is outside the scope of this book but it must be stressed that the chimney is an essential component of a system and needs to be expertly designed since it often has to be not only functional but also visually acceptable.

Plate 59 shows the boilerhouse of the coal fired Nottingham (UK) District Heating System. The chimney is made of concrete and is visually prominent and attractive. **Plate 58** shows a group of three 49 m high internally flanged multiflue mild steel clad chimneys. The exterior of the windshields is shotblast. **Plates 60** and **61** show two unusual chimney systems.

DETAILS FOR DESIGNER'S BRIEF

Equipment such as incinerators, boilers, turbines, air classifiers, pumps, hoggers and electric control panels must be protected from unauthorised interference and the effects of weather. Protection may also be required to inhibit the spread of fire into, or out of, the plant site. In certain areas precautions are necessary to prevent dust and dirt getting into equipment. The building enclosures for specific schemes must in fact provide adequate space for the plant within and they must fit in as appropriately as possible with the surrounding environment.

The designer, most commonly an architect, must be briefed concerning the following aspects:

Internal areas: these should suit the equipment they are intended to contain. Before finalising areas and heights the closest attention must be paid to the space necessary around the plant for safe and efficient maintenance procedures.

Access: this must be suitable for the installation and subsequent replacement of equipment. The needs for lifting beams, knock-out wall panels, platforms and ladders should be allowed for.

Drainage: the requirements may involve simple gullies, with or without covers and/or purpose made blowdown pits, traffic-bearing manholes, petrol or oil interceptors, sumps, condensation and drainage channels, roof drainage, toilet and washing facilities etc.

Leakages: safeguards against the spread of inflammable liquids such as petroleum or fuel oil may involve special enclosures and bung walls as well as suitable oil-proof (or other) rendering of the walls to partial or full height.

Weight of equipment: the need for heavy equipment on floors and/or roofs is likely to involve structural designs and facilities for lifting it into place and the design and weathering of the roof may need particular attention.

Plant base: special construction of foundations may be necessary to minimise noise, vibration and heat transmission and pits may be required for settlement or sediment purposes.

Gas meter room: this will require ventilation to specified standard; if a gas booster set is necessary, anti-vibration and sound proof enclosure should be considered.

Electrical: intake and switch rooms are generally required together with suitable cable entry pits and service ducts; standby emergency generator(s) may be required and will need fuel stores.

Communications: these may include telephones, paging, music, loudspeaker facilities etc, some of which will require special instructions to the builders.

Ventilation: buildings above ground with external walls may be suited by natural ventilation but mechanical ventilation provision may be required for other circumstances.

Water services: these involve the provision of water meter and stopcock facilities with the associated pits and ducts; many districts require substantial water storage capacity to be provided on site particularly where there is a sprinkler installation.

Explosion risks: relief doors may be required.

Duct and pipe services: it is preferable to pre-plan all major holes for the ducts and pipes which fit between the plant area and the other parts of the complex; this is particularly important where structural concrete walls are employed.

Traffic approach: the weight of wheeled traffic likely to use the yard must be ascertained and the civil engineering designed accordingly.

Fire engines: establish route for fire engines to the plant areas and ensure adequate turning circle for fire engines.

Fire hoses: agree and specify what types of hydraulic hose reels are required and where they are to be located; they will be needed for first-aid fire fighting before the Fire Brigade arrives at the premises: the reels are bulky and must be designed into the building.

Water hydrants: agree and specify types needed and where they are to be placed; points must be located outside buildings and should provide water supply connections for the fire fighters.

Fire alarms: an appropriate arrangement of fire alarm fittings and electric wiring will be required of a scope suitable to the hazard being protected and the use and occupation of the buildings.

Fire appliances: establish the nature of fire fighting facilities required; if pre-mixed foam, Halon gas sprinklers or similar installations are required allocate space and enclosures for the valve chamber, cylinders and associated equipment. Hand-held portable fire fighting appliances, usually of the foam type, will be required in locations specified by the fire fighters. Water buckets and sand buckets should be placed in appropriate positions.

Other fire precautions: wet fire fighting methods are not suitable in certain areas eg in rooms housing electronic equipment, electrical switchgear, computers, etc and smoke or heat detectors may have to be provided to raise the alarm there when dangerous conditions arise or to activate automatic fire fighting equipment (detection systems of this kind usually incorporate central visual display panels or consoles for quick identification of the danger area).

Emergency lighting: emergency lighting will be necessary to mark exit and escape routes, fire doors etc: such a system may consist of centralised battery-charged installations; alternatively self-contained rechargeable battery lights may be used.

Maintenance areas: provide adequate and appropriate on-site maintenance facilities, such as workshop, tool racks, spares store etc.

Staff facilities: provide appropriate toilets, washhand basins, showers, changing rooms, drying rooms, etc: these are particularly important for solid fuel fired boilerhouses.

Mechanical handling: establish requirements and allocate space for cranes, conveyors etc and design supports and foundations.

Chimneys: decide on method of construction, design foundations and establish guy points if needed (if chimneys are constructed of metal many local authorities insist on licensing them: this means that there will be an annual inspection for which facilities should be provided).

Lightning protection: establish requirements; a suitable location for the lightning conductor may be on a chimney terminating above its top; provide a base pit for the conductor.

Illumination: lighting levels must be adequate to permit safe operations within the plant areas (a minimum of 300 lux is recommended for this purpose); consider the need for flood lighting external to the building for operational, safety or display purposes or for preventing vandalism; if facilities for hand-held inspection lamps are required are these to be operated at low voltage?

Battery driven vehicles: if these are to be provided charging areas must be allocated with adequate access for the type of machinery to be employed.

Floor and wall finishes: these should be functional and easily cleaned and maintained.

Co-ordination: the architect has to work closely with the consulting engineers and with the other members of the team to establish and execute a fully detailed brief.
Major structures: may be required for the following: chimney; water tank towers; storage silos and supports; filter chambers and supports; gas washing plants and supports; crane gantries; lifting beams or frames; access gantries for overhead lights; platforms and gangways; fuel stores and delivery shutes; weighbridge. Special structures may also be required for uncommon applications.

LAYOUT AND APPEARANCE OF PLANT

Every endeavour must be made when laying out mechanical and electrical plant to ensure an arrangement which will permit orderly operation of the plant and ample access for maintenance of the equipment. Buildings and plant must also be arranged so as to permit the subsequent removal of large plant items without extensive disturbance.

Engineers often find themselves under intense pressure to minimise plant room spaces in the interest of economy but it is certainly a false economy to accept a design which will cramp the plant and commit the plant operator to future maintenance and operating difficulties. On the other hand if greatly excessive space is asked for and provided it will compromise the designer's reputation and cause unnecessary building expense. Decisions concerning plant space requirements must never therefore be rushed or decided off the cuff and must always be based on a thorough analysis of requirements and equipment. There is ample evidence to prove that a well-planned plant room will pay dividends by improving efficiency in operation and maintenance: it will be kept clean and the operators are likely to take pride in it as a showpiece.

Plate 62 shows the oil fired boilerhouse of a district heating system in North London. Note the packaged Allen Ygnis boilers and burners, the neat thermal insulation of the pipework, the trays under the oil burners, the tiled floor finish and the overhead lighting.

Plate 63 shows the coal fired boilers of the Billingham group heating scheme. This is particularly interesting because it illustrates the way in which a modern coal fired boiler plant can be arranged neatly and maintained spotlessly clean. Each of the three boilers is a GWB Vekos packaged unit of 4700 KW output and is fed from an overhead bunker of 70 ton capacity. Note the overhead coal feed from the bunker into the top of each boiler, the neat checker plate duct covers and the ash collection and grit recycling facility at the front of each boiler.

Plate 64 shows a pneumatic coal fired system with GWB boilers.

Plate 65 shows a typical screw-elevator feeding from a coal bunker into an anthracite burning cast iron sectional boiler such as may be used in a group heating scheme. The screw-elevator feeds the coal into a horizontal trough conveyor from whence it passes to the burner hoppers. It should be noted that the low ash content of the anthracite reduces clinker removal to a minimum (once in every twenty-four hours).

Plate 66 shows a Trianco gravity boiler installation. The boilers burn anthracite. Note the neat arrangement of coal hoppers and plant.

All the above pictures of boiler houses show economical plant layouts with generous (but not excessive) maintenance spaces around the equipment.

Where boiler houses or plant rooms are within a factory complex they are usually strictly functional in external appearance. However where they are located within the view of the general public or within a housing estate the architect should aim for a generally pleasing appearance. **Plate 68** shows a dramatic night view of the NCB District Heating Station at Edmonton (London, UK). Note the adjacent blocks of flats which are served off the system, the high level monorail coal-handling system, the ash silo for the pneumatic handling of ash and the 106 metre high chimney which contains two flues within its concrete windshield.

CHECK LIST

1 Choice of suitable fuel
2 Fuel storage capacity
3 Storage arrangements needed for fuel oil
4 Fire precuations needed in connection with oil storage
5 Codes of Practice relating to oil storage
6 Storage arrangements needed for solid fuel
7 Arrangements for coal delivery
8 Ash handling arrangements
9 Gas firing
10 Basic chimney requirements
11 Requirements relating to boiler houses, plant rooms and structures
12 Plant layout and appearance

REFERENCES

D. Kut, *Heating and Hot Water Services in Buildings*, Pergamon, 1968, pp 302–323

National Coal Board, 'Boiler Design for Solid Fuel', series of booklets, 1977

CHAPTER 14

Heat management of district heating installations

OBJECTIVE

The primary objective of heat management in a group or district heating system is to use the available resources of energy labour and materials as efficiently as possible to provide heat at the point of use at minimum cost in such a way that both consumer and operator receive satisfaction.

METHODS

To achieve the objective set out above the owner or operator (eg supply company, local authority, Housing Co-operative etc) of the district or group heating/cooling installation must decide on a suitable mode of operation in the light of the following major alternatives:

(a) *Employ own staff* to oversee the efficient performance of all sections of the installation, to monitor performance, order fuel and agree payments for it, provide routine maintenance, carry out repairs and provide replacements of worn out plant.

Advantage: The complete operation is under the control of the owner/operator; he can hire and fire staff, select sub-contractors, negotiate with the fuel suppliers: in other words the owner/operator appears to be in full control.

Disadvantage: The owner/operator has to set up an organisation which is outside his normal sphere of work, usually on a 'one-off' basis; he has to organise supervision, financial control and bureaucratic linkage between the heat management group and his main activity; he must provide standby labour and facilities to be prepared for emergencies.

(b) *Employ sub-contractors* to deal with specific aspects such as routine maintenance, fuel supply, maintenance of sophisticated equipment (eg central instrumentation); repair and replacement work will be carried out on the basis of estimates as and when such work becomes necessary.

Advantage: The owner/operator retains overall control and can negotiate costs of each major aspect.

Disadvantage: Staff must be employed to supervise the day-to-day operation and performance of the plant, select sub-contractors, order fuel, check the plant for any repairs and replacements which may be required, monitor the performance and costs of the sub-contractors; standby labour must also be provided for emergencies.

(c) *Enter into comprehensive long-term contract* with a heat management company for the complete operation of the plant on a continuous replacement basis.
Advantage: Once the terms of the heat management contract have been agreed the full responsibility for the management of the system is taken off the operator's shoulders; it is assumed that the contractors are a substantial company with numerous on-going contracts within the reasonable vicinity of the confines of the district system and that they will have a central control room to take care of emergencies: the operator is thus relieved of the need for provision of standby facilities; moreover the bulk supply of fuel may show a saving over the provision of fuel to one particular site only.
Disadvantage: Having fully delegated the heat management over a long term the operator loses the ability to control the individual functions and contracts: unless a fair agreement has been negotiated this solution may prove expensive.
Methods (a) and (b) are well-known. When applying either to a large engineering complex such as a district heating system the operator must employ experienced and reliable supervisory staff who will organise and administer comprehensive preventive and emergency maintenance and replacement procedures. Method (c) is not so well known. It is therefore described at some length below.

HEAT MANAGEMENT COMPANIES

Main functions

The primary objective of a heat management company is to satisfy the operator and consumer requirements on terms that enable it to make a profit. Its function will be to provide the necessary heat together with the supplies and services required in such a way that the operator can achieve his own objectives more conveniently or more cheaply than he could do if he were in sole or overall charge of them. In doing this the following secondary objectives, which depend on the nature of the company concerned, may also have to be taken into account:
Need to secure long-term tied outlets for a particular fuel without disturbing conventional marketing policies.
Need to employ a particular manufacturer's range of equipment.
The adoption of agreed maintenance procedures and/or standing orders.

The extent to which these factors will influence the main objective will be determined by the policy of the company itself and by whether heat management activities are seen as its primary function or as a means to some other end. Thus one company may wish to sell as much heat as possible in order to maximise fuel consumption. Another (perhaps unconcerned with fuel) will, where the system allows, be intent on providing the minimum heat necessary to satisfy an agreed performance standard so as to profit through economy. For these reasons each particular company will approach the question of heat management in its own particular way: obviously the emphasis will be on those supplies and services it feels best able to provide for the particular site.

Heat supply contracts

As a general rule, a heat management company will seek to enter into a heat supply contract with the owner/operator for a lengthy period of up to 30 years duration. In return for the charge agreed under the terms of this contract it will undertake to provide a guaranteed supply of heat in such a way as to satisfy a performance specification of some clearly defined kind. This will be most frequently determined by the design of the system itself and by whether control of the amount of heat supplied to buildings lies with the company or the occupiers.

Depending upon the circumstances of each case, appropriate specifications could be:

(a) Maintenance of specified internal air temperatures in specified hours during a basic heating season of agreed length.

(b) Maintenance of minimum flow temperatures at the boilerhouse wall. This specification could be employed in the case of a two-pipe distribution system supplying hot water to warm air units where (apart from a standing loss) heat input is controlled by each tenant by means of a manually operated thermostat. It would also apply to domestic hot water supplied from instantaneous or bulk storage calorifiers.

(c) Maintenance of an agreed combustion efficiency. This basis may be used where the agreement provides for the day to day operation of plant to remain in the hands of the owner's personnel. Such an arrangement could well arise for example in the heating of school or college buildings where the headmaster wishes to retain personal control or where the periods and standards of heating are subject to considerable variation and the plant itself is relatively small.

Choice of supplies and services offered

A heat supply company would normally carry out at its own expense a thorough investigation into the state of the existing or proposed plant concerned. Subject to the results its proposals might then cover all or part of the following extensive list of supplies and services:

Supply of all necessary fuel. The policy of one major company is to seek in each case the cheapest source of energy for the purpose; depending upon circumstances this could be coal, fuel oil, gas or quite conceivably in the future waste heat from power generation, refuse incineration or any other source capable of meeting the requirement.

Provision of supervisory staff responsible for planning and managing the

Associated Heat Services Limited

The alternative supply of heat from the first C.H.P. Scheme in the U.K., owned and operated by the C.E.G.B., will shortly be the responsibility of A.H.S.

Associated Heat Services Limited is the largest comprehensive Heat Service and Energy Conservation contractor in the U.K. Services include full responsibility for design, finance, operation, meter reading, account recording, etc for all types and sizes of District Heating and C.H.P. schemes fired by coal, gas, oil and waste materials.

Associated Heat Services Limited
15 Hobart Place, London S.W.1
Tel: 01 235 9161

No gimmicks.

But all the essentials.

NEW METRIC HANDBOOK

EDITED BY PATRICIA TUTT AND DAVID ADLER

512pp 1000 line drawings A4 Cloth £16·00 Paper £8·95

Available (remittance with order, please, plus 10% postage and packing) from Architectural Press, 9 Queen Anne's Gate, London SW1H 9BY.

operation and maintenance of the plant and equipment and for liaison with the owner's officers (**plate 67**).

Provision of boiler operators for the operation control and adjustment of the plant and equipment in sufficient number to ensure normal and continuous service in accordance with the terms agreed.

Running maintenance of plant and equipment: this comprises all work that can be carried out by men normally employed in operating boiler plant and that does not require specialist service or spare parts eg cleaning of boilerhouse plant equipment and premises, periodic inspections and checks in accordance with manufacturers' or installers' recommendations (testing greasing and lubrication together with the supply of oil, grease and rags etc), adjustment of automatic boilers and burners in order to maintain optimum efficiency, periodic cleaning and sweeping of boiler flueways and removal of soot to the outside of the boilerhouse (**plates 70, 71** and **72**).

Placing plant and equipment in a state of preservation at the end of the heating season eg de-scaling and cleaning boiler heating surfaces, spraying on of corrosion inhibiting products, drying out of combustion chambers, dismantling and cleaning the main burner parts, checking of mechanical parts, inspection and cleaning of electric motors, inspection of electric circuits, examination and repair of refractory brickwork, periodic visits to turn machinery during the period when it is not in use.

Other periodic maintenance eg cleaning and de-sludging of fuel oil tanks, removal of ash and clinker to the outside of the boilerhouse, sweeping of flues and chimneys as often as may be necessary and removal of soot to the outside of the boilerhouse.

Major maintenance of plant and equipment: this will include repair of equipment or parts of it and/or replacement of spares together with dismantling and assembling ancillary fittings if their failure is due to an accident or a fault on the part of the company. It would not normally include renewal or replacement of plant that has deteriorated as the result of physical aging.

Total guarantee contracts

The Total Guarantee contract is an alternative to the major maintenance facility described above (**plate 73**). It does in fact go further in that it covers major maintenance of plant and equipment, whatever the cause of their deterioration (including damage caused by fair wear and tear over a number of years). The effect of such a guarantee is that at the end of the contract period the client's plant is returned to him in good working condition and capable of satisfying the duties for which it was originally installed. The plant can be periodically inspected by an agreed independent organisation (usually a firm of consulting engineers nominated by the owner and agreed by the heat management company or an insurance company) which will report impartially upon the state of the plant and its maintenance.

Other facilities offered

Other supplies and services that may have to be provided to meet the needs of particular sites are:

Maintenance of heat distribution networks both underground and within buildings.

Operation and maintenance of air conditioning equipment.

Maintenance of other mechanical services particularly where these are sited in the boilerhouse area.

Feasibility studies: some companies can carry out initial studies to establish the economic viability of proposed group and district heating schemes or extensions to them and estimate the cost-in-use that could be expected from central plant as compared with the most likely alternative individual system.

Design and installation: an undertaking to provide these services may be given either for new schemes or for the renovation or improvement of existing plant; the nature of the heat management company concerned will determine whether it employs staff from its own organisation for this work or whether it retains the services of consulting engineers or the design services of installing contractors once an outline specification has been prepared.

Technical advice: to a limited extent this will be available free of charge and it may comprise preliminary plant and management surveys designed to establish present levels of heat cost in existing buildings so as to provide some yardstick against which to measure the extent and cost of any possible improvements.

Finance: the majority of companies are prepared to make varying amounts of capital available for new installations or for the replacement or renovation of old ones. The extent to which this service features varies from company to company.

Metering of heat to individual consumers and collection of heat charges: this may be undertaken by the company or sub-contracted to a specialised organisation.

Extent of responsibility undertaken

The extent of responsibility that a heat management company may undertake will vary according to requirements and the nature of the system which is to be operated. In its most comprehensive form it may well involve not only responsibility for the supply of heat but also for an entire system comprising a central boilerhouse and its plant together with the underground mains distribution system and the internal heating systems within the buildings (**plate 74**). The first and more usual arrangement is for the heat management company's responsibility to stop at the entry to the individual dwellings but to include pipework within blocks. The second alternative is for responsibility to terminate at the boiler house wall and for the remainder of the system to remain in the hands of the local authority.

Methods of charging

In return for its supplies and services the heat management company may charge for heat in a number of ways. The following are the most common:

Annual lump sum

This is the simplest and most attractive method of charging but it is only practicable where heat consumption can be forecast with reasonable accuracy and the control of heat input is in the hands of the heat management company. It is the recent practice in France for the annual lump sum to include charges for all domestic hot water supplied (commonly from bulk storage) without imposing any limitation on consumption.

Metered charge

Where it is not possible to forecast heat consumption with sufficient accuracy or where control is in the hands of individual tenants charges can be based upon heat consumption as recorded by heat or fuel meters. They may be expressed in the form of an inclusive flat rate per useful therm of heat or unit quantity of fuel supplied and may include all the supplies and services provided.

Multi-part tariff

In most cases the most equitable arrangement for both parties is a multi-part tariff consisting of an annual service charge in respect of fixed costs incurred regardless of the amount of heat supplied (eg for labour, materials and maintenance) plus a metered charge per useful therm covering the variable costs represented by fuel and some if not all of the electricity used. The benefit of this arrangement for both parties lies in the fact that it removes to a large extent the gamble inherent in forecasting heat consumption in consumer-controlled systems. Experience suggests that annual consumption can vary quite dramatically not only between different schemes but also for particular periods.

If an inclusive flat rate per useful therm is chosen as the method of charging, the heat management company may well fail to recover its fixed costs when heat consumption is below expectation. If on the other hand consumer demand exceeds the forecast, the owner may end up by paying far more for additional heat than would normally be the case. This is particularly important where finance is concerned and the heat charges include the recovery of capital and interest over the period of the contract. From the owner's point of view the annual lump sum is likely to be the most suitable method of charging where conditions are suitable. Where this is all embracing and includes provision for maintenance and any necessary replacement of plant during the contract period the owner can pass on the cost to his tenants directly without needing to make any reserve for the additional fuel expenditure which might otherwise be incurred in a severe winter and without needing to raise fresh loans to replace plant that has failed or has worn out.

Revision of contract charges

The proposed charges for heat are normally based on the prices ruling at the time of the proposal and the contract will provide for them to be put up or down from time to time in accordance with agreed formulae devised for that purpose.

One method is to divide the charge for heat into its three main components each of which may be revised in the following way:

(a) Fuel: by reference to variations in the net purchase price of a specific quantity of fuel.

(b) Electricity: by reference to variations in the net purchase price of a specific quantity of electricity under the most favourable tariff.

(c) Annual service charge: by reference to officially published national indices of the cost of labour and materials.

Reference values for each of these components can be established and stated in the contract and revised by comparing them with the current values at the time of review. If finance is involved, interest charges may be revised in a similar way in accordance with variations in the Bank Rate (or some other financial base rate or index): current indications however suggest that this traditional relationship may not continue in the future.

Functions outside the district heating field

The activities of heat management companies are not necessarily confined to group or district heating schemes. The same principles may be applied with equal success wherever heat or cooling is provided from central boiler plant for any large single building or complex of buildings such as blocks of flats or offices, schools, churches, civic buildings, baths, theatres, hospitals or commercial or industrial premises.

There is no particular minimum limit on the size of the plants that may be concerned but the extent to which a heat management company may be able to offer a service that is economically attractive must depend upon the amount of improvement it can make and the total amount of the annual heat bill after all the relevant costs have been considered. This may sound simple but it is in reality perhaps one of the most difficult tasks that heat management companies have to perform since outside the industries which rely on steam or heat for process work it is difficult to find anyone who knows with any certainty what he really pays for the heat he consumes. In the vast majority of cases the fuel cost is the only fact known and the buyer will recount his success in securing an additional rebate with evident pleasure: in doing so he does not apparently pay any regard to the efficiency with which that fuel may be burned or to the true costs of labour, maintenance and plant replacement (if these are known at all the totals are probably divided between a number of departmental budgets).

Heat management techniques

To justify its existence and make a profit the heat management company has to compete for the same business both with the client and with any other companies of its own kind. The problem is to find ways and means of providing a better and preferably cheaper service than the client can provide for himself or than can be offered by the competition.

The first and most obvious area of investigation is the fuel position since the fuel cost will invariably constitute the major part of the total heat cost. The following main points will have to receive consideration:

(a) Is the present fuel the cheapest source of energy?

(b) If so can the purchasing terms be improved?

(c) What opportunity exists to improve combustion efficiency?

(d) What opportunity exists to avoid wastage in distribution?

Particular attention will also have to be paid to the labour content of each job and its cost. Typical questions that have to be answered are:

(a) How many workers are required and for what kind of jobs?

(b) What skills are required in each case?

(c) Can they be provided from within the company's existing labour force or must they be recruited or brought in?

(d) At what frequency will these skills be required?

(e) To what extent can specialist skills be utilised elsewhere and the cost spread over a number of contracts?

Where the 'total guarantee' on plant and equipment has been undertaken (ie complete responsibility for maintenance and replacement during the contract period) the heat management company will have the most powerful incentive to develop and implement preventive maintenance programmes with a view to obtaining the maximum working life from the plant consistent with its continued reliability in securing optimum operating efficiencies.

The consumption of electricity is another area that will have to receive close scrutiny.

Thus a systematic appraisal has to be made for each item of cost so as to identify all practicable opportunities for improvements, the cost of implementing them and the time required to recover any investment involved. This analysis must be repeated at intervals during the life of each contract in order to ensure that no justifiable opportunity for improvement is overlooked and that the full benefit of any worthwhile technical innovation is obtained. During a contract period of say 25 years duration many changes can be expected to take place. The most obvious and important of these will be relative energy costs: if market conditions are favourable they may well justify a change of fuel. A company whose sole function is to sell heat and to provide a heat management service is favourably placed to take advantage of such opportunities but before it does so the condition of the plant at the time and the cost of any modifications which may be required must always receive consideration.

The need for an expert approach

Experience gained from surveys carried out in both public and private sectors to establish the operating efficiency of existing plant and the cost of heat obtained in a wide variety of installations has shown that in many instances considerable margin for improvement does in fact exist and that economies can result from a specialised approach which is also sufficiently comprehensive. The main reasons for this may be summarised as follows:

(a) The relatively low status which may be given to personnel employed on daily operation and maintenance of plant and the apparent reluctance to spend money on their instruction and training: unless the plant is big enough to warrant full-time attention its supervision may be regarded as an addition to general caretaking duties. This may have been acceptable in the days of the traditional hand fired coal boiler and lower standards of heating but in recent years manufacturers have vied with each other to produce higher performance boilers with automatic firing and comparatively sophisticated control systems: the emphasis in the labour requirement has therefore changed from quantity to quality and the consequences and cost of any failure have risen in proportion. Conversion to automatic firing may appear to solve one problem in that it is no longer necessary to find and keep staff who are prepared to shovel coal but it creates other needs: if these are not thoroughly considered they can give rise to new problems which will be no less expensive in terms of low combustion efficiency and premature plant failure.

(b) Reluctance to provide money for the upkeep of plant and equipment and the maintenance of good housekeeping standards: all too often a boilerhouse is found to be a neglected and depressing area black with soot and flooded from leaking glands on valves and pumps; in working conditions such as these, labourers have little incentive to take any pride in their work and it is not surprising that plant receives little care and less attention until the breakdowns start. Boilerhouse instrumentation can only be of value if it is correctly maintained and if the operator understands and is able to act upon the information it provides.

(c) General lack of appreciation of the cost-in-use of heat and the factors that affect it: efficient utilisation of the heat produced, effective length of plant life and current replacement costs are seldom fully taken into account.

Some evidence of acceptance of the benefits that are to be gained from a specialist approach may be seen in France where the concept of central plant operated, maintained and managed by specialist companies has now been well established for many years so that there are now over 130 different competing companies. Their Association now claims that the installations currently operated by its members represent the equivalent of just under 30 million KW of boiler capacity. Of this some 60 per cent is represented by housing (mostly owned or managed by local authorities and housing associations).

Advantages of employing specialised firms

The main advantages which can be gained by employing companies which specialise in heat management may be summarised as follows:

(a) A guaranteed supply of heat to satisfy the performance level specified as the basis of the heat supply contract.

(b) Release from the day to day problems normally associated with the production and distribution of heat on a large scale.

(c) The purchase of heat at a known cost: once agreed, the basic heat charges remain fixed throughout the contract period subject only to the operation of the rise and fall clauses incorporated in the contract to deal with variations in economic conditions which are outside the control of both parties. In formulating these charges the company will have taken into account the best fuel purchasing terms and optimum operating efficiencies. Its profit will therefore depend on maintaining or improving upon these figures. Any excess fuel consumed over and above that estimated would be the company's responsibility.

(d) The charge for heat in the form of an annual lump sum is not influenced by climatic conditions during the specified heating season: the company bases its calculations upon average weather conditions and will itself bear the cost of an exceptionally hard winter. This allows the owner to budget with confidence. Both the housing manager and his tenants can know in advance what their heating costs will be. This arrangement offers the additional advantage of reducing the cost of accounting and administration to a minimum since the annual charge may be collected in the form of equal monthly or bi-monthly instalments.

(e) The life of plant and equipment will be increased as the result of employing specialised labour for its operation and maintenance. Where the company accepts complete responsibility for maintenance and replacement of plant under the terms of a 'total guarantee', the owner will be relieved during the contract period of all concern for its maintenance and will not be faced with the need to raise fresh loans to replace plant which has failed or become worn out as the result of fair wear and tear.

(f) In the course of the long term operation and maintenance of all types of boiler plant burning a variety of fuels the heat management company acquires a very wide practical experience. It is therefore in a unique position to provide technical assistance and advice when new buildings are being planned or when old plant is to be replaced. In doing this its point of view is different from that of the installing contractor: if it is successful in securing a long term contract for the operation and maintenance of the new plant it must be prepared to undertake the responsibility for it and to bear the cost of any failure. The company's interest and that of the owner are thus identical in ensuring that the work is soundly designed and well executed. For this reason it is desirable that the heat management company should be consulted at the earliest stage of planning and thereafter play an active part in the design, commissioning and acceptance of plant.

(g) In purchasing heat on a comprehensive basis from a heat management company an authority will only have to deal with one organisation in place of the large number of specialists who might otherwise be involved.

CHECK LIST

1 Objectives of heat management
2 Methods of heat management and their advantages and disadvantages
3 Functions of a heat management company
4 Alternative services that may be provided by a heat management company
5 Total guarantee contracts
6 Extent of responsibility that may be undertakenby a heat management company
7 Alternative methods of charging for heat management services
8 Applications of heat management outside the district heating field
9 Heat management techniques
10 Summary given in this chapter of the benefits which may be afforded by employing a specialised heat management company

CHAPTER 15

Typical group and district heating schemes

A large number of group and district heating schemes have been installed and are in operation. The designer of a new installation always refers to previous ones and learns from past experience. Some relevant examples are therefore described in this chapter.

GROUP HEATING: UNITED KINGDOM

Rose Row scheme Aberdare

The Rose Row scheme at Cwmbach, Aberdare, South Wales was commissioned in 1974. It provides heating and hot water supply to 184 dwellings of mixed types (one-bedroom flats to three-bedroom houses). The system operates with a flow temperature of 82°C (180°F) and a return temperature of 71°C (160°F). It embodies a nitrogen pressurisation unit to maintain a system pressure of 4.14 bars. Two duplicate pump sets are provided (for winter and summer use respectively). Each dwelling has its own calorifier. The maximum winter load is 800 kW supplied by three Centrax horizontal steel boilers which are fuelled by semi-anthracite pea size coal supplied by the nearby Tower colliery. An inclined screw type elevator (one per boiler) feeds coal from the bunkers in front of the boilers into hoppers which supply the chain grates. The Centrax boilers cannot operate at loads below 293 kW without damaging the chain grates. Therefore Earlymil gravity feed burners are in use for the summer period. These burn the same grade of semi-anthracite coal as the chain grates (**15.1**). The ash and clinker are removed manually every 12 hours. This group heating scheme employs one full-time attendant and one relief attendant for evening and weekend work. The total coal consumption during 1975 was 500 tonnes, an average of 2717 kg per dwelling/year.

Rochdale (Lancashire)

In 1970 it was decided to improve council dwellings consisting of sem-detached, terraced and parlour type houses originally built in 1931. This exercise included the installation of whole-house heating. The scheme supplies heat to 963 dwellings, two shops and a primary school (**15.2**). In the initial stages this was supplied by a 1.76 MW temporary boiler plant. When the whole scheme has been completed, the boiler plant will have an output of 8.8 MW.

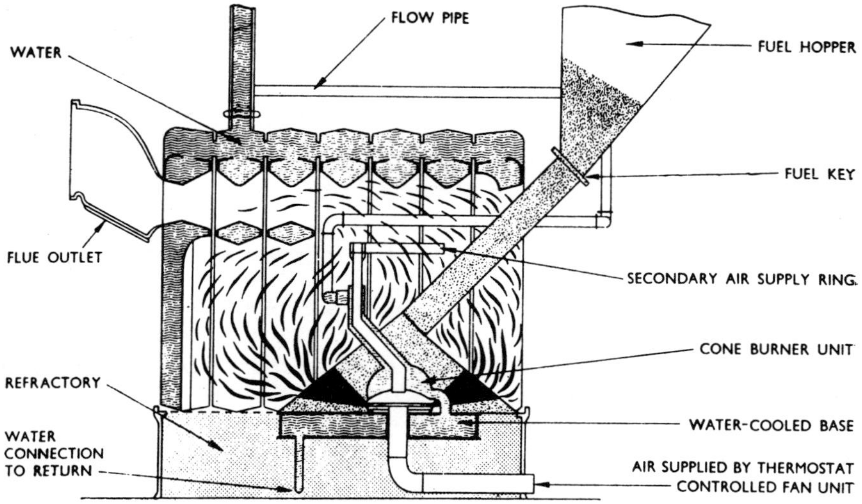

15.1 Earleymil burner arrangement

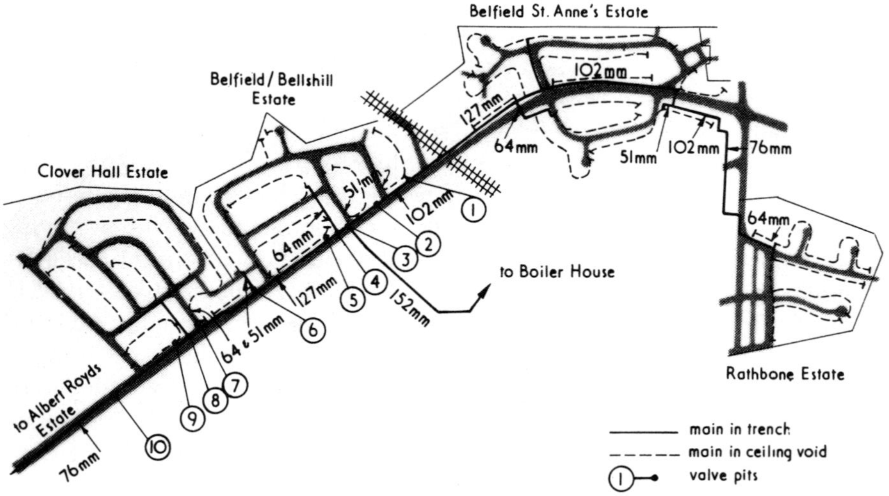

15.2 Layout of Rochdale group heating system

The hot water is circulated from the boilerhouse through mains which consist of steel pipes surrounded by polyurethane insulation and an outer plastic casing. The pH adjustment is by means of Houseman and Thompson Tri-so-tan No 2 water treatment equipment to a pH of 8.5 and is intended to prevent scale build-up and corrosion by acidic waters. Individual groups of dwellings are fed by branch mains off the trunk mains. These are run either under the floor (where houses are in pairs) or in the roof space (where dwellings are in groups of four or more). The branch mains are insulated with conventional pre-formed glass fibre sections secured with bands and supported from floor joists or roof trusses. The flow temperature varies between a minimum of 71°C (160°F) and a maximum of 93°C (200°F) depending on external ambient temperatures. The return is at 54°C (129°F).

The main boiler plant comprises three Parca Norrahammar 2.9 MW burner/boiler units each fired on 3500 seconds Redwood No 1 fuel oil. The oil is stored in three 54.5 cubic metre tanks.

The hot water is pumped by a Sulzer Bros pumping set, working against a total head of 79 m. The network is pressurised to maintain a static pressure of 4.15 bars at the boilers. The total working pressure of the system is 6.9 bars and pressure control devices are employed to limit the maximum pressure difference at any dwelling to 1.2 bars.

Within the dwellings, distribution loops to the Stelrad UB radiators are taken off the pre-insulated copper tubing below the floors. A combined packaged hot and cold water storage unit is installed in each cylinder cupboard. The cylinder return is at 57°C (135°F) and the space heating return at 54.5°C (130°F).

The radiator circuit is based on the use of Stelrad Unibelkon single-tapping radiators connected in series through U-B divertor valves. In the fully-on position the radiator valve diverts a set percentage of the total flow through the radiator: the balance bypasses the radiator. The flow to the radiators can be adjusted to compensate for falls in the supply temperature. The hot and cold water storage unit has an insulated casing, internal vents and drain cocks. Hot water storage tanks of 113 or 136 litre capacity are used (depending on the size of the dwelling); the primary coils are suitable for a supply pressure of 7 bars. The contents can be heated up completely in 2 to 2½ hours, when at the summer flow temperature of 66°C (151°F). Heating is measured by Clorius evaporative heat meters.

Totley Brook Sheffield

This system serves a residential development of 183 dwellings operated by the City of Sheffield Housing Committee (**15.3**). The total heat load is 2.1 MW supplied by four separate boiler houses using modular gas fired boilers. Distribution is by means of a twin-pipe low temperature pressurised system. All underground piping is pre-insulated with polyurethane foam and thermoplastic sheathing. Space heating in consumers' dwellings is by panel radiators each with a central connection and a limiter valve on the return line. Domestic hot water supply to each dwelling is by an individual temperature-controlled storage cylinder.

Four Norwich schemes

There are four separate City of Norwich group heating schemes:

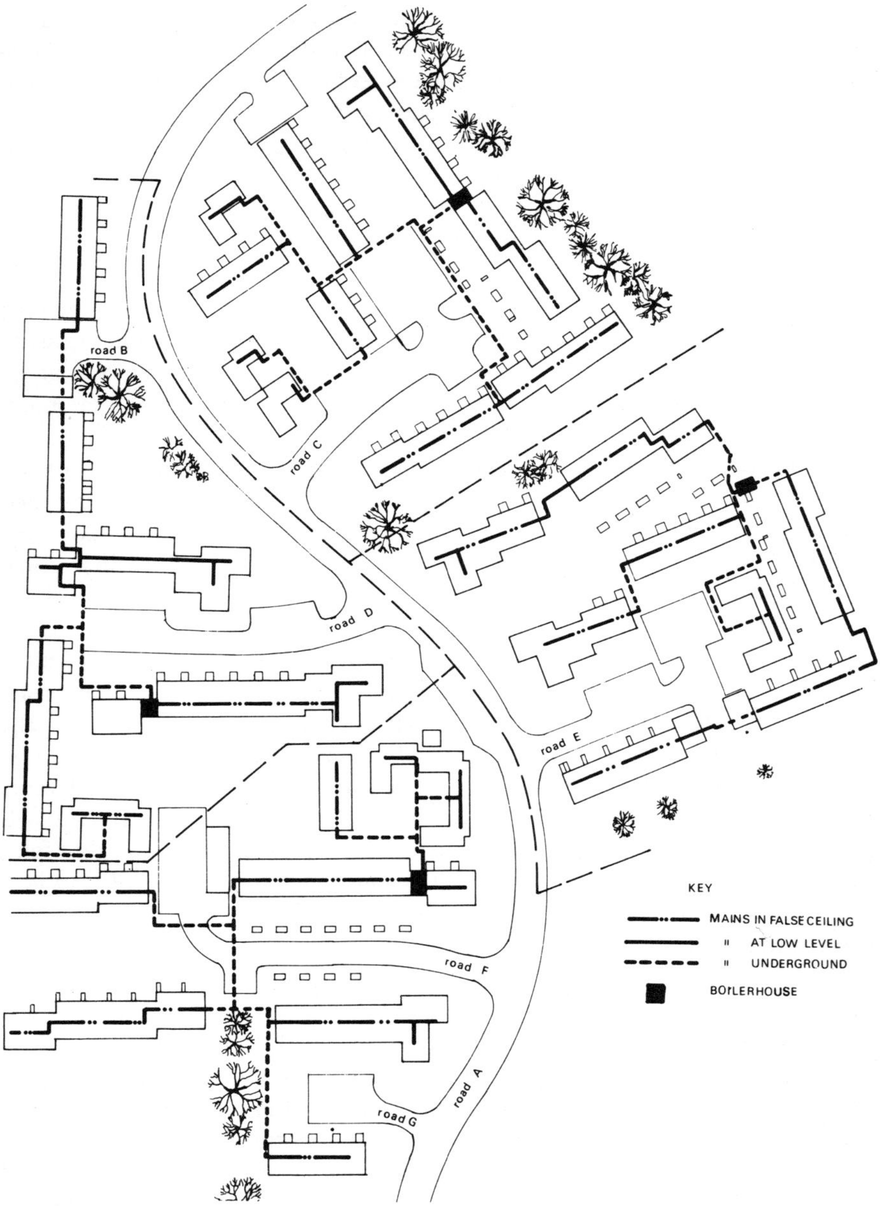

15.3 Layout of Totley Brook development, Sheffield

The Devonshire Street scheme serves 396 dwellings in two- and three-storey blocks. The boiler plant is located in a central free standing structure and houses three gas fired boilers with a combined capacity of 3.99 MW. Pro-

vision has been made for a further boiler to increase the output to 5.63 MW.
The Canterbury Place redevelopment consists currently of 128 dwellings in the form of two- and three-storey houses and flats with provision for the addition of 100 dwellings. The heat is supplied by two oil fired boilers with a combined output of 1.41 MW (to be increased to 2.24 MW). This plant is located in a free standing boiler room.
The Blackhorse Opening development comprises 64 dwellings and two hostels. The oil fired boiler plant is located on the ground floor of a three storey block. There are two boilers with a total output of 1.17 MW. A medium pressure system distributes hot water at 93°C (200°F) flow and at 66°C (151°F) return via roof spaces and under linking staircases; elsewhere there are pre-insulated underground mains covered with polyurethane foam and sheathed with PVC. Overall weather compensation temperature control is by means of a three-way automatic mixing valve. In the dwellings, space heating is by a twin-pipe distribution to steel panel radiators or skirting heating each equipped with thermostatic valves for individual temperature control. Domestic hot water is taken off an individual storage cylinder in each dwelling controlled by a temperature limiter.
The Mill Close development serves 22 flats within an old mill building and a 30-bed British Legion hostel; there is provision for a further 16 individual buildings. The boiler plant consisting of two oil fired boilers with a capacity of 0.59 MW is located in buildings that formerly housed a mill. Hot water is supplied at 93°C (200°F) flow temperature; return temperature is at 66°C (151°F). The pipes are installed at the flats in ceiling voids wherever this is practicable. Elsewhere polyurethane pre-insulated underground piping is used. The consumer services are similar to those in the Blackhorse Opening development.

Billingham

The Billingham group heating scheme became operational in 1964 (**15.4**). It supplies heat to an area of 25 hectares. There are 225 dwellings as well as a number of public, commercial and industrial buildings. Hot water is supplied by a central coal fired boiler plant of 14.1 MW capacity.

A major proportion of the consumers in the Billingham scheme are non-residential so that the peak heat demand always occurs early in the morning when the buildings are heated after night shut-down. Practical experience has shown that the heat demand falls linearly as the external temperature rises to 16°C (61°F) and then levels off. It was noted that every 1°C (1.8°F) drop in daily mean temperature below 16°C (61°F) increases the peak heating demand by 0.25 MW and the daily heat output by 21 GJ.

During the 12 months period between April 1972 and March 1973 the total output of the boile house was 109 TJ which corresponds to a mean output rate of 3.5 MW. All the main buildings in the scheme are connected to bulk heat meters and it is therefore possible to determine the heat con-

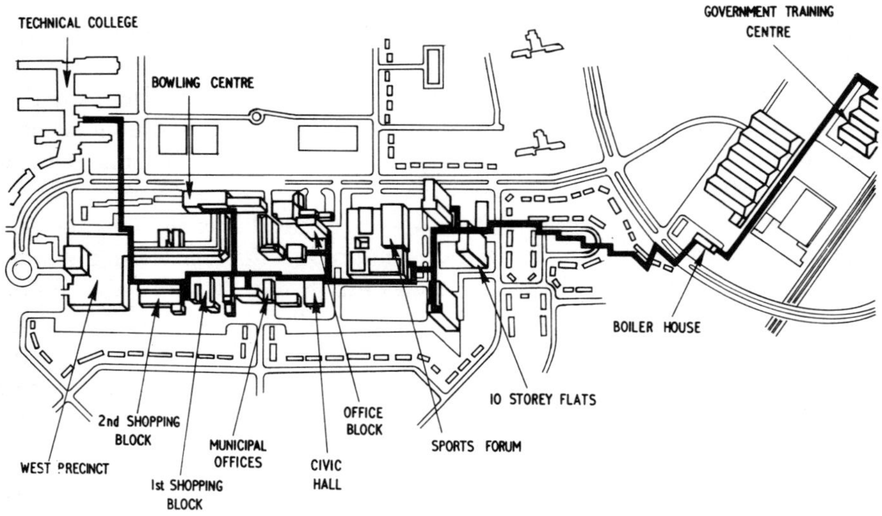

15.4 Billingham district heating scheme (route of mains in heavy line)

sumption and to compare this with the heat supply. The heat losses from this scheme were found to be high amounting to about 28 per cent of the total heat supply. They were caused by inadequately insulated underground mains. The boiler plant operates at an efficiency of 79 per cent: taking the mains heat losses into account the overall efficiency therefore falls to about 59 per cent.

The heat supplied to each flat is measured by an integrating mechanical heat meter, (Aquametro) and hot water meters are also fitted. Major difficulties have been caused by the omission of effective thermostatic controls on the radiators and the inadequacy of the heating mains: most of the latter had to be replaced.

Bonamy Street Southwark (London)

Commissioned in 1965 this scheme supplies 915 dwellings (mainly in four-storey blocks) from a centralised boiler plant capable of dual fuel firing (gas and 35 second oil). The three boilers can supply a total of 8.7 MW at a flow temperature of 90°C (194°F) and at a return temperature of 74°C (165°F). The mains are m.s. pipes with fibreglass insulation. The system covers a site area of 10 hectares.

There has been no direct measurement of the heat losses but it is suspected that these are high in certain sections because many of the heating mains were suspended over walkways underneath blocks of flats and have been severely vandalised (in that much of the thermal insulation has been ripped off). Elsewhere inadequate mains were used (as at Billingham) and it is considered that the heat losses are of the same order (or greater) than at Billingham.

In most European group heating systems outside Great Britain, including those which have been operational for many decades, heat losses seldom exceed 5 per cent of the total heat supply and pipeline maintenance costs are inevitably less than 0.5 per cent of the installed cost. The average heat loss from West German district heating systems (where only short distances are involved) was found to be 12.7 per cent for hot water networks and 11.1 percent for steam networks. The extensive 2600 km Moscow scheme has heat losses of only 5.44 per cent. The heat losses from district heating networks where the heat is produced cheaply from the condenser water of power stations) are nowhere near as costly as those from such group heating schemes as Billingham and Bonamy Street (where the heat is generated expensively in directly fired boilers). Neither of those two schemes has been a financial success and they have discouraged other authorities from considering group heating as a first stage towards district heating despite the fact that the major defects have been identified and can be avoided in well-designed schemes.

Rathenraw Estate, Antrim (Northern Ireland)

This development includes a hostel, shop premises, 465 houses (for 5 to 6 persons), 9 dwellings for the elderly and 157 flats. The complete scheme which mainly consists of traditional terrace type twin storey housing was completed in December 1976.

The boiler plant is located in an independent, centrally positioned services block and incorporates three boilers, two rated at 2.7 MW and one at 1.32 MW. The plant is fired by 35 seconds oil. The scheme is pressurised: variable speed circulating pumps and a constant speed pump meet fluctuations in seasonal loads. A boiler return water treatment plant is installed to prevent corrosion and scaling troubles. The plant can operate unattended, being equipped with a Datafonic remote information and control unit connected to the GPO telephone system.

A pressurised flow and return two-pipe system distributes hot water at a 95°C (203°F) flow and 65°C (149°F) return temperature in winter and at temperatures of 76°C (169°F) and 65°C (149°F) respectively in the summer. The distribution mains are routed through the roof spaces in terraced dwellings and are connected to external underground insulated pipes. The buried mains are pre-insulated in polyurethane hard foam covered by a pressure resistant high density polyethylene sheath. Isolating valves are positioned at the entry to each block.

The distribution to the consumer is by pre-insulated microbore piping connected to steel panel radiators. Domestic hot water is off individual indirect cylinders each with a temperature control valve and a remote sensor. Differential pressure control valves for balancing the system allow direct acting heating system controls to be used. Each individual system has a return water temperature limiter with automatic isolation of the heating system during summer periods when there is a drop in the operating temperature.

Aldershot (Hampshire)

This highly successful total energy group heating system serves the military camp of Aldershot (**15.5**). It comprises the supply of 8.5 MW of electricity, 27.8 MW of group heating capacity as hot water and 11.1 MW of steam at a pressure of 5 bars. The heating water is sent out at temperatures varying between 93°C (200°F) and 138°C (280°F) depending on ambient temperatures; return is at 71°C (160°F) (the same temperature as the condensate from the steam systems).

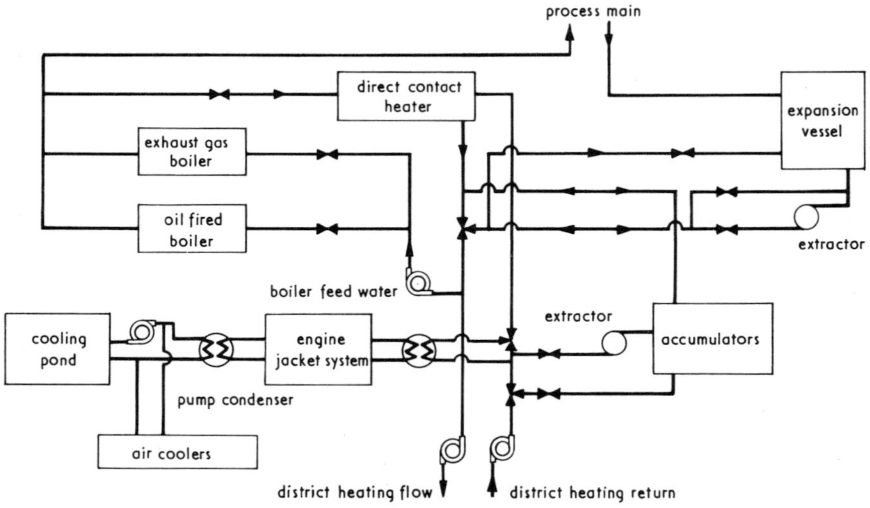

15.5 General layout of Aldershot district heating system

Much of the steam is supplied by four oil fired boilers. In addition use is made of the waste heat from the diesel engine prime power generators. The station is located about 800 metres from the main Stanhope Lines barracks and contains four Mirrlees diesel engines each developing 2.36 MW and four Richardson-Westgarth boilers each with a maximum hourly evaporation rate of 8.4 tonnes of steam at 4.8 bars.

The group heating system is supplied by a direct contact storage type steam to water heater at a maximum flow rate of 403 tonnes per hour. Steam enters through two control valves regulated by pressure recorder controllers which can adjust its pressure to within a range of 2.7 and 3.4 bars gauge.

The return water from the network at 71°C (160°F) is pumped through some or all the 24 spray nozzles and heated to the saturation temperature of the steam. Apart from heating the water this procedure also serves to expel dissolved gases thereby reducing the oxygen content to less than 10 ppm. The water then flows either through two direct acting pumps or through a parallel non-return valve. The pressure differential between the flow and return mains is governed by the configuration of the group heating load. But if appropriate adjustments are made to the pump speeds or discharge valve settings of the flow and return pumps, the pressure in both mains can be raised or lowered simultaneously.

The system incorporates four hot water accumulators, each with a capacity of 45.5 m^3 for the satisfaction of short term heating demands without the need to bring additional boilers on load. The temperature of the water in the accumulators is 143°C (270°F). The accumulators are connected in parallel to the direct contact heaters. To direct heat into the system some of the return flow is passed to the bottom sections of the accumulators causing hot water from the top to flow into the supply mains. Recharging of the accumulators is carried out by reversing this procedure. An expansion vessel accommodates the volume changes caused by the variations in temperature.

To safeguard the system a master relay trips under emergency conditions if the return main pressure falls below the minimum working setting or when the level in the direct contact heater falls outside the limits of the level control. The direct contact heater is then isolated to prevent it from being drained. If the pressure in the heater falls below the minimum setting a valve operates to reduce the steam demand by diverting water from the return main to the accumulators. In the event of the level in the expansion chamber falling below its minimum setting, the extraction pump is shut down automatically.

Use is made of the heat in the exhaust gas and of the jacket cooling heat from the diesel engines, which supply the electric power output to the system. There is one exhaust gas boiler for each engine which also serves as a partial engine silencer. These boilers discharge steam into the main boiler range and are designed for a normal working pressure of 4.8 bars (gauge) at 132°C (270°F). Their output varies between 490 kg/h when the diesel engines are running at half load and 1480 kg/h on full load. Heat is recovered from the cooling jackets of the engines in that the cooling water is injected under pressure. Steam is separated off under a pressure of about one bar gauge at a temperature of 121°C (250°F). On full load the temperature of the water in the cooling jackets of the engine is 127°C (261°F). The steam from the separators passes into a tubular condenser which is supplied with the return water from the return main at 71°C (160°F). The condensate returns by gravity flow via a collector to the cooling systems of the individual engines. The coolant flow in the condenser is regulated automatically to maintain a constant steam head. A dump condenser comes into operation to condense surplus steam if the main condenser cannot fully cope (ie when the return temperature from the system is too high or the flow rate is too low).

General observations on United Kingdom schemes

In view of space limitations only a few of the several hundred operational group heating schemes in the UK have been described. Those discussed have been chosen largely because published information is available about them. Little has however been published for most of the other systems in the UK. According to the 1976 District Heating Association Handbook 187 group heating schemes were completed between 1968 and 1976 (though this is probably an incomplete list). Most schemes have a connected capacity of around one megawatt and seem to be evenly divided as between oil, coal and gas.

The largest scheme in London is that at Thamesmead, Greenwich which will eventually supply heat to 5200 dwellings. When completed its connected capacity of 54 MW will be supplied by gas fired boilers. The next largest schemes are the 19 MW oil fired Ferrier scheme at Greenwich (serving 1908 dwellings) and the 13.34 MW Pepys scheme at Greenwich (serving 1334 dwellings).

Most of the more recent schemes have been reasonably successful and have not suffered the pipeline troubles commonly experienced in schemes started before 1965. This is largely because of the introduction of the modern water-resistant pipe-in-a-pipe mains systems.

GROUP HEATING: USA AND CANADA

Rochdale Village (New York)

Operated by the Brooklyn Union Gas Company this scheme comprises twenty 14-storey apartment blocks with 5840 dwelling units and two shopping areas with a total floor area of 28 000 m^2. Gas fired high pressure boilers supply steam at 44.5 bars gauge and 343°C (650°F) to two Westinghouse turbo-generators each capable of developing 7.5 MW of electricity. Two diesel gas engines are also included each with a capacity of 3.75 MW so that the system can develop a maximum of 22.5 MW which is sufficient for even peak powerload requirements (**15.6**). This scheme is not connected to the electric grid system.

The back pressure steam from the system combined with waste heat from the exhaust and cooling systems of the gas engines provides space heating and operates the Trane absorption chillers during the summer. The steam serves throughout the year for the provision of domestic hot water via heat exchangers. The exhaust steam is normally at a pressure of about 1.4 bars gauge.

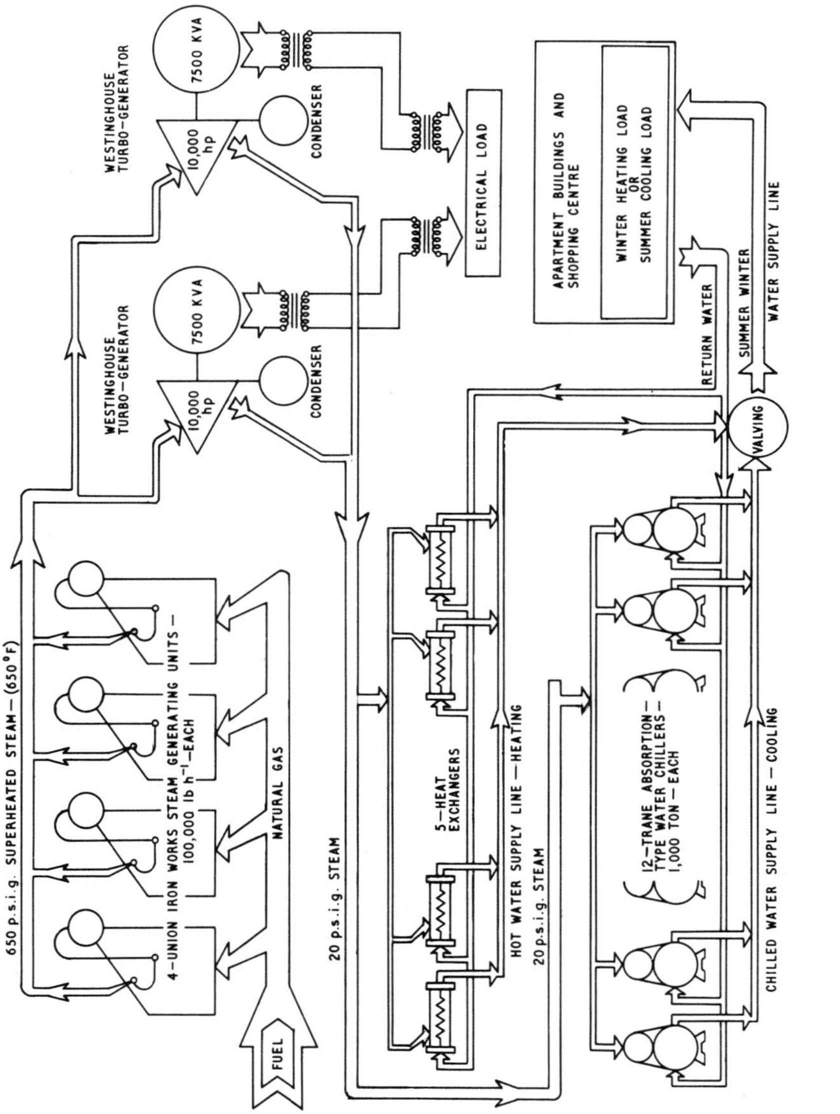

15.6 Brooklyn Union Rochdale Village district heating/cooling scheme (USA)

The cooling requirements amount to a maximum of 42 MW. All piping and the 25 000 fan coil units which are installed within the various apartment blocks are arranged for the twin duties of cooling in summer and heating in winter.

In addition to the gas engines and turbines, the system includes:

Four Iron Works packaged boilers capable of delivering 45.4 tonnes of high pressure steam per hour (one of these is on permanent standby)

Twenty oil tanks each with a capacity of 75 m³ to store grade 6 oil

Two Roots Connersville gas meters with a capacity of 9 500 m^3
Twelve Trane 3.5 MW absorption refrigeration units
Two Foster-Wheeler cooling towers each measuring 11.5 × 14.6 × 22 m with a capacity of 91 m^3 of water per minute.

Annual charges including interest, taxes and insurance were evaluated at 7.1 per cent and the system was amortized over a period of 30 years. Depending upon the season of the year the efficiency of utilisation of the incoming fuel is between 60 and 70 per cent.

To evaluate the viability of the system the total energy (CHP) plant was compared with an alternative system producing 43.6 GWh of electricity purchased by the operating authority: in the latter case no costs were involved for generators or their maintenance. The comparison indicated that the CHP system provided an annual financial saving of over 22 per cent.

Park Plaza shopping center Little Rock (Arkansas)

Park Plaza is on a 10 hectare site and is a two-storey structure with a total floor space of 22 000 m^2. The peak power demand for the area is 1.15 MW The scheme requires 5 tonnes of steam/hour for driving the air conditioning chillers in summer and 3.2 tonnes of steam/hour to provide heating in winter. The equipment installed for this system consists of one Ruston-Hornsby 800 kW gas turbine and two Waukesha reciprocating natural gas engines each with an approximate capacity of 400 kW. The performance of the system when on full load is:

Electricity	14.5%
Steam	47.25%
Heat losses	38.25%
	100.00%

The overall efficiency of the system is therefore 61.75 per cent. Most of the steam for the absorption chillers comes from the exhaust heat recovery boilers which utilise the waste heat from the gas engines and gas turbines. These exhaust boilers also act as silencers. The steam is at a pressure of about 1 bar gauge and it drives two 900 kW Carrier absorption refrigeration units during the summer. A standby boiler (capacity of 4.8 tonnes of steam/hour) is provided.

The fuel consumption was costed at only $10.6 per MWh of electricity provided yielding the associated steam and hot water at nil direct cost. In contrast the then comparative price for bought-in electricity was $18.6 per MWh without the benefit of waste heat. The cost of the CHP plant was $222.30 per installed kW. Since the plant is usually on load for a minimum period of 8000 hours/year it has proved to be eminently viable financially.

Dartmouth College

This small American University is located on the border between Vermont and New Hampshire. Steam and power is currently distributed to 90 buildings on the campus by four boilers, three of which are capable of producing saturated steam at 15 bars gauge each while one produces steam at 31 bars and 343°C (650°F). There are three turbines. Steam at 1.5 bars gauge is taken off the back pressure plants for heating purposes and some steam is passed at 4 bars gauge to the kitchens and to the campus hotel. All distribution pipes are located underground using a base concrete slab with a semi-annular precast concrete enclosure for the welded pre-insulated pipelines. The electric power of the College operates in parallel with the public utility supply (**15.7**).

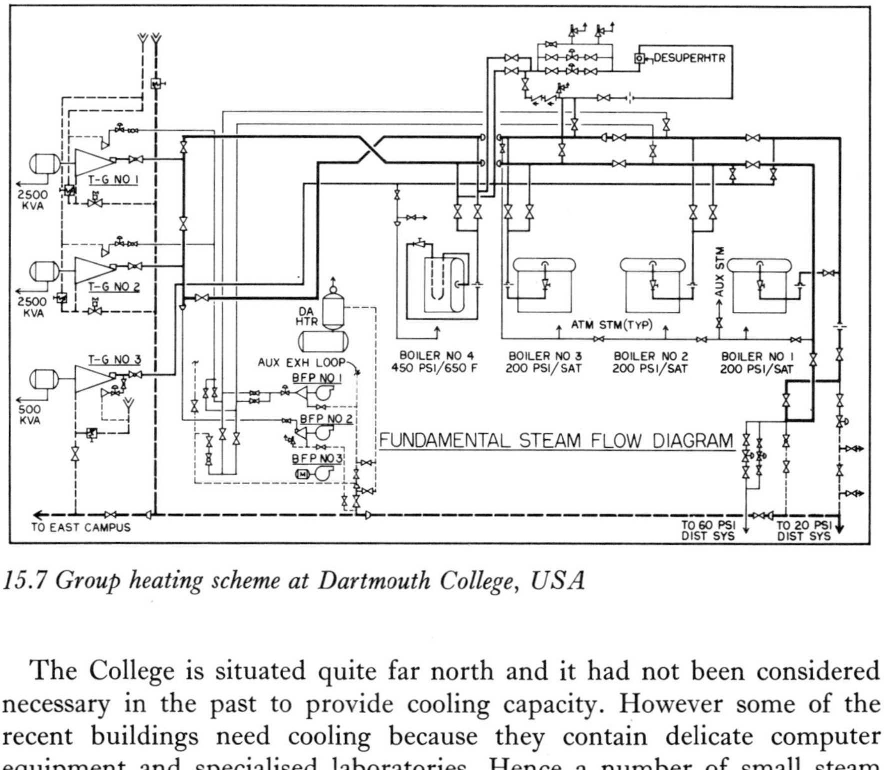

15.7 Group heating scheme at Dartmouth College, USA

The College is situated quite far north and it had not been considered necessary in the past to provide cooling capacity. However some of the recent buildings need cooling because they contain delicate computer equipment and specialised laboratories. Hence a number of small steam absorption units have been provided with an average capacity of 1.08 MW each fed by the back pressure steam.

The steam demand of the group heating network is about 900 tonnes/day of low pressure steam during the winter and 450 tonnes/day during the summer. The College is able to balance the power production in its own station with the amount of steam it requires. When the steam demand is low it purchases extra power from the local Utility Company. During the winter steam demand inside the campus is high and the College generates most of the electricity it needs. During the summer there is a low demand for steam inside the campus and the College purchases the external electricity at favourable summer rates.

Only one of the boilers (which can produce a steam output of 32 tonnes/hour at 31 bars) can generate electric power at a reasonable efficiency. The other boilers are generally only used for peak heat demand and standby. An increasing amount of electric power has had to be purchased from outside and some of the older underground mains have suffered from corrosion.

University of Guelph (Canada)

This University occupies a campus area of 445 hectares some 80 km to the west of Toronto, Canada. It has more than 9000 students and plans to accommodate 15 000 students by 1980. The central utility system was started in 1966 at a cost in excess of $6 million. Currently the boiler plant in the central utility building consists of the following:

One 23 tonnes per hour oil fired boiler
One 32 tonnes per hour oil or gas fired boiler
One 46 tonnes per hour oil or gas fired boiler
One 70 tonnes per hour oil or gas fired boiler.

Steam is distributed at a pressure of 11.4 bars gauge from the central utility station via steam mains in underground tunnels to the various buildings of the campus. These tunnels also accommodate the condensate returns, chilled water supply and return pipes, compressed air, deionised water, electrical distribution cables and signal circuits.

The central utilities plant includes 12.6 MW of chillers: these are driven by electricity supplied from the Guelph Hydro Utility System which provides all the power for the University via three 10 MW feeder cables.

Students' hostel, Ohio State University

This natural gas fired total energy system was described by J. T. Coan and B. I. Routh in the July 1968 issue of *Air Conditioning Heating and Ventilating* (**15.8**). The power plant for each of the 24-storey buildings consists of four 450 kW/4160 volt generators, each driven by a natural gas reciprocating engine. Heat is recovered as 0.83 bar (gauge) steam from the engine jackets and exhaust systems using ebullient cooling. The steam recovery system is supplemented by four 74.5 kW steam boilers located on the 24th floor which operate at the same pressure as the engines. The heat released from the lubricating oil is also recovered by means of shell and tube heat exchangers and is used for the domestic hot water supply. When heat recovery can be fully utilised a system efficiency of about 70 per cent can be achieved: close to that of a gas fired boiler producing heat alone and unlike the CHP system which provides both heat and high grade electric power. Nearly all the recoverable heat can usually be utilised, except during periods of short fluctuating loads which allow overshooting due to lag in the system.

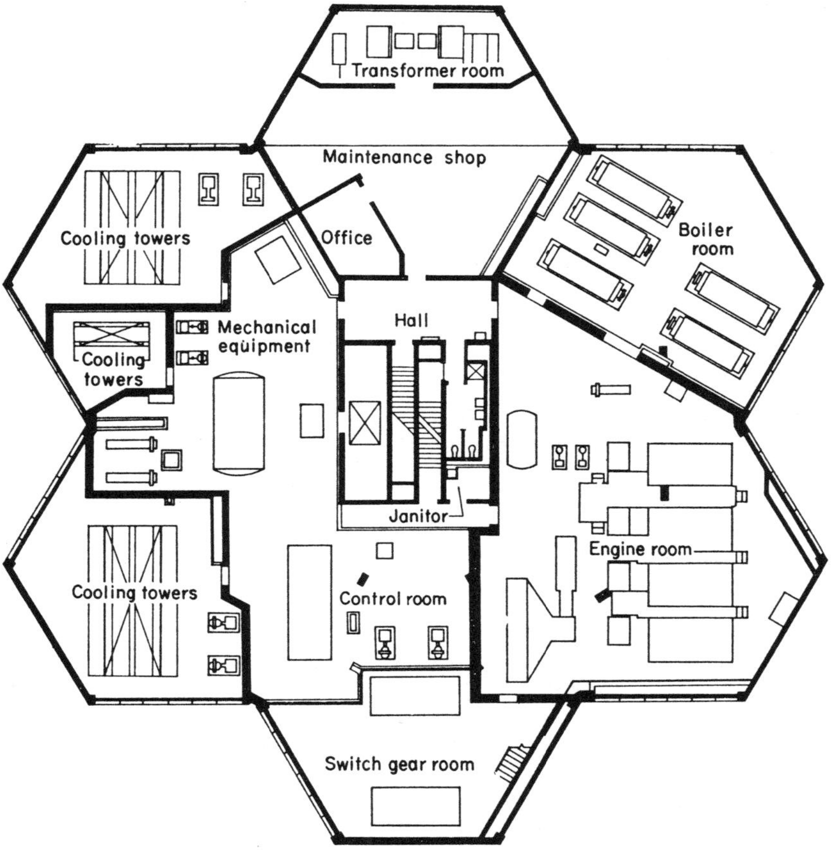

15.8 Plan view of the 24th storey of the Ohio State University Hostel showing location of CHP system

The heating/cooling system is of the terminal reheat type providing temperature and humidity control at practically no additional operating cost. Chilled water is supplied from a 1.6 MW engine driven centrifugal chiller and a 1.6 MW absorption chiller. Heat is recovered from the centrifugal chiller motor in the same way as from the generator engines. The two chillers are piped in series allowing the centrifugal unit to lead and the absorption unit to trim and supplement the load with chilled water at a constant temperature of 7°C (45°F) whenever the external temperature rises above 13°C (55°F).

In the domestic water heating system the lubricating oil heat recovery exchanger receives the lubricating oil cooling water at 71°C (160°F) and transfers this heat to the domestic water: any surplus heat passes to a cooling tower. If the domestic water temperature falls below 60°C (140°F) supplementary heating is provided by a steam generator and condenser which produce condensate for make-up water for the boiler feed and steam system and/or by a water/steam converter. These maintain the required storage tank temperature.

Figure 15.9 shows the steam side recovery system which utilises between 30 and 33 per cent of the engine input in the form of steam. The jacket water is heated to approximately 118°C (245°F) and delivered to a flash tank and separator to be flashed to steam. Heat is also recovered in an

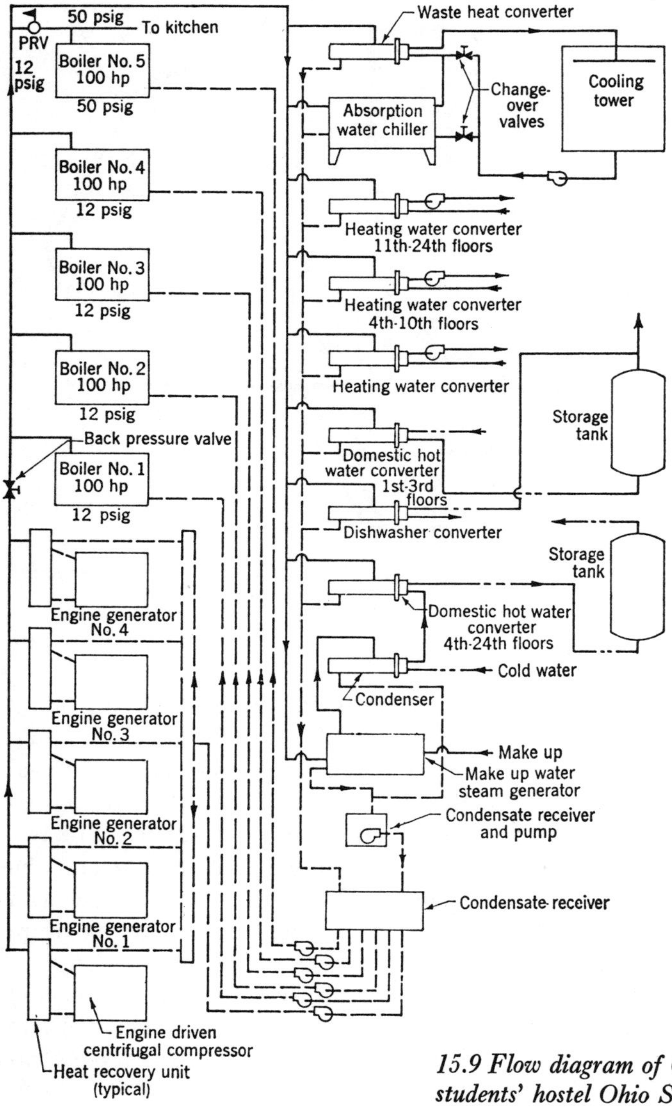

15.9 Flow diagram of CHP system used at students' hostel Ohio State University

exhaust heat recovery boiler in the same jacket as the steam separator. Steam from the engines is delivered to a header and then distributed to the various steam consuming systems.

A detailed cost analysis has indicated a payback period of less than 10 years on the initial investment (and about half that when allowing for inflation).

General observations on North American schemes

Group heating schemes are very popular in the USA for university and industrial precincts, office blocks, schools and apartment complexes. They are inevitably operated by private enterprise companies purchasing gas and oil directly. Intense competition exists between the gas companies and the power utilities which have in recent years streamlined and rationalised electricity production to meet the fierce competition of CHP stations in which heat and power are produced simultaneously.

Gas engines, gas turbines and back pressure steam turbines are the main prime movers used. A feature of US group heating is the importance of providing adequate cooling capacity. This has the enormous advantage of ensuring a very even and high heat take-off all the year round. Heat is usually distributed in the form of steam but high temperature hot water (at 300°C (572°F) and higher) is becoming increasingly popular.

There is ample documentation of American and Canadian group heating and cooling systems with or without the utilisation of the CHP principle. American systems are judged on whether they are individually viable and not (as is usually the case in Europe) on whether they can be first stage developments for large urban district heating systems. The latter do exist in USA but they are not quite as successful as in Europe because of different economic, political and social conditions. The systems described above have been selected because they illustrate the general principles employed in American group heating complexes.

GROUP HEATING IN WESTERN EUROPE

There are numerous group heating schemes all over western Europe but nearly all of them are regarded simply as first stages in the development of comprehensive district heating networks.

DISTRICT HEATING SCHEMES

Nottingham

This scheme is the largest or perhaps the only true district heating scheme in the UK (**15.10** and **15.11**). It has been hampered from its inception by the fact that under current British regulations it is impossible to use CHP turbines for heat generation (which would be the best way of operating a scheme of this kind).

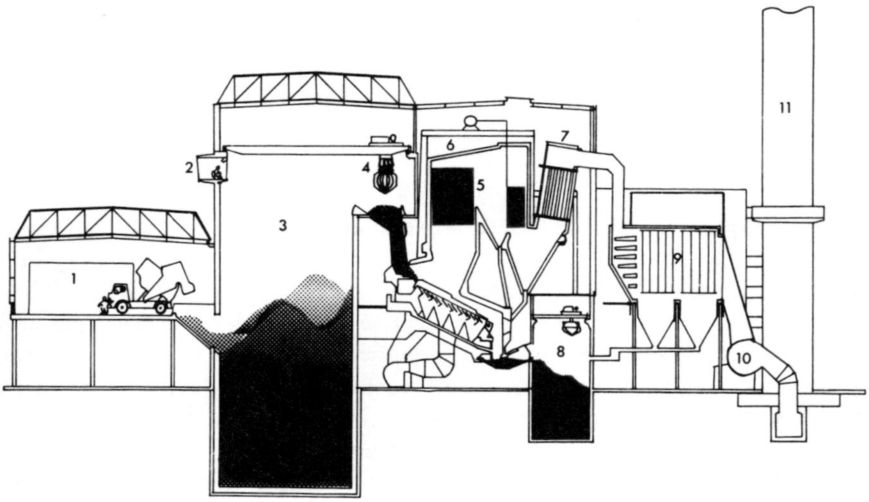

15.10 The Eastcroft incinerator, Nottingham district heating system. **Key**: **1** *Reception hall with truck at tipping bay*, **2** *Crane control cabin*, **3** *Refuse silo*, **4** *Grab crane*, **5** *and* **6** *Boiler with pendant superheaters*, **7** *Multi-tube waste heat boiler*, **8** *Residuals silo*, **9** *Electrostatic precipitator*, **10** *Induced draught fan*, **11** *Chimney*

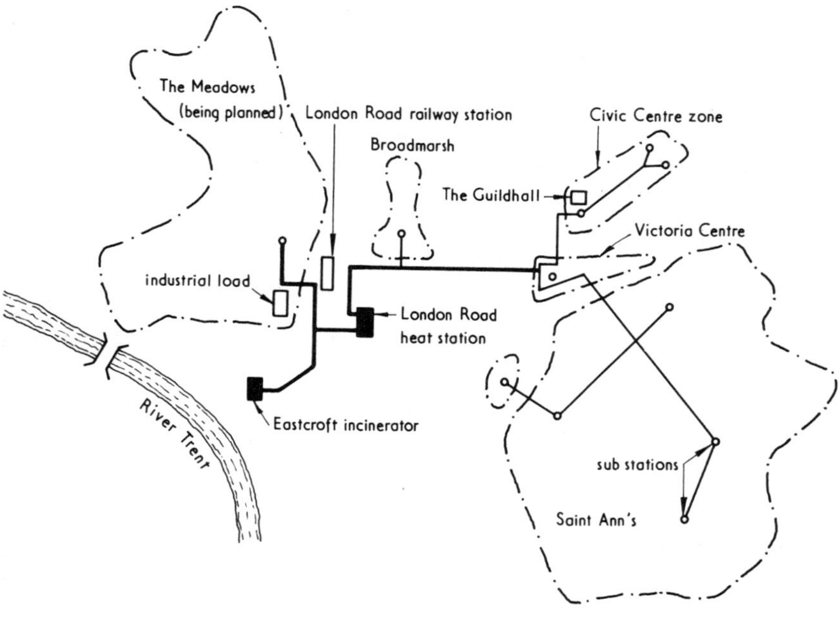

15.11 Layout of Nottingham district heating system

The scheme utilises two separate stations to provide heat and hot water for what will eventually be a total of 17 000 homes. The first of these stations, the Eastcroft refuse incineration plant, has an output of 180 000 tonnes of refuse per annum and is intended to dispose of the refuse generated by a population of 460 000 persons. The second, the London Road coal-fired heat station, used to be a private power station which was purchased by the Nottingham Corporation and then leased to the National Coal Board.

The incinerator station is housed in a concrete and steelwork building and it will eventually be extended to deal with 46 tonnes of refuse per hour. The silo compartment of the incinerator has a capacity of 13 500 m^3 representing 4250 tonnes of refuse. The mechanical stoker used is of the Martin type and has an inclined tumbling action grate. The boilers are essentially of water tube design: the hot combustion gases emerge at 800°C (1472°F) and transfer the heat by radiation to the water walls of the combustion chambers and to the superheaters. The balance of the heat is transferred by convection to the multi-tube waste heat boilers which are fitted at the rear of the plant. Each boiler arrangement produces steam at 27 bars and 371°C (700°F) at the rate of 7.6 kg/s. In the absence of refuse (through mechanical/electrical failure or industrial action) the incinerator can be fired by coal middlings. The flue gases are cleaned by Lodge-Cottrell electrostatic precipitators prior to exhaust to atmosphere via a 91.5 m high chimney.

The coal fired steam raising plant is sited at the London Road station and consists of three Babcock and Wilcox cross tube marine water tube boilers fired with smalls from the Gedling colliery (**15.12**). Each of the boilers is rated at a maximum of 5.7 kg/s of steam at 24 bars and 344°C (651°F). There is space to accommodate two additional boilers each with a capacity of 10 kg/s. Two back pressure steam sets deliver about 2.5 MW and supply all the electric power requirements of the plant. (This kind of operation is permitted in the UK: CHP turbines which sell electricity to external consumers are not.)

The ground floor of the main pump house serving both stations is situated at the London Road heat station and accommodates the main circulating pumps. Two direct contact steam heaters which generate primary hot water are overhead on the first floor of the same building. A nitrogen pressurisation vessel is provided which will cushion the pressure variations in the primary mains if the electricity supply to the circulating pumps fails. There are four air cooled dump condensers each fitted with a two-speed control: thus when the waste heat from the Eastcroft incinerator exceeds requirements the waste heat can be dumped to atmosphere at no less than eight different rates to balance heat demand instantaneously with heat supply (a vital feature of proper control). It is impracticable to adjust a coal or refuse fired plant quickly so as to cope with rapid variations in demand.

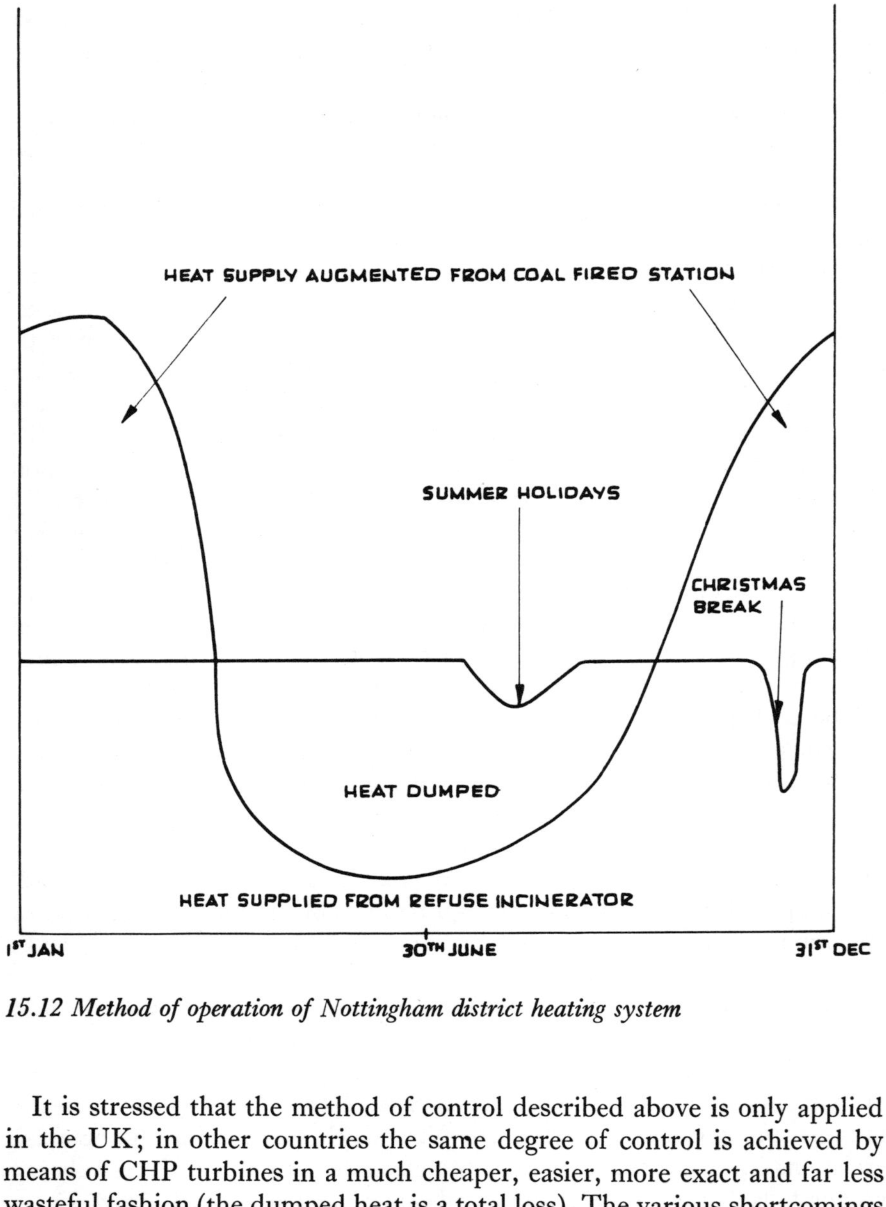

15.12 Method of operation of Nottingham district heating system

It is stressed that the method of control described above is only applied in the UK; in other countries the same degree of control is achieved by means of CHP turbines in a much cheaper, easier, more exact and far less wasteful fashion (the dumped heat is a total loss). The various shortcomings of the Nottingham plant have all been due to this inefficient control system.

The northern part of the initial development of the Nottingham district heating network is fed by 405 mm high pressure primary hot water mains. These are routed west along a canal and then turn north to the new Victoria Centre on the site of the old Great Central Railway station. On the way they feed a commercial load of 10.5 MW. The system serves four separate sections: the Victoria Centre, the St Ann's Redevelopment, the Civic Centre and the Meadows Development. The Victoria centre is a large enclosed shopping precinct with heated pedestrian malls and high rise flats and the connected load is 22.5 MW. The St Ann's Redevelopment at present consumes 42 MW which is to be expanded to about 70 MW. The Civic Centre has a loading of about 36 MW. The Meadows development area contains 2900 dwellings and utilises about 30 MW of heat.

There are usually severe problems in laying heating mains through urban areas where there are already numerous water, power, gas and sewage mains. In Nottingham the problem was eased by laying many mains along canals and railways.

In the northern section of the system, hot water was selected as the heating medium. It is distributed at a temperature of 140°C (284°F) pressurised to 6.9 bars and advantage is taken of the circulation pressure of the system to provide a margin against the risk of the water flashing into steam. At each substation the secondary mains water is heated by mixing the primary flow with some of the secondary return water to achieve the required temperature. The pressure is adjusted by means of control valves. The primary mains operate on a constant volume basis to maintain a stable pressure and the thermal inertia effect of the large volume of circulating water stabilises the temperature.

The southern section of the system has a potential industrial load: it was therefore decided to supply the heat as steam which is available at low or high pressure. Various types of pipelines are used to suit the ground conditions: in dry areas concrete conduits are used with the pipelines embedded in calcium silicate insulation (foamed concrete); in wet areas the pipeline systems are pipe-in-pipe with an all-welded outer steel casing protected against corrosion by a heavy duty wrapping of layers of fibreglass tissue impregnated with coal tar base enamel. For the secondary system where temperatures and pressures are lower the newer and cheaper techniques of pipeline laying are adopted, such as polyurethane insulation materials contained in PVC or GRP jackets.

The system is controlled centrally by a Bailey supervisory system which can receive and analyse up to 1000 transmissions in a three minute period. Analogue or status information can be given on all relevant parameters. The heat consumption of large point loads (supplied from their own sub-stations or plant rooms) is metered by mechanical integrating type meters read either monthly or quarterly. Domestic tenants pay on a fortnightly basis. Clorius evaporative meters are read annually and the tenants then receive either a rebate or an adjusting account.

The Nottingham district heating system is being expanded: it is anticipated that the connected capacity should be 132 MW by 1979 when the Eastcroft incinerator plant is scheduled to incinerate refuse at the rate of 200 000 tonnes per annum.

Vasteras (Sweden)

Vasteras is a typical Swedish industrial town which boasts two important companies and several back-up firms providing work for 120 000 inhabitants. The municipal district heating system was commenced in 1954 and during the first stage of expansion purchased the heat from the Swedish State Power Board's thermal power station in Vasteras and produced hot water in temporary boiler stations within the town. All types of new buildings in Vasteras were linked up to the various group heating schemes of the city. Gradually all the older buildings were also connected into the group heating complexes. Blocks 1 and 2 of the big Vasteras CHP station were commissioned in 1963 and provided 80 MW of electric power and 200 MW of back pressure heat. Blocks 3 and 4 yielding between 440 MW electrical power output plus 730 MW of heat output (at maximum winter load) and 500 MW electrical output (with zero heat output) were commissioned subsequently and were completed in 1973. The cheap back pressure power (economical in its use of fuel) from Vasteras power station supplements hydroelectric and nuclear power and all the electric energy consumed by the town of Vasteras is back pressure power from the thermal power station.

The economic advantages of a well developed district heating system such as that at Vasteras are saving in cost to the consumer in the form of cheaper heating and electric power (approximately 50 Swedish Kröne per inhabitant each year) and high profit to the operator (Vasteras makes an annual profit of around 6 million Kröne after all operating and capital costs have been met).

The Vasteras system provides its remarkably efficient production of heat and power in the following way:

Excellent thermal efficiency: combined power and heat production shows overall efficiencies of up to 80 per cent or even higher

Efficient use of labour: the entire system only employs 140 people

Considerable economy: the plant utilises low grade fuel which is purchased in bulk thus reducing transport costs.

The supply system has branches extending throughout the whole of Vasteras and consists mainly of conduits with insulated pipelines inside in a two ring main layout; at the end of 1973, the total length of the pipeline network was 292 km. Steel pipelines are used with bellows type stainless steel compensators. Pipe diameters up to 200 mm are insulated with glass wool or rock wool and are incorporated in asbestos-cement protective mantles. Pipes of larger dimensions than 200 mm are located inside concrete conduits with cellular concrete insulation. The network is maintained by some 20 operatives.

The water is normally delivered to the consumer at 85°C (185°F) but on cold days the temperature can be raised to a maximum of 120°C (248°F). The temperature of the water in the return main is 55 to 65°C (131°F to 149°F).

Malmo (Sweden)

The Rorsjo plant was the first CHP turbine plant used to supply heat to a district heating system in Sweden. It was completed in 1953 (**15.13**). A number of directly fired heating centrales were also built and the original CHP station was further expanded. By 1973 the system consisted of eight different plants servicing about 75 per cent of all households. Malmo is Sweden's third largest city; it has 258 800 inhabitants and covers an area of 152 km^2. In 1972 the technical division of the city's energy department had a connected district heating capacity of 915.1 MW, a pipeline network of 148 km and a total volume of water inside the network of 21 300 m^3, made up by storage containers to 36 000 m^3.

The system uses a total of 300 000 tonnes of oil each year and the biggest of the eight stations which contribute to the district heating system is the new and purpose built Heleneholm CHP plant with a gross electric power rating of 138 MW and a thermal power rating of 285 MW. The plant has no access to cooling water and the cooling which is required at high electrical and low thermal loads is by means of air cooled recoolers. If heat demand is low and the incoming district heating water has been overheated in the turbines it is cooled in the recoolers and then recirculated for renewed heating in the turbines.

Air cooled batteries also meet the cooling requirement of the alternators. The plant has four hot water storage tanks with a total capacity of 10 000 m^3 which corresponds to a storage of about 500 MW of thermal energy. CHP operation and storage of hot water combine to provide an even balance of operation which cannot be achieved with traditional condensing turbines.

The CHP stations of the system can supply a maximum of 150 MW of electricity to the city electric network. Additional electricity requirements are met from the grid system. The number of annual degree days in Malmo is 3020 on a Centigrade basis using a base temperature of 17°C (63°F): dividing the amount of heat delivered by the connection value, a peak utilisation of about 2000 hours per year is obtained. Dimensioning of the system is based on a minimum effective external temperature of −16°C (2°F).

Apart from the two CHP stations the network contains six smaller plants in which oil is burned at thermal efficiencies of 80 to 90 per cent to provide hot water for peak requirements. The CHP stations have varying efficiencies depending on their manner of use. Under full back pressure operation the total thermal efficiency of the Heleneholm and the older Rorsjo plant (thermal and electrical energy) is of the order of 86 per cent. When it is necessary to operate on a pure condensing cycle for peak electricity and zero heat production the thermal efficiency (electricity only) falls to 33.4 per cent.

Berlin

The Berlin system of district heating is operated by the Berliner Kraft und

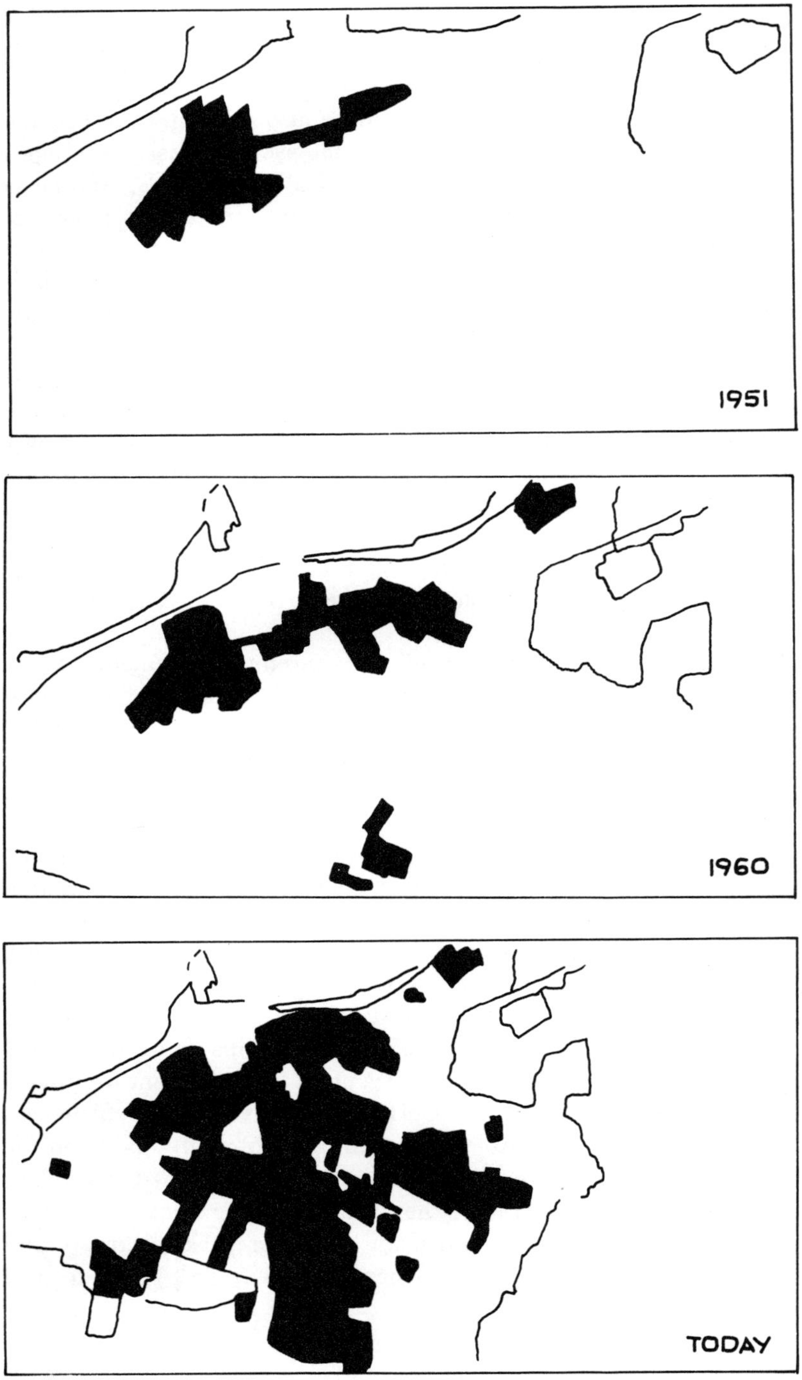

15.13 Development of district heating system at Malmo, Sweden

Licht AG (BEWAG) who have the concession of providing district heating and electricity to the Berlin area. The city heating system was begun in 1926 when the Charlottenburg power station was modified to supplying a district heating network with steam at about 2 bars pressure. A few years afterwards the area around the Steglitz power station was supplied with a hot water district heating network with 90°C (194°F) supply lines and 70°C (158°F) return lines. There were then many steam heated public buildings in Charlottenburg while the area around Steglitz consisted almost entirely of private dwellings which were heated with hot water. Steglitz was one of the first operational CHP stations. The systems were so well designed and constructed that even today after more than 50 years and one intense war most of the original pipelines are still in operation.

Subsequently, the Berlin system was greatly expanded, partly by amalgamation with other local systems and partly by the provision of CHP stations to serve additional areas. By 1974 the total building floor area heated by the BEWAG system was more than 7 million square metres and the investment in the plant amounted to 250 million Deutschmarks. The network is being steadily expanded: the annual rate of increase has remained over many years at between 7 and 8 per cent.

Traditionally the BEWAG company has always opted for an expensive and high quality district heating distribution system and this policy has been vindicated by the very long life which has been achieved and by the very low maintenance costs. The Berlin system uses almost exclusively a three-pipe distribution system with twin supply lines. The water is conveyed at a velocity of 3 m/s in the supply mains; there is a single common return line with a water velocity of 2 m/s. In winter both flow lines are in use, in summer only the smaller bore flow main.

The cost of the three-pipeline system was estimated in 1974 at between 4 and 4.5 million DM/km apportioned as follows:

Pipelines	34.1%
Surface works	26.4%
Concrete works	19.3%
Excavation	15.1%
Miscellaneous	5.1%
Total	100.0%

Since BEWAG is operated by the Berlin city authorities, it is possible for the layout of the pipelines to be arranged in conjunction with other utility services. Most of them are laid into in-situ cast concrete conduits rather than in pre-fabricated ones. Where there is adequate space, simple pipe expansion loops are installed; bellows expansion joints are used where space is limited.

At the end of 1974 the Berlin district heating system had an installed capacity of 1453.47 megawatts. The supply temperature is 110°C (230°F) in winter and 90°C (194°F) in summer. The common return is at 40 to 60°C (104 to 140°F). The heat is supplied by CHP turbines at seven power stations (Steglitz, Charlottenburg, Wilmersdorf, Moabit, Reuter, Lichterfelde and Rudow). Some 80 per cent of the fuel currently used is coal; it is anticipated that this may eventually reduce to 65 per cent. The cost of heat production at the CHP stations in 1979 was calculated at between DM 1.2 and 1.9 per GJ within the range of 2000 and 2500 hours of use per annum. It is estimated that CHP operation saves about 250 000 tonnes of coal equivalent in Berlin relative to individual group heating.

At Charlottenburg and Wilmersdorf gas turbines are now installed to meet the peak electricity and heating demands and offer the following advantages:

They can be started very quickly to meet instantaneous peak loads

They use low sulphur fuels (less than 0.5 per cent) and only need very low chimneys (advantageous in a densely developed urban area)

They require very little land area and can consequently be located close to the consumers.

The gas turbines used in conjunction with the Charlottenburg power station have an inlet temperature of 880°C (1616°F). The outlet temperature is 443°C (830°F) and the mass of gases blown off is 364 kg/s. With a design external temperature of 3°C (37.5°F) it is possible to extract 81.4 MW of heat from the flue gas heat exchanger. Two gas turbine heat exchangers enable the saleable heat from the Charlottenburg station to be raised from 427 MW to 590 MW: a third plant serves as standby. Heating is required for 4500 hours per annum (the average period during which the external temperature falls below 12°C (53°F)). The gas turbines are used as soon as the external temperature falls below 3°C (37.5°F) or whenever there is a peak demand for electric power.

Gas turbines are uneconomic in use unless the waste heat can be utilised. It therefore pays to use the waste heat from the gas turbines and reduce the heat loading on the CHP turbines. It is intended to utilise the waste heat from the gas turbines at Wilmersdorf for district heating purposes. In this case the inlet temperature is 1004°C (1839°F) and the outlet temperature 510°C (950°F). An additional 209 MW of heat can be drawn from this system. It is also planned to use these gas turbines for peak supply of both electricity and heat. The gas turbine plants at Charlottenburg cost 455 DM/kW (some 3 per cent of which was used to provide acoustic insulation, 2 per cent to prevent air pollution and 15 per cent to purchase specialised district heating equipment). By 1977 32 per cent of the installed electric capacity and 20 per cent of the installed heat capacity of the Berlin system was met by gas turbines.

Figure 15.14 shows another important West European district heating scheme.

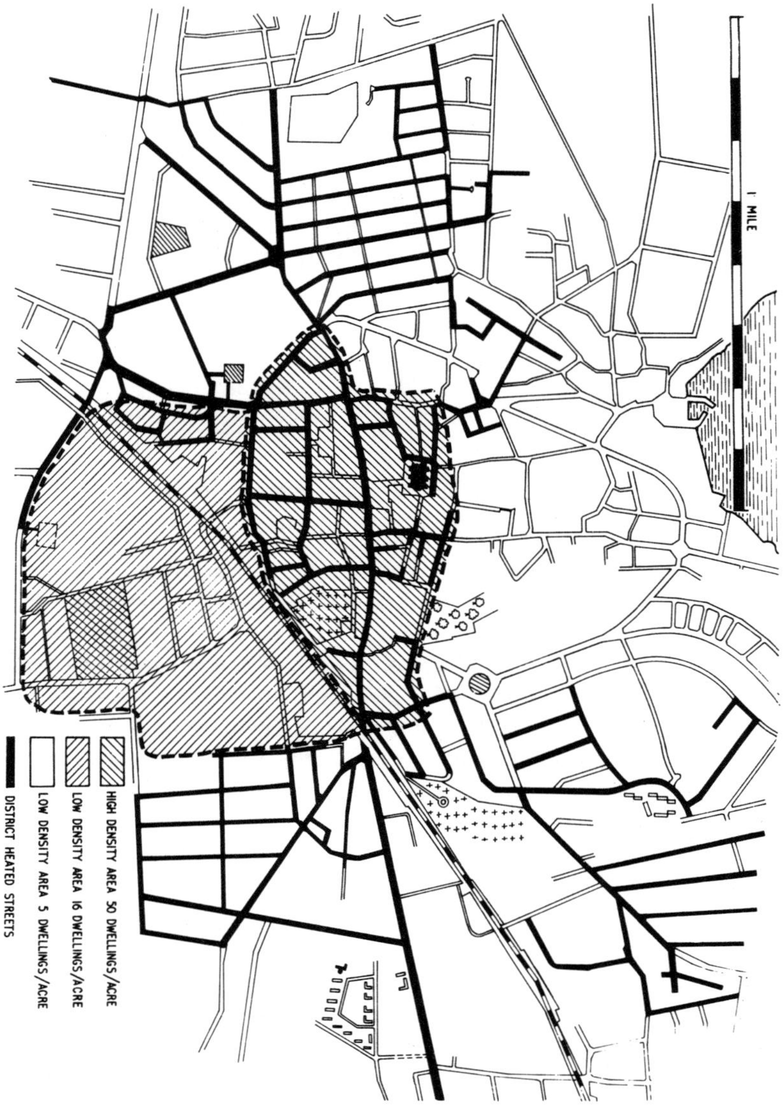

15.14 District heating system, Roskilde, Denmark

New York: Consolidated Edison system

This system is by far the largest in the USA: during 1975 it supplied some 23 million tonnes of steam to New York for space heating hot water supply, manufacturing processes and air conditioning (**15.15**). The steam is raised in back pressure turbine plants and in steam boilers from which it passes into a distribution grid of underground mains located under the city streets. The average supply pressure of the steam is at 8.5 bars gauge; there

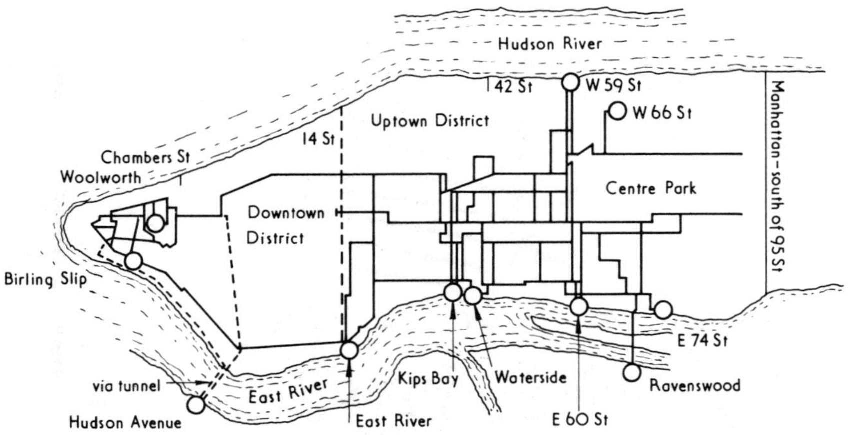

15.15 New York district heating system – main supply stations

are no condensate returns. The condensate is removed from the mains, cooled and discharged into the city sewers. The steam is produced in three types of stations:

Five electric generating stations are used to carry the general base load of the system: high pressure steam is fed into back pressure turbine sets and after passing through de-superheating and pressure regulation stations the exhaust steam is fed into the distribution stations. The system produces a maximum of 5784 tonnes of steam per hour though this figure may be slightly reduced if there is heavy electric loading.

Five straight boiler plants owned by the company are operated in winter and yield a total of 2000 tonnes of steam per hour.

Boiler plants leased by the company which are kept on standby for use on very cold days.

The steam mains consist of 190 km of single pipelines with diameters between 100 and 600 mm interconnected to form loop closures operating without pressure reducing stations. Steam is fed into the network at about 13 bars gauge and the pressure is allowed to find its own level between the source and the remote control points in the network. There are three long steam mains each of 600 mm diameter sized to supply 450 tonnes of steam per hour to the downtown system from very far distant stations and designed to operate at 27 bars gauge and at a temperature of 400°C (752°F). The pipes are of prefabricated design but use thicker than usual pipe walls and higher pressure-rated valves and other auxiliaries. The lower pressure mains use the conduit as the anchor for the pressure carrier pipe. The thrust forces developed with the 27 bars gauge pipe are about 100 tonnes so that the conduit method of anchorage is not practical; concrete slab anchors are used.

The steam pipelines are made in the form of prefabricated units. The unit pipe consists of the pressure pipe, thermal insulation, pipe support, collars and the steel conduit. Experience has shown that 12 m is the longest single length of pipeline that can be lowered into a trench. Each of the prefabricated units has a stainless steel bellows type expansion joint with welding ends, thermal insulation and a steel housing with steel end rings as pipe guides. The prefabricated unit inside the housing is inserted into the line and the ends of the joint welded to the steam main. Split insulation is placed over the joints at either end and welded sleeves are slipped into place and joined to the expansion joint housing and to the steel conduit of the adjoining steam main. The expansion joint is a rugged unit because the corrugated part has to be protected from damage during installation. It is prefabricated to the exact installed length and coated with bitumen in its entirety.

Prefabricated manholes are used: each consists of a precast concrete manhole including floor, walls and roof, necessary pipes, valves, fittings, thermal insulation and anchors. For ease of installation these manholes are cast with short lengths of conduit pipe in the end walls to permit the manhole piping to extend beyond the walls. The roof slab is integral with all the necessary openings for entrance heads and valve operating rods. When the manhole arrives on the site it is lifted off the truck, placed on the trench, lined up and welded to the adjoining pipe units. The packaged manhole is employed for all valves greater than 200 mm. None of these valves can be opened against the steam pressure: hence, the assembly includes bypass piping with two blow-offs to permit the relief of pressure in case of shutdown and also to blow off the air from a section of the main which is turned on. Since many of the valve points also function as condensate drain points the necessary valves and piping are included to permit connection to the traps which are installed in a separate manhole. The complete piping assembly can be constructed in an off-site workshop. It is lowered into the manhole and it then usually takes only two welds on to the smaller diameter piping to complete the manhole installation. No welding is required inside the manhole at the site. The packaged manhole design is also suitable at condensate drain locations and to house the street traps.

The New York Edison system has at the present time some 2500 individual customers in huge blocks of flats, commercial buildings and industrial premises. They are connected to the distribution mains by 32 km of service pipe, varying in diameter from 25 to 500 mm.

The steam services which enter consumers' premises are basically similar to the mains. New entries are inevitably of prefabricated design in units of 4 m in length. There are two valves, one at the main and one immediately inside the consumer's premises. Much steam is consumed by air conditioning equipment: the winter load is therefore sometimes smaller than the summer load in air conditioned buildings. When the building wall is at the

property line the service is installed through a sleeve to a point just inside the building. The service is standardised at a pressure of 8.5 bars and this is taken as the basis for all charges. Three types of steam meters are in use:
Condensate meter: this measures the weight of water resulting from the condensation of steam after use by the consumer. (This type is no longer being installed.)
Shuntflo meter: this type of meter is driven by flowing steam. There is an orifice in the line and the differential pressure causes flow through a shunt circuit. The steam passing through the shunt circuit turns a turbine rotor and the speed of the rotor shaft is reduced through suitable gearing to drive a magnetic coupling which actuates the counter gears. The counter is calibrated to measure the total flow. The Shuntflo meter is not accurate for very light loads. When there is a wide range in the load to be metered it is necessary to sequent the meters ie to use a small meter to read the initial load and to continue to register until it has reached close to its capacity, when a regulating valve brings the next meter into service and so on.
Bailey differential pressure meter: the pressure differential on either side of a fixed orifice operates a linkage recording the use on a flow chart and supplies information to an integrator registering flow on a counter. The Bailey meter is sequented to improve accuracy with varying loading.

The following general statistics about the Edison system may also be of interest. In 1974, 88 per cent of the total heat output was produced by the use of oil and 12 per cent by gas. The New York area environmental standards limit the sulphur content of the fuel to 2 per cent for Grade 2 oil and 3 per cent for other grades. Sulphur dioxide stack emission limits are 200 parts per million and particulate stack emission limits are 45 grammes of solid matter per GJ of heat produced. The annual systems load factor during 1974 was 38.16 per cent and the revenue from the sale of steam was $2.6 million. During 1974 coal cost 92 cents per GJ, oil 168 cents per GJ and gas 78.5 cents per GJ. The cost of steam varied between $2.40 and $13.50 per tonne sold with an overall average of $6.50. At a pressure of 8 bars steam has an available latent heat of 2.043 GJ per tonne so that the actual price paid by the consumer varied between 117.5 and 661 cents per GJ actual heat with an average of 318 cents.

Leningrad

The origins of this system date back to the years between 1910 and 1915. By 1931 the total length of the mains was 16 km and a CHP station had been built to provide heat to 94 apartment blocks. The CHP turbine used in those days had a power capacity of a mere 12 MW.

By the end of 1974 the Leningrad City power stations had a total capacity of 1160 MW of which 982 MW were taken from CHP power stations. The network had a total pipeline length of 1000 km with 10 223 connected buildings. The district heating network supplies heat to 2.08 million people comprising 50.2 per cent of all private dwellings and public buildings. The connected district heating load is 6680 MW. 59 million cubic metres of hot water were supplied during 1974. Hot water accumulators of varying capacities between 2000 and 10 000 m^3 are positioned at the various CHP stations (**15.16**). The newer CHP stations are all of far larger capacity than the original ones. The Severanaya station employs 500 MW turbine sets and it is planned to expand this to a total capacity of 1250 MW. The Yuzhnaya power station has a capacity of 1000 MW using two 500 MW CHP sets.

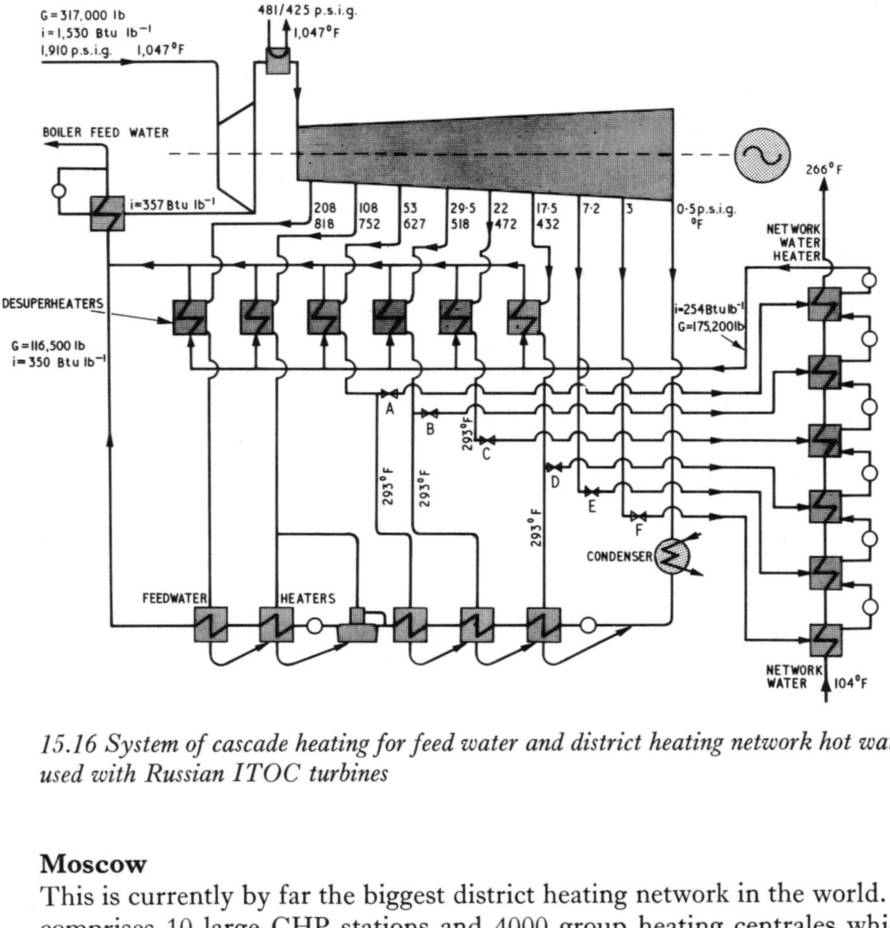

15.16 System of cascade heating for feed water and district heating network hot water used with Russian ITOC turbines

Moscow

This is currently by far the biggest district heating network in the world. It comprises 10 large CHP stations and 4000 group heating centrales which supply a total of no less than 17 422 MW of district heat to the network. 69 per cent of all houses, apartment blocks, public buildings etc in the city are connected to the district heating network. In 1973, 192 million GJ were supplied to the city from the CHP stations and 38 million GJ from heat-only boiler houses. 35 000 large blocks are heated by district heat and there are 2600 km of underground mains in use. The urban system also provides 30 per cent of the industrial heat requirement within the Moscow conurbation.

In the past the Moscow system included a number of condensing power stations. All these have now been converted to operate only on the CHP cycle. On average 271 kWh of electricity are produced with the development of 1 GJ of heat: the specific utilisation of coal in these new stations is only 2.55 grammes per kWh of electricity, far better than the utilisation offered by the old condensing stations which operated on an electricity-only basis.

All the Moscow stations use hot water for transmission. There are about 70 km of steamlines with maximum radii of 3 km. These are usually of 600 mm internal diameter at a working steam pressure of 6 to 7 bars gauge. Some employ condensate returns but the overall percentage condensate return is below 30 per cent. The heat losses in the Moscow district heating system are 5.44 per cent of the heat sent out. This low figure is due to the extremely good standards of pipeline insulation and it is indeed surprisingly small for a network of 2600 km.

Moscow is a fairly hilly city: pumping stations are therefore required at various locations. These are in general unmanned. Power expenditure for pumping purposes averages 4.44 kWh per GJ of heat supplied.

General observations on existing district heating networks

Highly efficient economical and well designed district heating systems can be found in virtually every major town and city in western and eastern Europe and in the Asian part of the Soviet Union. Comprehensive documentation is normally available and the various operating authorities always tend to be helpful in providing data. Almost inevitably the current basis of all these schemes is provided by CHP turbines which are becoming larger and more sophisticated. In many cases various urban areas are supplied with heat by a single joint system with feeder mains between them.

Because of the requirements for air conditioning, district heating systems in the USA are somewhat different in design from those in Europe: fuel has been cheaper in the USA and there is therefore less incentive to utilise as much waste heat as possible. The present changes in the energy supply situation may however press major changes in energy conservation.

CHECK LIST

1 Group heating schemes in UK, particularly the total energy group heating system serving Aldershot Military Camp
2 Group/CHP schemes in USA
3 District heating schemes in western Europe
4 District heating schemes in Soviet Russia.

References

E. A. Clymer and T. M. Loughery, 'The Philadelphia steam story', *District Heating*, 61 (4), 1976, pp 24–31

R. G. Courtney and F. J. Jackman, 'Study of three district heating schemes', *Heating and Ventilating Engineer*, March 1976, pp 6–14

J. Fox, 'Runcorn's district heating scheme', *Architects Journal*, 162, 2nd October 1975, pp 32–37

R. O. Klemstine, 'Chicago Union Station power plant', *District Heating*, 61 (4), 1976, pp 15–16

H. J. Knapp and others, 'Toronto district heating study', *District Heating*, 60 (1), 1974, pp 14–25

C. Mackenzie-Kennedy, 'District heating of a British residential area in Germany', *Heating and Ventilation Engineer*, October 1974, pp 129–132

"The Rochdale district heating scheme', *Heating and Air Conditioning Journal*, November 1975, pp 10–14

'The Rose Row district heating plant', *Heating and Air Conditioning Journal*, October 1974, pp 28–32

F. Tombs, 'Central Glasgow district heating study', Central Electricity Generating Board, March 1976

J. H. C. Vandermolen, 'University of Guelph central utility plant', *District Heating*, 60 (3), 1975, pp 22–36

'Waste heat recovery – Leeds total energy scheme', *Electrical Times*, 7th May 1976, pp 7–8

CHAPTER 16

District cooling

DISTRICT COOLING IN THE WORLD TODAY

A sustained and widespread demand for air conditioning such as that existing in many North American cities favours the adoption of a centralised system supplying chilled (refrigerated) water to the consumers on the district heating principle. The cooling centrale is usually either a high pressure steam driven centrifugal refrigerator, an absorption system, or a combination of both.

There are three major alternative ways of using such systems:

(a) Steam is generated locally and drives the central refrigeration plant

(b) The refrigeration plant is installed within the range of the district heating network and uses steam produced at the district heating centrale and pumped through the district heating steam system

(c) The water is chilled at a central location and pumped through insulated pipelines directly to the consumer.

In the United Kingdom and in other parts of Europe there is a tendency to use system (a) unless the air chilling load is appreciable. As can be seen from table 16.1 air conditioning demand is very heavy in many parts of the USA so that system (b) is more usually preferred there although the transition to system (c) is being made in some cases where the cooling demand is so extensive that it is economically feasible.

The cities on America's east coast such as New York require considerable air conditioning in the summer; cities on the west coast, such as San Francisco and Los Angeles, although much warmer, have a more equable climate and hence a lower air conditioning demand.

District cooling is only economically feasible in very densely concentrated central areas of offices, department stores, cinemas, dance halls etc which demand long periods of air conditioning during the year. It is doubtful whether such conditions would ever apply in the UK but there are many parts of Europe where conditions are favourable for district cooling and where systems are being installed to provide it.

Comprehensive district cooling necessitates the generation of refrigerated water at a central location using steam driven centrifugal or absorption type chillers. The chilled water is then pumped to the consumers at a temperature of roughly 4 to 5°C (39 to 41°F) in well insulated pipelines and returned to the plant at around 12 to 15°C (53 to 59°F).

District cooling is widely used in the USA. One of the biggest systems

TABLE 16.1 NUMBER OF DEGREE DAYS AND HOURS OF AIR CONDITIONING DEMAND IN VARIOUS AMERICAN CITIES

Town	Degree days (18.3°C base)	Hours of air conditioning demand per year
Baltimore (Maryland)	2659	1230
Boston (Massachusetts)	3217	820
Chicago (Illinois)	3505	1060
Cleveland (Ohio)	3337	820
Detroit (Michigan)	3558	740
Houston (Texas)	771	2770
Los Angeles (California)	1119	840
Louisville (Kentucky)	2466	1310
Memphis (Tennessee)	1743	1310
New York City	2772	1010
Philadelphia (Pennsylvania)	2703	1155
Rochester (New York)	3813	710
San Francisco (California)	1901	170
St. Louis (Missouri)	1901	1520

in the world outside Japan is located in Hartford, Connecticut where there is a connected cooling load of 52 MW. Another fifty systems of capacities between 11.6 and 70 MW are already installed or in progress in various North American cities.

The largest district cooling system in the Soviet Union is in Tashkent. It supplies chilled water to the air conditioning systems of a large number of buildings which include 1000 flats, offices, public buildings etc. Numerous smaller district cooling systems exist in Moscow and in other Russian cities.

Japan has been somewhat of a pioneer in the field of district cooling: a system of 130 MW capacity was provided at the 1970 EXPO in Osaka and district cooling is now being offered in conjunction with district heating systems in the majority of new installations. The largest scheme of its kind is in the Marounouchi area of Tokyo where there is a refrigeration capacity of 232 MW combined with a heating capacity of 263 MW (the heating and cooling capacitities being roughly of the same order). At Shinjuku New City Centre (another of Tokyo's new suburbs) the central system supplies 130 MW of cooling and 145 MW of heating.

Up to 1974 thirty district cooling systems were in operation in Japan and a large number planned. It is currently estimated that there will be about ninety different district cooling systems in operation by 1980.

In the UK district cooling is employed at the Heathrow international airport complex and also at a major shopping precinct in Chatham.

In West Germany and in France district cooling systems are becoming relatively common. They are operated by the various district heating companies within the framework of district heating supply and provide chilled water to hospitals, office blocks and public buildings.

REFRIGERATION

Refrigeration technology is based upon thermodynamics. There are two basic techniques of producing cold from heat:
(a) by compression refrigeration
(b) by absorption refrigeration.

Compression refrigeration

The first requirement in this technique is to change heat energy produced in a turbine system into mechanical energy. This can be done either by converting the mechanical energy produced into electricity (which can then be reconverted into mechanical energy at only a small loss) or by using the steam energy directly in steam driven rotary plant (**16.1**).

Whichever system is adopted, a refrigerant, either ammonia or one of the more modern Freons (table 16.2 below), is compressed using a pumping system. The hot compressed refrigerant is then cooled down as close to atmospheric temperature as possible in an efficient heat exchanger. The heat thus abstracted can either be utilised by a heat pump or (more usually) dissipated to the atmosphere. The compressed and liquified gas is allowed to expand and re-evaporate and, in so doing, it draws the latent heat of vaporisation from the heat exchange medium. The gas is then recycled to the compressor.

It can be demonstrated that the refrigeration effect (J/kg of fluid) is equal to the sum of the work done by the compressor and the heat which is given off: the refrigeration effect is therefore much higher than the work input to achieve it. The fraction $\frac{\text{refrigeration effect}}{\text{work done}}$ is termed the coefficient of performance. It is always above unity.

The coefficient of performance is closely related to the conditions conducing to the efficiency of steam or gas turbines and in any specific plant it depends upon the efficiency of thermal insulation, the nature of the refrigerant used and many other factors. The efficiency of a given refrigeration system can be stated in terms of the weight (kilogrammes) of steam required to achieve a given refrigeration effect or, conversely, in terms of the number of megajoules of refrigeration obtained from a kilogramme of steam. To establish the exact amount of heat which has been abstracted from the steam in such circumstances (megajoules) it is necessary to know the precise enthalpy content of the steam supplied and of the water condensed in the turbine section of the plant.

TABLE 16.2 PROPERTIES OF REFRIGERANTS AT 0°C (32°F)

Name	Formula	Enthalpy of vapour (kJ/kg)	Entropy of vapour (kJ/kgK)	Conductivity W/mK	Critical temperature °C (°F)
Ammonia	NH_3	1444.45	5.340	0.0493	132.4 (270)
Ethyl chloride	C_2H_5Cl	735.12	2.68	0.0952	187.2 (370)
Freon 11	CCl_3F	223.09	0.829	0.0981	198.3 (389)
Freon 12	CCl_2F_2	189.87	0.707	0.0764	112. (234)
Freon 13	$CClF_3$	n/a	n/a	n/a	29 (84)
Freon 14	CF_4	n/a	n/a	n/a	−45 (−49)
Freon 21	$CHCl_2F$	286.05	1.063	0.111	178.9 (354)
Freon 22	$CHClF_2$	n/a	n/a	0.101	97 (207)
Freon 23	CHF_3	n/a	n/a	n/a	33 (92)
Freon 113	CCl_2FCCl_2F	n/a	n/a	0.0997	n/a
Freon 114	$C_2H_2F_4$	n/a	n/a	0.0875	n/a
Freon 115	CF_3CClF_2	n/a	n/a	n/a	60 (140)
Freon 12 Bromine −1	$CBrClF_2$	n/a	n/a	n/a	154 (310)
Freon C 318	C_4F_8	n/a	n/a	n/a	117 (243)
Genetron 100	CH_3CHF_2	n/a	n/a	n/a	114 (237)
Genetron 101	CH_3CClF_2	n/a	n/a	n/a	138 (281)
Propane	C_3H_8	895.74	5.648	0.0151	96.8 (206)
Sulphur dioxide	SO_2	480.09	1.824	0.00865	157.2 (315)
Trichlorethylene	C_2HCl_3	274.28	1.009	n/a	

Note: n/a = not available

Example: It is claimed that a particular refrigeration plant requires 0.3 kg of steam at 200°C (392°F) and 15.548 bars pressure to produce 1 MJ of

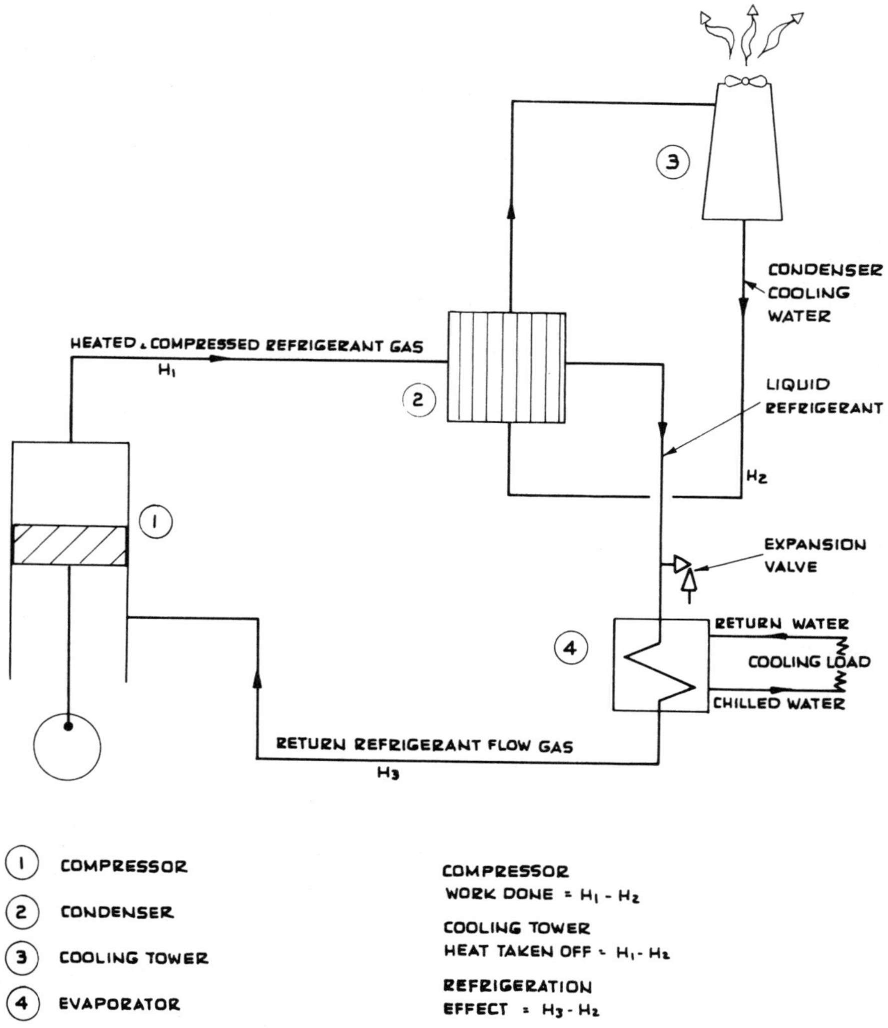

16.1 Schematic of operation of the compression refrigerator

refrigeration effect. The enthalpy content of the steam (the heat required to produce the steam) is given by the difference between the enthalpy content of steam at that temperature and pressure (ie 2.791 MJ/kg) and the enthalpy content of the return water. Assume that the feed water temperature to the boiler is 48°C (118°F) (an enthalpy of the feed water of 0.201 MJ/kg). The energy used in driving the refrigeration plant is therefore:

0.3 (2.791 − 0.201) MJ = 0.777 MJ

This produces a refrigeration effect of 1 MJ giving the practical coefficient of performance at $\underline{1.287}\left(\frac{1}{0.777}\right)$.

Absorption refrigeration

In this system cold is produced directly from heat without any arrangement for converting the latter to mechanical energy except for the use of small pumps (**16.2** and **16.3** and **plate 76**). The system has far fewer moving parts than compression refrigeration and is therefore less liable to break down. Numerous systems of absorption refrigeration are available and they all operate in a somewhat similar way.

The ammonia/water absorption plant is a typical system of this kind. It works as follows: a concentrated ammonia solution is heated in the generator still (A) and ammonia gas is driven off; still under considerable pressure it is then cooled down to atmospheric temperature in the cooler/condenser (B); the ammonia is liquified and the liquid then expands into the evaporator (C); in so doing it exerts its refrigeration effect to cool the circulating water; after the ammonia gas has again warmed to room temperature it is brought into contact with the cooled water from generator still (A): it then dissolves in this water to form concentrated ammonia solution which is recycled after which the process begins anew. The refrigeration effect in an absorption refrigeration system is equal to the sum of the heat supplied to the generator, and the various quantities of heat given off in gas cooler (B) and evaporator (D).

Units used in refrigeration practice

Nowadays SI units are employed in refrigeration practice as in the other energy fields. Heat abstracted is measured in joules and the capacity of a refrigerator is measured in J/S or watts. The traditional basis of specifying for refrigeration capacity is the heat which has to be abstracted to freeze one 'short' ton of 2000 pounds of water. This is equal to 288 000 British Thermal Units (Btu) or 303.86 megajoules of abstracted heat.

The term 'ton of refrigeration' is also used to express the capacity of a given refrigerator. It is assumed that a plant with a capacity of one ton is able to freeze 2000 pounds of water in 24 hours or that it can abstract $\frac{288\,000}{24} = 12\,000$ Btu/hr. Expressed in SI units such a plant has a capacity of 3.15 KW.

Less used is the British or Imperial ton of refrigeration equal to 340.320 MJ or representing a capacity of 3.938 89 KW.

The traditional Continental (European) measure for refrigeration expresses the amount of refrigeration in terms of calories and the rate of

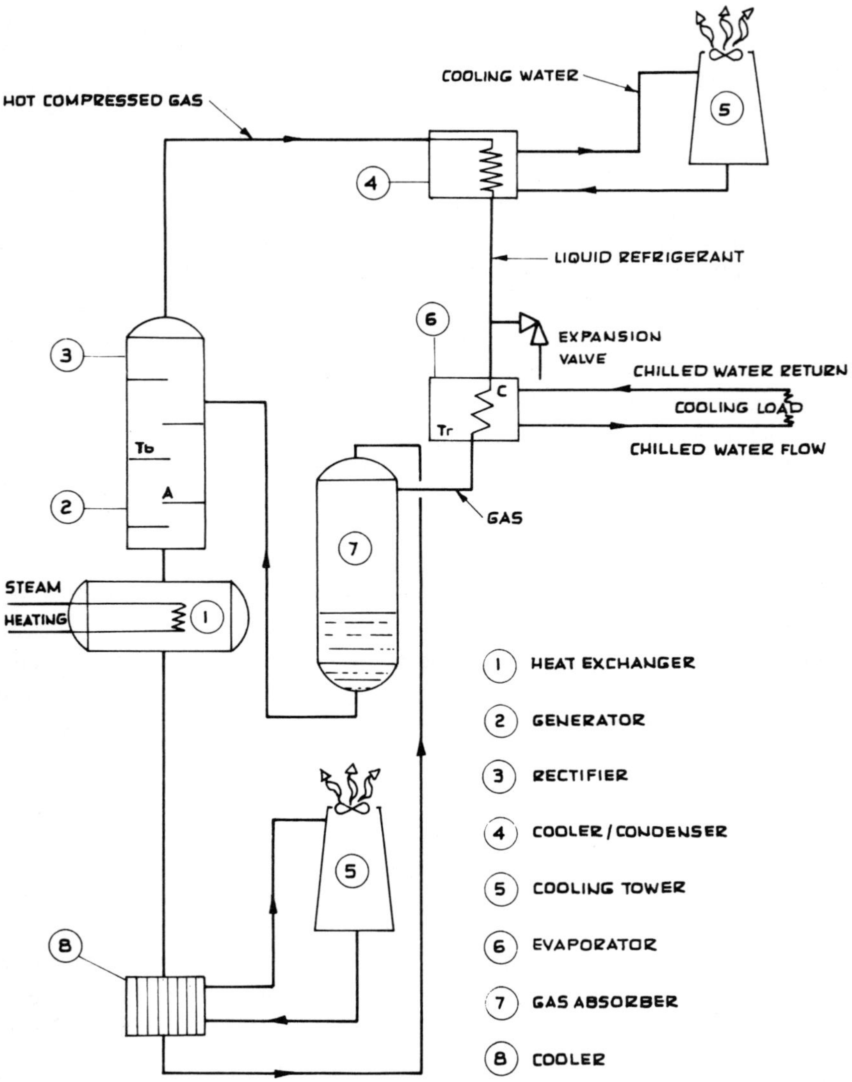

16.2 Schematic of operation of an absorption refrigerator

cooling in terms of calories/second. Conversion to SI units in both cases requires multiplication by the factor 4.186 8.

Summary

Short ton of refrigeration: Quantity = 303.857 28 MJ
Capacity = 3.516 866 kW

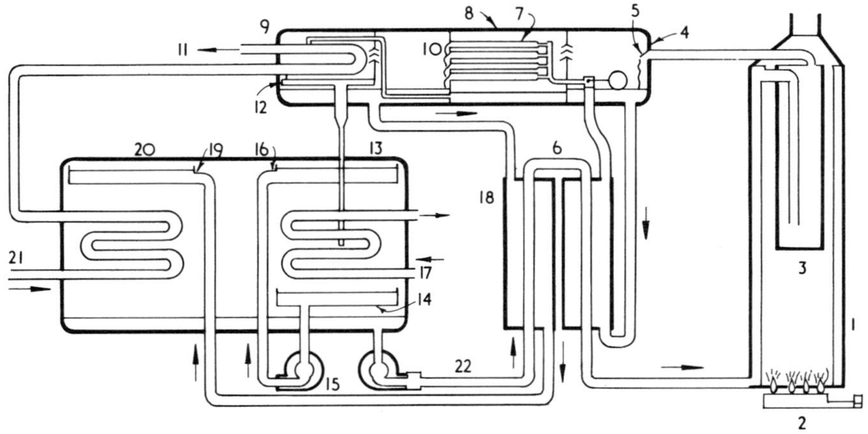

16.3 Larger absorption chiller (used by Philadelphia Electric Company)

Long ton of refrigeration: Quantity = 340.320 MJ
Capacity = 3.938 89 kW
Megacalorie of refrigeration = 4.186 8 MJ
Kilocalorie of refrigeration/second = 4.186 kW

REFRIGERATION PLANT USED IN DISTRICT COOLING

Centrifugal system

In this kind of plant high pressure steam drives a turbine which powers a standard compressor machine (**16.4**). Water from the district cooling circuit passes through the cooling coils of the refrigerator where it is cooled from 12°C (54°F) down to the circulating temperature of 3.3°C (38°F). The compressed ammonia is cooled to room temperature by cooling water drawn either from a nearby river or from a cooling tower. The water runs in series through the condensing coil of the steam turbine. One of the advantages of this plant is that the steam turbines can also power the system pumps.

Turbine-driven centrifugal machines operate at an efficiency of between 1.05 and 1.35 MJ/MJ of cooling effect and are likely to be the most efficient to use when the cooling load is in excess of 425 MW. Turbine-driven centrifugal machines are at their best when steam pressures are high and the periods of loading are long. This type of refrigeration machine is used for the main load at Hartford (USA) and it has operated there continuously at full load for long periods of time without giving major trouble.

Absorption systems

The absorption type of refrigeration circuit uses a heat operated generator to produce a temperature differential which progresses the processes of

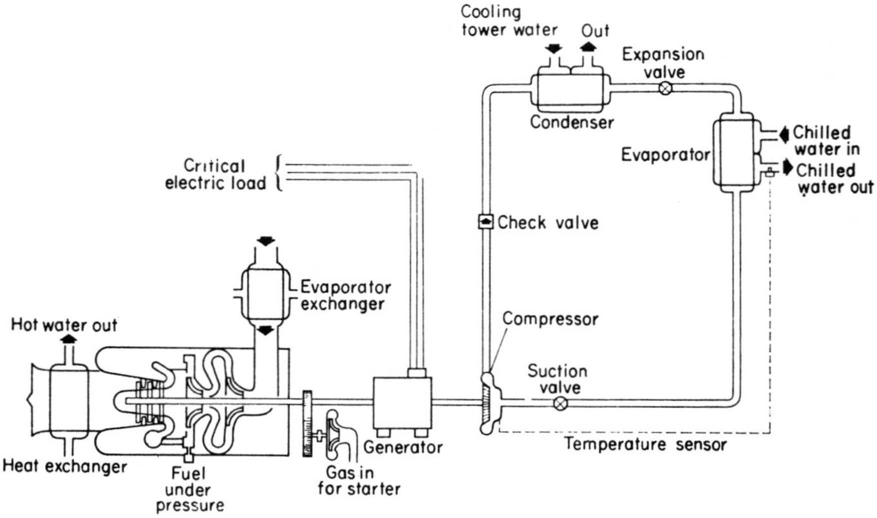

16.4 Layout of turbine driven absorption chilling systems

evaporation and condensation of the refrigerant (**16.5**). One of the most widely used plants of this type in the USA is the Swearingen absorption water chiller. The cycle uses a lithium bromide-water pair: the means of regenerating the absorbent and the refrigerant is a two-effect evaporator which achieves high degree of heat economy. The layout of the plant is shown in **figure 16.6.**

The liquid refrigerant (water) evaporates and heat is extracted from the evaporator coil which carries the chilled water. The refrigerant vapour is then absorbed by the rich lithium bromide solution. The absorbant solution has thus been exhausted, and the resultant spent solution drains to the bottom and is pumped through heat exchangers and through the first-effect generator. It is then discharged from the heater into the upper compartment of the upper left hand side of the vessel at about 6.9 k Pa (gauge) where the half-regenerated solution discharges the vapour. The vapour passes into the lower half of the vessel (the second-effect generator) condenses upon the coil and drops to the bottom. The other half of the absorbed refrigerant vapourises within the coil of the second-effect generator. The refrigerant vapour outside the coil condenses at 102°C (216°F) while, at the inside, the boiling point is 88°C (190°F) at the pressure of 11 k Pa (absolute) which prevails there. The rest of the refrigerant is thus able to distil off the lithium bromide solution. The liquid falls to the bottom and the vapour rises into the second-effect condenser and condenses on the cooling water coil. The refrigerant condensing in the second-effect generator flows into the second-effect condenser and the combined liquid refrigerant flows back to the

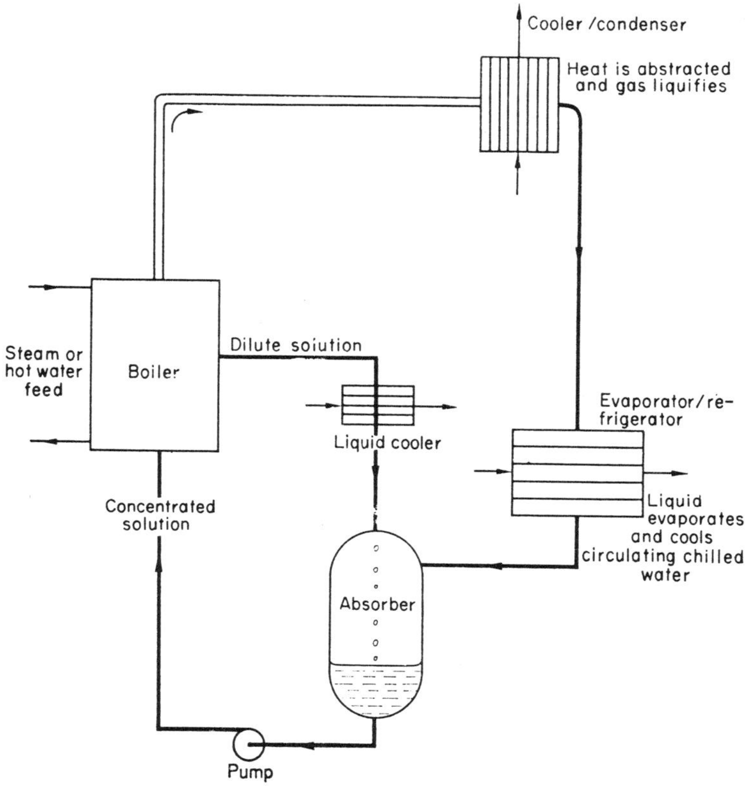

16.5 Flow sheet of ammonia refrigeration plant

evaporator. The only mechanical part of the machine is a totally enclosed pump.

The Swearingen absorption chiller uses only 0.94 MJ/MJ of refrigeration effect: a much better performance than that of most units of this type which require almost 1.5 MJ/MJ of refrigeration effect. Most absorption machines are within the 8.5 to 30 MW range and smaller although some steam powered absorption units are in service which are capable of producing 85 MW of refrigeration effect.

Absorption machines require larger cooling towers and condensers than compressor machines because their rate of heat rejection is greater. The extra costs involved are largely offset by lower requirements in starters, circuit breakers, transformers etc. Absorption plants occupy more space than compressor plants.

Combined absorption and centrifugal system

High pressure feed steam at a pressure of (say) 8 bars is introduced to drive the turbine of a centrifugal plant: when the pressure has been reduced to

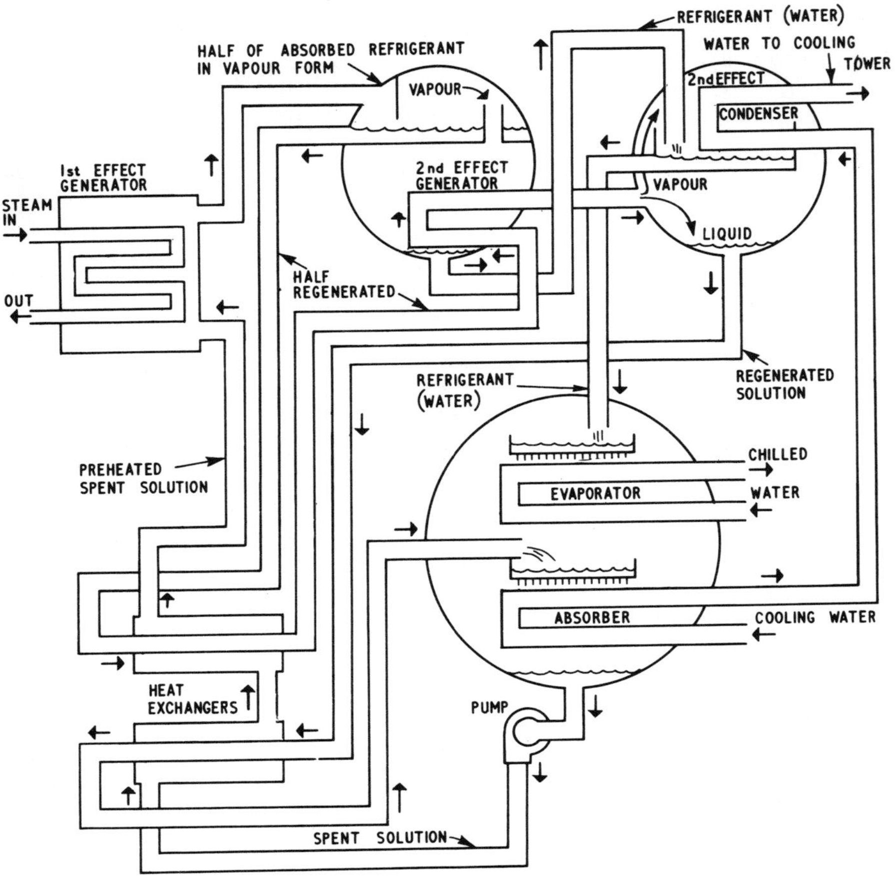

16.6 A Swearingen absorption chiller with two-effect generator

about 0.9 bars the residual heat content of the steam is used to operate an absorption plant. There are four basic conditions for the economic operation of such combined installations:
High pressure steam must be available
The output required from the chilling plant must be reasonably constant
The output required must exceed 40 per cent of the total capacity of the system for fairly long periods: at loads below 40 per cer't the entire capacity is served by the absorption machine only
The installation requires more than one chilling machine.
In general such a system is worthwhile when the load is greater than 42 MW.
system for fairly long periods: at loads below 40 per cent, the entire capacity
If steam at about 40 bars is available to the system a refrigeration effect of 0.825 MW/MW can be obtained: this is lower than with any other refrigeration system.

The Freon turbine

The Freon turbine utilises the favourable thermodynamic cycle of Freon and uses the sensible heat of water in excess of 50°C (122°F) so that it can function with the waste heat produced in CHP and back pressure turbines, total energy plants and other sources of low grade heat (**plate 77**). Such equipment is therefore superior to other types of refrigeration plant fed with expensive high pressure steam. Freon turbines are made by the Japanese Ishikawajima-Harima Heavy Industries Company. They are currently still at the pilot plant stage but future prospects for these plants in urban systems appear to be assured.

INSULATION OF PIPELINES CARRYING CHILLED WATER

The supply mains in district cooling systems carry chilled water at between 3 and 5°C (37 and 41°F). To conserve the cooling effect these mains must be insulated in a similar way to pipelines carrying hot water. The return mains usually carry water at about 12 to 15°C (54 to 59°F): whether they should be insulated or not depends on the ambient conditions, the operating costs of the plant and the capital cost of the thermal insulation. With major projects a feasibility study is required to arrive at a well founded decision.

CENTRAL COOLING SYSTEMS

Advantages

Centralised cooling systems have proved economical wherever it has been possible to supply a sufficient load of air conditioning without requiring excessive pipeline lengths. The same basic consideration also of course applies with district and group heating. Centralised cooling systems cannot however expect to save such a large margin of energy as district heating systems where economy is achieved by the difference in cost between waste heat produced centrally by power stations and refuse incinerators and heat from prime fuel burned locally at the consumer's premises.

The cost advantages of centralised cooling systems are:

(a) The total capacity of a centralised plant is less than the sum of individual plants at the consumer's premises because maximum cooling requirements for different types of consumers such as offices and dwellings are staggered: the diversity factor therefore operates in favour of the central plant

(b) Very large cooling plants are more economical both in their initial cost and in their running expenses than smaller plants: the need to provide standby equipment (costly for individual consumers) can be reduced; there are also savings in the cost of supervising maintenance and a better grade of engineer can be employed

(c) It is possible to use plant fed with low grade steam, thus enabling waste heat from CHP turbines to be utilised

(d) The refrigeration system can be operated as a heat pump utilising heat from that source to supplement the supply of heat to district heating networks: individual refrigeration plants in buildings waste most of the heat they produce.

It has been calculated that the running costs of centralised plants can in fact be as little as one third of those of individual air conditioning plants. The larger a given district cooling system is the more likely it is that it will be viable. High concentration of cooling load is needed to keep the pipeline network reasonably compact.

Centralised cooling can only be economic when it provides the cooling effect cheaper than cooling plant at the consumer's premises. The best results are achieved when district heating and district cooling are operated in conjunction with each other.

Turbo-refrigeration system

This system, shown in **figure 16.7,** uses high pressure steam in a back pressure turbine to drive a turbo-refrigerator which cools the circulating water from 8°C (46°F) down to the supply temperature which is around 5°C (41°F). The exhausted steam is then used in an absorption refrigeration system at approximately 160°C (320°F). A side-stream take-off of this steam is also effected in a heat exchanger/condenser to heat the circulating district heating water. The required balance between district heating and district cooling is adjusted by means of a valve. The condensate water is returned to the boiler which is incorporated with the plant.

Two cooling towers are required: one to operate in conjunction with the turbo-refrigerator, the other for the absorption plant. The heat content of the water emerging from these two plants is normally of too low an order for heating purposes. Where it is desirable to use low grade heat for space heating, the system can be operated so that the turbo-refrigerator and the absorption plant function as heat pumps providing waste heat at say 80°C (176°F): at this temperature the water can be used for space heating purposes. Very low temperature return water from the district heating network would be used having been cooled even further in cooling towers or condensers utilising river water.

Various modifications of the basic design shown in **figure 16.7** are used. For example, the layout can be coupled with an ITOC turbine, so that the plant can produce electricity during periods of low cooling demand and at the same time feed a controllable heat supply into the district heating system. It is advisable to design the centralised plant to retain the maximum flexibility in the proportion of primary energy which is being converted into (a) heat, (b) cold and (c) electricity.

Gas cooling system

A technique has been developed by Z. A. Melikyan (USSR) for supplying

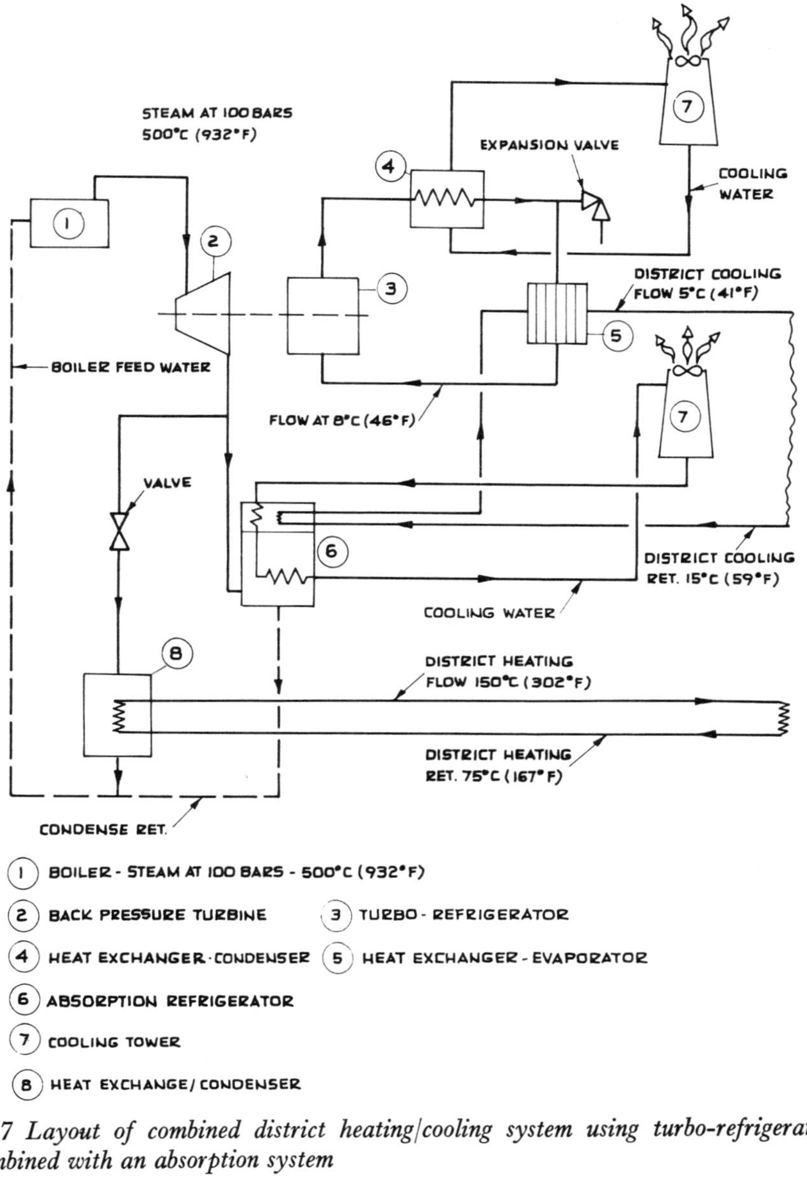

16.7 Layout of combined district heating/cooling system using turbo-refrigerators combined with an absorption system

district cooling by pumping compressed gas to the consumers: this overcomes the disadvantage caused by the need to pump large quantities of chilled water to them from the central cooling station. At the central station large compressors compress a refrigerant (such as Freon 12). The compressed gas is then transported by underground pipelines to the substations at the consumers' premises. Each building or group of buildings is equipped with a condenser in which the gases are cooled down to room temperature; they then pass through expansion valves installed in each individual room or public area of the building. The manual opening of these expansion valves permits close control over the amount of cooling provided in any particular room. On passing through the valve, the compressed gas is cooled down; this effect is then utilised by the use of heat exchangers to cool the incoming circulating air. Finally the expanded gas is returned to the plant for recompression.

The main advantages claimed for this system are that plant costs are some 30 to 40 per cent less than those of the competing systems in which water is the pipeline medium. The system can moreover be used in the heat pump mode during winter with the ground as the low-temperature heat source. The system is also suitable to supply centralised refrigeration to residences and to industry thereby eliminating the need for individual air conditioning or freeze refrigerators. The piped refrigerant can be used for food freezing via cooling coils mounted in well insulated food cupboards. To recover the loss of pressure in the return pipeline during the transport of the gas to the central compressors, use is made of auxiliary small scale compressors.

Figure 16.8 shows how the system works and **16.9** indicates the pressure-enthalpy relationship. In the latter figure:

Line AB shows the isentropic gas compression process in the main compressors of the station (it is necessary to compress to a higher level (line AB) so as to compensate for the pressure drop in the pipelines)

Line BD represents the loss in enthalpy during gas transit

Line DE represents the condensing process of the gas in the condensers

Line EF represents the pressure and enthalpy drop as the liquid refrigerant is conveyed from the condensers to the consumers

Line FG shows the expansion process in the valves

Line GH represents the evaporation process in the cooling coils of the air conditioning plant

Line HI is the pressure drop and enthalpy increase as the gas is transferred from the cooling coils to the compressor station

Line IA represents the action of the auxiliary compressor.

TWO EXISTING PLANTS

Hartford Connecticut

The combined district heating and cooling plant in Hartford, Connecticut was opened in June 1962 for the supply of heating and cooling to the large new Constitution Plaza (central city complex) which contains one office block of 20 storeys, one of 18 storeys, a 315 room hotel, a major shopping centre, 2100 flats in several large blocks, a 26-storey commercial tower block and several other minor buildings.

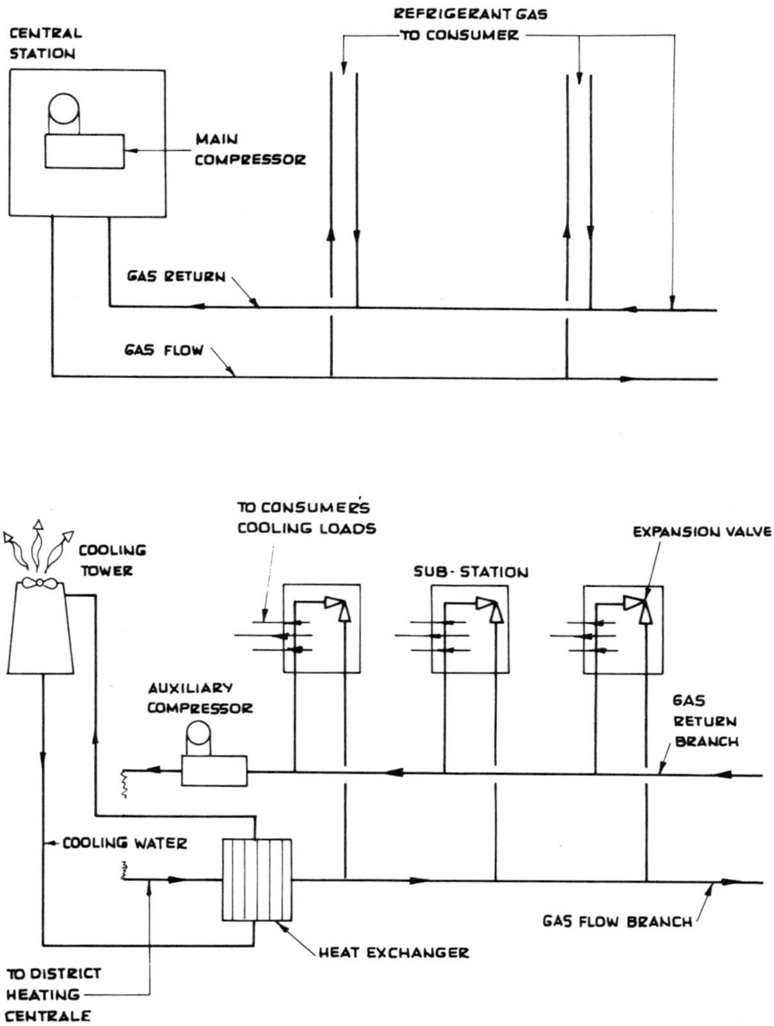

16.8 District gas cooling system. Above: general arrangement; below: plant at consumer's premises

The initial district heating capacity was 36.9 MW in 1964; at the time of writing this had grown to 152.8 MW. A steam plant of 102 tonnes per hour capacity supplies the complex with heat during the winter months. The steam is sold for heating and also for use in steam turbine driven absorption type refrigeration plant. This system has the advantage of providing a more even annual loading of the available steam plant. The boilers are fired by gas except during very severe winter conditions when fuel oil is used. The plant is situated adjacent to the Connecticut River from which water is

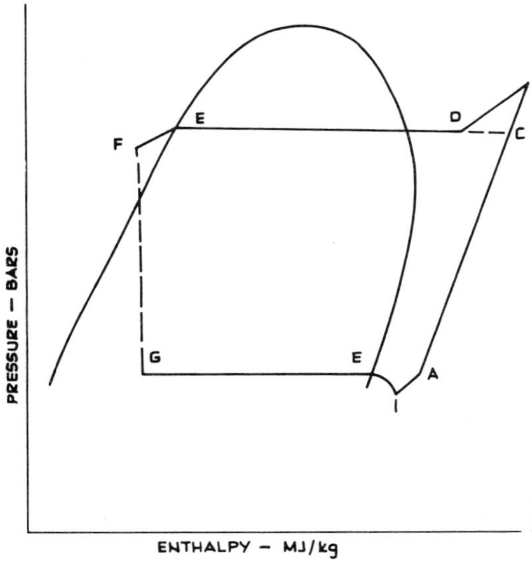

16.9 Pressure/enthalpy diagram of gas cooling system of district cooling

drawn for condensing purposes: the fact that a cooling medium is so readily available appears to have been most desirable since the alternative, which was not acceptable, would have been to build large cooling towers close to the densely developed central complex.

The central plant comprises about 2000 m² of floor space, containing three boilers, one centrifugal refrigerator of 10.6 MW capacity, two refrigerators of 15.3 MW capacity, one absorption type refrigerator of 1.8 MW capacity and a 15.8 MW capacity refrigeration plant (**16.10** and **16.11**). It delivers steam in winter and chilled water in summer. The steam supply is passed through a prefabricated conduit 650 mm in diameter which consists of a 300 mm carrier pipe, heavily insulated with Thermobestos, an air space, and a 10 gauge steel casing, surface-covered with bitumen.

The chilled water leaves the plant at 3.3°C (38°F) and passes through steel mains ranging from 400 to 600 mm in diameter which are insulated on the outside with 25 mm thick expanded polystyrene. The chilled water returns carry water at about 12.2°C (54°F) and are similar except that they are not insulated. The condensate return pipes for the steam supply are made of stainless steel 125 mm in diameter and are insulated with Gilsulate.

The chilled water is pumped by four turbine-driven pumps with one standby pump at a rate of 1136 litres/second at peak load. The quantity of water required for each MW of refrigeration capacity is 28.7 litres/second.

The condenser water pumps handle approximately 3.22 litres of water per KW of cooling effect. In practice at peak refrigeration demand, 1990 litres/second are pumped from the Connecticut River through a pipe 900 mm in diameter and 550 m long. The total length of the Hartford district heating/cooling system is 1585 m using four-pipe distribution.

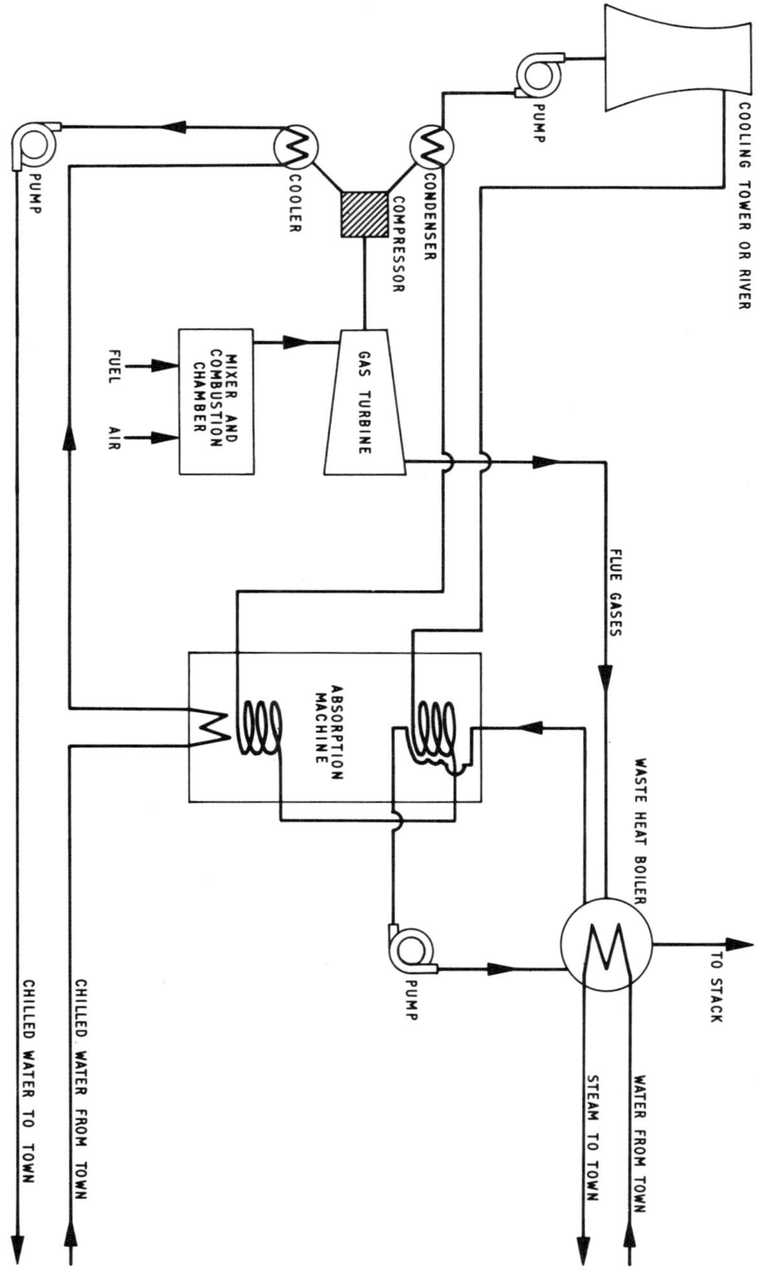

16.10 District cooling at Hartford, USA: general layout

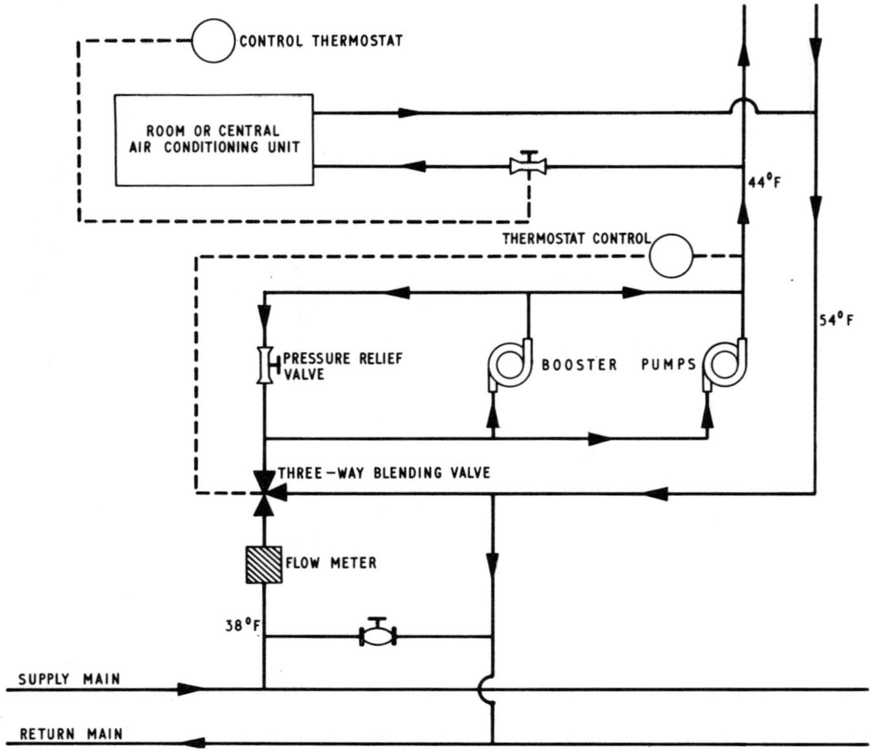

16.11 District cooling at Hartford, USA: consumer arrangement

Heathrow Airport

This scheme is currently by far the largest system of its kind in the UK. It comprises the air conditioning for a site which will eventually cover 3500 m² by refrigeration plant with a total capacity of 35 MW. In 1976 the cooling centrale had an installed capacity of 10.5 MW to serve Terminal 3 and the associated passenger traffic facilities (**16.12**).

The chilling system is based upon two open-drive two-stage centrifugal refrigeration machines with a capacity of 5.25 MW each. These are served by separate evaporators, condensers and cooling towers. Water circulates through these units into a single pressurised balance manifold and then returns to the primary system. The water produced by the chiller unit is at 5°C (41°F): it is circulated at a rate of 225 litres/s at a velocity of 2.13 m/s and with a head of 5.23 m (w.g.).

There are two condensers which are fed with 240 litres of water at a temperature of about 18°C (64°F) subsequently raised to 27°C (81°F), and a pressure of 4 m (w.g.). Air purge units are fitted to the condensers.

The compressors are driven by 1500 HP (1.12 MW) electric motors.

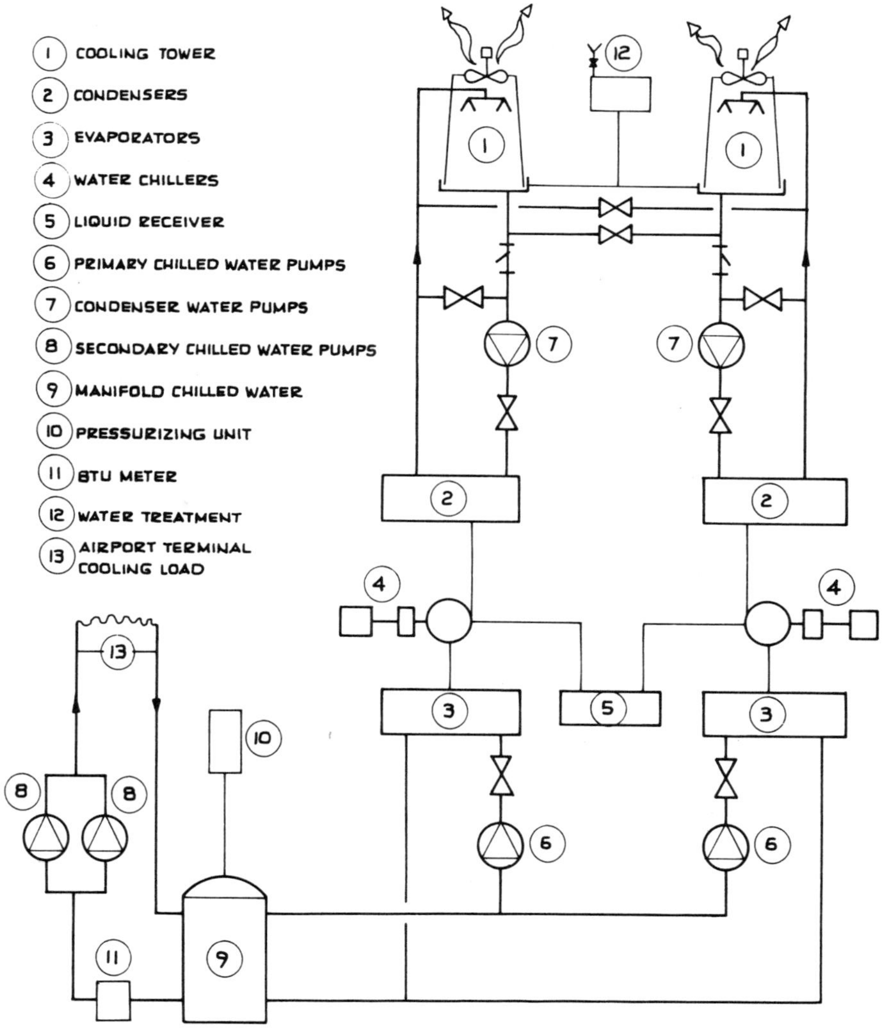

16.12 District cooling scheme at Heathrow Airport (London, UK)

The secondary system has an output of 225 litres/second, and is pressurised by nitrogen to about 6.5 bars.

Two cooling towers each capable of handling 240 litres of water per second are provided to cool down the circulating water from 27°C to 18°C (81 to 64°F). A 50 HP (37.3 kW) squirrel cage motor drives each of the 5.2 metre diameter fans.

The water used in the plant is treated by base exchange units capable of handling a flow rate of 1280 litres per second. The water is also treated by the addition of the following chemicals:

Dearcide 712 which is a phenate algicide and has to be added once a week to the tower ponds
Dearborn 909 which is an organo-zinc corrosion inhibitor and is injected directly into the condenser system
Dearborn 527 borate/nitrite solution which is also used as a corrosion inhibitor and is injected into the pressurised chilled water system.
Modern systems of automatic circulation control are installed.

FINAL POINTS

District heating has made much progress in recent years but air conditioning on a municipal scale is comparatively new. Centralised air conditioning has much the same advantages as centralised heating (**16.13**). The consumer does not need to finance refrigeration plant, cooling towers and other structures and he saves the labour required for operating and maintaining the plant. Air conditioning equipment is usually very noisy and special building construction is required if annoyance is to be avoided. Standby plant must be provided to protect the consumer in the event of plant break down. This can be achieved more cheaply at a centralised position than when each building is provided with separate air conditioning equipment. Finally, refrigeration plants, like furnaces and boilers, operate more economically when they are large.

As the pipe costs of central cooling plants are very appreciable, it is necessary for loads to be concentrated in a central area. According to experience in the USA the minimum load per metre of supply line is 7 kW of cooling capacity.

In a hot climate it is necessary to provide air conditioning to counteract the solar heat gain. It is also often necessary to provide air conditioning in

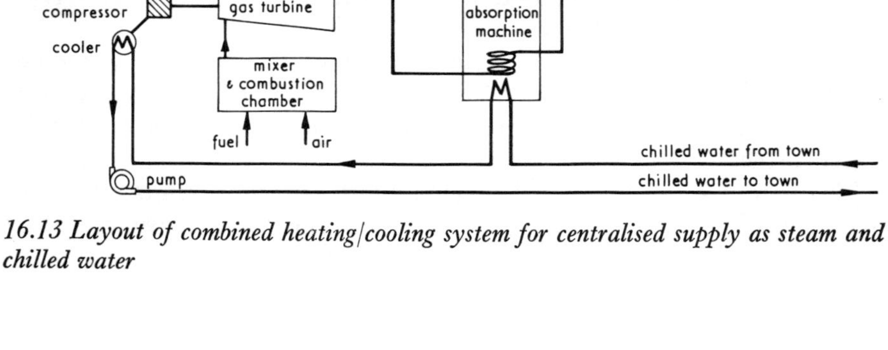

16.13 Layout of combined heating/cooling system for centralised supply as steam and chilled water

quite temperate areas to offset heat gains and/or to permit permanently fixed glazing to exclude external noise.

In the case of the district cooling scheme at Hartford, Connecticut (still one of the largest in the world) it was found that the average refrigeration loading figures were as follows:

Offices: about 12.5 kW per 100 m^2

Auditoria in theatres, cinemas, churches etc: about 1.8 kW of refrigeration per seat

Computer centres: about 38 kW of refrigeration capacity per 100 m^2

Shops and stores: about 19 kW of refrigeration capacity per 100 m^2.

It is very likely that in future district cooling will play a part as important as district heating does today. In the light of this prospect it is relevant to make the point that it is in general uneconomic to employ heat exchangers between the flowing chilled water and the room air cooler. For this reason it is normally necessary to preplan the district cooling system in one entity and to avoid a combination of long supply distances and high blocks. If a more flexible arrangement is needed it is probably better either to use steam lines to refrigeration plant positioned at substations or to employ a system like refrigerant gas distribution.

CHECK LIST

1 Advantages of district cooling systems in comparison with arrangements where cooling has to be provided by individual companies
2 Progress in respect of district cooling made in the USA, Japan and Europe
3 Summary of refrigeration theory as this applies to district cooling
4 Units of capacity in use in the field of refrigeration
5 Types of refrigeration plant suitable for district cooling
6 Thermal insulation needed for district cooling pipelines
7 Advantages of central cooling systems
8 Information given concerning the turbo-refrigeration system
9 Advantages of 'gas cooling' system
10 Possible applications for district cooling systems.

References

E. Gore and A. S. Qureshi, 'Design characteristics for engine driven chillers and heat pump systems', *Heating-Piping*, 49, August 1977, pp 59–64

G. G. Haselden, *Cryogenic fundamentals*, Academic Press, London, 1971

J. Hirsham, 'Natural cold water direct cooling systems', *The Specifying Engineer*, 39, pp 140–146

G. V. R. Holmes, 'Designing a central chilled water system', *Heating-Piping*, 50, September 1968, pp 111–122

K. J. Linton, 'Chilled water storage', *Heating-Piping*, 50, May 1978, pp 53–56

Z. A. Melikyan, 'Large scale centralised cooling systems', *Heating and Ventilation Engineer*, July 1974, pp 3–7
F. L. Schiebart, 'Improving the performance of absorption water chillers', *Plant Engineer*, 32, 2nd March 1978, pp 65–67
L. A. Stump, 'Absorption liquid chillers', *Plant Engineer*, 31, September 1977, pp 235–237, October 1977, p 13
R. W. Vance and W. M. Duke, *Applied cryogenic engineering*, Wiley, New York, 1969

APPENDIX A

Conversion factors and symbols

TABLE A.1 MULTIPLE AND SUBMULTIPLE DESIGNATIONS

Multiples and submultiples		Symbol
10^{12}	tera	T
10^{3}	giga	G
10^{9}	mega	M
10^{3}	kilo	k
10^{2}	hecto	h
10	deca	da
10^{-1}	deci	d
10^{-2}	centi	c
10^{-3}	milli	m
10^{-6}	micro	μ
10^{-9}	nano	n
10^{-12}	pico	p
10^{-15}	femto	f
10^{-18}	atto	a

The following table gives the conversions from imperial and non-SI metric units into the appropriate SI units. In each case the factor indicates by how much one has to multiply the imperial or metric non-SI values to obtain the SI values.

TABLE A.2 SIMPLE CONVERSION FACTORS

Imperial or non-SI metric unit	Factor	SI unit	Symbol
abampere	10	ampere	A
abcoulomb	10	coulomb	C (As)
abfarad	1	gigafarad	GF (As/V)
abhenry	1	nanohenry	nH (Vs/A)
abohm	1	nanoohm	nΩ (V/A)
abvolt	10	nanovolt	nV
acre	4 046.856	metre^2	m^2
angström	100	picometre	pm
atmosphere	1.013 25 (1 bar = 10^5 N/m^2)	bar	bar

Imperial or non-SI metric unit	Factor	SI unit	Symbol
barn (SI permitted unit)	10^{-28}	metre²	m²
barrel	0.158 987	metre³	m³
board foot	2.359 74	decimetre³	dm³
Btu (British thermal unit) IST	1.055 06	kilojoule	kJ
Btu (mean)	1.055 87	kilojoule	kJ
Btu (thermochemical)	1.054 35	kilojoule	kJ
Btu (39°F)	1.059 67	kilojoule	kJ
Btu (60°F)	1.054 68	kilojoule	kJ
bushel (UK)	36.368 7	decimetre³	dm³
bushel (US)	35.239 1	decimetre³	dm³
calorie (IST)	4.186 8	joule	J
calorie (mean)	4.190 0	joule	J
calorie (thermochemical)	4.184	joule	J
calorie (15°C)	4.185 8	joule	J
calorie (20°C)	4.181 9	joule	J
carat (new)	0.2	gramme	g
carat (old)	0.205 3	gramme	g
chain	20.116 8	metre	m
chaldron	1.309 27	metre³	m³
CHU	1.899 108	kilojoule	kJ
circular mil	506.707 5	picometre²	pm²
cord	3.624 56	metre³	m³
cup	236.588 2	millilitre	ml
curie (SI permitted unit)	3.7×10^{10}	disintegrations/second	Ci
degree (angle)	17.453 29	milliradian	mrad
dyne	10	micronewton	μN
electronvolt	0.160 21	attojoule	aJ
EMU of current, resistance, etc.	same as abampere, abohm, etc.		
ESU of capacitance	1.112 6	picofarad	pF
ESU of current	333.56	picoampere	pA
ESU of potential	299.79	volt	V
ESU of inductance	0.898 76	terahenry	TH
ESU of resistance	0.898 76	teraohm	TΩ
erg	100	nanojoule	nJ
Fahrenheit (degree)	$°C = (°F - 32) \times 5/9$ $K = (°F - 32) \times 5/9 + 273.15$		
faraday (Carbon 12)	96.487	kilocoulomb	kC
faraday (chemical)	96.495 7	kilocoulomb	kC
faraday (physical)	96.521 9	kilocoulomb	kC
fathom	1.828 8	metre	m
fluid drachm (UK)	3.551 63	centimetre³ (millilitre)	cm³

Imperial or non-SI metric unit	Factor	SI unit	Symbol
fluid dram (US)	3.697	centimetre^3	cm^3
fluid minim (UK)	59.193 9	millimetre^3	mm^3
fluid minim (US)	61.609 0	millimetre^3	mm^3
fluid ounce (UK)	28.413 1	centimetre^3	cm^3
fluid ounce (US)	29.573 5	centimetre^3	cm^3
foot	0.304 8	metre	m
foot^2	0.092 903	metre^2	m^2
foot^3	0.028 316 85	metre^3	m^3
gallon (UK)	4.546 087	decimetre^3 (litre)	dm^3
gallon (US dry)	4.404 884	decimetre^3	dm^3
gallon (US liquid)	3.785 412	decimetre^3	dm^3
gamma	1	nanotesla	nT
gauss	100	microtesla	μT
gill (UK)	142.065	centimetre^3	cm^3
gill (US)	118.294	centimetre^3	cm^3
grain	0.064 798 91	gramme	g
hand	101.6	millimetre	mm
hoppus foot	0.036 054	metre^3	m^3
horsepower (boiler)	9.809 5	kilowatt	kW
horsepower (electric)	0.746	kilowatt	kW
horsepower (550 ftlb/s)	0.745 699 9	kilowatt	kW
horsepower (metric)	0.735 499	kilowatt	kW
horsepower (UK)	0.745 7	kilowatt	kW
horsepower (water)	0.746 43	kilowatt	kW
hundredweight (UK)	50.802 35	kilogramme	kg
hundredweight (US)	45.359 24	kilogramme	kg
inch	25.4 (exactly)	millimetre	mm
inch^2	645.16	millimetre^2	mm^2
inch^3	16.387 06	centimetre^3	cm^3
iron	0.53	millimetre	mm
kilogramme force	9.806 65	newton	N
kip	4.448 222	kilonewton	kN
knot	0.514 444 4	metre/second	m/s
lambert	3.183 099	kilocandela/metre^2	kcd/m^2
langley	41.84	kilojoule/metre^2	kJ/m^2
league (international nautical)	5 556.000	metre	m
league (statute)	4 828.032	metre	m
league (UK nautical)	5 559.552	metre	m
lightyear (SI permitted unit)	9 460.55	terametre	Tm
link	0.201 168	metre	m
maxwell	10	nanoweber	nWb
micron	1	micrometre	μm
mil	25.4	micrometre	μm

Imperial or non-SI metric unit	**Factor**	**SI unit**	**Symbol**
mile (nautical UK)	1 853.184	metre	m
mile (nautical US)	1 852.000	metre	m
mile (statute)	1 609.344	metre	m
mile^2	2.589 988	kilometre^2	km^2
minute (angle)	290.888 2	microradian	μrad
oersted	79.577 47	ampere/metre	A/m
ounce force	0.278 013 9	newton	N
ounce mass (avoirdupois)	28.349 52	gramme	g
ounce troy	31.103 48	gramme	g
parsec (SI permitted unit)	$3.083\,74 \times 10^{16}$	metre	m
peck (UK)	9.092 18	decimetre^3	dm^3
peck (US)	8.809 768	decimetre^3	dm^3
pennyweight	1.555 174	gramme	g
perch (masonry)	0.701	metre^3	m^3
Petrograd standard (timber)	4.672 28	metre^3	m^3
phot	10	kilolumen/metre^2	klm/m^2
pica (printer's)	4.217 518	millimetre	mm
pint (UK)	0.568 261	decimetre^3	dm^3
pint (US dry)	0.550 610 5	decimetre^3	dm^3
pint (US liquid)	0.473 176 5	decimetre^3	dm^3
point (printer's)	0.351 459	millimetre	mm
point (silversmith's)	6.4	micrometre	μ
poise (SI permitted unit)	0.1	newton second/metre^2	N s/m^2
poundal	0.138 255	newton	N
pound (force)	4.448 222	newton	N
pound (mass)	0.453 592 4	kilogramme	kg
pound (troy)	0.373 241 7	kilogramme	kg
quart (UK)	1.136 52	decimetre^3	dm^3
quart (US dry)	1.101 221	decimetre^3	dm^3
quart (US liquid)	0.946 353	decimetre^3	dm^3
quarter (UK)	12.700 6	kilogramme	kg
Rankine degree	1.8	degree Kelvin	K
rod	5.029 2	metre	m
roentgen	0.257 976	millicoulomb/kilogramme	mC/kg
second (angle)	4.848 137	microradian	μrad
slug	14.593 9	kilogramme	kg
Statampere, statohm, etc., same as ESU of current, resistance, etc.			
stokes (SI permitted unit)	1	centimetre^2/second	cm^2/s
stone (UK)	6.350 29	kilogramme	kg
tablespoon	14.786 76	centimetre^3	cm^3

Imperial or non-SI metric unit	Factor	SI unit	Symbol
teaspoon	4.928 922	centimetre^3	cm^3
ton (assay)	29.166 67	gramme	g
ton (UK long)	1 016.047	kilogramme	kg
ton (US short)	907.184 7	kilogramme	kg
ton (nuclear)	4.20	gigajoule	GJ
ton (refrigeration)	3.516 85	kilowatt	kW
ton (register)	2.831 685	metre^3	m^3
ton (shipping UK)	1.189 3	metre^3	m^3
ton (shipping US)	1.132 67	metre^3	m^3
ton (displacement)	0.991 1	metre^3	m^3
torr	133.322	newton/metre^2	N/m^2
township (US)	93.239 57	kilometre^2	km^2
unit pole	125.663 7	nanoweber	nWb
yard	0.914 4	metre	m
yard^2	0.836 127 4	metre^2	m^2
yard^3	0.764 554 9	metre^3	m^3

Combined units

It is obviously quite impossible to give all the combinations of units which can be built up from the values in Table A.2. The required combined units can, however, easily be calculated from the simple units.

For example, let us determine the SI value of

$$\frac{\text{pound force}}{\text{inch}}$$

From Table A.2 we can read off that pound force = 4.448 222 newton and that inch = 0.025 4 metre.

$$\text{Therefore, } \frac{\text{pound force}}{\text{inch}} = \frac{4.448\,222}{0.025\,4}$$

$$= 175.126\,8 \text{ newton/metre}$$

Many of the more common compound conversions are given in the following tables.

TABLE A.3 MASS PER UNIT LENGTH, AREA AND VOLUME

Imperial or non-SI metric unit	Factor	SI unit	Symbol
ton (UK)/mile	0.631 342	kilogramme/ metre	kg/m
ton (US)/mile	0.563 698	kilogramme/ metre	kg/m

Imperial or non-SI metric unit	Factor	SI unit	Symbol
pound/yard	0.496 055	kilogramme/ metre	kg/m
pound/foot	1.488 16	kilogramme/ metre	kg/m
pound/inch	17.858	kilogramme/ metre	kg/m
ton (UK)/mile^2	392.298	kilogramme/ kilometre^2	kg/km^2
ton (US)/mile^2	350.266	kilogramme/ kilometre^2	kg/km^2
pound/foot^2	4.882 43	kilogramme/ metre^2	kg/m^2
ounce/foot^2	0.305 152	kilogramme/ metre^2	kg/m^2
ton (UK)/yard^3	1 328.94	kilogramme/ metre^3	kg/m^3
ton (US)/yard^3	1 186.55	kilogramme/ metre^3	kg/m^3
pound/foot^3	16.018 5	kilogramme/ metre^3	kg/m^3
pound/gallon (UK)	0.099 776	kilogramme/ decimetre^3	kg/dm^3
pound/gallon (US)	0.119 826 4	kilogramme/ decimetre^3	kg/dm^3
pound/inch^3	27 679.9	kilogramme/ metre^3	kg/m^3

TABLE A.4 FORCE PER UNIT LENGTH, MOMENTS AND MOMENTUM

Imperial or non-SI metric unit	Factor	SI unit	Symbol
ton (UK) force/foot	32.690 3	kilonewton/metre	kN/m
ton (US) force/foot	29.187 8	kilonewton/metre	kN/m
pound force/foot	14.593 9	newton/metre	N/m
pound force/inch	175.127	newton/metre	N/m
ton (UK) force foot	3.037 03	kilonewton metre	kNm
ton (US) force foot	2.711 63	kilonewton metre	kNm
pound force foot	1.355 82	newton metre	Nm
pound force inch	0.112 985	newton metre	Nm
pound foot^2	0.042 140	kilogramme metre^2	kgm^2
pound inch^2	2.926 4	kilogramme centimetre^2	kg cm^2
pound foot/second	0.138 255	kilogramme metre/second	kg m/s

TABLE A.5 PRESSURES

Imperial or non-SI metric unit	Factor	SI unit	Symbol
centimetre of mercury (0°C)	1.333 22	kilonewton/metre^2	kN/m^2
centimetre of water (4°C)	98.063 8	newton/metre^2	N/m^2
foot of water (39°F)	2.988 98	kilonewton/metre^2	kN/m^2
inch of mercury (32°F)	3.386 389	kilonewton/metre^2	kN/m^2
inch of water (39°F)	249.082	newton/metre^2	N/m^2
kilogramme force/metre^2	9.806 65	newton/metre^2	N/m^2
poundal/foot^2	1.488 164	newton/metre^2	N/m^2
pound force/foot^2	47.880 26	newton/metre^2	N/m^2
pound force/inch^2	6.894 757	kilonewton/metre^2	kN/m^2
ton (UK) force/foot^2	107.252	kilonewton/metre^2	kN/m^2
ton (US) force/foot^2	95.760 7	kilonewton/metre^2	kN/m^2
ton (UK) force/inch^2	15.444 3	newton/millimetre^2	N/mm^2 or MN/m^2
ton (US) force/inch^2	13.789 55	newton/millimetre^2	N/mm^2 or MN/m^2

TABLE A.6 HEAT AND HEAT TRANSFER

Imperial or non-SI metric unit	Factor	SI unit	Symbol
Btu/hour	0.293 071	watt	W (J/s)
Btu/pound	2.326	kilojoule/kilogramme	kJ/kg
Btu/foot^3	37.258 9	kilojoule/metre^3	kJ/m^3
therm/UK gallon	23.208 0	gigajoule/metre^3	GJ/m^3
therm/US gallon	27.871 7	gigajoule/metre^3	GJ/m^3
Btu/pound °F	4.186 8	kilojoule/kilogramme K	kJ/kg K
Btu/foot^3 °F	67.066 1	kilojoule/metre^3 K	kJ/m^3 K
Btu/foot^2 hour	3.154 59	watt/metre^2	W/m^2
Btu/foot^2 hour °F	5.678 26	watt/metre^2 deg C	W/m^2 deg C
kilocalorie/metre^2 hour °C	1.163	watt/metre^2 deg C	W/m^2 deg C
Btu/foot hour °F	1.730 073	watt/metre deg C	W/m deg C
Btu inch/foot^2 hour °F	0.144 228	watt/metre deg C	W/m deg C
Btu inch/foot^2 second °F	519.22	watt/metre deg C	W/m deg C
kilocalorie/metre hour °C	1.163	watt/metre deg C	W/m deg C
calorie/centimetre second °C	418.68	watt/metre deg C	W/m deg C

TABLE A.7 LIQUIDS

Imperial or non-SI metric unit	Factor	SI unit	Symbol
grain/100 foot³	0.022 883 5	gramme/metre³	g/m³
ounce/UK gallon	6.236 02	gramme/decimetre³	g/dm³
ounce/US gallon	7.489 142	gramme/decimetre³	g/dm³
pound force hour/foot²	0.172 369	meganewton second/metre²	MNs/m²
pound force second/foot²	47.880 3	newton second/metre²	Ns/m²
poundal second/foot²	1.488 16	newton second/metre²	Ns/m²
foot²/hour	25.806 4	millimetre²/second	mm²/s
foot²/second	0.092 903	metre²/second	m²/s
poise	0.1	newton second/metre²	Ns/m²
stoke	1	centimetre²/second	cm²/s

TABLE A.8 MISCELLANEOUS COMPOUND CONVERSIONS

Imperial or non-SI metric unit	Factor	SI unit	Symbol
gallon/mile (UK)	2.824 81	decimetre³/kilometre	dm³/km
gallon/mile (US)	2.352 15	decimetre³/kilometre	dm³/km
mile/gallon (UK)	0.354 006	kilometre/decimetre³	km/dm³
mile/gallon (US)	0.425 144	kilometre/decimetre³	km/dm³
ton (UK) mile	1.635 17	tonne kilometre	tonne km
ton (US) mile	1.459 973	tonne kilometre	tonne km
ton (UK) mile/gallon (UK)	359.687	kilogramme kilometre/decimetre³	kg km/dm³
ton (US) mile/gallon (US)	321.149	kilogramme kilometre/decimetre³	kg km/dm³
foot pound force/second	1.355 82	watt	W
foot candle	10.763 9	lumen/metre²	lm/m²
kilovolt/inch	39.370	kilovolt/metre	kV/m

APPENDIX B

Symbols for pipelines and fittings

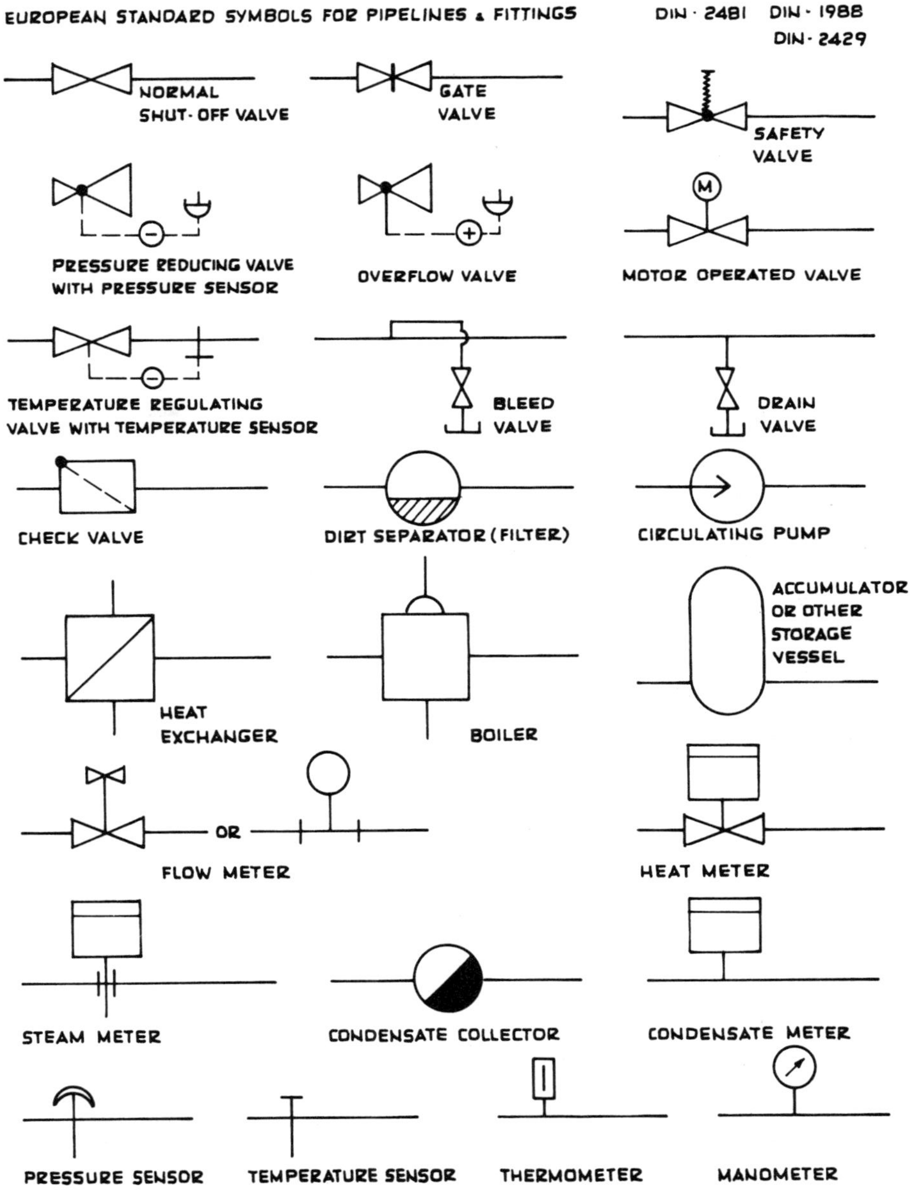

AMERICAN STANDARD SYMBOLS FOR PIPELINES & FITTINGS

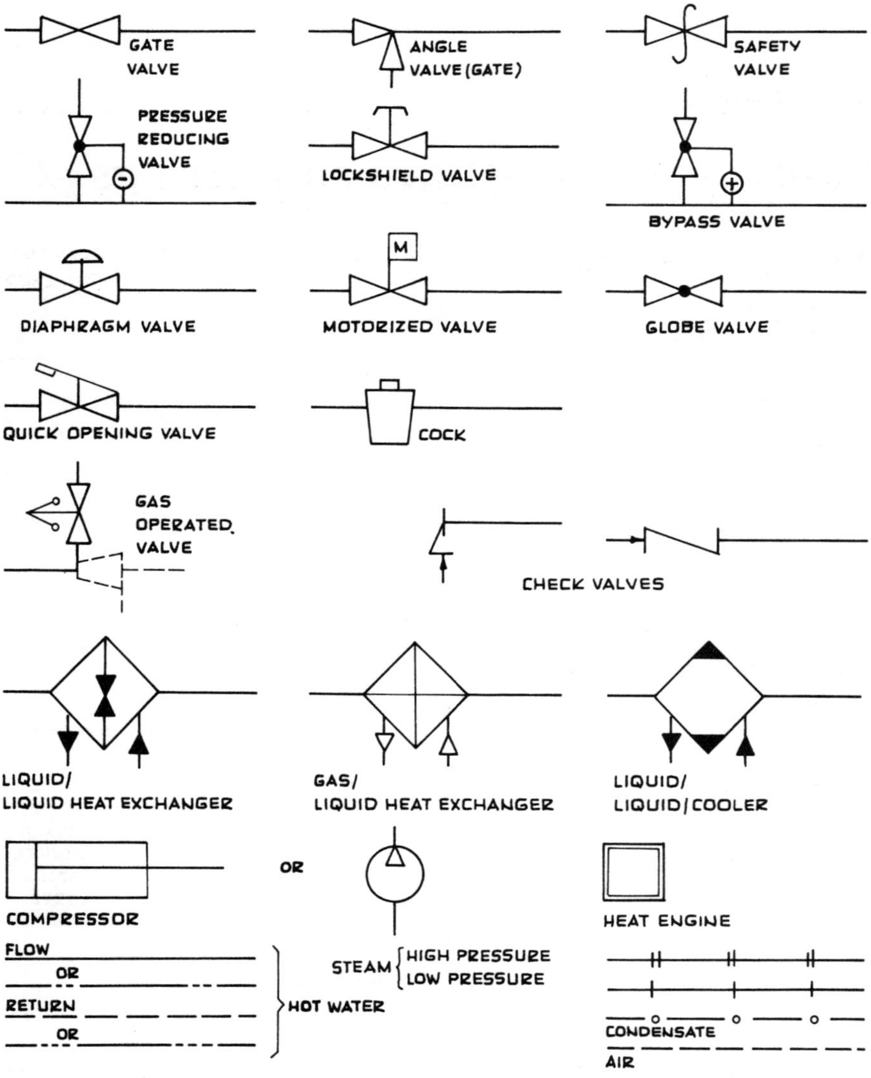

NOTE: IF OTHER FLUIDS ARE BEING PUMPED AN UNINTERUPTED LINE IS USED FOR FLOW & A DASHED LINE FOR RETURN WITH CODE LETTERS IN CENTRE AS UNDER

——— CHF ———
CHILLED WATER FLOW

——— CHR ———
CHILLED WATER RETURN

OTHER ABBREVIATIONS USED ARE:

B - BRINE FLOW	BR - BRINE RETURN	C - CONDENSER WATER
CR - CONDENSER RETURN	D - DRAIN	H - HUMIDIFICATION LINE
A - COMPRESSED AIR	RL - REFRIGERATION LIQUID	RD - REFRIGERATION DISCHARGE
FOF - FUEL OIL FLOW	FOR - FUEL OIL RETURN	FOV - FUEL OIL VENT
G - GAS	V - VACUUM	

BRITISH STANDARD SYMBOLS FOR PIPELINES & FITTINGS BS 2917 - BS 974 - BS M24

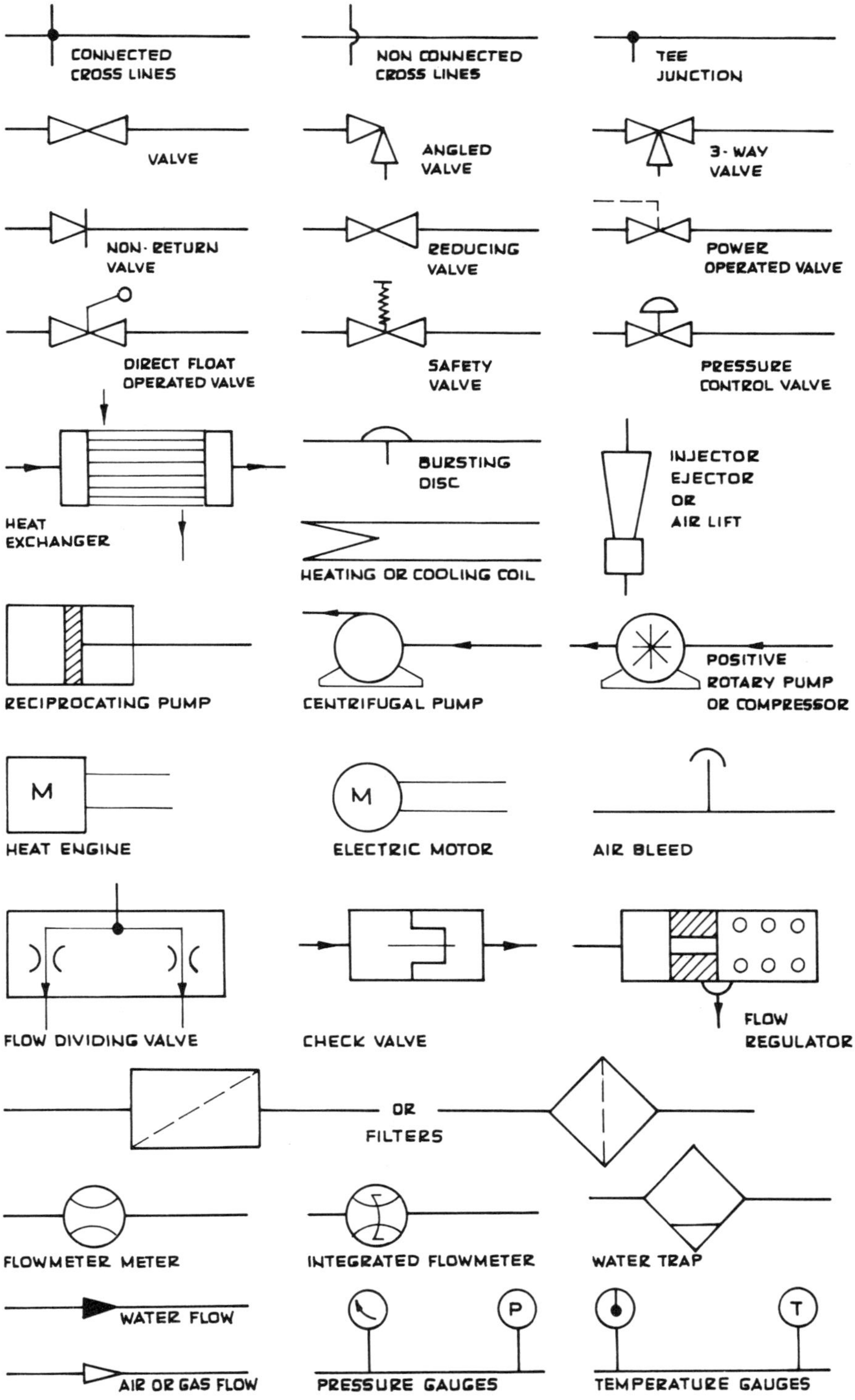

APPENDIX C

Steam tables

TABLE C.1 PROPERTIES OF SATURATED STEAM

Temperature		Pressure	Volume	Enthalpy		Entropy		Latent heat
				Steam	Water	Steam	Water	
°C	Kelvins	Bars	Litres/kg	MJ/kg		kJ/kg K		MJ/kg
0.00	273.15	0.006	206167.677	2.502	0.000	9.147	0.000	2.502
1.00	274.15	0.007	192604.570	2.503	0.004	9.124	0.015	2.499
2.00	275.15	0.007	179993.047	2.505	0.008	9.101	0.031	2.497
3.00	276.15	0.007	168268.553	2.507	0.013	9.078	0.046	2.494
4.00	277.15	0.008	157370.240	2.509	0.017	9.054	0.061	2.492
5.00	278.15	0.008	147240.820	2.511	0.021	9.031	0.076	2.490
6.00	279.15	0.009	137826.411	2.513	0.025	9.007	0.091	2.487
7.00	280.15	0.010	129076.395	2.514	0.029	8.983	0.106	2.485
8.00	281.15	0.010	120943.273	2.516	0.034	8.959	0.121	2.483
9.00	282.15	0.011	113382.532	2.518	0.038	8.935	0.136	2.480
10.00	283.15	0.012	106352.502	2.520	0.042	8.912	0.151	2.478
11.00	284.15	0.013	99814.229	2.522	0.046	8.888	0.166	2.476
12.00	285.15	0.014	93731.343	2.524	0.050	8.864	0.181	2.473
13.00	286.15	0.015	88069.934	2.525	0.055	8.840	0.195	2.471
14.00	287.15	0.016	82798.426	2.527	0.059	8.816	0.210	2.469
15.00	288.15	0.018	77887.460	2.529	0.063	8.792	0.224	2.466
16.00	289.15	0.019	73309.777	2.531	0.067	8.769	0.239	2.464
17.00	290.15	0.020	69040.105	2.533	0.071	8.745	0.253	2.461
18.00	291.15	0.022	65055.047	2.535	0.076	8.722	0.268	2.459
19.00	292.15	0.023	61332.976	2.536	0.080	8.699	0.282	2.457
19 00	292 15	0 023	61332 976	2 536	0 080	8.699	0.282	2.457
20.00	293.15	0.024	57853.931	2.538	0.084	8.675	0.296	2.454
21.00	294.15	0.026	54599.515	2.540	0.088	8.652	0.310	2.452
22.00	295.15	0.028	51552.801	2.542	0.092	8.630	0.324	2.450
23.00	296.15	0.029	48698.233	2.544	0.096	8.607	0.339	2.447
24.00	297.15	0.031	46021.541	2.546	0.101	8.584	0.352	2.445
25.00	298.15	0.033	43509.647	2.547	0.105	8.562	0.366	2.443
26.00	299.15	0.035	41150.585	2.549	0.109	8.540	0.380	2.440
27.00	300.15	0.037	38933.419	2.551	0.113	8.518	0.394	2.438
28.00	301.15	0.039	36848.161	2.553	0.117	8.496	0.408	2.435
29.00	302.15	0.041	34885.703	2.555	0.122	8.474	0.422	2.433
30.00	303.15	0.044	33037.739	2.556	0.126	8.453	0.436	2.431
31.00	304.15	0.046	31296.698	2.558	0.130	8.432	0.450	2.428
32.00	305.15	0.049	29655.681	2.560	0.134	8.411	0.464	2.426
33.00	306.15	0.051	28108.396	2.562	0.138	8.390	0.478	2.424
34.00	307.15	0.054	26649.100	2.564	0.142	8.369	0.492	2.421
35.00	308.15	0.057	25272.545	2.565	0.147	8.349	0.506	2.419
36.00	309.15	0.060	23973.921	2.567	0.151	8.329	0.521	2.416
37.00	310.15	0.063	22748.812	2.569	0.155	8.309	0.534	2.414

TABLE C.1 PROPERTIES OF SATURATED STEAM—*continued*

Temperature °C	Kelvins	Pressure Bars	Volume Litres/kg	Enthalpy Steam MJ/kg	Enthalpy Water MJ/kg	Entropy Steam kJ/kg K	Entropy Water kJ/kg K	Latent heat MJ/kg
38.00	311.15	0.067	21593.146	2.571	0.159	8.289	0.545	2.412
39.00	312.15	0.070	20503.152	2.573	0.163	8.270	0.559	2.409
40.00	313.15	0.074	19475.321	2.574	0.167	8.250	0.573	2.407
41.00	314.15	0.078	18506.373	2.576	0.172	8.231	0.586	2.405
42.00	315.15	0.082	17593.217	2.578	0.176	8.212	0.599	2.402
43.00	316.15	0.087	16732.923	2.580	0.180	8.194	0.613	2.400
44.00	317.15	0.091	15922.700	2.582	0.184	8.175	0.626	2.397
45.00	318.15	0.096	15159.863	2.583	0.188	8.157	0.639	2.395
46.00	319.15	0.101	14441.821	2.585	0.193	8.139	0.652	2.393
47.00	320.15	0.106	13766.054	2.587	0.197	8.121	0.664	2.390
48.00	321.15	0.111	13130.101	2.589	0.201	8.103	0.677	2.388
49.00	322.15	0.117	12531.549	2.590	0.205	8.086	0.690	2.385
50.00	323.15	0.123	11968.023	2.592	0.209	8.068	0.703	2.383
51.00	324.15	0.129	11437.184	2.594	0.213	8.051	0.716	2.381
52.00	325.15	0.136	10936.726	2.596	0.218	8.034	0.729	2.378
53.00	326.15	0.143	10464.379	2.597	0.222	8.017	0.741	2.376
54.00	327.15	0.150	10017.912	2.599	0.226	8.001	0.754	2.373
55.00	328.15	0.157	9595.141	2.601	0.230	7.984	0.767	2.371
56.00	329.15	0.165	9193.942	2.603	0.234	7.968	0.780	2.368
57.00	330.15	0.173	8812.267	2.604	0.238	7.952	0.793	2.366
58.00	331.15	0.181	8448.155	2.606	0.243	7.936	0.805	2.364
59.00	332.15	0.190	9099.761	2.608	0.247	7.920	0 818	2.361
60.00	333.15	0.199	7765.375	2.610	0.251	7.905	0.831	2.359
61.00	334.15	0.209	7443.449	2.611	0.255	7.889	0.844	2.356
62.00	335.15	0.318	7132.632	2.613	0.259	7.874	0.856	2.354
63.00	336.15	0.229	6831.798	2.615	0.264	7.859	0.869	2.351
64.00	337.15	0.239	6540.086	2.617	0.268	7.844	0.882	2.349
65.00	339.15	0.250	6256.940	2.618	0.272	7.829	0.894	2.346
66.00	339.15	0.261	5982.151	2.620	0.276	7.814	0.907	2.344
67.00	340.15	0.273	5715.904	2.622	0.280	7.799	0.919	2.341
68.00	341.15	0.285	5458.830	2.623	0.285	7.785	0.931	2.399
69.00	342.15	0.298	5212.051	2.625	0.289	7.770	0.944	2.336
70.00	343.15	0.311	4977.247	2.627	0.293	7.756	0.956	2.334
71.00	344.15	0.325	4756.706	2.629	0.297	7.742	0.968	2.332
72.00	345.15	0.339	4657.256	2.630	0.301	7.728	0.980	2.329
73.00	346.15	0.354	4475.217	2.632	0.305	7.714	0.992	2.327
74.00	347.15	0.370	4301.168	2.634	0.310	7.700	1.004	2.324
75.00	348.15	0.385	4134.779	2.635	0.314	7.686	1.016	2.322
76.00	349.15	0.402	3975.728	2.637	0.318	7.672	1.028	2.319
77.00	350.15	0.419	3823.702	2.639	0.322	7.659	1.039	2.316
78.00	351.15	0.436	3678.399	2.640	0.326	7.645	1.051	2.314
79.00	352.15	0.455	3539.521	2.642	0.331	7.632	1.063	2.311
80.00	353.15	0.474	3406.783	2.644	0.335	7.618	1.074	2.309
81.00	354.15	0.493	3279.907	2.645	0.339	7.605	1.086	2.306
82.00	355.15	0.513	3158.622	2.647	0.343	7.592	1.097	2.304
83.00	356.15	0.534	3042.667	2.649	0.347	7.579	1.109	2.301
84.00	357.15	0.556	2931.791	2.650	0.352	7.565	1.120	2.299

TABLE C.1 PROPERTIES OF SATURATED STEAM—*continued*

Temperature °C	Kelvins	Pressure Bars	Volume Litres/kg	Enthalpy Steam MJ/kg	Enthalpy Water	Entropy Steam kJ/kgK	Entropy Water	Latent heat MJ/kg
85.00	358.15	0.578	2825.749	2.652	0.356	7.553	1.132	2.296
86.00	359.15	0.601	2724.307	2.654	0.360	7.540	1.143	2.294
87.00	360.15	0.625	2627.236	2.655	0.364	7.527	1.155	2.291
88.00	361.15	0.650	2534.321	2.657	0.368	7.514	1.166	2.288
89.00	362.15	0.675	2445.350	2.649	0.373	7.501	1.178	2.286
90.00	363.15	0.701	2360.124	2.660	0.377	7.489	1.190	2.283
91.00	364.15	0.728	2278.450	2.662	0.381	7.476	1.201	2.281
92.00	365.15	0.756	2200.144	2.663	0.385	7.463	1.213	2.278
93.00	366.15	0.785	2125.032	2.665	0.390	7.451	1.225	2.275
94.00	367.15	0.815	2052.946	2.667	0.394	7.438	1.237	2.273
95.00	368.15	0.845	1983.730	2.668	0.398	7.426	1.249	2.270
96.00	369.15	0.877	1917.234	2.670	0.402	7.414	1.261	2.268
97.00	307.15	0.910	1853.317	2.671	0.406	7.401	1.273	2.265
98.00	371.15	0.943	1791.847	2.673	0.411	7.389	1.285	2.262
99.00	372.15	0.978	1732.702	2.675	0.415	7.377	1.297	2.260
100.00	373.15	1.013	1675.765	2.676	0.419	7.365	1.310	2.257
101.00	374.15	1.050	1620.931	2.678	0.423	7.352	1.322	2.254
102.00	375.15	1.088	1568.103	2.679	0.428	7.340	1.335	2.252
103.00	376.15	1.127	1517.190	2.681	0.432	7.328	1.348	2.249
104.00	377.15	1.167	1468.113	2.682	0.436	7.316	1.360	2.246
105.00	378.15	1.208	1420.801	2.684	0.440	7.304	1.373	2.244
106.00	379.15	1.251	1375.189	2.685	0.444	7.293	1.385	2.241
107.00	380.15	1.294	1331.223	2.687	0.449	7.281	1.398	2.238
108.00	381.15	1.339	1288.857	2.688	0.453	7.269	1.410	2.235
109.00	382.15	1.385	1248.053	2.690	0.457	7.257	1.423	2.233
110.00	383.15	1.433	1208.784	2.691	0.461	7.245	1.435	2.230
111.00	384.15	1.481	1171.028	2.693	0.466	7.234	1.446	2.227
112.00	385.15	1.532	1134.774	2.694	0.470	7.222	1.458	2.225
113.00	386.15	1.583	1100.018	2.696	0.474	7.210	1.469	2.222
114.00	387.15	1.636	1066.768	2.697	0.478	7.199	1.480	2.219
115.00	388.15	1.691	1035.036	2.699	0.483	7.187	1.490	2.216
116.00	389.15	1.746	1005.641	2.700	0.487	7.176	1.499	2.213
117.00	390.15	1.804	976.199	2.702	0.491	7.164	1.508	2.211
118.00	391.15	1.863	947.672	2.703	0.495	7.153	1.516	2.208
119.00	392.15	1.923	920.036	2.705	0.500	7.142	1.523	2.205
120.00	393.15	1.985	893.269	2.706	0.504	7.130	1.529	2.202
121.00	394.15	2.049	867.346	2.708	0.508	7.119	1.533	2.199
122.00	395.15	2.114	842.244	2.709	0.512	7.108	1.554	2.197
123.00	396.15	2.182	817.942	2.710	0.517	7.097	1.563	2.194
124.00	397.15	2.250	794.417	2.712	0.521	7.085	1.572	2.191
125.00	398.15	2.321	771.647	2.713	0.525	7.074	1.582	2.188
126.00	399.15	2.393	749.611	2.715	0.529	7.063	1.591	2.185
127.00	400.15	2.467	728.288	2.716	0.534	7.052	1.601	2.182
128.00	401.15	2.543	707.657	2.717	0.538	7.041	1.611	2.179
129.00	402.15	2.621	687.697	2.719	0.542	7.031	1.620	2.176
130.00	403.15	2.701	668.389	2.720	0.546	7.020	1.630	2.174
131.00	404.15	2.783	649.713	2.721	0.551	7.009	1.640	2.171

TABLE C.1 PROPERTIES OF SATURATED STEAM—*continued*

Temperature °C	Kelvins	Pressure Bars	Volume Litres/kg	Enthalpy Steam MJ/kg	Enthalpy Water MJ/kg	Entropy Steam kJ/kgK	Entropy Water kJ/kgK	Latent heat MJ/kg
132.00	405.15	2.867	631.649	2.723	0.555	6.998	1.650	2.168
133.00	406.15	2.953	614.180	2.724	0.559	6.988	1.661	2.165
134.00	407.15	3.041	597.285	2.725	0.564	6.977	1.671	2.162
135.00	408.15	3.131	580.947	2.727	0.568	6.966	1.681	2.159
136.00	409.15	3.223	565.148	2.728	0.572	6.956	1.691	2.156
137.00	410.15	3.317	549.870	2.729	0.576	6.946	1.702	2.153
138.00	411.15	3.414	535.096	2.731	0.581	6.935	1.712	2.150
139.00	412.15	3.513	520.810	2.732	0.585	6.925	1.722	2.147
140.00	413.15	3.614	506.994	2.733	0.589	6.915	1.733	2.144
141.00	414.15	3.717	493.633	2.734	0.594	6.904	1.743	2.141
142.00	415.15	3.823	480.711	2.736	0.598	6.894	1.754	2.138
143.00	416.15	3.931	468.212	2.737	0.602	6.884	1.765	2.135
144.00	417.15	4.042	456.121	2.738	0.606	6.874	1.775	2.132
145.00	418.15	4.155	444.424	2.739	0.611	6.864	1.786	2.129
146.00	419.15	4.271	433.107	2.741	0.615	6.854	1.797	2.126
147.00	420.15	4.389	422.154	2.742	0.619	6.845	1.807	2.123
148.00	421.15	4.510	411.554	2.743	0.624	6.835	1.818	2.119
149.00	422.15	4.634	401.291	2.744	0.628	6.825	1.829	2.116

TABLE C.2 PROPERTIES OF SUPERHEATED STEAM

Temperature °C	Kelvins	Entropy kJ/kgK					Enthalpy MJ/kg				
		Pressure in bars absolute									
		0.5	1.0	1.5	2.0	2.5	0.5	1.0	1.5	2.0	2.5
130	403.15	7.846	7.539	7.306	7.149	7.066	2.741	2.734	2.726	2.718	2.711
132	405.15	7.856	7.549	7.317	7.159	7.077	2.745	2.738	2.730	2.722	2.715
134	407.15	7.866	7.559	7.327	7.170	7.087	2.749	2.742	2.734	2.726	2.719
136	409.15	7.875	7.569	7.337	7.180	7.098	2.753	2.745	2.738	2.730	2.723
138	411.15	7.884	7.578	7.347	7.190	7.108	2.757	2.749	2.742	2.734	2.727
140	413.15	7.894	7.588	7.357	7.200	7.118	2.761	2.753	2.746	2.738	2.731
142	415.15	7.903	7.598	7.367	7.210	7.128	2.764	2.757	2.750	2.743	2.735
144	417.15	7.913	7.607	7.377	7.220	7.138	2.768	2.761	2.754	2.747	2.739
146	419.15	7.922	7.617	7.386	7.230	7.148	2.772	2.765	2.758	2.751	2.743
148	421.15	7.931	7.627	7.396	7.240	7.158	2.776	2.769	2.762	2.755	2.747
150	423.15	7.940	7.636	7.406	7.250	7.168	2.780	2.773	2.766	2.759	2.751
152	425.15	7.950	7.645	7.416	7.260	7.178	2.784	2.777	2.770	2.762	2.755
154	427.15	7.959	7.655	7.425	7.269	7.188	2.788	2.781	2.774	2.766	2.759
156	429.15	7.968	7.664	7.435	7.279	7.198	2.792	2.785	2.778	2.770	2.763
158	431.15	7.977	7.674	7.444	7.289	7.207	2.796	2.789	2.782	2.774	2.767
160	433.15	7.986	7.683	7.454	7.298	7.217	2.799	2.793	2.786	2.778	2.771
162	435.15	7.995	7.692	7.463	7.308	7.227	2.803	2.797	2.790	2.782	2.775

TABLE C.2 PROPERTIES OF SUPERHEATED STEAM—*continued*

Temperature °C	Kelvins	Entropy kJ/kgK					Enthalpy MJ/kg				
		Pressure in bars absolute									
		0.5	1.0	1.5	2.0	2.5	0.5	1.0	1.5	2.0	2.5
164	437.15	8.004	7.701	7.472	7.317	7.236	2.807	2.800	2.793	2.786	2.779
166	439.15	8.013	7.710	7.482	7.327	7.246	2.811	2.804	2.797	2.790	2.783
168	441.15	8.022	7.719	7.491	7.336	7.255	2.815	2.808	2.801	2.794	2.787
170	443.15	8.030	7.729	7.500	7.346	7.265	2.819	2.812	2.805	2.798	2.791
172	445.15	8.039	7.738	7.509	7.355	7.274	2.823	2.816	2.809	2.802	2.795
174	447.15	8.048	7.746	7.519	7.364	7.283	2.827	2.820	2.813	2.806	2.799
176	449.15	8.057	7.755	7.528	7.373	7.292	2.831	2.824	2.817	2.810	2.803
178	451.15	8.065	7.764	7.537	7.382	7.302	2.835	2.828	2.821	2.814	2.807
180	453.15	8.074	7.773	7.546	7.392	7.311	2.839	2.832	2.825	2.818	2.811
182	455.15	8.083	7.782	7.555	7.401	7.320	2.842	2.836	2.829	2.822	2.815
184	457.15	8.091	7.791	7.564	7.410	7.329	2.846	2.840	2.833	2.826	2.819
186	459.15	8.100	7.799	7.572	7.419	7.338	2.850	2.844	2.837	2.830	2.823
188	461.15	8.108	7.808	7.581	7.427	7.347	2.854	2.847	2.841	2.834	2.827
190	463.15	8.117	7.817	7.590	7.436	7.356	2.858	2.851	2.845	2.838	2.831
192	465.15	8.125	7.825	7.599	7.445	7.365	2.862	2.855	2.849	2.842	2.835
194	467.15	8.134	7.834	7.607	7.454	7.374	2.866	2.859	2.852	2.845	2.838
196	469.15	8.142	7.843	7.616	7.463	7.382	2.870	2.863	2.856	2.849	2.842
198	471.15	8.150	7.851	7.625	7.471	7.391	2.874	2.867	2.860	2.853	2.846
200	473.15	8.159	7.859	7.633	7.480	7.400	2.878	2.871	2.864	2.857	2.850
202	475.15	8.167	7.868	7.642	7.489	7.408	2.882	2.875	2.868	2.861	2.854
204	477.15	8.175	7.876	7.650	7.497	7.417	2.886	2.879	2.872	2.865	2.858
206	479.15	8.183	7.885	7.659	7.506	7.426	2.889	2.883	2.876	2.869	2.862
208	481.15	8.192	7.893	7.667	7.514	7.434	2.893	2.887	2.880	2.873	2.866
210	483.15	8.200	7.901	7.676	7.523	7.443	2.897	2.891	2.884	2.877	2.870
212	485.15	8.208	7.909	7.684	7.531	7.451	2.901	2.895	2.888	2.881	2.874
214	487.15	8.216	7.918	7.692	7.540	7.459	2.905	2.898	2.892	2.885	2.878
216	489.15	8.224	7.926	7.701	7.548	7.468	2.909	2.902	2.896	2.889	2.881
218	491.15	8.232	7.934	7.709	7.556	7.476	2.913	2.906	2.899	2.892	2.885
220	493.15	8.240	7.942	7.717	7.564	7.484	2.917	2.910	2.903	2.896	2.889
222	495.15	8.248	7.950	7.725	7.573	7.493	2.921	2.914	2.907	2.900	2.893
224	497.15	8.256	7.958	7.733	7.581	7.501	2.925	2.918	2.911	2.904	2.897
226	499.15	8.264	7.966	7.741	7.589	7.509	2.929	2.922	2.915	2.908	2.901
228	501.15	8.272	7.974	7.749	7.597	7.517	2.933	2.926	2.919	2.912	2.905

		Pressure in bars absolute									
		3	6	9	12	15	3	6	9	12	15
200	473.15	7.309	7.003	6.759	6.575	6.452	2.865	2.850	2.833	2.815	2.795
202	475.15	7.319	7.014	6.769	6.586	6.463	2.870	2.854	2.838	2.820	2.801
204	477.15	7.328	7.024	6.780	6.597	6.475	2.874	2.859	2.842	2.825	2.806
206	479.15	7.338	7.034	6.791	6.608	6.486	2.868	2.863	2.847	2.830	2.811

TABLE C.2 PROPERTIES OF SUPERHEATED STEAM—*continued*

Temperature °C	Kelvins	Entropy kJ/kgK					Enthalpy MJ/kg				
		Pressure in bars absolute									
		3	6	9	12	15	3	6	9	12	15
208	481.15	7.347	7.044	6.801	6.619	6.497	2.882	2.867	2.852	2.835	2.817
210	483.15	7.356	7.054	6.811	6.629	6.508	2.886	2.872	2.856	2.840	2.822
212	485.15	7.366	7.064	6.822	6.640	6.519	2.890	2.876	2.861	2.845	2.827
214	487.15	7.375	7.073	6.832	6.651	6.529	2.894	2.881	2.866	2.850	2.833
216	489.15	7.383	7.083	6.842	6.661	6.540	2.898	2.885	2.870	2.855	2.838
218	491.15	7.392	7.092	6.852	6.671	6.551	2.902	2.889	2.875	2.860	2.843
220	493.15	7.401	7.101	6.862	6.682	6.561	2.907	2.894	2.880	2.864	2.848
222	495.15	7.410	7.111	6.871	6.692	6.572	2.911	2.898	2.884	2.869	2.853
224	497.15	7.418	7.120	6.881	6.702	6.582	2.915	2.902	2.889	2.874	2.858
226	499.15	7.426	7.129	6.891	6.712	6.592	2.919	2.907	2.893	2.879	2.864
228	501.15	7.435	7.138	6.900	6.722	6.602	2.923	2.911	2.898	2.884	2.869
230	503.15	7.443	7.147	6.909	6.731	6.612	2.927	2.915	2.902	2.889	2.874
232	505.15	7.451	7.155	6.919	6.741	6.622	2.931	2.920	2.907	2.893	2.879
234	507.15	7.459	7.164	6.928	6.751	6.632	2.935	2.924	2.911	2.898	2.884
236	509.15	7.467	7.173	6.937	6.760	6.642	2.939	2.928	2.916	2.903	2.889
238	511.15	7.475	7.181	6.946	6.770	6.652	2.943	2.932	2.920	2.908	2.894
240	513.15	7.483	7.190	6.955	6.779	6.662	2.948	2.937	2.925	2.912	2.899
242	515.15	7.490	7.198	6.964	6.789	6.671	2.952	2.941	2.929	2.917	2.904
244	517.15	7.498	7.207	6.973	6.798	6.681	2.956	2.945	2.934	2.922	2.909
246	519.15	7.506	7.215	6.982	6.807	6.690	2.960	2.950	2.938	2.926	2.913
248	521.15	7.513	7.223	6.991	6.816	6.699	2.964	2.954	2.943	2.931	2.918
250	523.15	7.521	7.231	6.999	6.825	6.709	2.968	2.958	2.947	2.936	2.923
252	525.15	7.528	7.239	7.008	6.834	6.718	2.972	2.962	2.952	2.940	2.928
254	527.15	7.536	7.247	7.016	6.843	6.727	2.976	2.966	2.956	2.945	2.933
256	529.15	7.543	7.255	7.025	6.852	6.736	2.980	2.971	2.960	2.949	2.938
258	531.15	7.550	7.263	7.033	6.861	6.745	2.984	2.975	2.965	2.954	2.942
260	533.15	7.557	7.271	7.042	6.869	6.754	2.988	2.979	2.969	2.959	2.947
262	535.15	7.564	7.279	7.050	6.878	6.763	2.992	2.983	2.974	2.963	2.952
264	537.15	7.572	7.286	7.058	6.887	6.772	2.997	2.988	2.978	2.968	2.957
266	539.15	7.579	7.294	7.066	6.895	6.781	3.001	2.992	2.982	2.972	2.961
268	541.15	7.586	7.302	7.074	6.904	6.790	3.005	2.996	2.987	2.977	2.966
270	543.15	7.593	7.309	7.082	6.912	6.798	3.009	3.000	2.991	2.981	2.971
272	545.15	7.600	7.317	7.091	6.921	6.807	3.013	3.004	2.995	2.986	2.975
274	547.15	7.607	7.324	7.098	6.929	6.815	3.017	3.009	3.000	2.990	2.980
276	549.15	7.613	7.332	7.106	6.937	6.824	3.021	3.013	3.004	2.995	2.985
278	551.15	7.620	7.339	7.114	6.945	6.832	3.025	3.017	3.008	2.999	2.989
280	553.15	7.627	7.347	7.122	6.953	6.841	3.029	3.021	3.013	3.004	2.994
282	555.15	7.634	7.354	7.130	6.962	6.849	3.033	3.025	3.017	3.008	2.998
284	557.15	7.641	7.361	7.138	6.970	6.857	3.037	3.030	3.021	3.013	3.003
286	559.15	7.647	7.369	7.145	6.978	6.866	3.041	3.034	3.026	3.017	3.008
288	561.15	7.654	7.376	7.153	6.986	6.874	3.045	3.038	3.030	3.021	3.012
290	563.15	7.661	7.383	7.161	6.993	6.882	3.049	3.042	3.034	3.026	3.017
292	565.15	7.667	7.390	7.168	7.001	6.890	3.054	3.046	3.039	3.030	3.021
294	567.15	7.674	7.397	7.176	7.009	6.898	3.058	3.051	3.043	3.035	3.026
296	569.15	7.681	7.404	7.183	7.017	6.906	3.062	3.055	3.047	3.039	3.030
298	571.15	7.687	7.411	7.191	7.025	6.914	3.066	3.059	3.051	3.043	3.035

TABLE C.3 SPECIFIC DENSITY OF WATER AT VARYING TEMPERATURES

Temperature °C fens	Density in kg/dm³ Units									
	0	**1**	**2**	**3**	**4**	**5**	**6**	**7**	**8**	**9**
0	1.000	1.000	1.000	1.000	1.000	1.000	1.000	1.000	1.000	1.000
10	1.000	1.000	1.000	0.999	0.999	0.999	0.999	0.999	0.999	0.999
20	0.999	0.999	0.999	0.998	0.998	0.997	0.997	0.996	0.995	0.995
30	0.994	0.994	0.993	0.992	0.992	0.991	0.991	0.990	0.990	0.989
40	0.989	0.988	0.988	0.987	0.987	0.986	0.986	0.985	0.985	0.985
50	0.984	0.984	0.984	0.983	0.983	0.983	0.982	0.982	0.982	0.981
60	0.981	0.981	0.980	0.980	0.980	0.979	0.979	0.979	0.978	0.978
70	0.978	0.977	0.977	0.977	0.976	0.976	0.975	0.975	0.975	0.974
80	0.974	0.973	0.973	0.972	0.972	0.971	0.971	0.970	0.970	0.969
90	0.969	0.968	0.967	0.967	0.966	0.965	0.965	0.964	0.963	0.963
100	0.962	0.961	0.960	0.960	0.959	0.958	0.957	0.956	0.955	0.955
110	0.954	0.953	0.952	0.951	0.950	0.949	0.948	0.947	0.946	0.945
120	0.944	0.943	0.942	0.941	0.940	0.939	0.938	0.937	0.936	0.935
130	0.934	0.933	0.932	0.931	0.930	0.929	0.928	0.926	0.925	0.924
140	0.923	0.922	0.921	0.920	0.919	0.918	0.917	0.916	0.915	0.914
150	0.913	0.912	0.911	0.910	0.909	0.908	0.907	0.906	0.905	0.904
160	0.903	0.902	0.901	0.900	0.899	0.898	0.897	0.896	0.895	0.894
170	0.893	0.892	0.891	0.890	0.890	0.889	0.887	0.887	0.886	0.885
180	0.884	0.883	0.882	0.881	0.881	0.880	0.879	0.878	0.877	0.876
190	0.875	0.874	0.873	0.972	0.872	0.871	0.870	0.869	0.868	0.867
200	0.866	0.865	0.864	0.863	0.862	0.861	0.860	0.859	0.858	0.857
210	0.856	0.855	0.854	0.853	0.852	0.851	0.849	0.848	0.847	0.846
220	0.845	0.844	0.842	0.841	0.840	0.839	0.837	0.836	0.835	0.833
230	0.832	0.831	0.829	0.828	0.826	0.825	0.823	0.822	0.820	0.819
240	0.817	0.816	0.814	0.813	0.811	0.809	0.808	0.806	0.804	0.803
250	0.801	0.799	0.798	0.796	0.794	0.792	0.791	0.789	0.787	0.785
260	0.783	0.782	0.780	0.778	0.776	0.774	0.772	0.770	0.769	0.767
270	0.765	0.763	0.761	0.759	0.757	0.755	0.753	0.752	0.750	0.748
280	0.746	0.744	0.742	0.740	0.738	0.736	0.735	0.733	0.731	0.729
290	0.727	0.725	0.723	0.721	0.720	0.718	0.716	0.714	0.712	0.710
300	0.708	0.707	0.705	0.703	0.701	0.699	0.697	0.695	0.693	0.691
310	0.690	0.688	0.686	0.684	0.682	0.680	0.678	0.676	0.674	0.672
320	0.669	0.667	0.665	0.663	0.661	0.658	0.656	0.654	0.651	0.649
330	0.646	0.643	0.641	0.638	0.635	0.632	0.629	0.626	0.622	0.619
340	0.616	0.612	0.608	0.605	0.601	0.597	0.593	0.588	0.584	0.580
350	0.575	0.570	0.565	0.560	0.555	0.550	0.544	0.539	0.533	0.527
360	0.521	0.515	0.508	0.502	0.495	0.489	0.482	0.475	0.468	0.461

TABLE C.4 PROPERTIES OF WATER VAPOUR UNDER VACUUM CONDITIONS

Temperature °C	Volume cum/kg				Entropy kJ/kgK				Enthalpy MJ/kg		
	Pressure in kPa										
	1	3	5	7	1	3	5	7	1	3	7
40	144.5	48.1	28.9	20.6	9.184	8.675	8.444	8.283	2.576	2.575	2.574
42	145.4	48.4	29.0	20.7	9.196	8.687	8.456	8.295	2.580	2.579	2.578
44	146.3	48.7	29.2	20.9	9.208	8.699	8.468	8.307	2.583	2.583	2.582
46	147.3	49.1	29.4	21.0	9.220	8.711	8.480	8.318	2.587	2.587	2.586
48	148.2	49.4	29.6	21.1	9.231	8.723	8.491	8.330	2.591	2.590	2.590
50	149.1	49.7	29.8	21.3	9.243	8.734	8.503	8.342	2.595	2.594	2.593
52	150.0	50.0	30.0	21.4	9.254	8.746	8.515	8.353	2.598	2.598	2.597
54	151.0	50.3	30.2	21.5	9.266	8.757	8.526	8.365	2.602	2.602	2.601
56	151.9	50.6	30.3	21.7	9.277	8.769	8.538	8.376	2.606	2.605	2.605
58	152.8	50.9	30.5	21.8	9.289	8.780	8.549	8.388	2.610	2.609	2.608
60	153.7	51.2	30.7	21.9	9.300	8.791	8.560	8.399	2.613	2.613	2.612
62	154.7	51.5	30.9	22.1	9.311	8.803	8.572	8.411	2.617	2.617	2.616
64	155.6	51.8	31.1	22.2	9.322	8.814	8.583	8.422	2.621	2.620	2.620
66	156.5	52.1	31.3	22.3	9.333	8.825	8.594	8.433	2.625	2.624	2.624
68	157.4	52.4	31.5	22.5	9.345	8.836	8.605	8.444	2.628	2.628	2.627
70	158.3	52.8	31.6	22.6	9.356	8.847	8.616	8.455	2.632	2.632	2.631
72	159.3	53.1	31.8	22.7	9.366	8.858	8.627	8.466	2.636	2.636	2.635
74	160.2	53.4	32.0	22.9	9.377	8.869	8.638	8.477	2.640	2.639	2.639
76	161.1	53.7	32.2	23.0	9.388	8.880	8.649	8.488	2.643	2.643	2.642
78	162.0	54.0	32.4	23.1	9.399	8.891	8.660	8.499	2.647	2.647	2.646
80	163.0	54.3	32.6	23.3	9.410	8.901	8.671	8.509	2.651	2.651	2.650
82	163.9	54.6	32.8	23.4	9.420	8.912	8.681	8.520	2.655	2.654	2.654
84	164.8	54.9	32.9	23.5	9.431	8.923	8.692	8.531	2.658	2.658	2.658
86	165.7	55.2	33.1	23.6	9.441	8.933	8.702	8.541	2.662	2.662	2.661
88	166.7	55.5	33.3	23.8	9.452	8.944	8.713	8.552	2.666	2.666	2.665
90	167.6	55.8	33.5	23.9	9.462	8.954	8.723	8.562	2.670	2.669	2.669
92	168.5	56.1	33.7	24.0	9.473	8.965	8.734	8.573	2.673	2.673	2.673
94	169.4	56.5	33.9	24.2	9.483	8.975	8.744	8.583	2.677	2.677	2.677
96	170.4	56.8	34.0	24.3	9.493	8.985	8.755	8.593	2.681	2.681	2.680
98	171.3	57.1	34.2	24.4	9.503	8.996	8.765	8.604	2.685	2.685	2.684
100	172.2	57.4	34.4	24.6	9.514	9.006	8.775	8.614	2.689	2.688	2.688
102	173.1	57.7	34.6	24.7	9.524	9.016	8.785	8.624	2.692	2.692	2.692
104	174.0	58.0	34.8	24.8	9.534	9.026	8.795	8.634	2.696	2.696	2.695
106	175.0	58.3	35.0	25.0	9.544	9.036	8.805	8.644	2.700	2.700	2.699
108	175.9	58.6	35.2	25.1	9.554	9.046	8.815	8.654	2.704	2.704	2.703
110	176.8	58.9	35.3	25.2	9.564	9.056	8.825	8.664	2.708	2.707	2.707
112	177.7	59.2	35.5	25.4	9.574	9.066	8.835	8.674	2.711	2.711	2.711
114	178.7	59.5	35.7	25.5	9.583	9.076	8.845	8.684	2.715	2.715	2.715
116	179.6	59.8	35.9	25.6	9.593	9.085	8.855	8.694	2.719	2.719	2.718
118	180.5	60.2	36.1	25.8	9.603	9.095	8.865	8.704	2.723	2.723	2.722
120	181.4	60.5	36.3	25.9	9.613	9.105	8.874	8.713	2.727	2.726	2.726
122	182.4	60.8	36.5	26.0	9.622	9.115	8.884	8.723	2.730	2.730	2.730
124	183.3	61.1	36.6	26.2	9.632	9.124	8.894	8.733	2.734	2.734	2.734

TABLE C.4 PROPERTIES OF WATER VAPOUR UNDER VACUUM CONDITIONS—*continued*

Temperature °C	Volume cum/kg				Entropy kJ/kgK Pressure in kPa				Enthalpy MJ/kg		
	1	3	5	7	1	3	5	7	1	3	7
126	184.2	61.4	36.8	26.3	9.641	9.134	8.903	8.742	2.738	2.738	2.737
128	185.1	61.7	37.0	26.4	9.651	9.143	8.913	8.752	2.742	2.742	2.741
130	186.0	62.0	37.2	26.6	9.660	9.153	8.922	8.761	2.746	2.745	2.745
132	187.0	62.3	37.4	26.7	9.670	9.162	8.932	8.771	2.749	2.749	2.749
134	187.9	62.6	37.6	26.8	9.679	9.172	8.941	8.780	2.753	2.753	2.753
136	188.8	62.9	37.7	27.0	9.689	9.181	8.950	8.790	2.757	2.757	2.756
138	189.7	63.2	37.9	27.1	9.698	9.190	8.960	8.799	2.761	2.761	2.760

TABLE C.5 SPECIFIC VOLUME OF SUPERHEATED STEAM

Temperature °C	Volume in m^3/kg Pressure in bars absolute										
	0.50	0.75	1.00	1.25	1.50	2.00	2.50	3.00	4.00	5.00	6.00
90	3.324										
92	3.343										
94	3.362										
96	3.381										
98	3.399										
100	3.418	2.271	1.697								
102	3.437	2.283	1.706								
104	3.456	2.296	1.716								
106	3.475	2.309	1.726								
108	3.494	2.322	1.735								
110	3.513	2.334	1.745	1.391							
112	3.532	2.347	1.755	1.399							
114	3.551	2.360	1.764	1.407							
116	3.570	2.372	1.774	1.415							
118	3.588	2.385	1.783	1.422							
120	3.607	2.398	1.793	1.430	1.188						
122	3.626	2.410	1.803	1.438	1.195						
124	3.645	2.423	1.812	1.446	1.201						
126	3.664	2.436	1.822	1.453	1.208						
128	3.683	2.448	1.831	1.461	1.214						
130	3.701	2.461	1.841	1.469	1.221	0.910	0.724				
132	3.720	2.474	1.850	1.476	1.227	0.915	0.728	0.603			
134	3.739	2.486	1.860	1.484	1.234	0.920	0.732	0.607			
136	3.758	2.499	1.870	1.492	1.240	0.925	0.736	0.610			
138	3.777	2.512	1.879	1.500	1.247	0.930	0.740	0.614			
140	3.795	2.524	1.889	1.507	1.253	0.935	0.744	0.617	0.458		
142	3.814	2.537	1.898	1.515	1.259	0.940	0.748	0.620	0.460		
144	3.833	2.549	1.908	1.523	1.266	0.945	0.752	0.624	0.463		
146	3.852	2.562	1.917	1.530	1.272	0.950	0.756	0.627	0.466		
148	3.870	2.575	1.927	1.538	1.279	0.955	0.760	0.631	0.468		

TABLE C.5 SPECIFIC VOLUME OF SUPERHEATED STEAM—*continued*

Temperature °C	Volume in m^3/kg Pressure in bars absolute										
	0.50	0.75	1.00	1.25	1.50	2.00	2.50	3.00	4.00	5.00	6.00
150	3.889	2.587	1.936	1.546	1.285	0.960	0.764	0.634	0.471	0.373	0.308
152	3.908	2.600	1.946	1.553	1.292	0.965	0.768	0.637	0.473	0.375	0.309
154	3.927	2.612	1.955	1.561	1.298	0.969	0.772	0.641	0.476	0.377	0.311
156	3.945	2.625	1.965	1.569	1.305	0.974	0.776	0.644	0.479	0.379	0.313
158	3.964	2.638	1.974	1.576	1.311	0.979	0.780	0.647	0.481	0.381	0.315
160	3.983	2.650	1.984	1.584	1.317	0.984	0.784	0.651	0.484	0.384	0.317
162	4.001	2.663	1.993	1.592	1.324	0.989	0.788	0.654	0.486	0.386	0.318
164	4.020	2.675	2.003	1.599	1.330	0.994	0.792	0.657	0.489	0.388	0.320
166	4.039	2.688	2.012	1 607	1 337	0 999	0 796	0 661	0 491	0 390	0 322
168	4 058	2 700	2 022	1 614	1 343	1 004	0 800	0.664	0.494	0.392	0.324
170	4.076	2.713	2.031	1.622	1.349	1.008	0.804	0.667	0.497	0.394	0.326
172	4.095	2.725	2.041	1.630	1.356	1.013	0.808	0.671	0.499	0.396	0.327
174	4.114	2.738	2.050	1.637	1.362	1.018	0.812	0.674	0.502	0.398	0.329
176	4.132	2.750	2.059	1.645	1.368	1.023	0.815	0.677	0.504	0.400	0.331
178	4.151	2.763	2.069	1.652	1.375	1.028	0.819	0.680	0.507	0.402	0.333
180	4.170	2.775	2.078	1.660	1.381	1.033	0.823	0.684	0.509	0.404	0.335
182	4.188	2.788	2.088	1.668	1.388	1.037	0.827	0.687	0.512	0.407	0.336
184	4.207	2.800	2.097	1.675	1.394	1.042	0.831	0.690	0.514	0.409	0.338
186	4.226	2.813	2.107	1.683	1.400	1.047	0.835	0.694	0.517	0.411	0.340
188	4.244	2.825	2.116	1.690	1.407	1.052	0.839	0.697	0.519	0.413	0.342
190	4.263	2.838	2.125	1.698	1.413	1.057	0.843	0.700	0.522	0.415	0.343
192	4.282	2.850	2.135	1.706	1.419	1.061	0.847	0.703	0.524	0.417	0.345
194	4.300	2.863	2.144	1.713	1.426	1.066	0.851	0.707	0.527	0.419	0.347
196	4.319	2.875	2.154	1.721	1.432	1.071	0.854	0.710	0.529	0.421	0.348
198	4.337	2.888	2.163	1.728	1.438	1.076	0.858	0.713	0.532	0.423	0.350
200	4.356	2.900	2.172	1.736	1.445	1.081	0.862	0.716	0.534	0.425	0.352
202	4.375	2.913	2.182	1.743	1.451	1.085	0.866	0.720	0.537	0.427	0.354
204	4.393	2.925	2.191	1.751	1.457	1.090	0.870	0.723	0.539	0.429	0.355
206	4.412	2.938	2.201	1.758	1.464	1.095	0.874	0.726	0.542	0.431	0.357
208	4.431	2.950	2.210	1.766	1.470	1.100	0.878	0.729	0.544	0.433	0.359
210	4.449	2.963	2.219	1.773	1.476	1.104	0.881	0.733	0.547	0.435	0.360
212	4.468	2.975	2.229	1.781	1.482	1.109	0.885	0.736	0.549	0.437	0.362
214	4.486	2.988	2.238	1.789	1.489	1.114	0.889	0.739	0.552	0.439	0.364
216	4.505	3.000	2.248	1.796	1.495	1.119	0.893	0.742	0.554	0.441	0.366
218	4.524	3.012	2.257	1.804	1.501	1.123	0.897	0.746	0.557	0.443	0.367
220	4.542	3.025	2.266	1.811	1.508	1.128	0.901	0.749	0.559	0.445	0.369
222	4.561	3.037	2.276	1.819	1.514	1.133	0.904	0.752	0.561	0.447	0.371
224	4.579	3.050	2.285	1.826	1.520	1.138	0.908	0.755	0.564	0.449	0.372
226	4.598	3.062	2.294	1.834	1.526	1.142	0.912	0.758	0.566	0.451	0.374
228	4.617	3.075	2.304	1.841	1.533	1.147	0.916	0.762	0.569	0.453	0.376
230	4.635	3.087	2.313	1.849	1.539	1.152	0.920	0.765	0.571	0.455	0.377
232	4.654	3.100	2.322	1.856	1.545	1.157	0.924	0.768	0.574	0.457	0.379
234	4.672	3.112	2.332	1.864	1.552	1.161	0.927	0.771	0.576	0.459	0.381
236	4.619	3.124	2.341	1.871	1.558	1.166	0.931	0.774	0.578	0.461	0.382
238	4.709	3.137	2.350	1.879	1.564	1.171	0.935	0.778	0.581	0.463	0.384
240	4.728	3.149	2.360	1.886	1.570	1.176	0.939	0.781	0.583	0.465	0.386

TABLE C.5 SPECIFIC VOLUME OF SUPERHEATED STEAM—*continued*

Temperature °C	Volume in m^3/kg Pressure in bars absolute										
	0.50	0.75	1.00	1.25	1.50	2.00	2.50	3.00	4.00	5.00	6.00
242	4.747	3.162	2.369	1.894	1.577	1.180	0.943	0.784	0.586	0.467	0.387
244	4.765	3.174	2.378	1.901	1.583	1.185	0.946	0.787	0.588	0.469	0.389
246	4.784	3.186	2.388	1.909	1.589	1.190	0.950	0.790	0.591	0.471	0.391
248	4.802	3.199	2.397	1.916	1.595	1.195	0.954	0.794	0.583	0.473	0.392
250	4.821	3.211	2.406	1.924	1.602	1.199	0.958	0.797	0.595	0.475	0.394
252	4.839	3.224	2.416	1.931	1.608	1.204	0.962	0.800	0.598	0.477	0.396
254	4.858	3.236	2.425	1.939	1.614	1.209	0.965	0.803	0.600	0.478	0.397
256	4.876	3.248	2.434	1.946	1.620	1.213	0.969	0.806	0.603	0.480	0.399
258	4.895	3.261	2.444	1.954	1.627	1.218	0.973	0.809	0.605	0.482	0.401
260	4.913	3.273	2.453	1.961	1.633	1.223	0.977	0.813	0.607	0.484	0.402
262	4.932	3.286	2.462	1.968	1.639	1.227	0.980	0.816	0.610	0.486	0.404
264	4.951	3.298	2.472	1.976	1.645	1.232	0.984	0.819	0.612	0.488	0.405
266	4.969	3.310	2.481	1.983	1.652	1.237	0.988	0.822	0.615	0.490	0.407
268	4.988	3.323	2.490	1.991	1.658	1.242	0.992	0.825	0.617	0.492	0.409
270	5.006	3.335	2.500	1.998	1.664	1.246	0.996	0.828	0.619	0.494	0.410
272	5.025	3.348	2.509	2.006	1.670	1.251	0.999	0.832	0.622	0.496	0.412
274	5.043	3.360	2.518	2.013	1.677	1.256	1.003	0.835	0.624	0.498	0.414
276	5.062	3.372	2.528	2.021	1.683	1.260	1.007	0.838	0.627	0.500	0.415
278	5.080	3.385	2.537	2.028	1.689	1.265	1.011	0.841	0.629	0.502	0.417
280	5.099	3.397	2.546	2.036	1.695	1.270	1.014	0.844	0.631	0.504	0.418
282	5.117	3.409	2.555	2.043	1.701	1.274	1.018	0.847	0.634	0.506	0.420
284	5.136	3.422	2.565	2.051	1.708	1.279	1.022	0.851	0.636	0.507	0.422
286	5.154	3.434	2.574	2.058	1.714	1.284	1.026	0.854	0.639	0.509	0.423
288	5.173	3.447	2.583	2.065	1.720	1.288	1.029	0.857	0.641	0.511	0.425

APPENDIX D

District heating statistics

TABLE D.1 DISTRICT HEATING STATISTICS BY COUNTRY (1976)

Country	Number of systems	Number of CHP plants	Number of pure heating plants	Total heat capacity (MW)	Annual utilisation of CHP h/year	Notes
Australia	3	3	—	194	1440	Industrial use only
Bulgaria	38	25	13	6555	4760	Only pure heating plants in excess of 32 TJ/a are given here
Denmark	241	26	215	11 676	3440	Plants which do not belong to members of the Danish DH Assn. are excluded
Finland	104	83	21	12 315	3530	Only members of the Finnish DH Assn. are included
France	301	153	148	15 035	5910	Only public authority systems are included
Holland	5	5	—	985	3220	
Hungary	279	62	217	8109	4330	
Japan	349	349	—	68 200	5310	
Korea	11	11	—	481	—	No pure heating stations included
New Zealand	3	3	—	301	6590	CHP plants are all industrial
Poland	260	260	—	34 740	3940	

TABLE D.1 DISTRICT HEATING STATISTICS BY COUNTRY (1976)—*continued*

Country	Number of systems	Number of CHP plants	Number of pure heating plants	Total heat capacity (MW)	Annual utilisation of CHP h/year	Notes
Spain	100	100	—	—	2640	CHP plants are all industrial
Sweden	285	246	39	15 200	3890	Group heating schemes and industrial turbines without CHP excluded
United Kingdom	1701	179	1520	5818	4110	CHP plants except for 1 (Pimlico) are all industrial
USSR	1700	900	800	331 000	5050	Only heating systems in excess of 58 MW included
West Germany	—	106	378	120 800	—	

Source: *Fernwärme International* May 1978

TABLE D.2 DISTRICT HEATING STATISTICS FOR FINLAND (1978)

Group	Number of networks	Number of consumers	Capacity (MW)	Number of CHP stations	Heating capacity (MW)	Power capacity (MW)	Direct heating plants	
							Fixed	Mobile
1	22	9330	3824	27	3034	1192	36	26
2	37	1785	443	—	—	—	43	16
3	177	3010	518	—	—	—	—	283

TABLE D.3 DISTRICT HEATING STATISTICS FOR HELSINKI

Heating capacity requirement	1684 MW
% of city catered for by DH network	65%
Number of consumers	4143
Steam consumers	28
Total annual heat consumed	4.384 TWh
Income from sale of heat	230.9 million Finnmarks
% income from sale of heat in operation of CHP stations	38%
Ration of CHP to total electric power production	75%
Average price per GJ	14.72 Finnmarks

TABLE D.4 ANALYSIS OF DISTRICT HEATING USE IN HELSINKI BY CONSUMER GROUP

Consumer	Share of capacity	Average consumption
	%	kW
Private dwellings	53.1	266
Offices and public buildings	19.1	994
Industrial undertakings	11.1	1108
Schools	4.4	649
Hospitals	2.6	1162
Miscellaneous	9.7	—
	100.0	398

TABLE D.5 DISTRICT HEATING STATISTICS FOR THE UNITED KINGDOM (1979)

Name	District	Number of dwellings	Capacity (MW)	Fuel
1 London area				
198 schemes of between 0.3 and 54 MW capacity. The largest schemes are as follows:				
Thamesmead	Greenwich	5200	54.0	Gas
Roundshaw	Sutton	3771	37.7	Oil
Ferrier	Greenwich	1908	19.1	Oil
Grahame Park	Barnet	1809	18.1	Oil
Pepys	Greenwich	1334	13.3	Oil
Alton West	Wandsworth	1123	11.2	Oil
Kingshold	Hackney	771	7.7	Oil
Hanwell	Ealing	733	7.3	Oil
Mottingham	Bromley	499	5.0	Gas
Gloucester Grove	Southwark	1210	12.1	Oil
Alsen Road	Islington	925	9.3	Gas
Bartholomew Square	Islington	925	9.3	Gas
Brentford Dock	Hounslow	591	5.9	Gas
Phipps Bridge Road	Merton	596	6.0	Gas
Ethelred Estate	Lambeth	626	6.3	Oil

2 England—outside London

147 schemes of which the largest are:
Nottingham, Leicester, Sheffield, Rotherham, Mansfield, Newcastle-on-Tyne, Leeds, Greater Manchester, Oldham, Norwich.

TABLE D.5 DISTRICT HEATING STATISTICS FOR THE UNITED KINGDOM (1979)—*continued*

Name	District	Number of dwellings	Capacity (MW)	Fuel
3 Scotland				
10 schemes in the following towns:				
Glasgow, Edinburgh, Aberdeen, Greenock, Paisley, Aviemore and East Kilbride				
4 Wales				
11 schemes, mainly in Cardiff, Swansea, Aberdare and Flint				
5 Northern Ireland				
15 schemes, mainly in Belfast				

TABLE D.6 DISTRICT HEATING STATISTICS FOR THE UNITED STATES OF AMERICA AND CANADA (1975)

City and state	Company	Annual steam send out (million tonnes)	Maximum steam capacity (1000 tonnes/hour)
New York NY	Consolidated Edison Inc.	19.3	5.78
Philadelphia PA	Philadelphia Electric Co.	3.92	1.13
Detroit Mich.	Detroit Edison Co.	3.19	0.78
Boston Mass	Boston Edison Co.	2.84	0.84
Indianapolis Ind.	Indianapolis Power and Light Co.	2.16	0.59
Rochester NY	Rochester Gas and Electric Corp.	1.87	0.51
Milwaukee Wis	Wisconsin Electric Power Corp.	1.21	0.42

Baltimore Md.	Baltimore Gas and Electric Co.	1.14	0.31
Cleveland Ohio	Cleveland Electric Illum. Co.	1.18	0.41
Dayton Ohio	Dayton Power and Light Co.	0.89	0.23
Denver Colo.	Public Service Co. of Colorado	0.63	0.20
Pittsburgh Pa.	Allegheny County Steam Co.	0.74	0.30
Seattle Wash.	Seattle Steam Corp.	0.68	0.27
Toronto Ont. Canada	Toronto Hydro-electric system	0.66	0.20
Harrisburg Pa.	Pennsylvania Power and Light Co.	0.52	0.17
San Francisco Calif.	Pacific Gas and Electric Co.	0.48	0.15
Lansing Mich.	Board of water and light	0.42	0.101
Grand Rapids Mich.	Consumers Power Co.	0.353	0.103
Duluth Minn.	Duluth Steam Corp.	0.333	0.103
Akron Ohio	Ohio Edison Co.	0.356	0.093
Vancouver BC Canada	Central Heat Distribution Co.	0.307	0.118
Eugene Oregon	Eugene Water and Electric Board	0.264	0.064
Winnipeg Man. Canada	Winnipeg Hydro CSH Division	0.264	0.102
Spokane Wash.	Washington Waterpower Co.	0.238	0.127
Toledo Ohio	Toledo Edison Co.	0.239	0.086
Birmingham Ala.	Alabama Power Co.	0.276	0.092
Youngstown Ohio	Ohio Edison Co.	0.266	0.083
St Paul Minn.	Northern States Power Co.	0.225	0.109
Hartford Conn.	Hartford Steam Service Co.	0.157	0.070
Rochester Minn.	Rochester Public Utilities	0.130	0.041
Houston Texas	Houston Natural Gas Corp.	0.118	0.068
Wilkes-Barre Pa.	Wilkes-Barre Steam Heat Co.	0.125	0.072
Concord NH.	Concord Steam Corp.	0.071	0.030
Piqua Ohio	Municipal Power System	0.080	0.027
Springfield Ohio	Ohio Edison Co.	0.079	0.036

TABLE D.6 DISTRICT HEATING STATISTICS FOR THE UNITED STATES OF AMERICA AND CANADA (1975)—*continued*

City and state	Company	Annual steam send out (million tonnes)	Maximum steam capacity (1000 tonnes/hour)
St Joseph Mo.	St Joseph Light and Power Co.	0.078	0.036
Southfield Mich.	Dayton Hudson Props.	0.0472	0.0191
Nashville Tenn.	Nashville Thermal Transfer Co.	0.0372	0.0295
Rice Lake Wis.	Rice Lake Steam Corp.	0.0290	0.0130
San Antonio Texas	City Water Board	0.0295	0.0105

TABLE D.7 ADDITIONAL STATISTICS FOR THREE SELECTED USA SYSTEMS

	New York	Duluth Minn.	San Antonio Texas
Maximum hourly send-out (1000 tonnes/hour)	6.865	0.1392	0.0363
Annual system load factor (%)	38.16	36.90	32.00
Number of customers served	2541	260	28
Average revenue per tonne of steam sold (US $)	10.23	7.75	2.43
Number of meters installed:			
steam flow	4713	91	1
condensate	105	191	8
Total pipeline length (km)	173.3	22.25	1.525
Total radiation surface (1000 m^2)	223.0	10.72	1.282
Fuels used	No. 2 and No. 6 oil 88% Natural gas 12%	Bit. coal and natural gas	Natural gas

TABLE D.8 DISTRICT HEATING STATISTICS FOR WEST GERMANY (1978)

There are 111 companies operating district heating networks in West Germany, details of which are as follows:

Total connected capacity during 1978	27 220.5 MW
Space heating and hot water	24 892.0 MW
Industrial heating	2328.5 MW
Increase in capacity over 1977	941.6 MW
Total heat fed into network during 1978	1.836×10^{11} MJ
Maximum heat loading	16.182 GW
Period of maximum load	3146 hours
Ratio of average to maximum load	0.60

TABLE D.9 STATISTICS FOR THE LARGEST WEST GERMAN DISTRICT HEATING SCHEMES

City	Company	Connected capacity (MW)	Heat supplied to network (TJ/a)	Length of pipelines (km)
Hamburg	HEV	2266.6	17 098	342.3
Berlin	Bewag	1717.8	11 039	260
Munich	Stadtwerke	1671.0	12 795	288.6
Hamburg	Favorit GmbH	1108.8	6791	278.2
Essen	Steag	1038.0	7340	208.2
Wolfsburg	VW Werk GmbH	1135.4	6370	264.1
Mannheim	Rhein Neckar AG	793.9	5071	165.5
Dortmund	VEW	780.1	6018	78.7
Stuttgart	TWS	660.4	4151.3	103.0
Frankfurt/aM	Stadtwerke	632.7	4414.3	127.8
Kiel	Stadtwerke	620.0	3355.4	138.2
Cologne	GEW Werke AG	563.6	3800.6	149.4

TABLE D.10 COMPLETE STATISTICS FOR THREE TYPICAL GERMAN DISTRICT HEATING SYSTEMS

	Hanover	Erlangen	Fürth
Connected capacity (MW)	507.5	124.6	8.8
Increase 1977 (MW)	26.5	4.2	1.4
Maximum capacity (MW)	523.3	118.6	14.7
from own CHP (%)	82	41	nil
from own pure heating stations (%)	18	59	100
Loading ratio (%)	50	65	59
Power production (MW)	316	6.4	nil
Net annual power produced (MWh)	320.7	28.0	nil
Heat supplied to net (TJ/a)	2926.3	826.0	63.8
Period of maximum load (h/a)	3221	2847	3387
Annual degree days	3675	3187	3650
Length of network (km)	93.0	13.0	4.7
Number of sub-stations	978	99	50
Average length of pipeline per sub-station (m)	95	131	94

TABLE D.11 COSTS FOR CONSUMERS AT A SELECTION OF WEST GERMAN DISTRICT HEATING SCHEMES (MARCH 1979)

Scheme	Company	Cost in DM per GJ	Fuel
Hamburg	HEV	12.49 – 20.36	Coal in CHP stations
Hamburg	Favorit	12.67 – 20.79	Heavy oil in direct boiler plants
Berlin	Bewag	10.32 – 16.79	Coal in CHP stations
Essen	Steag	11.85 – 20.99	Gas in direct boiler plants
Fürth	STW AG	10.96 – 22.60	Gas in direct boiler plants
Duisburg	STW AG	10.49 – 14.14	Coal in CHP stations
Bochum	STW AG	13.16 – 13.41	Gas in CHP stations
Saarbrucken	Saarberg FW GmbH	13.28 – 18.53	Coal in direct boiler plants
Stuttgart	TWS	9.97 – 17.12	Heavy oil in CHP plants

APPENDIX E

Computer program to determine heat and pressure losses in district heating networks

Abstract

A Fortran computer program was compiled at the University of Salford, England to calculate heat and pressure losses in district heating networks. Standard theoretical and empirical equations were employed to deal with the numerous parameters involved. The program is of use for the design of new networks and for the maintenance of old ones.

Introduction

When designing a mathematical model of a district heating system for solution by means of a computer, one is faced by the problem that district heating distribution networks are so very variable in nature. Some pipeline systems are laid singly into the ground and others are buried as pairs. Pipelines are often carried along in the open while at other times they may be positioned along cellars or through railway tunnels. Some manufacturers of pipelines have carried out tests on the insulation properties of the various insulation layers and can quote the U-value to some degree of accuracy. In other cases no such information is available. Even the material from which the distribution network is constructed may differ. Mild steel is the most common but many systems use either plastic or copper to carry the distribution medium. The medium used may also vary: most district heating networks in Europe employ hot water as the heating medium, but there are still many steam-using systems. In such cases the steam may be condensed and the condensate recovered; alternatively the condensate may be discarded to the sewer as is common in the United States.

One of the authors has spent some years in developing two computer programs at the University of Salford, which can handle the huge numbers of parameters that come into play when calculating heat and pressure losses in district heating networks. The program used was FORTRAN; the computer was an ICL 1904 S operating at the University of Salford (**plate 78**). Two programs were developed

LABELS: this consists of about 320 statements and sorts out the input data into a form in which they can be readily understood and checked by the person setting the problem.

DIPROG: this is a fairly complex Fortran program of about 1250 state-

ments which tabulates the following data and results:
(a) Reference number of system studied
(b) Length of pipeline (metres)
(c) Net heat flow (kilowatts)
(d) Heat loss (kilowatts)
(e) U-value of pipeline (W/K per metre run)
(f) Heat loss as a percentage of the net heat flow
(g) The head loss due to friction (kilopascals)
(h) The pump work needed to pump the fluid against the pipeline friction.
Each calculation is carried out in subroutines attached to the main program. The computer requires approximately 11 seconds to calculate the data for each section of pipeline. Finally the various results are summated to produce the following information:
(a) Total pipeline length
(b) Total heat throughput
(c) Total heat loss
(d) Percentage heat loss
(e) Total pipeline head loss
(f) Total net pump work done.
In spite of the considerable complexity of the program and the numerous sophisticated equations used, the print-out comes in a very simple form which is read very easily.

General system parameters

As an introduction to the data supplied for each pipeline section, one requires to use a card which gives the following integers:
N (reference number of the system).
INDEX (classification of the type of pipeline system involved). Eleven such classifications are envisaged; they include single pipelines carrying either supply hot water or return water or steam which can be positioned above ground or buried underground. They also include twin pipelines in a common conduit which are buried in the soil. In some cases the manufacturer can quote the U-value of the pipeline or even the overall heat transfer coefficient. In other cases no such figure is available and the overall heat transfer coefficient has to be worked out from the given pipeline insulation material and the thickness data. The INDEX integer steers the computer to the correct subroutine for carrying out each of these separate and different computations.
IND (nature of pipeline). This instructs the computer whether the pipeline is a flow or return main or whether it is a twin main with both a flow and return pipe. It also gives the information as to whether the medium used is hot water or steam.
MAT (material of construction). This specifies whether the material from which the pipeline is constrcuted is steel, copper or plastic and thence

selects the correct equation to calculate the Fanning friction factor, which is needed to calculate head losses and pump work required.
ISH (sheltered position). This gives the condition in which the pipeline is being installed, in a sheltered position such as a cellar, a railway tunnel or other conduit etc. In such circumstances the computer automatically ignores the external windspeed and also substitutes a given temperature value for the normal external ambient temperature value used for the rest of the computations.
NSR (non-standard return temperature). This permits the substitution of a value for the return temperature of the circulated water which is not the standard one given at the beginning of the data file. In some cases the return water is used for some low temperature purpose, such as underfloor heating, so that its temperature afterwards is non-standard. In other cases variable return temperatures can be experienced when returning condensate from steam mains to the district heating station.

Primary data required

Apart from the various integers already mentioned, the computer needs to know the following:
(a) External temperature (degrees C)
(b) Average wind speed (m/s)
(c) Atmospheric pressure (bars)
(d) Machine efficiency of pumping systems employed
(e) Flow temperature of hot water pumped through system (°C)
(f) Return temperature of hot water from system (°C)
(g) Steam temperature if steam is used (°C)
(h) Steam pressure (bars).

Overall heat transfer coefficients

U-values of pipes known

The resistance to heat flow from the fluid flowing inside a pipeline to the ambient air can be subdivided into three parts:
(a) The thermal resistance of the internal boundary layer within the flowing fluid
(b) The resistance of the pipeline plus the insulation layers
(c) The resistance of the surface layer on the outside of the pipeline, if run in the open, or the earth resistance if buried.
If the thermal resistance of pipeline plus insulation is known, then one only needs to evaluate (a) and (c). The Dittus-Boelter equation (**1**) is used for the calculation of the boundary layer heat transfer coefficient from the body of the liquid to the wall of the pipeline. This can be quoted as:

$$h = A\pi\,(N_{Re})^{B} \times (N_{Pr})^{c} \times k_{(water)}\ \text{W/m K}, \qquad \textbf{(1)}$$

where
N_{Re} is the Reynold number of the flowing fluid (d1)
N_{Pr} is the Prandtl number of the flowing fluid (d1)

$k_{(water)}$ is the thermal conductivity of water at the temperature of the flowing fluid (W/m K)
A, B, and C are constants which depend upon the magnitudes of the Reynold and Prandtl numbers and are worked out automatically by the computer before insertion into the equation. In the case of steam, differences between the temperature of the flowing steam and the wall surface are negligibly small and are ignored.

Surface heat transfer coefficients of open pipelines
Heat losses from the surface of pipelines which are run in the open are two-fold: by convective heat transfer and radiation heat transfer. The two are commonly bulked to give h_o, the external boundary layer coefficient (W/m K). To determine the convection heat transfer from the surface of the insulated pipeline when a wind is blowing, one uses the Hilpert equation which is as follows:

$$h_c = \frac{A\, k_{air}\, (N_{Re})^B}{2\, r_{ext}}\ \mathrm{W/m^2\,K} \tag{2}$$

where
h_c is the convection heat transfer coefficient from the surface
k_{air} is the thermal conductivity of the air at the temperature in question (W/m K)
N_{Re} is the Reynold number of the air (d1)
r_{ext} is the external radius of pipeline plus insulation (metres).
A and B are constants depending upon the magnitude of the Reynold number, which are worked out automatically by the computer.

If the air velocity is less than 0.1 m/s the Hilpert equation fails and it is then necessary to use a modified Koch equation:

$$h_c = 0.66\,(t_{ins} - t_{ext})^{0.25}\,(r_{ext})^{-0.25}\ \mathrm{W/m^2\,K} \tag{3}$$

where
t_{ins} is the insulation surface temperature (°C)
t_{ext} is the external ambient temperature (°C)
r_{ext} is the external radius (metres).
This happens when the pipeline is laid in sheltered conditions. The heat loss due to convection is worked out as follows:
$Q_c = 2\,\pi\, r_{ext}\, h_c$ W/metre run
The computer automatically selects the correct equation and the correct constants on the basis of the data fed in.

The radiation heat transfer is calculated according to the Stefan Boltzmann equation. This is written as:

$$Q_{rad} = 5.673 \times 10^{-8}\, C\,(T^4_{ins} - T^4_{ext}) \tag{4}$$

where
T_{ins} is the insulation surface temperature (kelvins) (°C + 273.15)
T_{ext} is the external temperature (kelvins)

Q'_{rad} is the heat transfer (W/m^2) from the surface of the pipe by radiation agency only. To obtain Q_{rad} (which is the heat loss due to radiation) we write:
$Q'_{rad} = Q'_{rad} \times 2 \pi r_{ext}$ W/metre run
C is the emissivity constant of the insulation surface (d1)
Q'_{rad} and Q_c are bulked and equated as Q, which is the total rate of heat transfer from the centre of the fluid to the external air. The computer now carries out complex loop operations to determine the temperature conditions at the inside of the pipeline and also at the surface of the insulation layer so the condition of a uniform flow of heat through all three portions (namely the internal surface layer, the insulation layer and the external surface layer) is obeyed. Once a balance has been achieved the heat loss can be quoted in terms of watts per metre run of pipe.

Buried pipelines

Where the pipeline is buried in the soil, one obviously does not evaluate the surface conductivity from the insulation layer; instead, it becomes necessary to determine the thermal resistance of the earth cover, using the Louden equation:

$$R_{earth} = \frac{1}{2 \pi k_{earth}} (1n\ h/r_{ext}) (1 + (1 - (r_{ext}/h)^2)^{\frac{1}{2}}\ mK/W \qquad (5)$$

where
R is the resistance of the earth to heat flow (m K/W)
k_{earth} is the thermal conductivity of the soil (W/m K)
h is the depth of soil between the centre of the pipe and the surface (metres)
k_{earth} can be fed into the computer in two ways:
(a) It can be given as an input data, provided that it is already known
(b) It can be worked out by the computer from the dry density of the soil and its moisture content. Both these can be evaluated easily by simple tests in non-specialised laboratories. A regression analysis equation was developed in which the thermal conductivity of dry soil is given in terms of its density. A second regression analysis correlates this with the moisture content of the soil. This regression analysis works well at low moisture contents, but when the soil is very wet it approximates the conductivity of convection-free water, ie it approaches a value of 1.9642 W/m K. This value is inserted into the program as the limiting value of the thermal conductivity of the soil.

It is assumed that the surface of the soil has the same temperature as the external ambient air. A multiple loop program calculates the heat flows through the various sections of the pipeline system, which are now the internal boundary layer within the pipe, the flow of heat through pipeline and insulation, and the flow of heat from the pipeline insulation to the surface of the earth cover. The temperatures at the inside surface of the pipe and at the outside surface of the insulation are adjusted until the heat flows equate.

Insulated pipelines with unknown thermal conductance

In many cases the thermal conductance values of the pipeline insulation are unknown. The computer can compute the U-value of a pipeline insulated with up to 5 different layers of insulation materials. At times, air spaces are kept between layers of insulation materials, using metal tie-rods or flanges to ensure stability. The computer has been programmed to allow for such bridged air gaps.

The thermal resistance of a concentric pipeline insulation system is evaluated from the following equation:

$$Res_{pipe} = (1n(r_1/r_p))/2\,\pi\,k_p + (1n(r_2/r_1))/2\,\pi\,k_1 + (1n(r_3/r_2))/2\,\pi\,k_2 \ldots \text{ etc Km/W} \quad \textbf{(6)}$$

The thermal conductance is equal to $1/Res_{pipe}$ W/m K

where

r_p, r_1, r_2, r_3 etc are the radii (m) of pipeline and layers of insulation material as measured from the centre of the pipe, while

k_p, k_1, k_2 etc are the thermal conductivity values of the various materials of construction of pipeline and insulation layers (W/m K).

Bridged air gap

Heat is transmitted across such an air gap by conduction, convection and radiation. Conduction takes place along the metal or other tie bars; convection across the air space by movement of the air from the hotter to the colder surface; radiation by heat transfer from the heated to the cooler surface in the form of infra-red rays.

The equation which applies to convection heat transfer across the air gap is a modified Fishenden and Saunders equation which can be quoted as follows:

$$h_c = \frac{A\,(N_{Gr})^B\,(NPr)^B}{d}\ \text{W/mK} \quad \textbf{(7)}$$

where

N_{Gr} is the Grasshof number of the air within the gap (d1)

N_{Pr} is the Prandtl number of the air within the gap (d1)

d is the width of gap (m)

A and B are constants which depend upon the magnitude of the Grasshof number; they are evaluated automatically by the computer.

The heat transfer by radiation is evaluated by the Christiansen equation:

$$Q^{-}_{rad} = \frac{5.673 \times 10^{-8}\,(T_{ig}^4 - T_{og}^4)}{1/e_{ig} + (1/e_{og} - 1)}\ \text{W/m}^2 \quad \textbf{(8)}$$

where

T_{ig} is the temperature on the inside of the gap (K)

T_{og} is the temperature on the outside of the gap (K)
e_{ig} is the emissivity on the inside of the gap (d1)
e_{og} is the emissivity on the outside of the gap (d1)
Q'_{rad} is heat transferred per m^2 pipeline surface
The heat transported by convection (Q_c) is evaluated from the following equation:
$Q_c = h_c (t_{ig} - t_{og})$ W/m run, while
$Q'_{rad} = Q'_{rad} \times 2 \pi r_{ext}$ W/m run,
when r_{ext} is the average radius of the air gap (metres) the sum $Q'_{rad} + Q_c$ gives the heat which travels across the unbridged air gap, per metre run.
When there is a bridged air gap, one employs the following equation:
$h'_{br} = h'_{unbr} (1 - f_{br}) + k_{br} f_{br}/d$ W/mK **(9)**
where
h'_{br} is the coefficient of thermal conductance of the bridged air gap (W/m^2K
h'_{unbr} is the coefficient of thermal conductance of the unbridged air gap (W/m^2K)
k_{br} is the thermal conductivity of the cold bridge (W/m K)
f_{br} is the fraction of air gap which is bridged (d1)
d is the thickness of the air gap (metres)
To convert h'_{br} and h'_{unbr} to thermal conductances per metre pipelinelength, these values must be multiplied by $2\pi\, r_{ave}$, where r_{ave} is the average radius of the air gap (metres).

The computer carries out all these calculations, provided the data are fed in correctly. Using the loop technique previously described the heat transfer across the air gap is equated with the remaining heat transfers to arrive at the overall figure.

Twin pipelines

When twin pipelines are buried in the soil, one uses the mathematical approach developed by Eneborg, coupled with the calculation of the earth resistance as calculated by the Louden equation (**5**). The pipeline insulation system is divided into four sections, and the heat transfer is calculated separately for each section. After summation, the thermal resistance is added to the thermal resistance of the soil to give the total thermal resistance of the pipeline. Twin pipelines can be calculated with up to three different widths, heights and thermal conductivities of insulation materials covering the twin pipes. It is also necessary to specify the distance between the pipes. If the twin pipelines are positioned in a conduit with an air gap it is better to use one of the standard equations instead and to treat the twin line as two single lines.

Regression analyses

In order to evaluate many of the parameters needed for calculating the heat flow, the heat loss and the pressure head loss, a number of regression analyses had to be carried out and the results obtained substituted in the main program, using the UMRCC (University of Manchester regional computer centre) software library. The computer examines a large number of pairs of data fed in and develops an empirical equation in the form of:

$y = Ax + Bx^2 + Cx^3 + Dx^4 \ldots$ etc

where A, B, C, D etc are constants.

The series is convergent so that increased accuracy is achieved with an increasing number of terms. The following table gives the x and y terms of the regression analyses which were carried out for the main program. The numbers ('Data' column) are those of sources in the list of references at the end of this appendix.

x	y	Data
water temperature	specific heat of water	10
water temperature	thermal conductivity of water	11
water temperature	viscosity of water	11
water temperature	specific volume of water	11
water temperature	enthalpy of water	11
density of soil	dry conductivity of soil	12
% moisture content of soil	moisture factor of soil	10
temperature of air	density of air	12
steam pressure	saturated steam temperature	11
steam pressure	enthalpy of steam	11
steam pressure	specific heat of steam	11
steam pressure	density of steam	11

Calculation of net heat transported

The enthalpy of water at any specific temperature is calculated by a regression analysis. The product of the flow rate and the enthalpy of water in the flow mains is given as a positive value; the product of the flow rate and enthalpy of water in the return mains is given a negative value, irrespective of whether the flowing water passes through single pipelines or combined flow and return mains.

In the case of steam this calculation is somewhat more complicated as steam transfers its heat in three different ways:

(a) The difference in temperature between the actual steam temperature and the saturation temperature appropriate to the steam pressure (the superheat) is calculated by means of a regression analysis equation. The superheat is the product of steam flow, specific heat and temperature difference (W)

(b) The enthalpy of steam is its heat content, assuming that it is condensed and cooled to 0°C. This is calculated by multiplying the enthalpy of steam by the steam flow (kg/s)

(c) As steam is not cooled down to 0°C, but either to the standard water return temperature or, if the NSR integer is used, to a specific water return temperature, the enthalpy represented by the condensed water must be deducted from that of the steam. In other words, the computer adds (a) and (b), but deducts (c). The total heat supplied is quoted in both cases in terms of kilowatts.

Pressure losses and pump work

The program as at present constituted only evaluates pressure losses due to friction and not those which are due to differences in altitude. This is done on the assumption that the return water returns to the starting point of the supply water so that level differences cancel out.

A distinction is made in the program between mild steel pipelines, for which the modified von Karman equation is:

$1/f^{\frac{1}{2}} = 3.2 \log (N_{Re}\, f^{\frac{1}{2}}) + 1.2$ **(11)**

and either copper or plastic pipelines for which the von Karman equation for smooth pipelines applies:

$1/f^{\frac{1}{2}} = 4.0 \log (N_{Re}\, f^{\frac{1}{2}}) - 0.4$ **(12)**

where

f is the Fanning friction factor (d1)

N_{re} is the Reynold number for the fluid (d1)

The choice of integer MAT automatically selects the correct equation, which calculates f by means of a loop technique.

Pressure head losses are then worked out from the following equation:

$$\text{Pr. hd.} = \frac{f\, w^2\, L_{adj}}{2\, V\, d} \text{ pascals} \qquad \text{E}(\mathbf{13})$$

where

Pr. hd. is the pressure head (pascals: N/m^2)

w is the mass flow rate of the fluid (kg/s)

V is the flow rate of the fluid (m/s)

d is the internal diameter of the pipe (metres)

L_{adj} is the pipeline length (metres), adjusted to account for such items as constrictions, bends, expansion joints, valves etc.

The pump load is then calculated by multiplying the pressure head by the cross-sectional area of the pipe and dividing by the pump efficiency index (% efficiency/100).

Limitations of program

Although every step in the calculations is based upon theoretical and empirical equations which have been tested and established for some time, the program obviously deals with ideal systems, ie with standardised conditions. Many of the equations were adapted from US publications and were given originally in foot-pound-second units. These have all been converted to SI. In many cases information needed for the program may not be available or may vary during the operation of the pipeline system: cases in point might be variations in wind speed and in external ambient temperatures. Average values have then to be given. The program is capable of accepting alternative information. For example mass flow rates can be substituted for fluid flow velocities and there are also choices available for giving data on soil conductivity and pipeline conductances.

Practical applications

The programs as developed would appear to have two distinct practical applications:

(a) When designing a new network, or sections of a new network, it is possible to use the program to obtain some guide as to heat losses and pressure losses to be anticipated when employing different design parameters, thus avoiding the expense of laying experimental piping. It is therefore an extremely cheap way of making use of a computer to avoid errors which would otherwise be very costly.

(b) In existing networks it would be possible to calculate the theoretical heat losses and pressure losses for sections of the network and compare these to the actual values obtained. Should such comparison throw up marked discrepancies then these would point to some irregular conditions: eg a heat loss which is much higher than that calculated by the computer may indicate a waterlogged or missing insulation layer, while an abnormally high pressure drop may indicate a blockage in the pipeline. The program can therefore also serve as a tool to the maintenance engineer.

Although the above program can deal with a large number of different types of pipeline systems and parameters, it cannot cope with all the systems in existence eg the Berlin district heating system is almost unique in that it employs a triple pipeline system (which is not included in the program). However the program design is such that it can be readily expanded to deal with additional systems by including further sub-routines: one of the advantages of FORTRAN which was used in the compilation of this program.

REFERENCES

1 F. W. Dittus and L. M. K. Boelter, *Engineering 2*, p 443, University of California Publications, 1930

2 J. F. Malloy, *Thermal insulation*, van Nostrand-Reinhold, New York, 1969

3 W. Koch, *Gesundheitsingenieur*, 22 (1), 1927, pp 1–27 (in German)

4 R. M. E. Diamant, *Insulation deskbook*, H & V Publications, Croydon, 1977

TABLE E.1 TYPICAL PRINT-OUT OF RESULTS USING SALFORD UNIVERSITY COMPUTER PROGRAM DIPROG

1	4.69	1630.5	45.59	1517.6	0.358	71.19	0.406
2	3.87	1630.5	62.32	−854.5	0.489	33.11	0.577
3	5.00	705.5	52.13	1809.4	0.312	90.43	1.192
4	4.69	1853.0	47.71	−682.2	0.375	31.97	0.332
5	3.87	1853.0	71.27	1741.9	0.541	67.47	0.339
6	9.59	1030.0	240.53	104.3	0.391	10.00	0.122
7	7.67	905.0	29.74	909.7	0.235	69.76	0.967
8	17.16	893.0	0.63	594.2	0.057	101.99	0.789
9	18.44	893.0	4.36	594.2	0.176	109.56	0.760
10	14.74	876.0	17.28	−501.8	0.259	73.98	0.523
11	5.99	630.5	14.10	1040.5	0.460	62.33	0.612
212							
12	11.45	983.0	28.74	252.0	0.289	28.85	0.182
13	11.23	800.0	14.94	−485.9	0.117	54.56	1.550

Total pipeline length	14683.0 metres
Total heat throughout	6039.4 kilowatts
Total heat loss	805.2 kilowatts
Percentage heat loss	13.3 per cent
Total pipeline head loss	629.3 kilopascals
Total net pumpwork	4.06 kilowatts

5 A. G. Loudon, 'Heat losses from underground heating mains', *IHVE Journal*, 1957, pp 196–203

6 W. H. McAdams, *Heat transmission*, McGraw-Hill, New York, Third edition 1954

7 M. Fishenden and O. A. Saunders, *An introduction to heat transfer*, Oxford Press, New York, 1950

8 F. B. Rowley and A. B. Algren, *Transactions ASHVE*, 35, 1929, pp 165–181

9 D. H. Stevens in *Heating and Ventilation Engineer*, February 1977, pp 20–21

10 I. Eneborg, 'Stora eller små värmecentraler', SNFB Stockholm No 34, 1960

11 E. Schmidt, 'Properties of water and steam in SI units', Springer Verlag, Berlin, 1964

12 *Handbook of chemistry and physics*, Chemical Rubber Co, Cleveland, 52nd edition, 1972

13 T. von Karman, 'On laminar and turbulent friction', NACA technical memorandum No 1092, 1946

14 R. H. Perry and C. H. Chilton, *Chemical engineers handbook*, chapter 5, McGraw-Hill, New York, 5th edition 1973

15 R. M. E. Diamant and C. L. Rice, 'The use of a computer to solve insulation problems', *Heating and Ventilation Engineer*, February 1978, pp 6–9

GENERAL BIBLIOGRAPHY

Specialised journals

DHA Journal, published by District Heating Association, Bedford House, Stafford Road, Caterham, Surrey, CR3 6JA

District Heating, published by the International District Heating Association 5940 Baum Square, Pittsburgh, Pennsylvania 15206, USA

Fernwärme International, published by VDEW, 6000 Frankfurt (Main) 70, Stresemannallee 23, West Germany

Handbooks

District Heating Association handbook, annual publication by International District Heating Association (see above)

District heating: a survey of practice in Europe and America, National Coal Board, London 1970

I. G. C. Dryden, *Efficient use of energy*, IPC, London, 1976

Articles in English

A. E. Haseler, numerous articles in such journals as *Heating and Ventilation Engineer*, *Steam* and *Heating Engineer* etc; also author of various government publications on the subject of district heating

D. Manifold, 'District heating and the forces of change', *Clean Air* 6, Autumn 1976, pp 27–32

'The history of district heating', *District Heating*, July–September 1976, pp 18–23

Articles in German (all in *Ferwärme International*)

K. Ahrens, 'Cities and district heating', 4 (1975) 6, pp 194–199

H. P. Winkens, 'General matters regarding district heating', 4 (1975) 6, pp 194–199

H. P. Winkens, 'Comparison of methods to save energy', 5 (1976) 4, pp 123–128

P. Schenk, 'The 1977 Dusseldorf District Heating Conference', 6 (1977) 2, pp 21–22

Conference Proceedings

Total Energy Conference at Institute of Energy, Brighton, 29th November–1st December 1971

International District Heating Association Conference at Pittsburgh, 3rd–5th June 1975

International District Heating Conference at Warsaw, 1976

IDHC, Unichal and IDHA Conference at Stockholm, 14th–18th May 1979

INDEX

References to the pages on which figures appear are in italics. References to tables are by table number and page number (e.g. Table 4.2 44)